Quantum Gravity
&
Gravitational Waves

" Ripples in The Curvature of Space-Time "

Edited by Paul F. Kisak

Contents

Chapter 1

Quantum gravity

Quantum gravity (QG) is a field of theoretical physics that seeks to describe the force of gravity according to the principles of quantum mechanics.

The current understanding of gravity is based on Albert Einstein's general theory of relativity, which is formulated within the framework of classical physics. On the other hand, the nongravitational forces are described within the framework of quantum mechanics, a radically different formalism for describing physical phenomena based on probability.[1] The necessity of a quantum mechanical description of gravity follows from the fact that one cannot consistently couple a classical system to a quantum one.[2]

Although a quantum theory of gravity is needed in order to reconcile general relativity with the principles of quantum mechanics, difficulties arise when one attempts to apply the usual prescriptions of quantum field theory to the force of gravity.[3] From a technical point of view, the problem is that the theory one gets in this way is not renormalizable and therefore cannot be used to make meaningful physical predictions. As a result, theorists have taken up more radical approaches to the problem of quantum gravity, the most popular approaches being string theory and loop quantum gravity.[4] A recent development is the theory of causal fermion systems which gives quantum mechanics, general relativity, and quantum field theory as limiting cases.[5][6][7][8][9][10]

Strictly speaking, the aim of quantum gravity is only to describe the quantum behavior of the gravitational field and should not be confused with the objective of unifying all fundamental interactions into a single mathematical framework. While any substantial improvement into the present understanding of gravity would aid further work towards unification, study of quantum gravity is a field in its own right with various branches having different approaches to unification. Although some quantum gravity theories, such as string theory, try to unify gravity with the other fundamental forces, others, such as loop quantum gravity, make no such attempt; instead, they make an effort to quantize the gravitational field while it is kept separate from the

other forces. A theory of quantum gravity that is also a grand unification of all known interactions is sometimes referred to as a theory of everything (TOE).

One of the difficulties of quantum gravity is that quantum gravitational effects are only expected to become apparent near the Planck scale, a scale far smaller in distance (equivalently, far larger in energy) than what is currently accessible at high energy particle accelerators. As a result, quantum gravity is a mainly theoretical enterprise, although there are speculations about how quantum gravity effects might be observed in existing experiments.[11]

1.1 Overview

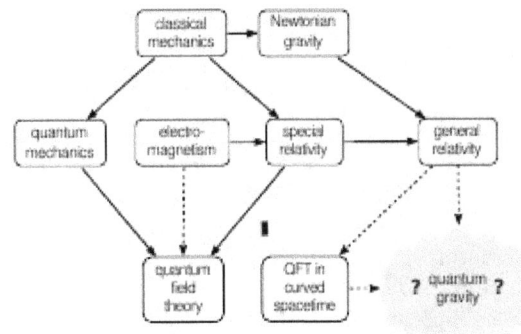

Diagram showing where quantum gravity sits in the hierarchy of physics theories

Much of the difficulty in meshing these theories at all energy scales comes from the different assumptions that these theories make on how the universe works. Quantum field theory depends on particle fields embedded in the flat spacetime of special relativity. General relativity models gravity as a curvature within space-time that changes as a gravitational mass moves. Historically, the most obvious way of combining the two (such as treating gravity as simply another particle field) ran quickly into what is known as the

1

renormalization problem. In the old-fashioned understanding of renormalization, gravity particles would attract each other and adding together all of the interactions results in many infinite values which cannot easily be cancelled out mathematically to yield sensible, finite results. This is in contrast with quantum electrodynamics where, given that the series still do not converge, the interactions sometimes evaluate to infinite results, but those are few enough in number to be removable via renormalization.

1.1.1 Effective field theories

Quantum gravity can be treated as an effective field theory. Effective quantum field theories come with some high-energy cutoff, beyond which we do not expect that the theory provides a good description of nature. The "infinities" then become large but finite quantities depending on this finite cutoff scale, and correspond to processes that involve very high energies near the fundamental cutoff. These quantities can then be absorbed into an infinite collection of coupling constants, and at energies well below the fundamental cutoff of the theory, to any desired precision; only a finite number of these coupling constants need to be measured in order to make legitimate quantum-mechanical predictions. This same logic works just as well for the highly successful theory of low-energy pions as for quantum gravity. Indeed, the first quantum-mechanical corrections to graviton-scattering and Newton's law of gravitation have been explicitly computed[12] (although they are so infinitesimally small that we may never be able to measure them). In fact, gravity is in many ways a much better quantum field theory than the Standard Model, since it appears to be valid all the way up to its cutoff at the Planck scale.

While confirming that quantum mechanics and gravity are indeed consistent at reasonable energies, it is clear that near or above the fundamental cutoff of our effective quantum theory of gravity (the cutoff is generally assumed to be of the order of the Planck scale), a new model of nature will be needed. Specifically, the problem of combining quantum mechanics and gravity becomes an issue only at very high energies, and may well require a totally new kind of model.

1.1.2 Quantum gravity theory for the highest energy scales

The general approach to deriving a quantum gravity theory that is valid at even the highest energy scales is to assume that such a theory will be simple and elegant and, accordingly, to study symmetries and other clues offered by current theories that might suggest ways to combine them into a comprehensive, unified theory. One problem with this approach is that it is unknown whether quantum grav-

ity will actually conform to a simple and elegant theory, as it should resolve the dual conundrums of special relativity with regard to the uniformity of acceleration and gravity, and general relativity with regard to spacetime curvature.

Such a theory is required in order to understand problems involving the combination of very high energy and very small dimensions of space, such as the behavior of black holes, and the origin of the universe.

1.2 Quantum mechanics and general relativity

Gravity Probe B (GP-B) has measured spacetime curvature near Earth to test related models in application of Einstein's general theory of relativity.

1.2.1 The graviton

Main article: Graviton

At present, one of the deepest problems in theoretical physics is harmonizing the theory of general relativity, which describes gravitation, and applications to large-scale structures (stars, planets, galaxies), with quantum mechanics, which describes the other three fundamental forces acting on the atomic scale. This problem must be put in the proper context, however. In particular, contrary to the popular claim that quantum mechanics and general relativity are fundamentally incompatible, one can demonstrate that the structure of general relativity essentially follows inevitably from the quantum mechanics of interacting theoretical spin-2 massless particles (called gravitons).[13][14][15][16][17]

While there is no concrete proof of the existence of gravitons, quantized theories of matter may necessitate their existence. Supporting this theory is the observation that all fundamental forces except gravity have one or more known messenger particles, leading researchers to believe that at least one most likely does exist; they have dubbed this hypothetical particle the *graviton*. The predicted find would result in the classification of the graviton as a "force particle" similar to the photon of the electromagnetic field. Many of the accepted notions of a unified theory of physics since the 1970s assume, and to some degree depend upon, the existence of the graviton. These include string theory, superstring theory, M-theory, and loop quantum gravity. Detection of gravitons is thus vital to the validation of various lines of research to unify quantum mechanics and relativity theory.

1.2.2 The dilaton

Main article: Dilaton

The dilaton made its first appearance in Kaluza–Klein theory, a five-dimensional theory that combined gravitation and electromagnetism. Generally, it appears in string theory. More recently, however, it's become central to the lower-dimensional many-bodied gravity problem[18] based on the field theoretic approach of Roman Jackiw. The impetus arose from the fact that complete analytical solutions for the metric of a covariant N-body system have proven elusive in general relativity. To simplify the problem, the number of dimensions was lowered to *(1+1)*, i.e., one spatial dimension and one temporal dimension. This model problem, known as $R=T$ theory[19] (as opposed to the general $G=T$ theory) was amenable to exact solutions in terms of a generalization of the Lambert W function. It was also found that the field equation governing the dilaton (derived from differential geometry) was the Schrödinger equation and consequently amenable to quantization.[20]

Thus, one had a theory which combined gravity, quantization, and even the electromagnetic interaction, promising ingredients of a fundamental physical theory. It is worth noting that this outcome revealed a previously unknown and already existing *natural link* between general relativity and quantum mechanics. However, this theory lacks generalization to the *(2+1)* or *(3+1)* dimensions. In principle, the field equations are amenable to such generalization (as shown with the inclusion of a one-graviton process[21]) and yield the correct Newtonian limit in d dimensions but only if if a dilaton is included. Furthermore, it is not yet clear what the fully generalized field equation governing the dilaton in (3+1) dimensions should be. The fact that gravitons can propagate in *(3+1)* dimensions implies that gravitons and dilatons do exist in the real world. Nonetheless, detection of the dilaton is expected to be even more elusive than the graviton. But since this simplified approach combines gravitational, electromagnetic and quantum effects, their coupling could potentially lead to a means of vindicating the theory, through cosmology and even, perhaps, experimentally.

1.2.3 Nonrenormalizability of gravity

Further information: Renormalization

General relativity, like electromagnetism, is a classical field theory. One might expect that, as with electromagnetism, the gravitational force should also have a corresponding quantum field theory.

However, gravity is perturbatively nonrenormalizable.[22][23] For a quantum field theory to be well-defined according to this understanding of the subject, it must be asymptotically free or asymptotically safe. The theory must be characterized by a choice of *finitely many* parameters, which could, in principle, be set by experiment. For example, in quantum electrodynamics these parameters are the charge and mass of the electron, as measured at a particular energy scale.

On the other hand, in quantizing gravity there are *infinitely many independent parameters* (counterterm coefficients) needed to define the theory. For a given choice of those parameters, one could make sense of the theory, but since it's impossible to conduct infinite experiments to fix the values of every parameter, we do not have a meaningful physical theory:

- At low energies, the logic of the renormalization group tells us that, despite the unknown choices of these infinitely many parameters, quantum gravity will reduce to the usual Einstein theory of general relativity.

- On the other hand, if we could probe very high energies where quantum effects take over, then *every one* of the infinitely many unknown parameters would begin to matter, and we could make no predictions at all.

If we treat QG as an effective field theory, there is a way around this problem.

That is, the meaningful theory of quantum gravity (that makes sense and is predictive at all energy levels) inherently implies some deep principle that reduces the infinitely many unknown parameters to a finite number that can then be measured:

- One possibility is that normal perturbation theory is not a reliable guide to the renormalizability of the theory, and that there really *is* a UV fixed point for gravity. Since this is a question of non-perturbative quantum field theory, it is difficult to find a reliable answer, but some people still pursue this option.

- Another possibility is that there are new unfound symmetry principles that constrain the parameters and reduce them to a finite set. This is the route taken by string theory, where all of the excitations of the string essentially manifest themselves as new symmetries.

1.2.4   QG as an effective field theory

Main article: Effective field theory

In an effective field theory, all but the first few of the infinite set of parameters in a non-renormalizable theory are suppressed by huge energy scales and hence can be neglected when computing low-energy effects. Thus, at least in the low-energy regime, the model is indeed a predictive quantum field theory.[12] (A very similar situation occurs for the very similar effective field theory of low-energy pions.) Furthermore, many theorists agree that even the Standard Model should really be regarded as an effective field theory as well, with "nonrenormalizable" interactions suppressed by large energy scales and whose effects have consequently not been observed experimentally.

Recent work[12] has shown that by treating general relativity as an effective field theory, one can actually make legitimate predictions for quantum gravity, at least for low-energy phenomena. An example is the well-known calculation of the tiny first-order quantum-mechanical correction to the classical Newtonian gravitational potential between two masses.

1.2.5   Spacetime background dependence

Main article: Background independence

A fundamental lesson of general relativity is that there is no fixed spacetime background, as found in Newtonian mechanics and special relativity; the spacetime geometry is dynamic. While easy to grasp in principle, this is the hardest idea to understand about general relativity, and its consequences are profound and not fully explored, even at the classical level. To a certain extent, general relativity can be seen to be a relational theory,[24] in which the only physically relevant information is the relationship between different events in space-time.

On the other hand, quantum mechanics has depended since its inception on a fixed background (non-dynamic) structure. In the case of quantum mechanics, it is time that is given and not dynamic, just as in Newtonian classical mechanics. In relativistic quantum field theory, just as in classical field theory, Minkowski spacetime is the fixed background of the theory.

String theory

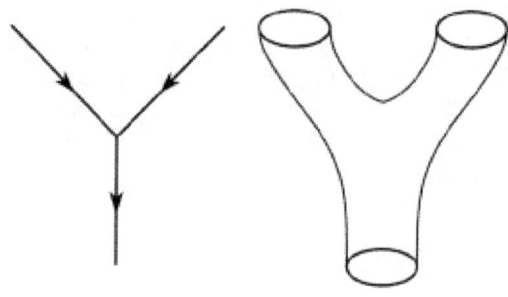

Interaction in the subatomic world: world lines of point-like particles in the Standard Model or a world sheet swept up by closed strings in string theory

String theory can be seen as a generalization of quantum field theory where instead of point particles, string-like objects propagate in a fixed spacetime background, although the interactions among closed strings give rise to space-time in a dynamical way. Although string theory had its origins in the study of quark confinement and not of quantum gravity, it was soon discovered that the string spectrum contains the graviton, and that "condensation" of certain vibration modes of strings is equivalent to a modification of the original background. In this sense, string perturbation theory exhibits exactly the features one would expect of a perturbation theory that may exhibit a strong dependence on asymptotics (as seen, for example, in the AdS/CFT correspondence) which is a weak form of background dependence.

Background independent theories

Loop quantum gravity is the fruit of an effort to formulate a background-independent quantum theory.

Topological quantum field theory provided an example of background-independent quantum theory, but with no local degrees of freedom, and only finitely many degrees of freedom globally. This is inadequate to describe gravity in 3+1 dimensions, which has local degrees of freedom according to general relativity. In 2+1 dimensions, however, gravity is a topological field theory, and it has been successfully quantized in several different ways, including spin networks.

1.2.6   Semi-classical quantum gravity

Quantum field theory on curved (non-Minkowskian) backgrounds, while not a full quantum theory of gravity, has shown many promising early results. In an analogous way to the development of quantum electrodynamics in the early part of the 20th century (when physicists considered quantum mechanics in classical electromagnetic fields), the consideration of quantum field theory on a curved background has led to predictions such as black hole radiation.

Phenomena such as the Unruh effect, in which particles exist in certain accelerating frames but not in stationary ones, do not pose any difficulty when considered on a curved background (the Unruh effect occurs even in flat Minkowskian backgrounds). The vacuum state is the state with the least energy (and may or may not contain particles). See Quantum field theory in curved spacetime for a more complete discussion.

1.2.7   Points of tension

There are other points of tension between quantum mechanics and general relativity.

- First, classical general relativity breaks down at singularities, and quantum mechanics becomes inconsistent with general relativity in the neighborhood of singularities (however, no one is certain that classical general relativity applies near singularities in the first place).

- Second, it is not clear how to determine the gravitational field of a particle, since under the Heisenberg uncertainty principle of quantum mechanics its location and velocity cannot be known with certainty. The

resolution of these points may come from a better understanding of general relativity.[25]

- Third, there is the problem of time in quantum gravity. Time has a different meaning in quantum mechanics and general relativity and hence there are subtle issues to resolve when trying to formulate a theory which combines the two.[26]

1.3   Candidate theories

There are a number of proposed quantum gravity theories.[27] Currently, there is still no complete and consistent quantum theory of gravity, and the candidate models still need to overcome major formal and conceptual problems. They also face the common problem that, as yet, there is no way to put quantum gravity predictions to experimental tests, although there is hope for this to change as future data from cosmological observations and particle physics experiments becomes available.[28][29]

1.3.1   String theory

Main article: String theory
One suggested starting point is ordinary quantum field the-

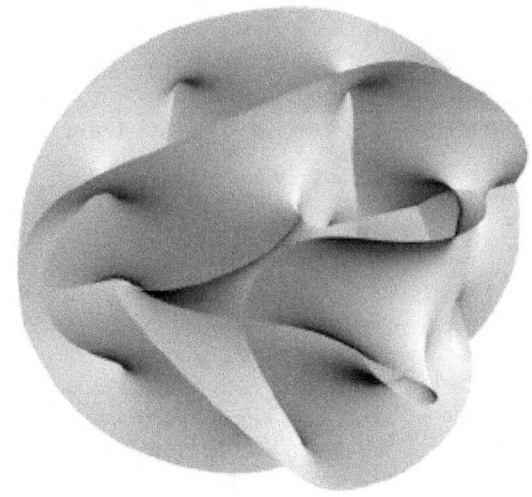

Projection of a Calabi–Yau manifold, one of the ways of compactifying the extra dimensions posited by string theory

ories which, after all, are successful in describing the other three basic fundamental forces in the context of the standard model of elementary particle physics. However, while this leads to an acceptable effective (quantum) field theory of gravity at low energies,[30] gravity turns out to be much more problematic at higher energies. For ordinary field theories such as quantum electrodynamics, a technique known as renormalization is an integral part of deriving predictions which take into account higher-energy contributions,[31] but gravity turns out to be nonrenormalizable: at high energies, applying the recipes of ordinary quantum field theory yields models that are devoid of all predictive power.[32]

One attempt to overcome these limitations is to replace ordinary quantum field theory, which is based on the classical concept of a point particle, with a quantum theory of one-dimensional extended objects: string theory.[33] At the energies reached in current experiments, these strings are indistinguishable from point-like particles, but, crucially, different modes of oscillation of one and the same type of fundamental string appear as particles with different (electric and other) charges. In this way, string theory promises to be a unified description of all particles and interactions.[34] The theory is successful in that one mode will always correspond to a graviton, the messenger particle of gravity; however, the price of this success are unusual features such as six extra dimensions of space in addition to the usual three for space and one for time.[35]

In what is called the second superstring revolution, it was conjectured that both string theory and a unification of general relativity and supersymmetry known as supergravity[36] form part of a hypothesized eleven-dimensional model known as M-theory, which would constitute a uniquely defined and consistent theory of quantum gravity.[37][38] As presently understood, however, string theory admits a very large number (10^{500} by some estimates) of consistent vacua, comprising the so-called "string landscape". Sorting through this large family of solutions remains a major challenge.

1.3.2   Loop quantum gravity

Main article: Loop quantum gravity
 Loop quantum gravity seriously considers general relativity's insight that spacetime is a dynamical field and is therefore a quantum object. Its second idea is that the quantum discreteness that determines the particle-like behavior of other field theories (for instance, the photons of the electromagnetic field) also affects the structure of space.

The main result of loop quantum gravity is the derivation of a granular structure of space at the Planck length. This is derived from following considerations: In the case of electromagnetism, the quantum operator representing the energy of each frequency of the field has a discrete spectrum. Thus the energy of each frequency is quantized, and the quanta are the photons. In the case of gravity, the operators representing the area and the volume of each sur-

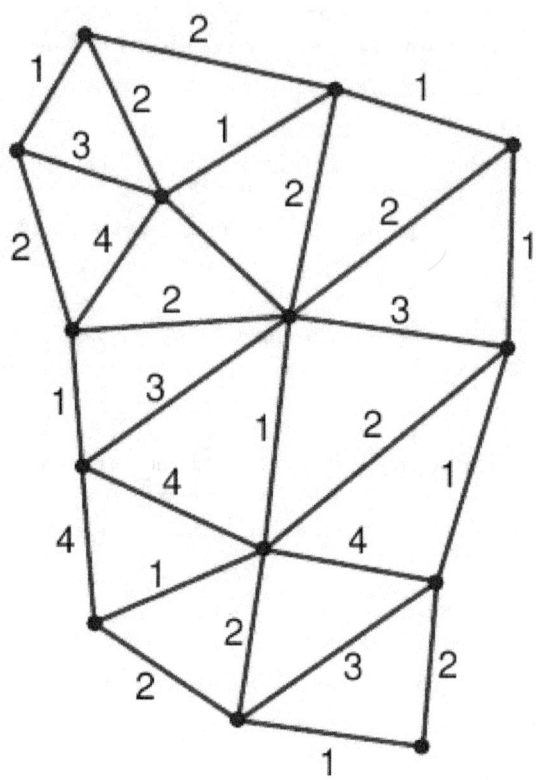

Simple spin network of the type used in loop quantum gravity

face or space region likewise have discrete spectrum. Thus area and volume of any portion of space are also quantized, where the quanta are elementary quanta of space. It follows, then, that spacetime has an elementary quantum granular structure at the Planck scale, which cuts off the ultraviolet infinities of quantum field theory.

The quantum state of spacetime is described in the theory by means of a mathematical structure called spin networks. Spin networks were initially introduced by Roger Penrose in abstract form, and later shown by Carlo Rovelli and Lee Smolin to derive naturally from a non-perturbative quantization of general relativity. Spin networks do not represent quantum states of a field in spacetime: they represent directly quantum states of spacetime.

The theory is based on the reformulation of general relativity known as Ashtekar variables, which represent geometric gravity using mathematical analogues of electric and magnetic fields.[39][40] In the quantum theory, space is represented by a network structure called a spin network, evolving over time in discrete steps.[41][42][43][44]

The dynamics of the theory is today constructed in several versions. One version starts with the canonical quantization of general relativity. The analogue of the Schrödinger equation is a Wheeler–DeWitt equation, which can be defined within the theory.[45] In the covariant, or spinfoam formulation of the theory, the quantum dynamics is obtained via a sum over discrete versions of spacetime, called spinfoams. These represent histories of spin networks.

1.3.3 Scale relativity

Main article: Scale relativity

Most quantum gravity theories assume quantum laws as

Schrödinger's flower. Morphogenesis of a flower-like structure, solution of a growth process equation that takes the form of a Schrödinger equation under fractal conditions.

a starting point. However, in the framework of scale relativity, this is not needed.[46] The theory is an extension of special and general relativity, including the relativity of scale transformations. It thus takes a geometrical approach to the problem, where quantum phenomena became a manifestation of the fractality of spacetime. This is similar to the geometrical interpretation of gravitation in general relativity, where gravitation become a manifestation of spacetime curvature instead of a force. Although much remains to be developed, validated predictions have already been obtained in physics, astrophysics and cosmology.

1.3.4 Other approaches

There are a number of other approaches to quantum gravity. The approaches differ depending on which features of general relativity and quantum theory are accepted unchanged, and which features are modified.[47][48] Examples include:

- Acoustic metric and other analog models of gravity
- Asymptotic safety in quantum gravity
- Euclidean quantum gravity

- Causal dynamical triangulation[49]
- Causal fermion systems,[5][6][7][8][9][10] giving quantum mechanics, general relativity and quantum field theory as limiting cases.
- Causal sets[50]
- Covariant Feynman path integral approach
- Group field theory[51]
- Wheeler-DeWitt equation
- Geometrodynamics
- Hořava–Lifshitz gravity
- MacDowell–Mansouri action
- Noncommutative geometry.
- Path-integral based models of quantum cosmology[52]
- Regge calculus
- String-nets giving rise to gapless helicity ±2 excitations with no other gapless excitations[53]
- Superfluid vacuum theory a.k.a. theory of BEC vacuum
- Supergravity
- Twistor theory[54]
- Canonical quantum gravity
- E8 Theory
- Quantum holonomy theory[55]

1.4 Weinberg–Witten theorem

In quantum field theory, the Weinberg–Witten theorem places some constraints on theories of composite gravity/emergent gravity. However, recent developments attempt to show that if locality is only approximate and the holographic principle is correct, the Weinberg–Witten theorem would not be valid.

1.5 Experimental tests

As was emphasized above, quantum gravitational effects are extremely weak and therefore difficult to test. For this reason, the possibility of experimentally testing quantum gravity had not received much attention prior to the late 1990s. However, in the past decade, physicists have realized that evidence for quantum gravitational effects can guide the development of the theory. Since theoretical development has been slow, the field of phenomenological quantum gravity, which studies the possibility of experimental tests, has obtained increased attention.[56][57]

The most widely pursued possibilities for quantum gravity phenomenology include violations of Lorentz invariance, imprints of quantum gravitational effects in the cosmic microwave background (in particular its polarization), and decoherence induced by fluctuations in the space-time foam.

The BICEP2 experiment detected what was initially thought to be primordial B-mode polarization caused by gravitational waves in the early universe. If truly primordial, these waves were born as quantum fluctuations in gravity itself. Cosmologist Ken Olum (Tufts University) stated: "I think this is the only observational evidence that we have that actually shows that gravity is quantized....It's probably the only evidence of this that we will ever have."[58]

1.6 See also

1.7 References

[1] Griffiths, David J. (2004). *Introduction to Quantum Mechanics*. Pearson Prentice Hall. OCLC 803860989.

[2] Wald, Robert M. (1984). *General Relativity*. University of Chicago Press. p. 382. OCLC 471881415.

[3] Zee, Anthony (2010). *Quantum Field Theory in a Nutshell* (2nd ed.). Princeton University Press. p. 172. OCLC 659549695.

[4] Penrose, Roger (2007). *The road to reality : a complete guide to the laws of the universe*. Vintage. p. 1017. OCLC 716437154.

[5] F. Finster, J. Kleiner, Causal Fermion Systems as a Candidate for a Unified Physical Theory, http://arxiv.org/abs/1502.03587

[6] F. Finster, The Principle of the Fermionic Projector, hep-th/0001048, hep-th/0202059, hep- th/0210121, AMS/IP Studies in Advanced Mathematics, vol. **35**, American Mathematical Society, Providence, RI, 2006.

[7] F. Finster, A formulation of quantum field theory realizing a sea of interacting Dirac particles, arXiv:0911.2102 [hep-th], Lett. Math. Phys. **97** (2011), no. 2, 165–183.

[8] F. Finster, An action principle for an interacting fermion system and its analysis in the continuum limit, arXiv:0908.1542 [math-ph] (2009).

[9] F. Finster, The continuum limit of a fermion system involving neutrinos: Weak and gravitational interactions, arXiv: 1211.3351 [math-ph] (2012).

[10] F. Finster, Perturbative quantum field theory in the framework of the fermionic projector, arXiv:1310.4121 [math-ph], J. Math. Phys. **55** (2014), no. 4, 042301.

[11] Quantum effects in the early universe might have an observable effect on the structure of the present universe, for example, or gravity might play a role in the unification of the other forces. Cf. the text by Wald cited above.

[12] Donoghue (1995). "Introduction to the Effective Field Theory Description of Gravity". arXiv:gr-qc/9512024. (verify against ISBN 9789810229085)

[13] Kraichnan, R. H. (1955). "Special-Relativistic Derivation of Generally Covariant Gravitation Theory". *Physical Review* **98** (4): 1118–1122. Bibcode:1955PhRv...98.1118K. doi:10.1103/PhysRev.98.1118.

[14] Gupta, S. N. (1954). "Gravitation and Electromagnetism". *Physical Review* **96** (6): 1683–1685. Bibcode:1954PhRv...96.1683G. doi:10.1103/PhysRev.96.1683.

[15] Gupta, S. N. (1957). "Einstein's and Other Theories of Gravitation". *Reviews of Modern Physics* **29** (3): 334–336. Bibcode:1957RvMP...29..334G. doi:10.1103/RevModPhys.29.334.

[16] Gupta, S. N. (1962). "Quantum Theory of Gravitation". *Recent Developments in General Relativity*. Pergamon Press. pp. 251–258.

[17] Deser, S. (1970). "Self-Interaction and Gauge Invariance". *General Relativity and Gravitation* **1**: 9–18. arXiv:gr-qc/0411023. Bibcode:1970GReGr...1....9D. doi:10.1007/BF00759198.

[18] Ohta, Tadayuki; Mann, Robert (1996). "Canonical reduction of two-dimensional gravity for particle dynamics". *Classical and Quantum Gravity* **13** (9): 2585–2602. arXiv:gr-qc/9605004. Bibcode:1996CQGra..13.2585O. doi:10.1088/0264-9381/13/9/022.

[19] Sikkema, A E; Mann, R B (1991). "Gravitation and cosmology in (1+1) dimensions". *Classical and Quantum Gravity* **8**: 219–235. Bibcode:1991CQGra...8..219S. doi:10.1088/0264-9381/8/1/022.

[20] Farrugia; Mann; Scott (2007). "N-body Gravity and the Schroedinger Equation". *Classical and Quantum Gravity* **24** (18): 4647–4659. arXiv:gr-qc/0611144.

Bibcode:2007CQGra..24.4647F.   doi:10.1088/0264-9381/24/18/006.

[21] Mann, R B; Ohta, T (1997). "Exact solution for the metric and the motion of two bodies in (1+1)-dimensional gravity". *Physical Review D* **55** (8): 4723–4747. arXiv:gr-qc/9611008.   Bibcode:1997PhRvD..55.4723M. doi:10.1103/PhysRevD.55.4723.

[22] Feynman, R. P.; Morinigo, F. B.; Wagner, W. G.; Hatfield, B. (1995). *Feynman lectures on gravitation*. Addison-Wesley. ISBN 0-201-62734-5.

[23] Hamber, H. W. (2009). *Quantum Gravitation - The Feynman Path Integral Approach*. Springer Publishing. ISBN 978-3-540-85292-6.

[24] Smolin, Lee (2001). *Three Roads to Quantum Gravity*. Basic Books. pp. 20–25. ISBN 0-465-07835-4. Pages 220–226 are annotated references and guide for further reading.

[25] Hunter Monroe (2005). "Singularity-Free Collapse through Local Inflation". arXiv:astro-ph/0506506.

[26] Edward Anderson (2010). "The Problem of Time in Quantum Gravity". arXiv:1009.2157 [gr-qc]. (also published as chapter 4 of ISBN 9781611229578)

[27] A timeline and overview can be found in Rovelli, Carlo (2000). "Notes for a brief history of quantum gravity". arXiv:gr-qc/0006061. (verify against ISBN 9789812777386)

[28] Ashtekar, Abhay (2007). "Loop Quantum Gravity: Four Recent Advances and a Dozen Frequently Asked Questions". *11th Marcel Grossmann Meeting on Recent Developments in Theoretical and Experimental General Relativity*. p. 126. arXiv:0705.2222. Bibcode:2008mgm..conf..126A. doi:10.1142/9789812834300_0008.

[29] Schwarz, John H. (2007). "String Theory: Progress and Problems". *Progress of Theoretical Physics Supplement* **170**: 214–226. arXiv:hep-th/0702219.   Bibcode:2007PThPS.170..214S. doi:10.1143/PTPS.170.214.

[30] Donoghue, John F. (editor) (1995). "Introduction to the Effective Field Theory Description of Gravity". In Cornet, Fernando. *Effective Theories: Proceedings of the Advanced School, Almunecar, Spain, 26 June–1 July 1995*. Singapore: World Scientific. arXiv:gr-qc/9512024. ISBN 981-02-2908-9.

[31] Weinberg, Steven (1996). "Chapters 17–18". *The Quantum Theory of Fields II: Modern Applications*. Cambridge University Press. ISBN 0-521-55002-5.

[32] Goroff, Marc H.; Sagnotti, Augusto; Sagnotti, Augusto (1985). "Quantum gravity at two loops". *Physics Letters B* **160**: 81–86. Bibcode:1985PhLB..160...81G. doi:10.1016/0370-2693(85)91470-4.

[33] An accessible introduction at the undergraduate level can be found in Zwiebach, Barton (2004). *A First Course in String Theory*. Cambridge University Press. ISBN 0-521-83143-1., and more complete overviews in Polchinski, Joseph (1998). *String Theory Vol. I: An Introduction to the Bosonic String*. Cambridge University Press. ISBN 0-521-63303-6. and Polchinski, Joseph (1998b). *String Theory Vol. II: Superstring Theory and Beyond*. Cambridge University Press. ISBN 0-521-63304-4.

[34] Ibanez, L. E. (2000). "The second string (phenomenology) revolution". *Classical & Quantum Gravity* **17** (5): 1117–1128. arXiv:hep-ph/9911499. Bibcode:2000CQGra..17.1117I.   doi:10.1088/0264-9381/17/5/321.

[35] For the graviton as part of the string spectrum, e.g. Green, Schwarz & Witten 1987, sec. 2.3 and 5.3; for the extra dimensions, ibid sec. 4.2.

[36] Weinberg, Steven (2000). "Chapter 31". *The Quantum Theory of Fields II: Modern Applications*. Cambridge University Press. ISBN 0-521-55002-5.

[37] Townsend, Paul K. (1996). *Four Lectures on M-Theory*. ICTP Series in Theoretical Physics. p. 385. arXiv:hep-th/9612121. Bibcode:1997hepcbconf..385T.

[38] Duff, Michael (1996). "M-Theory (the Theory Formerly Known as Strings)". *International Journal of Modern Physics A* **11** (32): 5623–5642. arXiv:hep-th/9608117.   Bibcode:1996IJMPA..11.5623D. doi:10.1142/S0217751X96002583.

[39] Ashtekar, Abhay (1986). "New variables for classical and quantum gravity". *Physical Review Letters* **57** (18): 2244–2247. Bibcode:1986PhRvL..57.2244A. doi:10.1103/PhysRevLett.57.2244. PMID 10033673.

[40] Ashtekar, Abhay (1987). "New Hamiltonian formulation of general relativity". *Physical Review D* **36** (6): 1587–1602. Bibcode:1987PhRvD..36.1587A. doi:10.1103/PhysRevD.36.1587.

[41] Thiemann, Thomas (2006). "Loop Quantum Gravity: An Inside View". *Approaches to Fundamental Physics*. Lecture Notes in Physics **721**: 185. arXiv:hep-th/0608210.   Bibcode:2007LNP...721..185T. doi:10.1007/978-3-540-71117-9_10.   ISBN 978-3-540-71115-5.

[42] Rovelli, Carlo (1998). "Loop Quantum Gravity". *Living Reviews in Relativity* **1**. Retrieved 2008-03-13.

[43] Ashtekar, Abhay; Lewandowski, Jerzy (2004). "Background Independent Quantum Gravity: A Status Report". *Classical & Quantum Gravity* **21** (15): R53–R152. arXiv:gr-qc/0404018.   Bibcode:2004CQGra..21R..53A. doi:10.1088/0264-9381/21/15/R01.

[44] Thiemann, Thomas (2003). "Lectures on Loop Quantum Gravity". *Lecture Notes in Physics*. Lecture

Notes in Physics **631**: 41–135. arXiv:gr-qc/0210094. Bibcode:2003LNP...631...41T. doi:10.1007/978-3-540-45230-0_3. ISBN 978-3-540-40810-9.

[45] Rovelli, Carlo (2004). *Quantum Gravity*. Cambridge University Press. ISBN 0521715962.

[46] Nottale, L. (2011). *Scale Relativity and Fractal Space-Time: A New Approach to Unifying Relativity and Quantum Mechanics*. World Scientific Publishing Company. ISBN 1848166508.:p. 458

[47] Isham, Christopher J. (1994). "Prima facie questions in quantum gravity". In Ehlers, Jürgen; Friedrich, Helmut. *Canonical Gravity: From Classical to Quantum*. Springer. arXiv:gr-qc/9310031. ISBN 3-540-58339-4.

[48] Sorkin, Rafael D. (1997). "Forks in the Road, on the Way to Quantum Gravity". *International Journal of Theoretical Physics* **36** (12): 2759–2781. arXiv:gr-qc/9706002. Bibcode:1997IJTP...36.2759S. doi:10.1007/BF02435709.

[49] Loll, Renate (1998). "Discrete Approaches to Quantum Gravity in Four Dimensions". *Living Reviews in Relativity* **1**: 13. arXiv:gr-qc/9805049. Bibcode:1998LRR.....1...13L. doi:10.12942/lrr-1998-13. Retrieved 2008-03-09.

[50] Sorkin, Rafael D. (2005). "Causal Sets: Discrete Gravity". In Gomberoff, Andres; Marolf, Donald. *Lectures on Quantum Gravity*. Springer. arXiv:gr-qc/0309009. ISBN 0-387-23995-2.

[51] See Daniele Oriti and references therein.

[52] Hawking, Stephen W. (1987). "Quantum cosmology". In Hawking, Stephen W.; Israel, Werner. *300 Years of Gravitation*. Cambridge University Press. pp. 631–651. ISBN 0-521-37976-8.

[53] Wen 2006

[54] See ch. 33 in Penrose 2004 and references therein.

[55] "Quantum Holonomy Theory by J. Aastrup and J. M. Grimstrup" (PDF).

[56] Hossenfelder, Sabine (2011). "Experimental Search for Quantum Gravity". In V. R. Frignanni. *Classical and Quantum Gravity: Theory, Analysis and Applications*. Chapter 5: Nova Publishers. ISBN 978-1-61122-957-8.

[57] Hossenfelder, Sabine (2010-10-17). V. R. Frignanni. ed. "Experimental Search for Quantum Gravity". *Classical and Quantum Gravity: Theory, Analysis and Applications* (Nova Publishers) **5** (2011). arXiv:1010.3420. Bibcode:2010arXiv1010.3420H. |chapter= ignored (help)

[58] Camille Carlisle. "First Direct Evidence of Big Bang Inflation". SkyandTelescope.com. Retrieved March 18, 2014.

1.8 Further reading

- Ahluwalia, D. V. (2002). "Interface of Gravitational and Quantum Realms". *Modern Physics Letters A* **17** (15–17): 1135. arXiv:gr-qc/0205121. Bibcode:2002MPLA...17.1135A. doi:10.1142/S021773230200765X.

- Ashtekar, Abhay (2005). "The winding road to quantum gravity" (PDF). *Current Science* **89**: 2064–2074.

- Carlip, Steven (2001). "Quantum Gravity: a Progress Report". *Reports on Progress in Physics* **64** (8): 885–942. arXiv:gr-qc/0108040. Bibcode:2001RPPh...64..885C. doi:10.1088/0034-4885/64/8/301.

- Herbert W. Hamber (2009). *Quantum Gravitation*. Springer Publishing. doi:10.1007/978-3-540-85293-3. ISBN 978-3-540-85292-6.

- Kiefer, Claus (2007). *Quantum Gravity*. Oxford University Press. ISBN 0-19-921252-X.

- Kiefer, Claus (2005). "Quantum Gravity: General Introduction and Recent Developments". *Annalen der Physik* **15**: 129–148. arXiv:gr-qc/0508120. Bibcode:2006AnP...518..129K. doi:10.1002/andp.200510175.

- Lämmerzahl, Claus, ed. (2003). *Quantum Gravity: From Theory to Experimental Search*. Lecture Notes in Physics. Springer. ISBN 3-540-40810-X.

- Rovelli, Carlo (2004). *Quantum Gravity*. Cambridge University Press. ISBN 0-521-83733-2.

- Quantum gravity Carlo Rovelli, Scholarpedia, 3(5):7117. doi:10.4249/scholarpedia.7117

- Trifonov, Vladimir (2008). "GR-friendly description of quantum systems". *International Journal of Theoretical Physics* **47** (2): 492–510. arXiv:math-ph/0702095. Bibcode:2008IJTP...47..492T. doi:10.1007/s10773-007-9474-3.

Chapter 2

String theory

For the study of strings of characters, see Concatenation theory.
For a more accessible and less technical introduction to this topic, see Introduction to M-theory.

In physics, **string theory** is a theoretical framework in which the point-like particles of particle physics are replaced by one-dimensional objects called strings. String theory describes how these strings propagate through space and interact with each other. On distance scales larger than the string scale, a string looks just like an ordinary particle, with its mass, charge, and other properties determined by the vibrational state of the string. In string theory, one of the many vibrational states of the string corresponds to the graviton, a quantum mechanical particle that carries gravitational force. Thus string theory is a theory of quantum gravity.

String theory is a broad and varied subject that attempts to address a number of deep questions of fundamental physics. String theory has been applied to a variety of problems in black hole physics, early universe cosmology, nuclear physics, and condensed matter physics, and it has stimulated a number of major developments in pure mathematics. Because string theory potentially provides a unified description of gravity and particle physics, it is a candidate for a theory of everything, a self-contained mathematical model that describes all fundamental forces and forms of matter. Despite much work on these problems, it is not known to what extent string theory describes the real world or how much freedom the theory allows to choose the details.

String theory was first studied in the late 1960s as a theory of the strong nuclear force, before being abandoned in favor of quantum chromodynamics. Subsequently, it was realized that the very properties that made string theory unsuitable as a theory of nuclear physics made it a promising candidate for a quantum theory of gravity. The earliest version of string theory, bosonic string theory, incorporated only the class of particles known as bosons. It later developed into superstring theory, which posits a connection called supersymmetry between bosons and the class of particles called fermions. Five consistent versions of superstring theory were developed before it was conjectured in the mid-1990s that they were all different limiting cases of a single theory in eleven dimensions known as M-theory. In late 1997, theorists discovered an important relationship called the AdS/CFT correspondence, which relates string theory to another type of physical theory called a quantum field theory.

One of the challenges of string theory is that the full theory does not yet have a satisfactory definition in all circumstances. Another issue is that the theory is thought to describe an enormous landscape of possible universes, and this has complicated efforts to develop theories of particle physics based on string theory. These issues have led some in the community to criticize these approaches to physics and question the value of continued research on string theory unification.

2.1   Fundamentals

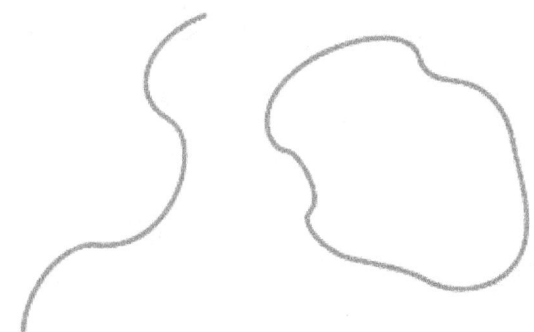

The fundamental objects of string theory are open and closed strings.

In the twentieth century, two theoretical frameworks emerged for formulating the laws of physics. One of these frameworks was Albert Einstein's general theory of relativity, a theory that explains the force of gravity and the structure of space and time. The other was quantum mechanics, a radically different formalism for describing physical phenomena using probability. By the late 1970s, these two frameworks had proven to be sufficient to explain most of the observed features of the universe, from elementary particles to atoms to the evolution of stars and the universe as a whole.[1]

In spite of these successes, there are still many problems that remain to be solved. One of the deepest problems in modern physics is the problem of quantum gravity.[1] The general theory of relativity is formulated within the framework of classical physics, whereas the other fundamental forces are described within the framework of quantum mechanics. A quantum theory of gravity is needed in order to reconcile general relativity with the principles of quantum mechanics, but difficulties arise when one attempts to apply the usual prescriptions of quantum theory to the force of gravity.[2] In addition to the problem of developing a consistent theory of quantum gravity, there are many other fundamental problems in the physics of atomic nuclei, black holes, and the early universe.[lower-alpha 1]

String theory is a theoretical framework that attempts to address these questions and many others. The starting point for string theory is the idea that the point-like particles of particle physics can also be modeled as one-dimensional objects called strings. String theory describes how strings propagate through space and interact with each other. In a given version of string theory, there is only one kind of string, which may look like a small loop or segment of ordinary string, and it can vibrate in different ways. On distance scales larger than the string scale, a string will look just like an ordinary particle, with its mass, charge, and other properties determined by the vibrational state of the string. In this way, all of the different elementary particles may be viewed as vibrating strings. In string theory, one of the vibrational states of the string gives rise to the graviton, a quantum mechanical particle that carries gravitational force. Thus string theory is a theory of quantum gravity.[3]

One of the main developments of the past several decades in string theory was the discovery of certain "dualities", mathematical transformations that identify one physical theory with another. Physicists studying string theory have discovered a number of these dualities between different versions of string theory, and this has led to the conjecture that all consistent versions of string theory are subsumed in a single framework known as M-theory.[4]

Studies of string theory have also yielded a number of results on the nature of black holes and the gravitational interaction. There are certain paradoxes that arise when one attempts to understand the quantum aspects of black holes, and work on string theory has attempted to clarify these issues. In late 1997 this line of work culminated in the discovery of the anti-de Sitter/conformal field theory correspondence or AdS/CFT.[5] This is a theoretical result which relates string theory to other physical theories which are better understood theoretically. The AdS/CFT correspondence has implications for the study of black holes and quantum gravity, and it has been applied to other subjects, including nuclear[6] and condensed matter physics.[7][8]

Since string theory incorporates all of the fundamental interactions, including gravity, many physicists hope that it fully describes our universe, making it a theory of everything. One of the goals of current research in string theory is to find a solution of the theory that reproduces the observed spectrum of elementary particles, with a small cosmological constant, containing dark matter and a plausible mechanism for cosmic inflation. While there has been progress toward these goals, it is not known to what extent string theory describes the real world or how much freedom the theory allows to choose the details.[9]

One of the challenges of string theory is that the full theory does not yet have a satisfactory definition in all circumstances. The scattering of strings is most straightforwardly defined using the techniques of perturbation theory, but it is not known in general how to define string theory nonperturbatively.[10] It is also not clear whether there is any principle by which string theory selects its vacuum state, the physical state that determines the properties of our universe.[11] These problems have led some in the community to criticize these approaches to the unification of physics and question the value of continued research on these problems.[12]

2.1.1 Strings

Main article: String (physics)

The application of quantum mechanics to physical objects such as the electromagnetic field, which are extended in space and time, is known as quantum field theory. In particle physics, quantum field theories form the basis for our understanding of elementary particles, which are modeled as excitations in the fundamental fields.[13]

In quantum field theory, one typically computes the probabilities of various physical events using the techniques of perturbation theory. Developed by Richard Feynman and others in the first half of the twentieth century, perturbative quantum field theory uses special diagrams called Feynman diagrams to organize computations. One imagines that these diagrams depict the paths of point-like particles and their interactions.[13]

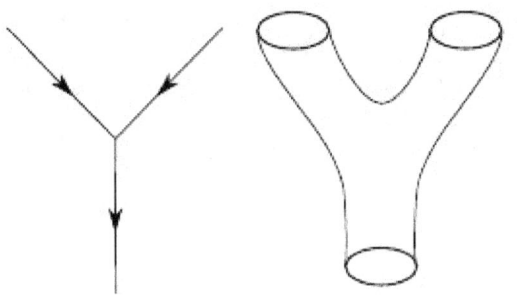

Interaction in the quantum world: worldlines of point-like particles or a worldsheet swept up by closed strings in string theory.

The starting point for string theory is the idea that the point-like particles of quantum field theory can also be modeled as one-dimensional objects called strings.[14] The interaction of strings is most straightforwardly defined by generalizing the perturbation theory used in ordinary quantum field theory. At the level of Feynman diagrams, this means replacing the one-dimensional diagram representing the path of a point particle by a two-dimensional surface representing the motion of a string.[15] Unlike in quantum field theory, string theory does not yet have a full non-perturbative definition, so many of the theoretical questions that physicists would like to answer remain out of reach.[16]

In theories of particle physics based on string theory, the characteristic length scale of strings is assumed to be on the order of the Planck length, or 10^{-35} meters, the scale at which the effects of quantum gravity are believed to become significant.[15] On much larger length scales, such as the scales visible in physics laboratories, such objects would be indistinguishable from zero-dimensional point particles, and the vibrational state of the string would determine the type of particle. One of the vibrational states of a string corresponds to the graviton, a quantum mechanical particle that carries the gravitational force.[3]

The original version of string theory was bosonic string theory, but this version described only bosons, a class of particles which transmit forces between the matter particles, or fermions. Bosonic string theory was eventually superseded by theories called superstring theories. These theories describe both bosons and fermions, and they incorporate a theoretical idea called supersymmetry. This is a mathematical relation that exists in certain physical theories between the bosons and fermions. In theories with supersymmetry, each boson has a counterpart which is a fermion, and vice versa.[17]

There are several versions of superstring theory: type I, type IIA, type IIB, and two flavors of heterotic string theory ($SO(32)$ and $E_8 \times E_8$). The different theories allow different types of strings, and the particles that arise at low energies exhibit different symmetries. For example, the type I theory includes both open strings (which are segments with endpoints) and closed strings (which form closed loops), while types IIA and IIB include only closed strings.[18]

2.1.2 Extra dimensions

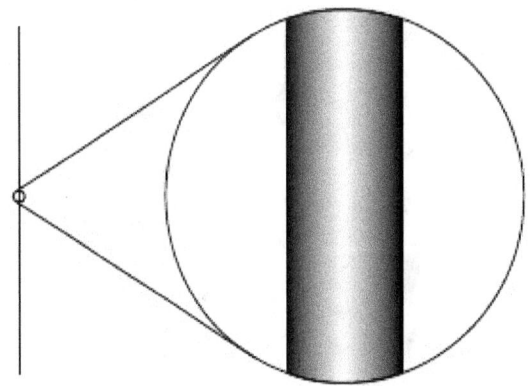

An example of compactification: At large distances, a two dimensional surface with one circular dimension looks one-dimensional.

In everyday life, there are three familiar dimensions of space: height, width and length. Einstein's general theory of relativity treats time as a dimension on par with the three spatial dimensions; in general relativity, space and time are not modeled as separate entities but are instead unified to a four-dimensional spacetime. In this framework, the phenomenon of gravity is viewed as a consequence of the geometry of spacetime.[19]

In spite of the fact that the universe is well described by four-dimensional spacetime, there are several reasons why physicists consider theories in other dimensions. In some cases, by modeling spacetime in a different number of dimensions, a theory becomes more mathematically tractable, and one can perform calculations and gain general insights more easily.[lower-alpha 2] There are also situations where theories in two or three spacetime dimensions are useful for describing phenomena in condensed matter physics.[20] Finally, there exist scenarios in which there could actually be more than four dimensions of spacetime which have nonetheless managed to escape detection.[21]

One notable feature of string theories is that these theories require extra dimensions of spacetime for their mathematical consistency. In bosonic string theory, spacetime is 26-dimensional, while in superstring theory it is ten-dimensional. In order to describe real physical phenomena using string theory, one must therefore imagine scenarios in which these extra dimensions would not be observed in experiments.[22]

A cross section of a quintic Calabi–Yau manifold

Compactification is one way of modifying the number of dimensions in a physical theory. In compactification, some of the extra dimensions are assumed to "close up" on themselves to form circles.[23] In the limit where these curled up dimensions become very small, one obtains a theory in which spacetime has effectively a lower number of dimensions. A standard analogy for this is to consider a multidimensional object such as a garden hose. If the hose is viewed from a sufficient distance, it appears to have only one dimension, its length. However, as one approaches the hose, one discovers that it contains a second dimension, its circumference. Thus, an ant crawling on the surface of the hose would move in two dimensions.[24]

Compactification can be used to construct models in which spacetime is effectively four-dimensional. However, not every way of compactifying the extra dimensions produces a model with the right properties to describe nature. In a viable model of particle physics, the compact extra dimensions must be shaped like a Calabi–Yau manifold.[23] A Calabi–Yau manifold is a special space which is typically taken to be six-dimensional in applications to string theory. It is named after mathematicians Eugenio Calabi and Shing-Tung Yau.[25]

Another approach to reducing the number of dimensions is the so-called brane-world scenario. In this approach, physicists assume that the observable universe is a four-dimensional subspace of a higher dimensional space. In such models, the force-carrying bosons of particle physics arise from open strings with endpoints attached to the

four-dimensional subspace, while gravity arises from closed strings propagating through the larger ambient space. This idea plays an important role in attempts to develop models of real world physics based on string theory, and it provides a natural explanation for the weakness of gravity compared to the other fundamental forces.[26]

2.1.3 Dualities

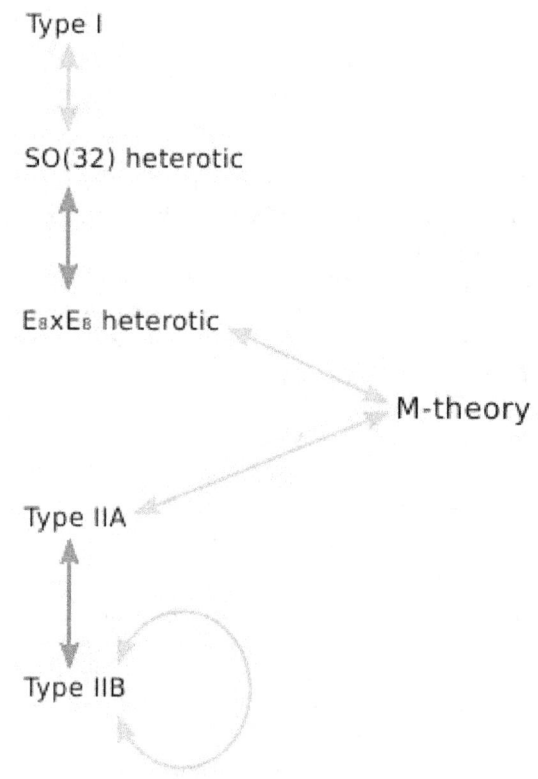

A diagram of string theory dualities. Yellow arrows indicate S-duality. Blue arrows indicate T-duality.

Main articles: S-duality and T-duality

One notable fact about string theory is that the different versions of the theory all turn out to be related in highly nontrivial ways. One of the relationships that can exist between different string theories is called S-duality. This is a relationship which says that a collection of strongly interacting particles in one theory can, in some cases, be viewed as a collection of weakly interacting particles in a completely different theory. Roughly speaking, a collection of particles is said to be strongly interacting if they combine and decay often and weakly interacting if they do so infrequently. Type I string theory turns out to be equivalent by S-duality to the $SO(32)$ heterotic string theory. Similarly, type IIB string theory is related to itself in a nontrivial way by S-duality.[27]

Another relationship between different string theories is T-duality. Here one considers strings propagating around a circular extra dimension. T-duality states that a string propagating around a circle of radius R is equivalent to a string propagating around a circle of radius $1/R$ in the sense that all observable quantities in one description are identified with quantities in the dual description. For example, a string has momentum as it propagates around a circle, and it can also wind around the circle one or more times. The number of times the string winds around a circle is called the winding number. If a string has momentum p and winding number n in one description, it will have momentum n and winding number p in the dual description. For example, type IIA string theory is equivalent to type IIB string theory via T-duality, and the two versions of heterotic string theory are also related by T-duality.[27]

In general, the term *duality* refers to a situation where two seemingly different physical systems turn out to be equivalent in a nontrivial way. Two theories related by a duality need not be string theories. For example, Montonen–Olive duality is example of an S-duality relationship between quantum field theories. The AdS/CFT correspondence is example of a duality which relates string theory to a quantum field theory. If two theories are related by a duality, it means that one theory can be transformed in some way so that it ends up looking just like the other theory. The two theories are then said to be *dual* to one another under the transformation. Put differently, the two theories are mathematically different descriptions of the same phenomena.[28]

2.1.4   Branes

Main article: Brane

In string theory and related theories, a brane is a physi-

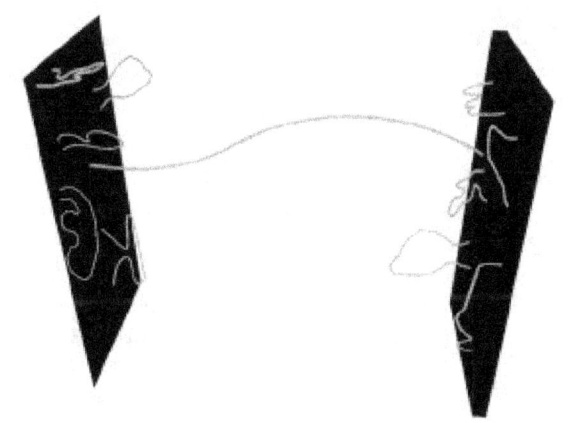

Open strings attached to a pair of D-branes

cal object that generalizes the notion of a point particle to higher dimensions. For example, a point particle can be viewed as a brane of dimension zero, while a string can be viewed as a brane of dimension one. It is also possible to consider higher-dimensional branes. In dimension p, these are called p-branes. The word brane comes from the word "membrane" which refers to a two-dimensional brane.[29]

Branes are dynamical objects which can propagate through spacetime according to the rules of quantum mechanics. They have mass and can have other attributes such as charge. A p-brane sweeps out a $(p+1)$-dimensional volume in spacetime called its *worldvolume*. Physicists often study fields analogous to the electromagnetic field which live on the worldvolume of a brane.[29]

In string theory, D-branes are an important class of branes that arise when one considers open strings. As an open string propagates through spacetime, its endpoints are required to lie on a D-brane. The letter "D" in D-brane refers to a certain mathematical condition on the system known as the Dirichlet boundary condition. The study of D-branes in string theory has led to important results such as the AdS/CFT correspondence, which has shed light on many problems in quantum field theory.[30]

Branes are also frequently studied from a purely mathematical point of view. Mathematically, branes can be described as objects of certain categories, such as the derived category of coherent sheaves on a complex algebraic variety, or the Fukaya category of a symplectic manifold.[31] The connection between the physical notion of a brane and the mathematical notion of a category has led to important mathematical insights in the fields of algebraic and symplectic geometry[32] and representation theory.[33]

2.2   M-theory

Main article: M-theory

Prior to 1995, theorists believed that there were five consistent versions of superstring theory (type I, type IIA, type IIB, and two versions of heterotic string theory). This understanding changed in 1995 when Edward Witten suggested that the five theories were just special limiting cases of an eleven-dimensional theory called M-theory. Witten's conjecture was based on the work of a number of other physicists, including Ashoke Sen, Chris Hull, Paul Townsend, and Michael Duff. His announcement led to a flurry of research activity now known as the second superstring revolution.[34]

2.2.1   Unification of superstring theories

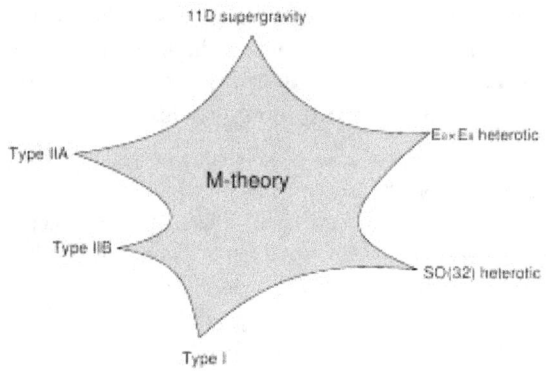

A schematic illustration of the relationship between M-theory, the five superstring theories, and eleven-dimensional supergravity. The shaded region represents a family of different physical scenarios that are possible in M-theory. In certain limiting cases corresponding to the cusps, it is natural to describe the physics using one of the six theories labeled there.

In the 1970s, many physicists became interested in supergravity theories, which combine general relativity with supersymmetry. Whereas general relativity makes sense in any number of dimensions, supergravity places an upper limit on the number of dimensions.[35] In 1978, work by Werner Nahm showed that the maximum spacetime dimension in which one can formulate a consistent supersymmetric theory is eleven.[36] In the same year, Eugene Cremmer, Bernard Julia, and Joel Scherk of the École Normale Supérieure showed that supergravity not only permits up to eleven dimensions but is in fact most elegant in this maximal number of dimensions.[37][38]

Initially, many physicists hoped that by compactifying eleven-dimensional supergravity, it might be possible to construct realistic models of our four-dimensional world. The hope was that such models would provide a unified description of the four fundamental forces of nature: electromagnetism, the strong and weak nuclear forces, and gravity. Interest in eleven-dimensional supergravity soon waned as various flaws in this scheme were discovered. One of the problems was that the laws of physics appear to distinguish between clockwise and counterclockwise, a phenomenon known as chirality. Edward Witten and others observed this chirality property cannot be readily derived by compactifying from eleven dimensions.[38]

In the first superstring revolution in 1984, many physicists turned to string theory as a unified theory of particle physics and quantum gravity. Unlike supergravity theory, string theory was able to accommodate the chirality of the standard model, and it provided a theory of gravity consistent with quantum effects.[38] Another feature of string theory

that many physicists were drawn to in the 1980s and 1990s was its high degree of uniqueness. In ordinary particle theories, one can consider any collection of elementary particles whose classical behavior is described by an arbitrary Lagrangian. In string theory, the possibilities are much more constrained: by the 1990s, physicists had argued that there were only five consistent supersymmetric versions of the theory.[38]

Although there were only a handful of consistent superstring theories, it remained a mystery why there was not just one consistent formulation.[38] However, as physicists began to examine string theory more closely, they realized that these theories are related in intricate and nontrivial ways. They found that a system of strongly interacting strings can, in some cases, be viewed as a system of weakly interacting strings. This phenomenon is known as S-duality. It was studied by Ashoke Sen in the context of heterotic strings in four dimensions[39][40] and by Chris Hull and Paul Townsend in the context of the type IIB theory.[41] Theorists also found that different string theories may be related by T-duality. This duality implies that strings propagating on completely different spacetime geometries may be physically equivalent.[42]

At around the same time, as many physicists were studying the properties of strings, a small group of physicists was examining the possible applications of higher dimensional objects. In 1987, Eric Bergshoeff, Ergin Sezgin, and Paul Townsend showed that eleven-dimensional supergravity includes two-dimensional branes.[43] Intuitively, these objects look like sheets or membranes propagating through the eleven-dimensional spacetime. Shortly after this discovery, Michael Duff, Paul Howe, Takeo Inami, and Kellogg Stelle considered a particular compactification of eleven-dimensional supergravity with one of the dimensions curled up into a circle.[44] In this setting, one can imagine the membrane wrapping around the circular dimension. If the radius of the circle is sufficiently small, then this membrane looks just like a string in ten-dimensional spacetime. In fact, Duff and his collaborators showed that this construction reproduces exactly the strings appearing in type IIA superstring theory.[45]

Speaking at a string theory conference in 1995, Edward Witten made the surprising suggestion that all five superstring theories were in fact just different limiting cases of a single theory in eleven spacetime dimensions. Witten's announcement drew together all of the previous results on S- and T-duality and the appearance of higher dimensional branes in string theory.[46] In the months following Witten's announcement, hundreds of new papers appeared on the Internet confirming different parts of his proposal.[47] Today this flurry of work is known as the second superstring revolution.[48]

Initially, some physicists suggested that the new theory was a fundamental theory of membranes, but Witten was skeptical of the role of membranes in the theory. In a paper from 1996, Hořava and Witten wrote "As it has been proposed that the eleven-dimensional theory is a supermembrane theory but there are some reasons to doubt that interpretation, we will non-committally call it the M-theory, leaving to the future the relation of M to membranes."[49] In the absence of an understanding of the true meaning and structure of M-theory, Witten has suggested that the *M* should stand for "magic", "mystery", or "membrane" according to taste, and the true meaning of the title should be decided when a more fundamental formulation of the theory is known.[50]

2.2.2   Matrix theory

Main article: Matrix theory (physics)

In mathematics, a matrix is a rectangular array of numbers or other data. In physics, a matrix model is a particular kind of physical theory whose mathematical formulation involves the notion of a matrix in an important way. A matrix model describes the behavior of a set of matrices within the framework of quantum mechanics.[51]

One important example of a matrix model is the BFSS matrix model proposed by Tom Banks, Willy Fischler, Stephen Shenker, and Leonard Susskind in 1997. This theory describes the behavior of a set of nine large matrices. In their original paper, these authors showed, among other things, that the low energy limit of this matrix model is described by eleven-dimensional supergravity. These calculations led them to propose that the BFSS matrix model is exactly equivalent to M-theory. The BFSS matrix model can therefore be used as a prototype for a correct formulation of M-theory and a tool for investigating the properties of M-theory in a relatively simple setting.[51]

The development of the matrix model formulation of M-theory has led physicists to consider various connections between string theory and a branch of mathematics called noncommutative geometry. This subject is a generalization of ordinary geometry in which mathematicians define new geometric notions using tools from noncommutative algebra.[52] In a paper from 1998, Alain Connes, Michael R. Douglas, and Albert Schwarz showed that some aspects of matrix models and M-theory are described by a noncommutative quantum field theory, a special kind of physical theory in which spacetime is described mathematically using noncommutative geometry.[53] This established a link between matrix models and M-theory on the one hand, and noncommutative geometry on the other hand. It quickly led to the discovery of other important links between noncommutative geometry and various physical

theories.[54][55]

2.3   Black holes

In general relativity, a black hole is defined as a region of spacetime in which the gravitational field is so strong that no particle or radiation can escape. In the currently accepted models of stellar evolution, black holes are thought to arise when massive stars undergo gravitational collapse, and many galaxies are thought to contain supermassive black holes at their centers. Black holes are also important for theoretical reasons, as they present profound challenges for theorists attempting to understand the quantum aspects of gravity. String theory has proved to be an important tool for investigating the theoretical properties of black holes because it provides a framework in which theorists can study their thermodynamics.[56]

2.3.1   Bekenstein–Hawking formula

In the branch of physics called statistical mechanics, entropy is a measure of the randomness or disorder of a physical system. This concept was studied in the 1870s by the Austrian physicist Ludwig Boltzmann, who showed that the thermodynamic properties of a gas could be derived from the combined properties of its many constituent molecules. Boltzmann argued that by averaging the behaviors of all the different molecules in a gas, one can understand macroscopic properties such as volume, temperature, and pressure. In addition, this perspective led him to give a precise definition of entropy as the natural logarithm of the number of different states of the molecules (also called *microstates*) that give rise to the same macroscopic features.[57]

In the twentieth century, physicists began to apply the same concepts to black holes. In most systems such as gases, the entropy scales with the volume. In the 1970s, the physicist Jacob Bekenstein suggested that the entropy of a black hole is instead proportional to the *surface area* of its event horizon, the boundary beyond which matter and radiation is lost to its gravitational attraction.[58] When combined with ideas of the physicist Stephen Hawking,[59] Bekenstein's work yielded a precise formula for the entropy of a black hole. The formula expresses the entropy S as

$$S = \frac{c^3 kA}{4\hbar G}$$

where c is the speed of light, k is Boltzmann's constant, $\hbar$ is the reduced Planck constant, G is Newton's constant, and A is the surface area of the event horizon.[60]

Like any physical system, a black hole has an entropy defined in terms of the number of different microstates that lead to the same macroscopic features. The Bekenstein–Hawking entropy formula gives the expected value of the entropy of a black hole, but by the 1990s, physicists still lacked a derivation of this formula by counting microstates in a theory of quantum gravity. Finding such a derivation of this formula was considered an important test of the viability of any theory of quantum gravity such as string theory.[61]

2.3.2 Derivation within string theory

In a paper from 1996, Andrew Strominger and Cumrun Vafa showed how to derive the Beckenstein–Hawking formula for certain black holes in string theory.[62] Their calculation was based on the observation that D-branes—which look like fluctuating membranes when they are weakly interacting—become dense, massive objects with event horizons when the interactions are strong. In other words, a system of strongly interacting D-branes in string theory is indistinguishable from a black hole. Strominger and Vafa analyzed such D-brane systems and calculated the number of different ways of placing D-branes in spacetime so that their combined mass and charge is equal to a given mass and charge for the resulting black hole. Their calculation reproduced the Bekenstein–Hawking formula exactly, including the factor of 1/4.[63] Subsequent work by Strominger, Vafa, and others refined the original calculations and gave the precise values of the "quantum corrections" needed to describe very small black holes.[64][65]

The black holes that Strominger and Vafa considered in their original work were quite different from real astrophysical black holes. One difference was that Strominger and Vafa considered only extremal black holes in order to make the calculation tractable. These are defined as black holes with the lowest possible mass compatible with a given charge.[66] Strominger and Vafa also restricted attention to black holes in five-dimensional spacetime with unphysical supersymmetry.[67]

Although it was originally developed in this very particular and physically unrealistic context in string theory, the entropy calculation of Strominger and Vafa has led to a qualitative understanding of how black hole entropy can be accounted for in any theory of quantum gravity. Indeed, in 1998, Strominger argued that the original result could be generalized to an arbitrary consistent theory of quantum gravity without relying on strings or supersymmetry.[68] In collaboration with several other authors in 2010, he showed that some results on black hole entropy could be extended to non-extremal astrophysical black holes.[69][70]

2.4 AdS/CFT correspondence

Main article: AdS/CFT correspondence

One approach to formulating string theory and studying its properties is provided by the anti-de Sitter/conformal field theory (AdS/CFT) correspondence. This is a theoretical result which implies that string theory is in some cases equivalent to a quantum field theory. In addition to providing insights into the mathematical structure of string theory, the AdS/CFT correspondence has shed light on many aspects of quantum field theory in regimes where traditional calculational techniques are ineffective.[6] The AdS/CFT correspondence was first proposed by Juan Maldacena in late 1997.[71] Important aspects of the correspondence were elaborated in articles by Steven Gubser, Igor Klebanov, and Alexander Markovich Polyakov,[72] and by Edward Witten.[73] By 2010, Maldacena's article had over 7000 citations, becoming the most highly cited article in the field of high energy physics.[lower-alpha 3]

2.4.1 Overview of the correspondence

In the AdS/CFT correspondence, the geometry of spacetime is described in terms of a certain vacuum solution of Einstein's equation called anti-de Sitter space.[74] In very elementary terms, anti-de Sitter space is a mathematical model of spacetime in which the notion of distance between points (the metric) is different from the notion of distance in ordinary Euclidean geometry. It is closely related to hyperbolic space, which can be viewed as a disk as illustrated on the left.[75] This image shows a tessellation of a disk by triangles and squares. One can define the distance between points of this disk in such a way that all the triangles and squares are the same size and the circular outer boundary is infinitely far from any point in the interior.[76]

One can imagine a stack of hyperbolic disks where each disk represents the state of the universe at a given time. The resulting geometric object is three-dimensional anti-de Sitter space.[75] It looks like a solid cylinder in which any cross section is a copy of the hyperbolic disk. Time runs along the vertical direction in this picture. The surface of this cylinder plays an important role in the AdS/CFT correspondence. As with the hyperbolic plane, anti-de Sitter space is curved in such a way that any point in the interior is actually infinitely far from this boundary surface.[76]

This construction describes a hypothetical universe with only two space dimensions and one time dimension, but it can be generalized to any number of dimensions. Indeed, hyperbolic space can have more than two dimensions and one can "stack up" copies of hyperbolic space to get higher-dimensional models of anti-de Sitter space.[75]

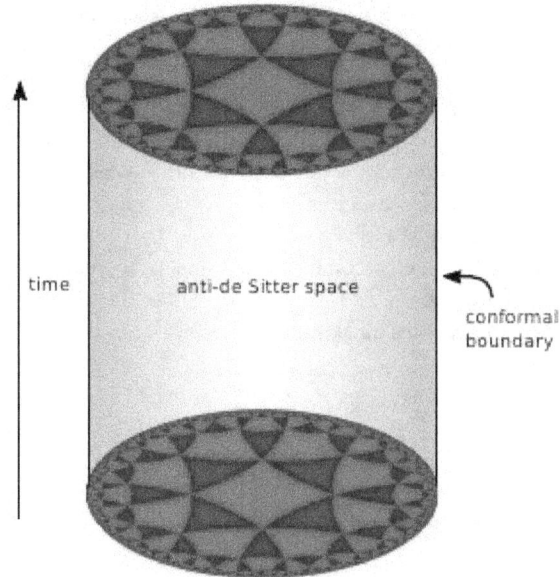

Three-dimensional anti-de Sitter space is like a stack of hyperbolic disks, each one representing the state of the universe at a given time. The resulting spacetime looks like a solid cylinder.

An important feature of anti-de Sitter space is its boundary (which looks like a cylinder in the case of three-dimensional anti-de Sitter space). One property of this boundary is that, within a small region on the surface around any given point, it looks just like Minkowski space, the model of spacetime used in nongravitational physics.[77] One can therefore consider an auxiliary theory in which "spacetime" is given by the boundary of anti-de Sitter space. This observation is the starting point for AdS/CFT correspondence, which states that the boundary of anti-de Sitter space can be regarded as the "spacetime" for a quantum field theory. The claim is that this quantum field theory is equivalent to a gravitational theory, such as string theory, in the bulk anti-de Sitter space in the sense that there is a "dictionary" for translating entities and calculations in one theory into their counterparts in the other theory. For example, a single particle in the gravitational theory might correspond to some collection of particles in the boundary theory. In addition, the predictions in the two theories are quantitatively identical so that if two particles have a 40 percent chance of colliding in the gravitational theory, then the corresponding collections in the boundary theory would also have a 40 percent chance of colliding.[78]

2.4.2 Applications to quantum gravity

The discovery of the AdS/CFT correspondence was a major advance in physicists' understanding of string theory and quantum gravity. One reason for this is that the corresponding-

dence provides a formulation of string theory in terms of quantum field theory, which is well understood by comparison. Another reason is that it provides a general framework in which physicists can study and attempt to resolve the paradoxes of black holes.[56]

In 1975, Stephen Hawking published a calculation which suggested that black holes are not completely black but emit a dim radiation due to quantum effects near the event horizon.[59] At first, Hawking's result posed a problem for theorists because it suggested that black holes destroy information. More precisely, Hawking's calculation seemed to conflict with one of the basic postulates of quantum mechanics, which states that physical systems evolve in time according to the Schrödinger equation. This property is usually referred to as unitarity of time evolution. The apparent contradiction between Hawking's calculation and the unitarity postulate of quantum mechanics came to be known as the black hole information paradox.[79]

The AdS/CFT correspondence resolves the black hole information paradox, at least to some extent, because it shows how a black hole can evolve in a manner consistent with quantum mechanics in some contexts. Indeed, one can consider black holes in the context of the AdS/CFT correspondence, and any such black hole corresponds to a configuration of particles on the boundary of anti-de Sitter space.[80] These particles obey the usual rules of quantum mechanics and in particular evolve in a unitary fashion, so the black hole must also evolve in a unitary fashion, respecting the principles of quantum mechanics.[81] In 2005, Hawking announced that the paradox had been settled in favor of information conservation by the AdS/CFT correspondence, and he suggested a concrete mechanism by which black holes might preserve information.[82]

2.4.3 Applications to quantum field theory

Main articles: AdS/QCD correspondence and AdS/CMT correspondence

In addition to its applications to theoretical problems in quantum gravity, the AdS/CFT correspondence has been applied to a variety of problems in quantum field theory. One physical system that has been studied using the AdS/CFT correspondence is the quark–gluon plasma, an exotic state of matter produced in particle accelerators. This state of matter arises for brief instants when heavy ions such as gold or lead nuclei are collided at high energies. Such collisions cause the quarks that make up atomic nuclei to deconfine at temperatures of approximately two trillion kelvins, conditions similar to those present at around 10^{-11} seconds after the Big Bang.[83]

The physics of the quark–gluon plasma is governed by a theory called quantum chromodynamics, but this the-

A magnet levitating above a high-temperature superconductor. Today some physicists are working to understand high-temperature superconductivity using the AdS/CFT correspondence.[7]

ory is mathematically intractable in problems involving the quark–gluon plasma.[lower-alpha 4] In an article appearing in 2005, Đàm Thanh Sơn and his collaborators showed that the AdS/CFT correspondence could be used to understand some aspects of the quark–gluon plasma by describing it in the language of string theory.[84] By applying the AdS/CFT correspondence, Sơn and his collaborators were able to describe the quark gluon plasma in terms of black holes in five-dimensional spacetime. The calculation showed that the ratio of two quantities associated with the quark–gluon plasma, the shear viscosity and volume density of entropy, should be approximately equal to a certain universal constant. In 2008, the predicted value of this ratio for the quark–gluon plasma was confirmed at the Relativistic Heavy Ion Collider at Brookhaven National Laboratory.[85][86]

The AdS/CFT correspondence has also been used to study aspects of condensed matter physics. Over the decades, experimental condensed matter physicists have discovered a number of exotic states of matter, including superconductors and superfluids. These states are described using the formalism of quantum field theory, but some phenomena are difficult to explain using standard field theoretic techniques. Some condensed matter theorists including Subir Sachdev hope that the AdS/CFT correspondence will make it possible to describe these systems in the language of string theory and learn more about their behavior.[85]

So far some success has been achieved in using string theory methods to describe the transition of a superfluid to an insulator. A superfluid is a system of electrically neutral atoms that flows without any friction. Such systems are often produced in the laboratory using liquid helium, but recently experimentalists have developed new ways of producing artificial superfluids by pouring trillions of cold atoms into a lattice of criss-crossing lasers. These atoms

initially behave as a superfluid, but as experimentalists increase the intensity of the lasers, they become less mobile and then suddenly transition to an insulating state. During the transition, the atoms behave in an unusual way. For example, the atoms slow to a halt at a rate that depends on the temperature and on Planck's constant, the fundamental parameter of quantum mechanics, which does not enter into the description of the other phases. This behavior has recently been understood by considering a dual description where properties of the fluid are described in terms of a higher dimensional black hole.[87]

2.5 Phenomenology

Main article: String phenomenology

In addition to being an idea of considerable theoretical interest, string theory provides a framework for constructing models of real world physics that combine general relativity and particle physics. Phenomenology is the branch of theoretical physics in which physicists construct realistic models of nature from more abstract theoretical ideas. String phenomenology is the part of string theory that attempts to construct realistic models based on string theory.

Partly because of theoretical and mathematical difficulties and partly because of the extremely high energies needed to test these theories experimentally, there is so far no experimental evidence that would unambiguously point to any of these models being a correct fundamental description of nature. This has led some in the community to criticize these approaches to unification and question the value of continued research on these problems.[12]

2.5.1 Particle physics

The currently accepted theory describing elementary particles and their interactions is known as the standard model of particle physics. This theory provides a unified description of three of the fundamental forces of nature: electromagnetism and the strong and weak nuclear forces. Despite its remarkable success in explaining a wide range of physical phenomena, the standard model cannot be a complete description of reality. This is because the standard model fails to incorporate the force of gravity and because of problems such as the hierarchy problem and the inability to explain the structure of fermion masses or dark matter.

String theory has been used to construct a variety of models of particle physics going beyond the standard model. Typically, such models are based on the idea of compactification. Starting with the ten- or eleven-dimensional spacetime of string or M-theory, physicists postulate a shape for the extra dimensions. By choosing this shape appropriately, they can construct models roughly similar to the standard model of particle physics, together with additional undiscovered particles.[88] One popular way of deriving realistic physics from string theory is to start with the heterotic theory in ten dimensions and assume that the six extra dimensions of spacetime are shaped like a six-dimensional Calabi–Yau manifold. Such compactifications offer many ways of extracting realistic physics from string theory. Other similar methods can be used to construct realistic models of our four-dimensional world based on M-theory.[89]

2.5.2 Cosmology

Main article: String cosmology

The Big Bang theory is the prevailing cosmological model

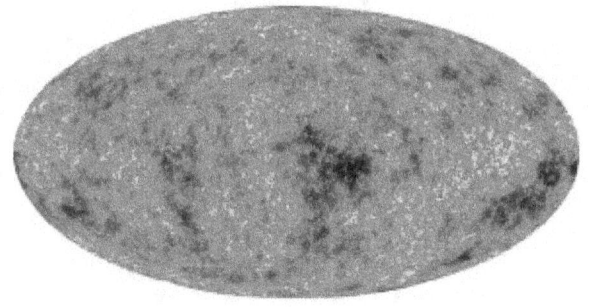

A map of the cosmic microwave background produced by the Wilkinson Microwave Anisotropy Probe

for the universe from the earliest known periods through its subsequent large-scale evolution. Despite its success in explaining many observed features of the universe including galactic redshifts, the relative abundance of light elements such as hydrogen and helium, and the existence of a cosmic microwave background, there are several questions that remain unanswered. For example, the standard Big Bang model does not explain why the universe appears to be same in all directions, why it appears flat on very large distance scales, or why certain hypothesized particles such as magnetic monopoles are not observed in experiments.[90]

Currently, the leading candidate for a theory going beyond the Big Bang is the theory of cosmic inflation. Developed by Alan Guth and others in the 1980s, inflation postulates a period of extremely rapid accelerated expansion of the universe prior to the expansion described by the standard Big Bang theory. The theory of cosmic inflation preserves the successes of the Big Bang while providing a natural explanation for some of the mysterious features of the universe.[91] The theory has also received striking support from observations of the cosmic microwave background,

the radiation that has filled the sky since around 380,000 years after the Big Bang.[92]

In the theory of inflation, the rapid initial expansion of the universe is caused by a hypothetical particle called the inflaton. The exact properties of this particle are not fixed by the theory but should ultimately be derived from a more fundamental theory such as string theory.[93] Indeed, there have been a number of attempts to identify an inflaton within the spectrum of particles described by string theory and to study inflation using string theory. While these approaches might eventually find support in observational data such as measurements of the cosmic microwave background, the application of string theory to cosmology is still in its early stages.[94]

2.6 Connections to mathematics

In addition to influencing research in theoretical physics, string theory has stimulated a number of major developments in pure mathematics. Like many developing ideas in theoretical physics, string theory does not at present have a mathematically rigorous formulation in which all of its concepts can be defined precisely. As a result, physicists who study string theory are often guided by physical intuition to conjecture relationships between the seemingly different mathematical structures that are used to formalize different parts of the theory. These conjectures are later proved by mathematicians, and in this way, string theory serves as a source of new ideas in pure mathematics.[95]

2.6.1 Mirror symmetry

Main article: Mirror symmetry (string theory)

After Calabi–Yau manifolds had entered physics as a way to compactify extra dimensions in string theory, many physicists began studying these manifolds. In the late 1980s, several physicists noticed that given such a compactification of string theory, it is not possible to reconstruct uniquely a corresponding Calabi–Yau manifold.[96] Instead, two different versions of string theory, type IIA and type IIB, can be compactified on completely different Calabi–Yau manifolds giving rise to the same physics. In this situation, the manifolds are called mirror manifolds, and the relationship between the two physical theories is called mirror symmetry.[97]

Regardless of whether Calabi–Yau compactifications of string theory provide a correct description of nature, the existence of the mirror duality between different string theories has significant mathematical consequences. The Calabi–Yau manifolds used in string theory are of interest in pure mathematics, and mirror symmetry allows math-

The Clebsch cubic is an example of a kind of geometric object called an algebraic variety. A classical result of enumerative geometry states that there are exactly 27 straight lines that lie entirely on this surface.

ematicians to solve problems in enumerative geometry, a branch of mathematics concerned with counting the numbers of solutions to geometric questions.[31][98]

Enumerative geometry studies a class of geometric objects called algebraic varieties which are defined by the vanishing of polynomials. For example, the Clebsch cubic illustrated on the right is an algebraic variety defined using a certain polynomial of degree three in four variables. A celebrated result of nineteenth-century mathematicians Arthur Cayley and George Salmon states that there are exactly 27 straight lines that lie entirely on such a surface.[99]

Generalizing this problem, one can ask how many lines can be drawn on a quintic Calabi–Yau manifold, such as the one illustrated above, which is defined by a polynomial of degree five. This problem was solved by the nineteenth-century German mathematician Hermann Schubert, who found that there are exactly 2,875 such lines. In 1986, geometer Sheldon Katz proved that the number of curves, such as circles, that are defined by polynomials of degree two and lie entirely in the quintic is 609,250.[100]

By the year 1991, most of the classical problems of enumerative geometry had been solved and interest in enumerative geometry had begun to diminish.[101] The field was reinvigorated in May 1991 when physicists Philip Candelas, Xenia de la Ossa, Paul Green, and Linda Parks showed that mirror symmetry could be used to translate difficult mathematical questions about one Calabi–Yau manifold into easier questions about its mirror.[102] In particular, they used mirror symmetry to show that a six-dimensional Calabi–Yau

manifold can contain exactly 317,206,375 curves of degree three.[101] In addition to counting degree-three curves, Candelas and his collaborators obtained a number of more general results for counting rational curves which went far beyond the results obtained by mathematicians.[103]

Originally, these results of Candelas were justified on physical grounds. However, mathematicians generally prefer rigorous proofs that do not require an appeal to physical intuition. Inspired by physicists' work on mirror symmetry, mathematicians have therefore constructed their own arguments proving the enumerative predictions of mirror symmetry.[lower-alpha 5] Today mirror symmetry is an active area of research in mathematics, and mathematicians are working to develop a more complete mathematical understanding of mirror symmetry based on physicists' intuition.[104] Major approaches to mirror symmetry include the homological mirror symmetry program of Maxim Kontsevich[32] and the SYZ conjecture of Andrew Strominger, Shing-Tung Yau, and Eric Zaslow.[105]

2.6.2 Monstrous moonshine

Main article: Monstrous moonshine
Group theory is the branch of mathematics that studies the

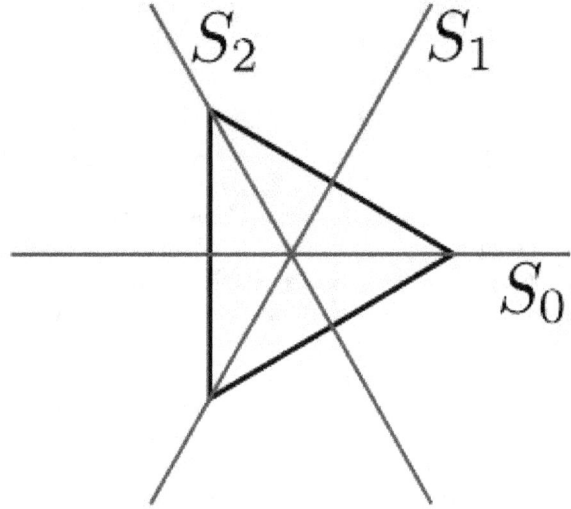

An equilateral triangle can be rotated through 120°, 240°, or 360°, or reflected in any of the three lines pictured without changing its shape.

concept of symmetry. For example, one can consider a geometric shape such as an equilateral triangle. There are various operations that one can perform on this triangle without changing its shape. One can rotate it through 120°, 240°, or 360°, or one can reflect in any of the lines labeled S_0, S_1, or S_2 in the picture. Each of these operations is called a *symmetry*, and the collection of these symmetries satisfies certain technical properties making it into what mathematicians call a group. In this particular example, the group is known as the dihedral group of order 6 because it has six elements. A general group may describe finitely many or infinitely many symmetries; if there are only finitely many symmetries, it is called a finite group.[106]

Mathematicians often strive for a classification (or list) of all mathematical objects of a given type. It is generally believed that finite groups are too diverse to admit a useful classification. A more modest but still challenging problem is to classify all finite *simple* groups. These are finite groups which may be used as building blocks for constructing arbitrary finite groups in the same way that prime numbers can be used to construct arbitrary whole numbers by taking products.[lower-alpha 6] One of the major achievements of contemporary group theory is the classification of finite simple groups, a mathematical theorem which provides a list of all possible finite simple groups.[107]

This classification theorem identifies several infinite families of groups as well as 26 additional groups which do not fit into any family. The latter groups are called the "sporadic" groups, and each one owes its existence to a remarkable combination of circumstances. The largest sporadic group, the so-called monster group, has over 10^{53} elements, more than a thousand times the number of atoms in the Earth.[108]

A graph of the j-function in the complex plane

A seemingly unrelated construction is the *j*-function of number theory. This object belongs to a special class of functions called modular functions, whose graphs form a certain kind of repeating pattern.[109] Although this function appears in a branch of mathematics which seems very different from the theory of finite groups, the two subjects turn out to be intimately related. In the late 1970s, mathematicians John McKay and John Thompson noticed that certain numbers arising in the analysis of the monster group (namely, the dimensions of its irreducible representations) are related to numbers that appear in a formula for the *j*-

function (namely, the coefficients of its Fourier series).[110] This relationship was further developed by John Horton Conway and Simon Norton[111] who called it monstrous moonshine because it seemed so far fetched.[112]

In 1992, Richard Borcherds constructed a bridge between the theory of modular functions and finite groups and, in the process, explained the observations of McKay and Thompson.[113][114] Borcherds' work used ideas from string theory in an essential way, extending earlier results of Igor Frenkel, James Lepowsky, and Arne Meurman, who had realized the monster group as the symmetries of a particular version of string theory.[115] In 1998, Borcherds was awarded the Fields medal for his work.[116]

Since the 1990s, the connection between string theory and moonshine has led to further results in mathematics and physics.[108] In 2010, physicists Tohru Eguchi, Hirosi Ooguri, and Yuji Tachikawa discovered connections between a different sporadic group, the Mathieu group M_{24}, and a certain version of string theory.[117] Miranda Cheng, John Duncan, and Jeffrey A. Harvey proposed a generalization of this moonshine phenomenon called umbral moonshine,[118] and their conjecture was proved mathematically by Duncan, Michael Griffin, and Ken Ono.[119] Witten has also speculated that the version of string theory appearing in monstrous moonshine might be related to a certain simplified model of gravity in three spacetime dimensions.[120]

2.7 History

Main article: History of string theory

2.7.1 Early results

Some of the structures reintroduced by string theory arose for the first time much earlier as part of the program of classical unification started by Albert Einstein. The first person to add a fifth dimension to a theory of gravity was Gunnar Nordström in 1914, who noted that gravity in five dimensions describes both gravity and electromagnetism in four. Nordström attempted to unify electromagnetism with his theory of gravitation, which was however superseded by Einstein's general relativity in 1919. Thereafter, German mathematician Theodor Kaluza combined the fifth dimension with general relativity, and only Kaluza is usually credited with the idea. In 1926, the Swedish physicist Oskar Klein gave a physical interpretation of the unobservable extra dimension—it is wrapped into a small circle. Einstein introduced a non-symmetric metric tensor, while much later Brans and Dicke added a scalar component to gravity. These ideas would be revived within string theory, where they are demanded by consistency conditions.

Leonard Susskind

String theory was originally developed during the late 1960s and early 1970s as a never completely successful theory of hadrons, the subatomic particles like the proton and neutron that feel the strong interaction. In the 1960s, Geoffrey Chew and Steven Frautschi discovered that the mesons make families called Regge trajectories with masses related to spins in a way that was later understood by Yoichiro Nambu, Holger Bech Nielsen and Leonard Susskind to be the relationship expected from rotating strings. Chew advocated making a theory for the interactions of these trajectories that did not presume that they were composed of any fundamental particles, but would construct their interactions from self-consistency conditions on the S-matrix. The S-matrix approach was started by Werner Heisenberg in the 1940s as a way of constructing a theory that did not rely on the local notions of space and time, which Heisenberg believed break down at the nuclear scale. While the scale was off by many orders of magnitude, the approach he advocated was ideally suited for a theory of quantum gravity.

Working with experimental data, R. Dolen, D. Horn and C. Schmid developed some sum rules for hadron exchange. When a particle and antiparticle scatter, virtual particles can

be exchanged in two qualitatively different ways. In the s-channel, the two particles annihilate to make temporary intermediate states that fall apart into the final state particles. In the t-channel, the particles exchange intermediate states by emission and absorption. In field theory, the two contributions add together, one giving a continuous background contribution, the other giving peaks at certain energies. In the data, it was clear that the peaks were stealing from the background—the authors interpreted this as saying that the t-channel contribution was dual to the s-channel one, meaning both described the whole amplitude and included the other.

Gabriele Veneziano

The result was widely advertised by Murray Gell-Mann, leading Gabriele Veneziano to construct a scattering amplitude that had the property of Dolen-Horn-Schmid duality, later renamed world-sheet duality. The amplitude needed poles where the particles appear, on straight line trajectories, and there is a special mathematical function whose poles are evenly spaced on half the real line— the Gamma function— which was widely used in Regge theory. By manipulating combinations of Gamma functions, Veneziano was able to find a consistent scattering amplitude with poles on straight lines, with mostly positive residues, which obeyed duality and had the appropriate Regge scaling at high energy. The amplitude could fit near-beam scattering data as well as other Regge type fits, and had a suggestive integral representation that could be used for generalization.

Over the next years, hundreds of physicists worked to complete the bootstrap program for this model, with many surprises. Veneziano himself discovered that for the scattering amplitude to describe the scattering of a particle that appears in the theory, an obvious self-consistency condition, the lightest particle must be a tachyon. Miguel Virasoro and Joel Shapiro found a different amplitude now understood to be that of closed strings, while Ziro Koba and Holger Nielsen generalized Veneziano's integral representation to multiparticle scattering. Veneziano and Sergio Fubini introduced an operator formalism for computing the scattering amplitudes that was a forerunner of world-sheet conformal theory, while Virasoro understood how to remove the poles with wrong-sign residues using a constraint on the states. Claud Lovelace calculated a loop amplitude, and noted that there is an inconsistency unless the dimension of the theory is 26. Charles Thorn, Peter Goddard and Richard Brower went on to prove that there are no wrong-sign propagating states in dimensions less than or equal to 26.

In 1969, Yoichiro Nambu, Holger Bech Nielsen, and Leonard Susskind recognized that the theory could be given a description in space and time in terms of strings. The scattering amplitudes were derived systematically from the action principle by Peter Goddard, Jeffrey Goldstone, Claudio Rebbi, and Charles Thorn, giving a space-time picture to the vertex operators introduced by Veneziano and Fubini and a geometrical interpretation to the Virasoro conditions.

John Schwarz

In 1970, Pierre Ramond added fermions to the model,

which led him to formulate a two-dimensional supersymmetry to cancel the wrong-sign states. John Schwarz and André Neveu added another sector to the fermi theory a short time later. In the fermion theories, the critical dimension was 10. Stanley Mandelstam formulated a world sheet conformal theory for both the bose and fermi case, giving a two-dimensional field theoretic path-integral to generate the operator formalism. Michio Kaku and Keiji Kikkawa gave a different formulation of the bosonic string, as a string field theory, with infinitely many particle types and with fields taking values not on points, but on loops and curves.

In 1974, Tamiaki Yoneya discovered that all the known string theories included a massless spin-two particle that obeyed the correct Ward identities to be a graviton. John Schwarz and Joel Scherk came to the same conclusion and made the bold leap to suggest that string theory was a theory of gravity, not a theory of hadrons. They reintroduced Kaluza–Klein theory as a way of making sense of the extra dimensions. At the same time, quantum chromodynamics was recognized as the correct theory of hadrons, shifting the attention of physicists and apparently leaving the bootstrap program in the dustbin of history.

String theory eventually made it out of the dustbin, but for the following decade all work on the theory was completely ignored. Still, the theory continued to develop at a steady pace thanks to the work of a handful of devotees. Ferdinando Gliozzi, Joel Scherk, and David Olive realized in 1976 that the original Ramond and Neveu Schwarz-strings were separately inconsistent and needed to be combined. The resulting theory did not have a tachyon, and was proven to have space-time supersymmetry by John Schwarz and Michael Green in 1981. The same year, Alexander Polyakov gave the theory a modern path integral formulation, and went on to develop conformal field theory extensively. In 1979, Daniel Friedan showed that the equations of motions of string theory, which are generalizations of the Einstein equations of General Relativity, emerge from the Renormalization group equations for the two-dimensional field theory. Schwarz and Green discovered T-duality, and constructed two superstring theories—IIA and IIB related by T-duality, and type I theories with open strings. The consistency conditions had been so strong, that the entire theory was nearly uniquely determined, with only a few discrete choices.

2.7.2 First superstring revolution

In the early 1980s, Edward Witten discovered that most theories of quantum gravity could not accommodate chiral fermions like the neutrino. This led him, in collaboration with Luis Alvarez-Gaumé to study violations of the conservation laws in gravity theories with anomalies, conclud-

Edward Witten

Joseph Polchinski

ing that type I string theories were inconsistent. Green and Schwarz discovered a contribution to the anomaly that Witten and Alvarez-Gaumé had missed, which restricted the gauge group of the type I string theory to be SO(32). In coming to understand this calculation, Edward Witten became convinced that string theory was truly a consistent theory of gravity, and he became a high-profile advocate. Following Witten's lead, between 1984 and 1986, hundreds of physicists started to work in this field, and this is sometimes called the first superstring revolution.

During this period, David Gross, Jeffrey Harvey, Emil Martinec, and Ryan Rohm discovered heterotic strings. The gauge group of these closed strings was two copies of E8, and either copy could easily and naturally include the standard model. Philip Candelas, Gary Horowitz, Andrew Strominger and Edward Witten found that the Calabi–Yau manifolds are the compactifications that preserve a realistic amount of supersymmetry, while Lance Dixon and others worked out the physical properties of orbifolds, distinctive geometrical singularities allowed in string theory. Cumrun Vafa generalized T-duality from circles to arbitrary manifolds, creating the mathematical field of mirror symmetry. Daniel Friedan, Emil Martinec and Stephen Shenker further developed the covariant quantization of the superstring using conformal field theory techniques. David Gross and Vipul Periwal discovered that string perturbation theory was divergent. Stephen Shenker showed it diverged much faster than in field theory suggesting that new nonperturbative objects were missing.

In the 1990s, Joseph Polchinski discovered that the theory requires higher-dimensional objects, called D-branes

and identified these with the black-hole solutions of supergravity. These were understood to be the new objects suggested by the perturbative divergences, and they opened up a new field with rich mathematical structure. It quickly became clear that D-branes and other p-branes, not just strings, formed the matter content of the string theories, and the physical interpretation of the strings and branes was revealed—they are a type of black hole. Leonard Susskind had incorporated the holographic principle of Gerardus 't Hooft into string theory, identifying the long highly excited string states with ordinary thermal black hole states. As suggested by 't Hooft, the fluctuations of the black hole horizon, the world-sheet or world-volume theory, describes not only the degrees of freedom of the black hole, but all nearby objects too.

2.7.3 Second superstring revolution

In 1995, at the annual conference of string theorists at the University of Southern California (USC), Edward Witten gave a speech on string theory that in essence united the five string theories that existed at the time, and giving birth to a new 11-dimensional theory called M-theory. M-theory was also foreshadowed in the work of Paul Townsend at approximately the same time. The flurry of activity that began at this time is sometimes called the second superstring revolution.[34]

Juan Maldacena

During this period, Tom Banks, Willy Fischler, Stephen Shenker and Leonard Susskind formulated matrix theory, a full holographic description of M-theory using IIA D0 branes.[51] This was the first definition of string theory that was fully non-perturbative and a concrete mathematical realization of the holographic principle. It is an example of a gauge-gravity duality and is now understood to be a special case of the AdS/CFT correspondence. Andrew Strominger and Cumrun Vafa calculated the entropy of certain configurations of D-branes and found agreement with the semi-classical answer for extreme charged black holes.[62] Petr Hořava and Witten found the eleven-dimensional formulation of the heterotic string theories, showing that orbifolds solve the chirality problem. Witten noted that the effective description of the physics of D-branes at low energies is by a supersymmetric gauge theory, and found geometrical interpretations of mathematical structures in gauge theory that he and Nathan Seiberg had earlier discovered in terms of the location of the branes.

In 1997, Juan Maldacena noted that the low energy excitations of a theory near a black hole consist of objects close to the horizon, which for extreme charged black holes looks like an anti-de Sitter space.[71] He noted that in this limit the gauge theory describes the string excitations near the branes. So he hypothesized that string theory on a near-horizon extreme-charged black-hole geometry, an anti-deSitter space times a sphere with flux, is equally well described by the low-energy limiting gauge theory, the N = 4 supersymmetric Yang–Mills theory. This hypothesis, which is called the AdS/CFT correspondence, was further developed by Steven Gubser, Igor Klebanov and Alexander Polyakov,[72] and by Edward Witten,[73] and it is now well-accepted. It is a concrete realization of the holographic principle, which has far-reaching implications for black holes, locality and information in physics, as well as the nature of the gravitational interaction.[56] Through this relationship, string theory has been shown to be related to gauge theories like quantum chromodynamics and this has led to more quantitative understanding of the behavior of hadrons, bringing string theory back to its roots.[84]

2.8 Criticism

2.8.1 Number of solutions

To construct models of particle physics based on string theory, physicists typically begin by specifying a shape for the extra dimensions of spacetime. Each of these different shapes corresponds to a different possible universe, or "vacuum state", with a different collection of particles and forces. String theory as it is currently understood has an enormous number of vacuum states, typically estimated to be around 10^{500}, and these might be sufficiently diverse to accommodate almost any phenomena that might be observed at low energies.[121]

Many critics of string theory have expressed concerns about the large number of possible universes described by string theory. In his book *Not Even Wrong*, Peter Woit, a lecturer in the mathematics department at Columbia University, has argued that the large number of different physical scenarios renders string theory vacuous as a framework for constructing models of particle physics. According to Woit,

> The possible existence of, say, 10^{500} consistent different vacuum states for superstring theory probably destroys the hope of using the theory to predict anything. If one picks among this large set just those states whose properties agree with present experimental observations, it is likely there still will be such a large number of these that one can get just about whatever value one wants for the results of any new observation.[122]

Some physicists believe this large number of solutions is actually a virtue because it may allow a natural anthropic explanation of the observed values of physical constants, in particular the small value of the cosmological constant.[122] The anthropic principle is the idea that some of the numbers appearing in the laws of physics are not fixed by any fundamental principle but must be compatible with the evolution of intelligent life. In 1987, Steven Weinberg published an article in which he argued that the cosmological constant could not have been too large, or else galaxies and intelligent life would not have been able to develop.[123] Weinberg suggested that there might be a huge number of possible consistent universes, each with a different value of the cosmological constant, and observations indicate a small value of the cosmological constant only because humans happen to live in a universe that has allowed intelligent life, and hence observers, to exist.[124]

String theorist Leonard Susskind has argued that string theory provides a natural anthropic explanation of the small value of the cosmological constant.[125] According to Susskind, the different vacuum states of string theory might be realized as different universes within a larger multiverse. The fact that the observed universe has a small cosmological constant is just a tautological consequence of the fact that a small value is required for life to exist.[126] Many prominent theorists and critics have disagreed with Susskind's conclusions.[127] According to Woit, "in this case [anthropic reasoning] is nothing more than an excuse for failure. Speculative scientific ideas fail not just when they make incorrect predictions, but also when they turn out to be vacuous and incapable of predicting anything."[128]

2.8.2   Background independence

Main article: Background independence

One of the fundamental principles of Einstein's general theory of relativity is the idea that the laws of physics should be background independent. This means that the geometry of spacetime is not specified from the outset but is instead determined dynamically by the theory. In general relativity, the geometry of spacetime can evolve in time, responding to whatever matter is present.[129]

One of the older criticisms of string theory is that it is not manifestly background independent. In string theory, one must typically specify a fixed reference geometry for spacetime, and all other possible geometries are described as perturbations of this fixed one. In his book *The Trouble With Physics*, physicist Lee Smolin of the Perimeter Institute for Theoretical Physics claims that this is the principal weakness of string theory as a theory of quantum gravity, saying that string theory has failed to incorporate this important insight from general relativity.[130]

Others have disagreed with Smolin's characterization of string theory. In a review of Smolin's book, string theorist Joseph Polchinski writes

> [Smolin] is mistaking an aspect of the mathematical language being used for one of the physics being described. New physical theories are often discovered using a mathematical language that is not the most suitable for them... In string theory it has always been clear that the physics is background-independent even if the language being used is not, and the search for more suitable language continues. Indeed, as Smolin belatedly notes, [AdS/CFT] provides a solution to this problem, one that is unexpected and powerful.[131]

Polchinski notes that an important open problem in quantum gravity is to develop holographic descriptions of gravity which do not require the gravitational field to be asymptotically anti-de Sitter.[131]

2.8.3   Sociological issues

Since the superstring revolutions of the 1980s and 1990s, string theory has become the dominant paradigm of high energy theoretical physics.[132] Some string theorists have expressed the view that there does not exist an equally successful alternative theory addressing the deep questions of fundamental physics. In an interview from 1987, Nobel laureate David Gross made the following controversial comments about the reasons for the popularity of string theory:

> The most important [reason] is that there are no other good ideas around. That's what gets most people into it. When people started to get interested in string theory they didn't know anything about it. In fact, the first reaction of most people is that the theory is extremely ugly and unpleasant, at least that was the case a few years ago when the understanding of string theory was much less developed. It was difficult for people to learn about it and to be turned on. So I think the real reason why people have got attracted by it is because there is no other game in town. All other approaches of constructing grand unified theories, which were more conservative to begin with, and only gradually became more and more radical, have failed, and this game hasn't failed yet.[133]

Several other high profile theorists and commentators have expressed similar views, suggesting that there are no viable alternatives to string theory.[134]

Many critics of string theory have commented on this state of affairs. In his book criticizing string theory, Peter Woit views the status of string theory research as unhealthy and detrimental to the future of fundamental physics. He argues that the extreme popularity of string theory among theoretical physicists is partly a consequence of the financial structure of academia and the fierce competition for scarce resources.[135] In his book *The Road to Reality*, mathematical physicist Roger Penrose expresses similar views, stating "The often frantic competitiveness that this ease of communication engenders leads to 'bandwagon' effects, where researchers fear to be left behind if they do not join in."[136] Penrose also claims that the technical difficulty of modern physics forces young scientists to rely on the preferences of established researchers, rather than forging new paths of their own.[137] Lee Smolin expresses a slightly different position in his critique, claiming that string theory grew out of a tradition of particle physics which discourages speculation about the foundations of physics, while his preferred approach, loop quantum gravity, encourages more radical thinking. According to Smolin,

> String theory is a powerful, well-motivated idea and deserves much of the work that has been devoted to it. If it has so far failed, the principal reason is that its intrinsic flaws are closely tied to its strengths—and, of course, the story is unfinished, since string theory may well turn out to be part of the truth. The real question is not why we have expended so much energy on string theory but why we haven't expended nearly enough on alternative approaches.[138]

Smolin goes on to offer a number of prescriptions for how scientists might encourage a greater diversity of approaches to quantum gravity research.[139]

2.9   References

2.9.1   Notes

[1] For example, physicists are still working to understand the phenomenon of quark confinement, the paradoxes of black holes, and the origin of dark energy.

[2] For example, in the context of the AdS/CFT correspondence, theorists often formulate and study theories of gravity in unphysical numbers of spacetime dimensions.

[3] "Top Cited Articles during 2010 in hep-th". Retrieved 25 July 2013.

[4] More precisely, one cannot apply the methods of perturbative quantum field theory.

[5] Two independent mathematical proofs of mirror symmetry were given by Givental 1996, 1998 and Lian, Liu, Yau 1997, 1999, 2000.

[6] More precisely, a nontrivial group is called *simple* if its only normal subgroups are the trivial group and the group itself. The Jordan–Hölder theorem exhibits finite simple groups as the building blocks for all finite groups.

2.9.2   Citations

[1] Becker, Becker, and Schwarz 2007, p. 1

[2] Zwiebach 2009, p. 6

[3] Becker, Becker, and Schwarz 2007, pp. 2–3

[4] Becker, Becker, and Schwarz 2007, pp. 9–12

[5] Becker, Becker, and Schwarz 2007, pp. 14–15

[6] Klebanov and Maldacena 2009

[7] Merali 2011

[8] Sachdev 2013

[9] Becker, Becker, and Schwarz 2007, pp. 3, 15–16

[10] Becker, Becker, and Schwarz 2007, p. 8

[11] Becker, Becker, and Schwarz 13–14

[12] Woit 2006

[13] Zee 2010

[14] Becker, Becker, and Schwarz 2007, p. 2

[15] Becker, Becker, and Schwarz 2007, p. 6

[16] Zwiebach 2009, p. 12

[17] Becker, Becker, and Schwarz 2007, p. 4

[18] Zwiebach 2009, p. 324

[19] Wald 1984, p. 4

[20] Zee 2010, Parts V and VI

[21] Zwiebach 2009, p. 9

[22] Zwiebach 2009, p. 8

[23] Yau and Nadis 2010, Ch. 6

[24] Greene 2000, p. 186

[25] Yau and Nadis 2010, p. ix

[26] Randall and Sundrum 1999

[27] Becker, Becker, and Schwarz 2007

[28] Zwiebach 2009, p. 376

[29] Moore 2005, p. 214

[30] Moore 2005, p. 215

[31] Aspinwall et al. 2009

[32] Kontsevich 1995

[33] Kapustin and Witten 2007

[34] Duff 1998

[35] Duff 1998, p. 64

[36] Nahm 1978

[37] Cremmer, Julia, and Scherk 1978

[38] Duff 1998, p. 65

[39] Sen 1994a

[40] Sen 1994b

[41] Hull and Townsend 1995

[42] Duff 1998, p. 67

[43] Bergshoeff, Sezgin, and Townsend 1987

[44] Duff et al. 1987

[45] Duff 1998, p. 66

[46] Witten 1995

[47] Duff 1998, pp. 67–68

[48] Becker, Becker, and Schwarz 2007, p. 296

[49] Hořava and Witten 1996

[50] Duff 1996, sec. 1

[51] Banks et al. 1997

[52] Connes 1994

[53] Connes, Douglas, and Schwarz 1998

[54] Nekrasov and Schwarz 1998

[55] Seiberg and Witten 1999

[56] de Haro et al. 2013, p. 2

[57] Yau and Nadis 2010, p. 187–188

[58] Bekenstein 1973

[59] Hawking 1975

[60] Wald 1984, p. 417

[61] Yau and Nadis 2010, p. 189

[62] Strominger and Vafa 1996

[63] Yau and Nadis 2010, pp. 190–192

[64] Maldacena, Strominger, and Witten 1997

[65] Ooguri, Strominger, and Vafa 2004

[66] Yau and Nadis 2010, pp. 192–193

[67] Yau and Nadis 2010, pp. 194–195

[68] Strominger 1998

[69] Guica et al. 2009

[70] Castro, Maloney, and Strominger 2010

[71] Maldacena 1998

[72] Gubser, Klebanov, and Polyakov 1998

[73] Witten 1998

[74] Klebanov and Maldacena 2009, p. 28

[75] Maldacena 2005, p. 60

[76] Maldacena 2005, p. 61

[77] Zwiebach 2009, p. 552

[78] Maldacena 2005, pp. 61–62

[79] Susskind 2008

[80] Zwiebach 2009, p. 554

[81] Maldacena 2005, p. 63

[82] Hawking 2005

[83] Zwiebach 2009, p. 559

[84] Kovtun, Son, and Starinets 2001

[85] Merali 2011, p. 303

[86] Luzum and Romatschke 2008

[87] Sachdev 2013, p. 51

[88] Candelas et al. 1985

[89] Yau and Nadis 2010, pp. 147–150

[90] Becker, Becker, and Schwarz 2007, pp. 530–531

[91] Becker, Becker, and Schwarz 2007, p. 531

[92] Becker, Becker, and Schwarz 2007, p. 538

[93] Becker, Becker, and Schwarz 2007, p. 533

[94] Becker, Becker, and Schwarz 2007, pp. 539–543

[95] Deligne et al. 1999, p. 1

[96] Hori et al. 2003, p. xvii

[97] Aspinwall et al. 2009, p. 13

[98] Hori et al. 2003

[99] Yau and Nadis 2010, p. 167

[100] Yau and Nadis 2010, p. 166

[101] Yau and Nadis 2010, p. 169

[102] Candelas et al. 1991

[103] Yau and Nadis 2010, p. 171

[104] Hori et al. 2003, p. xix

[105] Strominger, Yau, and Zaslow 1996

[106] Dummit and Foote 2004

[107] Dummit and Foote 2004, pp. 102–103

[108] Klarreich 2015

[109] Gannon 2006, p. 2

[110] Gannon 2006, p. 4

[111] Conway and Norton 1979

[112] Gannon 2006, p. 5

[113] Gannon 2006, p. 8

[114] Borcherds 1992

[115] Frenkel, Lepowsky, and Meurman 1988

[116] Gannon 2006, p. 11

[117] Eguchi, Ooguri, and Tachikawa 2010

[118] Cheng, Duncan, and Harvey 2013

[119] Duncan, Griffin, and Ono 2015

[120] Witten 2007

[121] Woit 2006, pp. 240–242

[122] Woit 2006, p. 242

[123] Weinberg 1987

[124] Woit 2006, p. 243

[125] Susskind 2005

[126] Woit 2006, pp. 242–243

[127] Woit 2006, p. 240

[128] Woit 2006, p. 249

[129] Smolin 2006, p. 81

[130] Smolin 2006, p. 184

[131] Polchinski 2007

[132] Penrose 2004, p. 1017

[133] Woit 2006, pp. 224–225

[134] Woit 2006, Ch. 16

[135] Woit 2006, p. 239

[136] Penrose 2004, p. 1018

[137] Penrose 2004, pp. 1019–1020

[138] Smolin 2006, p. 349

[139] Smolin 2006, Ch. 20

2.9.3 Bibliography

- Aspinwall, Paul; Bridgeland, Tom; Craw, Alastair; Douglas, Michael; Gross, Mark; Kapustin, Anton; Moore, Gregory; Segal, Graeme; Szendröi, Balázs; Wilson, P.M.H., eds. (2009). *Dirichlet Branes and Mirror Symmetry*. American Mathematical Society. ISBN 978-0-8218-3848-8.

- Banks, Tom; Fischler, Willy; Schenker, Stephen; Susskind, Leonard (1997). "M theory as a matrix model: A conjecture". *Physical Review D* **55** (8): 5112. arXiv:hep-th/9610043. Bibcode:1997PhRvD..55.5112B. doi:10.1103/physrevd.55.5112.

- Becker, Katrin; Becker, Melanie; Schwarz, John (2007). *String theory and M-theory: A modern introduction*. Cambridge University Press. ISBN 978-0-521-86069-7.

- Bekenstein, Jacob (1973). "Black holes and entropy". *Physical Review D* **7** (8): 2333. Bibcode:1973PhRvD...7.2333B. doi:10.1103/PhysRevD.7.2333.

- Bergshoeff, Eric; Sezgin, Ergin; Townsend, Paul (1987). "Supermembranes and eleven-dimensional supergravity". *Physics Letters B* **189** (1): 75–78. Bibcode:1987PhLB..189...75B. doi:10.1016/0370-2693(87)91272-X.

- Borcherds, Richard (1992). "Monstrous moonshine and Lie superalgebras". *Inventiones mathematicae* **109** (1): 405–444. Bibcode:1992InMat.109..405B. doi:10.1007/BF01232032.

- Candelas, Philip; de la Ossa, Xenia; Green, Paul; Parks, Linda (1991). "A pair of Calabi–Yau manifolds as an exactly soluble superconformal field theory". *Nuclear Physics B* **359** (1): 21–74. Bibcode:1991NuPhB.359...21C. doi:10.1016/0550-3213(91)90292-6.

- Candelas, Philip; Horowitz, Gary; Strominger, Andrew; Witten, Edward (1985). "Vacuum configurations for superstrings". *Nuclear Physics B* **258**: 46–74. Bibcode:1985NuPhB.258...46C. doi:10.1016/0550-3213(85)90602-9.

- Castro, Alejandra; Maloney, Alexander; Strominger, Andrew (2010). "Hidden conformal symmetry of the Kerr black hole". *Physical Review D* **82** (2). arXiv:1004.0996. Bibcode:2010PhRvD..82b4008C. doi:10.1103/PhysRevD.82.024008.

- Cheng, Miranda; Duncan, John; Harvey, Jeffrey (2013). "Umbral Moonshine". arXiv:1204.2779.

- Connes, Alain (1994). *Noncommutative Geometry*. Academic Press. ISBN 978-0-12-185860-5.

- Connes, Alain; Douglas, Michael; Schwarz, Albert (1998). "Noncommutative geometry and matrix theory". *Journal of High Energy Physics*. 19981 (2): 003. arXiv:hep-th/9711162. Bibcode:1998JHEP...02..003C. doi:10.1088/1126-6708/1998/02/003.

- Conway, John; Norton, Simon (1979). "Monstrous moonshine". *Bull. London. Math. Soc.* **11** (3): 308–339.

- Cremmer, Eugene; Julia, Bernard; Scherk, Joel (1978). "Supergravity theory in eleven dimensions". *Physics Letters B* **76** (4): 409–412. Bibcode:1978PhLB...76..409C. doi:10.1016/0370-2693(78)90894-8.

- de Haro, Sebastian; Dieks, Dennis; 't Hooft, Gerard; Verlinde, Erik (2013). "Forty Years of String Theory Reflecting on the Foundations". *Foundations of Physics* **43** (1): 1–7. Bibcode:2013FoPh...43....1D. doi:10.1007/s10701-012-9691-3.

- Deligne, Pierre; Etingof, Pavel; Freed, Daniel; Jeffery, Lisa; Kazhdan, David; Morgan, John; Morrison, David; Witten, Edward, eds. (1999). *Quantum Fields and Strings: A Course for Mathematicians* **1**. American Mathematical Society. ISBN 978-0821820124.

- Duff, Michael (1996). "M-theory (the theory formerly known as strings)". *International Journal of Modern Physics A* **11** (32): 6523–41. arXiv:hep-th/9608117. Bibcode:1996IJMPA..11.5623D. doi:10.1142/S0217751X96002583.

- Duff, Michael (1998). "The theory formerly known as strings". *Scientific American* **278** (2): 64–9. doi:10.1038/scientificamerican0298-64.

- Duff, Michael; Howe, Paul; Inami, Takeo; Stelle, Kellogg (1987). "Superstrings in D=10 from supermembranes in D=11". *Nuclear Physics B* **191** (1): 70–74. Bibcode:1987PhLB..191...70D. doi:10.1016/0370-2693(87)91323-2.

- Dummit, David; Foote, Richard (2004). *Abstract Algebra*. Wiley. ISBN 978-0-471-43334-7.

- Duncan, John; Griffin, Michael; Ono, Ken (2015). "Proof of the Umbral Moonshine Conjecture". arXiv:1503.01472.

- Eguchi, Tohru; Ooguri, Hirosi; Tachikawa, Yuji (2011). "Notes on the K3 surface and the Mathieu group M_{24}". *Experimental Mathematics* **20** (1): 91–96. doi:10.1080/10586458.2011.544585.

- Frenkel, Igor; Lepowsky, James; Meurman, Arne (1988). *Vertex Operator Algebras and the Monster*. Pure and Applied Mathematics **134**. Academic Press. ISBN 0-12-267065-5.

- Gannon, Terry. *Moonshine Beyond the Monster: The Bridge Connecting Algebra, Modular Forms, and Physics*. Cambridge University Press.

- Givental, Alexander (1996). "Equivariant Gromov-Witten invariants". *International Mathematics Research Notices* **1996** (13): 613–663. doi:10.1155/S1073792896000414.

- Givental, Alexander (1998). "A mirror theorem for toric complete intersections". *Topological field theory, primitive forms and related topics*: 141–175. doi:10.1007/978-1-4612-0705-4_5. ISBN 978-1-4612-6874-1.

- Gubser, Steven; Klebanov, Igor; Polyakov, Alexander (1998). "Gauge theory correlators from non-critical string theory". *Physics Letters B* **428**: 105–114. arXiv:hep-th/9802109. Bibcode:1998PhLB..428..105G. doi:10.1016/S0370-2693(98)00377-3.

- Guica, Monica; Hartman, Thomas; Song, Wei; Strominger, Andrew (2009). "The Kerr/CFT Correspondence". *Physical Review D* **80** (12). arXiv:0809.4266. Bibcode:2009PhRvD..80l4008G. doi:10.1103/PhysRevD.80.124008.

- Hawking, Stephen (1975). "Particle creation by black holes". *Communications in mathematical physics* **43** (3): 199–220. Bibcode:1975CMaPh..43..199H. doi:10.1007/BF02345020.

- Hawking, Stephen (2005). "Information loss in black holes". *Physical Review D* **72** (8). arXiv:hep-th/0507171. Bibcode:2005PhRvD..72h4013H. doi:10.1103/PhysRevD.72.084013.

- Hořava, Petr; Witten, Edward (1996). "Heterotic and Type I string dynamics from eleven dimensions". *Nuclear Physics B* **460** (3): 506–524. arXiv:hep-th/9510209. Bibcode:1996NuPhB.460..506H. doi:10.1016/0550-3213(95)00621-4.

- Hori, Kentaro; Katz, Sheldon; Klemm, Albrecht; Pandharipande, Rahul; Thomas, Richard; Vafa, Cumrun; Vakil, Ravi; Zaslow, Eric, eds. (2003). *Mirror Symmetry* (PDF). American Mathematical Society. ISBN 0-8218-2955-6.

- Hull, Chris; Townsend, Paul (1995). "Unity of superstring dualities". *Nuclear Physics B* **4381** (1): 109–137. arXiv:hep-th/9410167. Bibcode:1995NuPhB.438..109H. doi:10.1016/0550-3213(94)00559-W.

- Kapustin, Anton; Witten, Edward (2007). "Electric-magnetic duality and the geometric Langlands program". *Communications in Number Theory and Physics* **1** (1): 1–236. arXiv:hep-th/0604151. Bibcode:2007CNTP....1....1K. doi:10.4310/cntp.2007.v1.n1.a1.

- Klarreich, Erica. "Mathematicians chase moonshine's shadow". *Quanta Magazine*. Retrieved March 2015.

- Klebanov, Igor; Maldacena, Juan (2009). "Solving Quantum Field Theories via Curved Spacetimes" (PDF). *Physics Today* **62**: 28. Bibcode:2009PhT....62a..28K. doi:10.1063/1.3074260. Retrieved May 2013.

- Kontsevich, Maxim (1995). "Homological algebra of mirror symmetry". *Proceedings of the International Congress of Mathematicians*: 120–139. arXiv:alg-geom/9411018. Bibcode:1994alg.geom.11018K.

- Kovtun, P. K.; Son, Dam T.; Starinets, A. O. (2001). "Viscosity in strongly interacting quantum field theories from black hole physics". *Physical review letters* **94** (11): 111601. arXiv:hep-th/0405231. Bibcode:2005PhRvL..94k1601K. doi:10.1103/PhysRevLett.94.111601. PMID 15903845.

- Lian, Bong; Liu, Kefeng; Yau, Shing-Tung (1997). "Mirror principle, I". *Asian Journal of Math* **1**: 729–763. arXiv:alg-geom/9712011. Bibcode:1997alg.geom.12011L.

- Lian, Bong; Liu, Kefeng; Yau, Shing-Tung (1999a). "Mirror principle, II". *Asian Journal of Math* **3**: 109–146. arXiv:math/9905006. Bibcode:1999math......5006L.

- Lian, Bong; Liu, Kefeng; Yau, Shing-Tung (1999b). "Mirror principle, III". *Asian Journal of Math* **3**: 771–800. arXiv:math/9912038. Bibcode:1999math.....12038L.

- Lian, Bong; Liu, Kefeng; Yau, Shing-Tung (2000). "Mirror principle, IV". *Surveys in Differential Geometry*: 475–496. arXiv:math/0007104. Bibcode:2000math......7104L.

- Luzum, Matthew; Romatschke, Paul (2008). "Conformal relativistic viscous hydrodynamics: Applications to RHIC results at $\sqrt{s_{NN}}=200$ GeV". *Physical Review C* **78** (3). arXiv:0804.4015. doi:10.1103/PhysRevC.78.034915.

- Maldacena, Juan (1998). "The Large N limit of superconformal field theories and supergravity". *Advances in Theoretical and Mathematical Physics* **2**: 231–252. arXiv:hep-th/9711200. Bibcode:1998AdTMP...2..231M. doi:10.1063/1.59653.

- Maldacena, Juan (2005). "The Illusion of Gravity" (PDF). *Scientific American* **293** (5): 56–63. Bibcode:2005SciAm.293e..56M. doi:10.1038/scientificamerican1105-56. PMID 16318027. Retrieved July 2013.

- Maldacena, Juan; Strominger, Andrew; Witten, Edward (1997). "Black hole entropy in M-theory". *Journal of High Energy Physics* **1997** (12).

- Merali, Zeeya (2011). "Collaborative physics: string theory finds a bench mate". *Nature* **478** (7369): 302–304. Bibcode:2011Natur.478..302M. doi:10.1038/478302a. PMID 22012369.

- Moore, Gregory (2005). "What is ... a Brane?" (PDF). *Notices of the AMS* **52**: 214. Retrieved June 2013.

- Nahm, Walter (1978). "Supersymmetries and their representations". *Nuclear Physics B* **135** (1): 149–166. Bibcode:1978NuPhB.135..149N. doi:10.1016/0550-3213(78)90218-3.

- Nekrasov, Nikita; Schwarz, Albert (1998). "Instantons on noncommutative $\mathbf{R}^4$ and (2,0) superconformal six dimensional theory". *Communications in Mathematical Physics* **198** (3): 689–703. arXiv:hep-th/9802068. Bibcode:1998CMaPh.198..689N. doi:10.1007/s002200050490.

- Ooguri, Hirosi; Strominger, Andrew; Vafa, Cumrun (2004). "Black hole attractors and the topological string". *Physical Review D* **70** (10).

- Polchinski, Joseph (2007). "All Strung Out?". *American Scientist*. Retrieved April 2015.

- Penrose, Roger (2005). *The Road to Reality: A Complete Guide to the Laws of the Universe*. Knopf. ISBN 0-679-45443-8.

- Randall, Lisa; Sundrum, Raman (1999). "An alternative to compactification". *Physical Review Letters* **83** (23): 4690. arXiv:hep-th/9906064. Bibcode:1999PhRvL..83.4690R. doi:10.1103/PhysRevLett.83.4690.

- Sachdev, Subir (2013). "Strange and stringy". *Scientific American* **308** (44): 44. Bibcode:2012SciAm.308a..44S. doi:10.1038/scientificamerican0113-44.

- Seiberg, Nathan; Witten, Edward (1999). "String Theory and Noncommutative Geometry". *Journal of High Energy Physics* **1999** (9): 032. arXiv:hep-th/9908142. Bibcode:1999JHEP...09..032S. doi:10.1088/1126-6708/1999/09/032.

- Sen, Ashoke (1994a). "Strong-weak coupling duality in four-dimensional string theory". *International Journal of Modern Physics A* **9** (21): 3707–3750. arXiv:hep-th/9402002. Bibcode:1994IJMPA...9.3707S. doi:10.1142/S0217751X94001497.

- Sen, Ashoke (1994b). "Dyon-monopole bound states, self-dual harmonic forms on the multi-monopole moduli space, and $SL(2,\mathbf{Z})$ invariance in string theory". *Physics Letters B* **329** (2): 217–221. arXiv:hep-th/9402032. Bibcode:1994PhLB..329..217S. doi:10.1016/0370-2693(94)90763-3.

- Smolin, Lee (2006). *The Trouble with Physics: The Rise of String Theory, the Fall of a Science, and What Comes Next*. New York: Houghton Mifflin Co. ISBN 0-618-55105-0.

- Strominger, Andrew (1998). "Black hole entropy from near-horizon microstates". *Journal of High Energy Physics* **1998** (2). arXiv:hep-th/9712251. Bibcode:1998JHEP...02..009S. doi:10.1088/1126-6708/1998/02/009.

- Strominger, Andrew; Vafa, Cumrun (1996). "Microscopic origin of the Bekenstein–Hawking entropy". *Physics Letters B* **379** (1): 99–104. arXiv:hep-th/9601029. Bibcode:1996PhLB..379...99S. doi:10.1016/0370-2693(96)00345-0.

- Strominger, Andrew; Yau, Shing-Tung; Zaslow, Eric (1996). "Mirror symmetry is T-duality". *Nuclear Physics B* **479** (1): 243–259. arXiv:hep-th/9606040. Bibcode:1996NuPhB.479..243S. doi:10.1016/0550-3213(96)00434-8.

- Susskind, Leonard (2005). *The Cosmic Landscape: String Theory and the Illusion of Intelligent Design*. Back Bay Books. ISBN 978-0316013338.

- Susskind, Leonard (2008). *The Black Hole War: My Battle with Stephen Hawking to Make the World Safe for Quantum Mechanics*. Little, Brown and Company. ISBN 978-0-316-01641-4.

- Wald, Robert (1984). *General Relativity*. University of Chicago Press. ISBN 978-0-226-87033-5.

- Weinberg, Steven (1987). *Anthropic bound on the cosmological constant* **59** (22). Physical Review Letters. p. 2607.

- Witten, Edward (1995). "String theory dynamics in various dimensions". *Nuclear Physics B* **443** (1): 85–126. arXiv:hep-th/9503124. Bibcode:1995NuPhB.443...85W. doi:10.1016/0550-3213(95)00158-O.

- Witten, Edward (1998). "Anti-de Sitter space and holography". *Advances in Theoretical and Mathematical Physics* **2**: 253–291. arXiv:hep-th/9802150. Bibcode:1998AdTMP...2..253W.

- Witten, Edward (2007). "Three-dimensional gravity revisited". arXiv:0706.3359 [hep-th].

- Woit, Peter (2006). *Not Even Wrong: The Failure of String Theory and the Search for Unity in Physical Law*. Basic Books. p. 105. ISBN 0-465-09275-6.

- Yau, Shing-Tung; Nadis, Steve (2010). *The Shape of Inner Space: String Theory and the Geometry of the Universe's Hidden Dimensions*. Basic Books. ISBN 978-0-465-02023-2.

- Zee, Anthony (2010). *Quantum Field Theory in a Nutshell* (2nd ed.). Princeton University Press. ISBN 978-0-691-14034-6.

- Zwiebach, Barton (2009). *A First Course in String Theory*. Cambridge University Press. ISBN 978-0-521-88032-9.

2.10 Further reading

2.10.1 Popularizations

General

- Greene, Brian (2003). *The Elegant Universe: Superstrings, Hidden Dimensions, and the Quest for the Ultimate Theory*. New York: W.W. Norton & Company. ISBN 0-393-05858-1.

- Greene, Brian (2004). *The Fabric of the Cosmos: Space, Time, and the Texture of Reality*. New York: Alfred A. Knopf. ISBN 0-375-41288-3.

Critical

- Penrose, Roger (2005). *The Road to Reality: A Complete Guide to the Laws of the Universe*. Knopf. ISBN 0-679-45443-8.

- Smolin, Lee (2006). *The Trouble with Physics: The Rise of String Theory, the Fall of a Science, and What Comes Next*. New York: Houghton Mifflin Co. ISBN 0-618-55105-0.

- Woit, Peter (2006). *Not Even Wrong: The Failure of String Theory And the Search for Unity in Physical Law*. London: Jonathan Cape &: New York: Basic Books. ISBN 978-0-465-09275-8.

2.10.2 Textbooks

For physicists

- Becker, Katrin; Becker, Melanie; Schwarz, John (2007). *String Theory and M-theory: A Modern Introduction*. Cambridge University Press. ISBN 978-0-521-86069-7.

- Green, Michael; Schwarz, John; Witten, Edward (2012). *Superstring theory. Vol. 1: Introduction*. Cambridge University Press. ISBN 978-1107029118.

- Green, Michael; Schwarz, John; Witten, Edward (2012). *Superstring theory. Vol. 2: Loop amplitudes, anomalies and phenomenology*. Cambridge University Press. ISBN 978-1107029132.

- Polchinski, Joseph (1998). *String Theory Vol. 1: An Introduction to the Bosonic String*. Cambridge University Press. ISBN 0-521-63303-6.

- Polchinski, Joseph (1998). *String Theory Vol. 2: Superstring Theory and Beyond*. Cambridge University Press. ISBN 0-521-63304-4.

- Zwiebach, Barton (2009). *A First Course in String Theory*. Cambridge University Press. ISBN 978-0-521-88032-9.

For mathematicians

- Deligne, Pierre; Etingof, Pavel; Freed, Daniel; Jeffery, Lisa; Kazhdan, David; Morgan, John; Morrison, David; Witten, Edward, eds. (1999). *Quantum Fields and Strings: A Course for Mathematicians, Vol. 2*. American Mathematical Society. ISBN 978-0821819883.

2.11 External links

- *The Elegant Universe*—A three-hour miniseries with Brian Greene by *NOVA* (original PBS Broadcast Dates: October 28, 8–10 p.m. and November 4, 8–9 p.m., 2003). Various images, texts, videos and animations explaining string theory.

- Not Even Wrong—A blog critical of string theory

- The Official String Theory Web Site

- Why String Theory—An introduction to string theory.

Chapter 3

Loop quantum gravity

Loop quantum gravity (**LQG**) is a theory that attempts to describe the quantum properties of the universe and gravity. It is also a theory of quantum spacetime because, according to general relativity, gravity is a manifestation of the geometry of spacetime. LQG is an attempt to merge quantum mechanics and general relativity. The main output of the theory is a physical picture of space where space is granular. The granularity is a direct consequence of the quantization. It has the same nature as the granularity of the photons in the quantum theory of electromagnetism and the discrete levels of the energy of the atoms. Here, it is space itself that is discrete. In other words, there is a minimum distance possible to travel through it.

More precisely, space can be viewed as an extremely fine fabric or network "woven" of finite loops. These networks of loops are called spin networks. The evolution of a spin network over time is called a spin foam. The predicted size of this structure is the Planck length, which is approximately 10^{-35} meters. According to the theory, there is no meaning to distance at scales smaller than the Planck scale. Therefore, LQG predicts that not just matter, but space itself, has an atomic structure.

Today LQG is a vast area of research, developing in several directions, which involves about 30 research groups worldwide.[1] They all share the basic physical assumptions and the mathematical description of quantum space. The full development of the theory is being pursued in two directions: the more traditional canonical loop quantum gravity, and the newer covariant loop quantum gravity, more commonly called spin foam theory.

Research into the physical consequences of the theory is proceeding in several directions. Among these, the most well-developed is the application of LQG to cosmology, called loop quantum cosmology (LQC). LQC applies LQG ideas to the study of the early universe and the physics of the Big Bang. Its most spectacular consequence is that the evolution of the universe can be continued beyond the Big Bang. The Big Bang appears thus to be replaced by a sort of cosmic Big Bounce.

3.1 History

Main article: History of loop quantum gravity

In 1986, Abhay Ashtekar reformulated Einstein's general relativity in a language closer to that of the rest of fundamental physics. Shortly after, Ted Jacobson and Lee Smolin realized that the formal equation of quantum gravity, called the Wheeler–DeWitt equation, admitted solutions labelled by loops, when rewritten in the new Ashtekar variables, and Carlo Rovelli and Lee Smolin defined a nonperturbative and background-independent quantum theory of gravity in terms of these loop solutions. Jorge Pullin and Jerzy Lewandowski understood that the intersections of the loops are essential for the consistency of the theory, and the theory should be formulated in terms of intersecting loops, or graphs.

In 1994, Rovelli and Smolin showed that the quantum operators of the theory associated to area and volume have a discrete spectrum. That is, geometry is quantized. This result defines an explicit basis of states of quantum geometry, which turned out to be labelled by Roger Penrose's spin networks, which are graphs labelled by spins.

The canonical version of the dynamics was put on firm ground by Thomas Thiemann, who defined an anomaly-free Hamiltonian operator, showing the existence of a mathematically consistent background-independent theory. The covariant or spinfoam version of the dynamics developed during several decades, and crystallized in 2008, from the joint work of research groups in France, Canada, UK, Poland, and Germany, lead to the definition of a family of transition amplitudes, which in the classical limit can be shown to be related to a family of truncations of general relativity.[2] The finiteness of these amplitudes was proven in 2011.[3][4] It requires the existence of a positive cosmological constant, and this is consistent with observed acceleration in the expansion of the Universe.

3.2 General covariance and back-ground independence

Main articles: General covariance, background-independent and diffeomorphism

In theoretical physics, general covariance is the invariance of the form of physical laws under arbitrary differentiable coordinate transformations. The essential idea is that coordinates are only artifices used in describing nature, and hence should play no role in the formulation of fundamental physical laws. A more significant requirement is the principle of general relativity that states that the laws of physics take the same form in all reference systems. This is a generalization of the principle of special relativity which states that the laws of physics take the same form in all inertial frames.

In mathematics, a diffeomorphism is an isomorphism in the category of smooth manifolds. It is an invertible function that maps one differentiable manifold to another, such that both the function and its inverse are smooth. These are the defining symmetry transformations of General Relativity since the theory is formulated only in terms of a differentiable manifold.

In general relativity, general covariance is intimately related to "diffeomorphism invariance". This symmetry is one of the defining features of the theory. However, it is a common misunderstanding that "diffeomorphism invariance" refers to the invariance of the physical predictions of a theory under arbitrary coordinate transformations; this is untrue and in fact every physical theory is invariant under coordinate transformations this way. Diffeomorphisms, as mathematicians define them, correspond to something much more radical; intuitively a way they can be envisaged is as simultaneously dragging all the physical fields (including the gravitational field) over the bare differentiable manifold while staying in the same coordinate system. Diffeomorphisms are the true symmetry transformations of general relativity, and come about from the assertion that the formulation of the theory is based on a bare differentiable manifold, but not on any prior geometry — the theory is background-independent (this is a profound shift, as all physical theories before general relativity had as part of their formulation a prior geometry). What is preserved under such transformations are the coincidences between the values the gravitational field take at such and such a "place" and the values the matter fields take there. From these relationships one can form a notion of matter being located with respect to the gravitational field, or vice versa. This is what Einstein discovered: that physical entities are located with respect to one another only and not with respect to the spacetime manifold. As Carlo Rovelli puts it: "No more fields on

spacetime: just fields on fields.".[5] This is the true meaning of the saying "The stage disappears and becomes one of the actors"; space-time as a "container" over which physics takes place has no objective physical meaning and instead the gravitational interaction is represented as just one of the fields forming the world. This is known as the relationalist interpretation of space-time. The realization by Einstein that general relativity should be interpreted this way is the origin of his remark "Beyond my wildest expectations".

In LQG this aspect of general relativity is taken seriously and this symmetry is preserved by requiring that the physical states remain invariant under the generators of diffeomorphisms. The interpretation of this condition is well understood for purely spatial diffeomorphisms. However, the understanding of diffeomorphisms involving time (the Hamiltonian constraint) is more subtle because it is related to dynamics and the so-called "problem of time" in general relativity.[6] A generally accepted calculational framework to account for this constraint has yet to be found.[7][8] A plausible candidate for the quantum hamiltonian constraint is the operator introduced by Thiemann.[9]

LQG is formally background independent. The equations of LQG are not embedded in, or dependent on, space and time (except for its invariant topology). Instead, they are expected to give rise to space and time at distances which are large compared to the Planck length. The issue of background independence in LQG still has some unresolved subtleties. For example, some derivations require a fixed choice of the topology, while any consistent quantum theory of gravity should include topology change as a dynamical process.

3.3 Constraints and their Poisson bracket algebra

Main articles: Poisson bracket and Hamiltonian constraint

3.3.1 The constraints of classical canonical general relativity

Main article: Lie derivative

In the Hamiltonian formulation of ordinary classical mechanics the Poisson bracket is an important concept. A "canonical coordinate system" consists of canonical position and momentum variables that satisfy canonical Poisson-bracket relations,

$$\{q_i, p_j\} = \delta_{ij}$$

where the Poisson bracket is given by

$$\{f,g\} = \sum_{i=1}^{N} \left(\frac{\partial f}{\partial q_i} \frac{\partial g}{\partial p_i} - \frac{\partial f}{\partial p_i} \frac{\partial g}{\partial q_i} \right).$$

for arbitrary phase space functions $f(q_i, p_j)$ and $g(q_i, p_j)$. With the use of Poisson brackets, the Hamilton's equations can be rewritten as,

$$\dot{q}_i = \{q_i, H\},$$

$$\dot{p}_i = \{p_i, H\}.$$

These equations describe a "flow" or orbit in phase space generated by the Hamiltonian H. Given any phase space function $F(q, p)$, we have

$$\tfrac{d}{dt} F(q_i, p_i) = \{F, H\}.$$

Let us consider constrained systems, of which General relativity is an example. In a similar way the Poisson bracket between a constraint and the phase space variables generates a flow along an orbit in (the unconstrained) phase space generated by the constraint. There are three types of constraints in Ashtekar's reformulation of classical general relativity:

$SU(2)$ Gauss gauge constraints

The Gauss constraints

$$G_j(x) = 0.$$

This represents an infinite number of constraints one for each value of x. These come about from re-expressing General relativity as an $SU(2)$ Yang–Mills type gauge theory (Yang–Mills is a generalization of Maxwell's theory where the gauge field transforms as a vector under Gauss transformations, that is, the Gauge field is of the form $A_a^i(x)$ where i is an internal index. See Ashtekar variables). These infinite number of Gauss gauge constraints can be smeared with test fields with internal indices, $\lambda^j(x)$,

$$G(\lambda) = \int d^3 x G_j(x) \lambda^j(x).$$

which we demand vanish for any such function. These smeared constraints defined with respect to a suitable space of smearing functions give an equivalent description to the original constraints.

In fact Ashtekar's formulation may be thought of as ordinary $SU(2)$ Yang–Mills theory together with the following special constraints, resulting from diffeomorphism invariance, and a Hamiltonian that vanishes. The dynamics of such a theory are thus very different from that of ordinary Yang–Mills theory.

Spatial diffeomorphisms constraints

The spatial diffeomorphism constraints

$$C_a(x) = 0$$

can be smeared by the so-called shift functions $\vec{N}(x)$ to give an equivalent set of smeared spatial diffeomorphism constraints,

$$C(\vec{N}) = \int d^3 x C_a(x) N^a(x).$$

These generate spatial diffeomorphisms along orbits defined by the shift function $N^a(x)$.

Hamiltonian constraints

The Hamiltonian

$$H(x) = 0$$

can be smeared by the so-called lapse functions $N(x)$ to give an equivalent set of smeared Hamiltonian constraints,

$$H(N) = \int d^3 x H(x) N(x).$$

These generate time diffeomorphisms along orbits defined by the lapse function $N(x)$.

In Ashtekar formulation the gauge field $A_a^i(x)$ is the configuration variable (the configuration variable being analogous to q in ordinary mechanics) and its conjugate momentum is the (densitized) triad (electrical field) $E_i^a(x)$. The constraints are certain functions of these phase space variables.

We consider the action of the constraints on arbitrary phase space functions. An important notion here is the Lie derivative, $\mathcal{L}_V$, which is basically a derivative operation that infinitesimally "shifts" functions along some orbit with tangent vector V.

3.3.2   The Poisson bracket algebra

Of particular importance is the Poisson bracket algebra formed between the (smeared) constraints themselves as it completely determines the theory. In terms of the above smeared constraints the constraint algebra amongst the Gauss' law reads,

$$\{G(\lambda), G(\mu)\} = G([\lambda, \mu])$$

where $[\lambda, \mu]^k = \lambda_i \mu_j \epsilon^{ijk}$. And so we see that the Poisson bracket of two Gauss' law is equivalent to a single Gauss' law evaluated on the commutator of the smearings. The Poisson bracket amongst spatial diffeomorphisms constraints reads

$$\{C(\vec{N}), C(\vec{M})\} = C(\mathcal{L}_{\vec{N}} \vec{M})$$

and we see that its effect is to "shift the smearing". The reason for this is that the smearing functions are not functions of the canonical variables and so the spatial diffeomorphism does not generate diffeomorphims on them. They do however generate diffeomorphims on everything else. This is equivalent to leaving everything else fixed while shifting the smearing .The action of the spatial diffeomorphism on the Gauss law is

$$\{C(\vec{N}), G(\lambda)\} = G(\mathcal{L}_{\vec{N}}\lambda) \,,$$

again, it shifts the test field λ. The Gauss law has vanishing Poisson bracket with the Hamiltonian constraint. The spatial diffeomorphism constraint with a Hamiltonian gives a Hamiltonian with its smearing shifted,

$$\{C(\vec{N}), H(M)\} = H(\mathcal{L}_{\vec{N}}M) \,.$$

Finally, the poisson bracket of two Hamiltonians is a spatial diffeomorphism,

$$\{H(N), H(M)\} = C(K)$$

where K is some phase space function. That is, it is a sum over infinitesimal spatial diffeomorphisms constraints where the coefficients of proportionality are not constants but have non-trivial phase space dependence.

A (Poisson bracket) Lie algebra, with constraints C_I, is of the form

$$\{C_I, C_J\} = f_{IJ}^K C_K$$

where f_{IJ}^K are constants (the so-called structure constants). The above Poisson bracket algebra for General relativity does not form a true Lie algebra as we have structure functions rather than structure constants for the Poisson bracket between two Hamiltonians. This leads to difficulties.

3.3.3 Dirac observables

The constraints define a constraint surface in the original phase space. The gauge motions of the constraints apply to all phase space but have the feature that they leave the constraint surface where it is, and thus the orbit of a point in the hypersurface under gauge transformations will be an orbit entirely within it. Dirac observables are defined as phase space functions, O, that Poisson commute with all the constraints when the constraint equations are imposed,

$$\{G_j, O\}_{G_j = C_a = H = 0} = \{C_a, O\}_{G_j = C_a = H = 0} = \{H, O\}_{G_j = C_a = H = 0} = 0 \,,$$

that is, they are quantities defined on the constraint surface that are invariant under the gauge transformations of the theory.

Then, solving only the constraint $G_j = 0$ and determining the Dirac observables with respect to it leads us back to the ADM phase space with constraints H, C_a. The dynamics of general relativity is generated by the constraints, it can be shown that six Einstein equations describing time evolution (really a gauge transformation) can be obtained by calcu-

lating the Poisson brackets of the three-metric and its conjugate momentum with a linear combination of the spatial diffeomorphism and Hamiltonian constraint. The vanishing of the constraints, giving the physical phase space, are the four other Einstein equations.[10]

3.4 Quantization of the constraints – the equations of quantum general relativity

3.4.1 Pre-history and Ashtekar new variables

Main articles: Frame fields in general relativity, Ashtekar variables and Self-dual Palatini action

Many of the technical problems in canonical quantum gravity revolve around the constraints. Canonical general relativity was originally formulated in terms of metric variables, but there seemed to be insurmountable mathematical difficulties in promoting the constraints to quantum operators because of their highly non-linear dependence on the canonical variables. The equations were much simplified with the introduction of Ashtekars new variables. Ashtekar variables describe canonical general relativity in terms of a new pair canonical variables closer to that of gauge theories. The first step consists of using densitized triads $\tilde{E}_i^a$ (a triad E_i^a is simply three orthogonal vector fields labeled by $i = 1, 2, 3$ and the densitized triad is defined by $\tilde{E}_i^a = \sqrt{\det(q)}E_i^a$) to encode information about the spatial metric,

$$\det(q)q^{ab} = \tilde{E}_i^a \tilde{E}_j^b \delta^{ij} \,.$$

(where δ^{ij} is the flat space metric, and the above equation expresses that q^{ab}, when written in terms of the basis E_i^a, is locally flat). (Formulating general relativity with triads instead of metrics was not new.) The densitized triads are not unique, and in fact one can perform a local in space rotation with respect to the internal indices i. The canonically conjugate variable is related to the extrinsic curvature by $K_a^i = K_{ab}\tilde{E}^{ai}/\sqrt{\det(q)}$. But problems similar to using the metric formulation arise when one tries to quantize the theory. Ashtekar's new insight was to introduce a new configuration variable,

$$A_a^i = \Gamma_a^i - iK_a^i$$

that behaves as a complex SU(2) connection where Γ_a^i is related to the so-called spin connection via $\Gamma_a^i = \Gamma_{ajk}\epsilon^{jki}$. Here A_a^i is called the chiral spin connection. It defines a covariant derivative $\mathcal{D}_a$. It turns out that $\tilde{E}_i^a$ is the conjugate momentum of A_a^i, and together these form Ashtekar's new variables.

The expressions for the constraints in Ashtekar variables: the Gauss's law, the spatial diffeomorphism constraint and the (densitized) Hamiltonian constraint then read:

$$G^i = \mathcal{D}_a \tilde{E}_i^a = 0$$

$$C_a = \tilde{E}_i^b F_{ab}^i - A_a^i (\mathcal{D}_b \tilde{E}_i^b) = V_a - A_a^i G^i = 0 ,$$

$$\tilde{H} = \epsilon_{ijk} \tilde{E}_i^a \tilde{E}_j^b F_{ab}^i = 0$$

respectively, where F_{ab}^i is the field strength tensor of the connection A_a^i and where V_a is referred to as the vector constraint. The above-mentioned local in space rotational invariance is the original of the SU(2) gauge invariance here expressed by the Gauss law. Note that these constraints are polynomial in the fundamental variables, unlike as with the constraints in the metric formulation. This dramatic simplification seemed to open up the way to quantizing the constraints. (See the article Self-dual Palatini action for a derivation of Ashtekar's formulism).

With Ashtekar's new variables, given the configuration variable A_a^i , it is natural to consider wavefunctions $\Psi(A_a^i)$. This is the connection representation. It is analogous to ordinary quantum mechanics with configuration variable q and wavefunctions $\psi(q)$. The configuration variable gets promoted to a quantum operator via:

$$\hat{A}_a^i \Psi(A) = A_a^i \Psi(A) ,$$

(analogous to $\hat{q}\psi(q) = q\psi(q)$) and the triads are (functional) derivatives,

$$\hat{\tilde{E}}_i^a \Psi(A) = -i \frac{\delta \Psi(A)}{\delta A_a^i} .$$

(analogous to $\hat{p}\psi(q) = -i\hbar d\psi(q)/dq$). In passing over to the quantum theory the constraints become operators on a kinematic Hilbert space (the unconstrained SU(2) Yang–Mills Hilbert space). Note that different ordering of the A 's and $\tilde{E}$'s when replacing the $\tilde{E}$'s with derivatives give rise to different operators - the choice made is called the factor ordering and should be chosen via physical reasoning. Formally they read

$$\hat{G}_j |\psi\rangle = 0$$

$$\hat{C}_a |\psi\rangle = 0$$

$$\hat{\tilde{H}} |\psi\rangle = 0 .$$

There are still problems in properly defining all these equations and solving them. For example the Hamiltonian constraint Ashtekar worked with was the densitized version instead of the original Hamiltonian, that is, he worked with $\tilde{H} = \sqrt{\det(q)} H$. There were serious difficulties in promoting this quantity to a quantum operator. Moreover, although Ashtekar variables had the virtue of simplifying the Hamiltonian, they are complex. When one quantizes the theory, it is difficult to ensure that one recovers real general relativity as opposed to complex general relativity.

3.4.2 Quantum constraints as the equations of quantum general relativity

We now move on to demonstrate an important aspect of the quantum constraints. We consider Gauss' law only. First we state the classical result that the Poisson bracket of the smeared Gauss' law $G(\lambda) = \int d^3x \lambda^j (D_a E^a)^j$ with the connections is

$$\{G(\lambda), A_a^i\} = \partial_a \lambda^i + g\epsilon^{ijk} A_a^j \lambda^k = (D_a \lambda)^i.$$

The quantum Gauss' law reads

$$\hat{G}_j \Psi(A) = -i D_a \frac{\delta \lambda \Psi[A]}{\delta A_a^i} = 0.$$

If one smears the quantum Gauss' law and study its action on the quantum state one finds that the action of the constraint on the quantum state is equivalent to shifting the argument of Ψ by an infinitesimal (in the sense of the parameter λ small) gauge transformation,

$$\left[1 + \int d^3x \lambda^j(x) \hat{G}_j \right] \Psi(A) = \Psi[A + D\lambda] = \Psi[A],$$

and the last identity comes from the fact that the constraint annihilates the state. So the constraint, as a quantum operator, is imposing the same symmetry that its vanishing imposed classically: it is telling us that the functions $\Psi[A]$ have to be gauge invariant functions of the connection. The same idea is true for the other constraints.

Therefore the two step process in the classical theory of solving the constraints $C_I = 0$ (equivalent to solving the admissibility conditions for the initial data) and looking for the gauge orbits (solving the `evolution' equations) is replaced by a one step process in the quantum theory, namely looking for solutions Ψ of the quantum equations $\hat{C}_I \Psi = 0$. This is because it obviously solves the constraint at the quantum level and it simultaneously looks for states that are gauge invariant because $\hat{C}_I$ is the quantum generator of gauge transformations (gauge invariant functions are constant along the gauge orbits and thus characterize them).[11] Recall that, at the classical level, solving the admissibility conditions and evolution equations was equivalent to solving all of Einstein's field equations, this underlines the central role of the quantum constraint equations in canonical quantum gravity.

3.4.3 Introduction of the loop representation

Main articles: Holonomy, Wilson loop and Knot invariant

It was in particular the inability to have good control over the space of solutions to the Gauss' law and spacial diffeomorphism constraints that led Rovelli and Smolin to consider a new representation - the loop representation in gauge theories and quantum gravity.[12]

We need the notion of a holonomy. A holonomy is a measure of how much the initial and final values of a spinor or vector differ after parallel transport around a closed loop; it is denoted

$$h_\gamma[A] \; .$$

Knowledge of the holonomies is equivalent to knowledge of the connection, up to gauge equivalence. Holonomies can also be associated with an edge; under a Gauss Law these transform as

$$(h'_e)_{\alpha\beta} = U^{-1}_{\alpha\gamma}(x)(h_e)_{\gamma\sigma} U_{\sigma\beta}(y) \; .$$

For a closed loop $x = y$ if we take the trace of this, that is, putting $\alpha = \beta$ and summing we obtain

$$(h'_e)_{\alpha\alpha} = U^{-1}_{\alpha\gamma}(x)(h_e)_{\gamma\sigma} U_{\sigma\alpha}(x) =$$
$$[U_{\sigma\alpha}(x)U^{-1}_{\alpha\gamma}(x)](h_e)_{\gamma\sigma} = \delta_{\sigma\gamma}(h_e)_{\gamma\sigma} = (h_e)_{\gamma\gamma}$$

or

$$\mathrm{Tr}\, h'_\gamma = \mathrm{Tr}\, h_\gamma \; . .$$

The trace of an holonomy around a closed loop is written

$$W_\gamma[A]$$

and is called a Wilson loop. Thus Wilson loops are gauge invariant. The explicit form of the Holonomy is

$$h_\gamma[A] = \mathcal{P} \exp\left\{ - \int_{\gamma_0}^{\gamma_1} ds \dot\gamma^a A^i_a(\gamma(s)) T_i \right\}$$

where γ is the curve along which the holonomy is evaluated, and s is a parameter along the curve, $\mathcal{P}$ denotes path ordering meaning factors for smaller values of s appear to the left, and T_i are matrices that satisfy the SU(2) algebra

$$[T^i, T^j] = 2i\epsilon^{ijk}T^k \; .$$

The Pauli matrices satisfy the above relation. It turns out that there are infinitely many more examples of sets of matrices that satisfy these relations, where each set comprises $(N+1) \times (N+1)$ matrices with $N = 1, 2, 3, \ldots$, and where none of these can be thought to 'decompose' into two or more examples of lower dimension. They are called different irreducible representations of the SU(2) algebra. The most fundamental representation being the Pauli matrices. The holonomy is labelled by a half integer $N/2$ according to the irreducible representation used.

The use of Wilson loops explicitly solves the Gauss gauge constraint. To handle the spatial diffeomorphism constraint we need to go over to the loop representation. As Wilson loops form a basis we can formally expand any Gauss gauge invariant function as,

$$\Psi[A] = \sum_\gamma \Psi[\gamma] W_\gamma[A] \; .$$

This is called the loop transform. We can see the analogy with going to the momentum representation in quantum mechanics(see Position and momentum space). There

one has a basis of states $\exp(ikx)$ labelled by a number k and one expands

$$\psi[x] = \int dk \psi(k) \exp(ikx) \; .$$

and works with the coefficients of the expansion $\psi(k)$.

The inverse loop transform is defined by

$$\Psi[\gamma] = \int [dA]\Psi[A] W_\gamma[A] \; .$$

This defines the loop representation. Given an operator $\hat{O}$ in the connection representation,

$$\Phi[A] = \hat{O}\Psi[A] \qquad Eq\ 1 \; .$$

one should define the corresponding operator $\hat{O}'$ on $\Psi[\gamma]$ in the loop representation via,

$$\Phi[\gamma] = \hat{O}'\Psi[\gamma] \qquad Eq\ 2 \; ,$$

where $\Phi[\gamma]$ is defined by the usual inverse loop transform,

$$\Phi[\gamma] = \int [dA]\Phi[A] W_\gamma[A] \qquad Eq\ 3 \; . .$$

A transformation formula giving the action of the operator $\hat{O}'$ on $\Psi[\gamma]$ in terms of the action of the operator $\hat{O}$ on $\Psi[A]$ is then obtained by equating the R.H.S. of Eq 2 with the R.H.S. of Eq 3 with Eq 1 substituted into Eq 3 , namely

$$\hat{O}'\Psi[\gamma] = \int [dA] W_\gamma[A] \hat{O}\Psi[A] \; ,$$

or

$$\hat{O}'\Psi[\gamma] = \int [dA](\hat{O}^\dagger W_\gamma[A])\Psi[A] \; ,$$

where by $\hat{O}^\dagger$ we mean the operator $\hat{O}$ but with the reverse factor ordering (remember from simple quantum mechanics where the product of operators is reversed under conjugation). We evaluate the action of this operator on the Wilson loop as a calculation in the connection representation and rearranging the result as a manipulation purely in terms of loops (one should remember that when considering the action on the Wilson loop one should choose the operator one wishes to transform with the opposite factor ordering to the one chosen for its action on wavefunctions $\Psi[A]$). This gives the physical meaning of the operator $\hat{O}'$. For example if $\hat{O}^\dagger$ corresponded to a spatial diffeomorphism, then this can be thought of as keeping the connection field A of $W_\gamma[A]$ where it is while performing a spatial diffeomorphism on γ instead. Therefore the meaning of $\hat{O}'$ is a spatial diffeomorphism on γ , the argument of $\Psi[\gamma]$.

In the loop representation we can then solve the spatial diffeomorphism constraint by considering functions of loops $\Psi[\gamma]$ that are invariant under spatial diffeomorphisms of the loop γ . That is, we construct what mathematicians call knot invariants. This opened up an unexpected connection between knot theory and quantum gravity.

What about the Hamiltonian constraint? Let us go back to the connection representation. Any collection of non-intersecting Wilson loops satisfy Ashtekar's quantum

Hamiltonian constraint. This can be seen from the following. With a particular ordering of terms and replacing $\hat{E}_i^a$ by a derivative, the action of the quantum Hamiltonian constraint on a Wilson loop is

$$\hat{H}^\dagger W_\gamma[A] = -\epsilon_{ijk}\hat{F}_{ab}^k \frac{\delta}{\delta A_a^i} \frac{\delta}{\delta A_b^j} W_\gamma[A] .$$

When a derivative is taken it brings down the tangent vector, $\dot{\gamma}^a$, of the loop. γ. So we have something like

$$\hat{F}_{ab}^i \dot{\gamma}^a \dot{\gamma}^b .$$

However, as F_{ab}^i is anti-symmetric in the indices a and b this vanishes (this assumes that γ is not discontinuous anywhere and so the tangent vector is unique). Now let us go back to the loop representation.

We consider wavefunctions $\Psi[\gamma]$ that vanish if the loop has discontinuities and that are knot invariants. Such functions solve the Gauss law, the spatial diffeomorphism constraint and (formally) the Hamiltonian constraint. Thus we have identified an infinite set of exact (if only formal) solutions to all the equations of quantum general relativity![12] This generated a lot of interest in the approach and eventually led to LQG.

3.4.4 Geometric operators, the need for intersecting Wilson loops and spin network states

The easiest geometric quantity is the area. Let us choose coordinates so that the surface Σ is characterized by $x^3 = 0$. The area of small parallelogram of the surface Σ is the product of length of each side times $\sin\theta$ where θ is the angle between the sides. Say one edge is given by the vector $\vec{u}$ and the other by $\vec{v}$ then,

$$A = \|\vec{u}\|\|\vec{v}\|\sin\theta = \sqrt{\|\vec{u}\|^2\|\vec{v}\|^2(1-\cos^2\theta)} = \sqrt{\|\vec{u}\|^2\|\vec{v}\|^2 - (\vec{u}\cdot\vec{v})^2}$$

In the space spanned by x^1 and x^2 we have an infinitesimal parallelogram described by $\vec{u} = \vec{e}_1 dx^1$ and $\vec{v} = \vec{e}_2 dx^2$. Using $q_{AB}^{(2)} = \vec{e}_A \cdot \vec{e}_B$ (where the indices A and B run from 1 to 2), we get the area of the surface Σ to be given by

$$A_\Sigma = \int_\Sigma dx^1 dx^2 \sqrt{\det(q^{(2)})}$$

where $\det(q^{(2)}) = q_{11}q_{22} - q_{12}^2$ and is the determinant of the metric induced on Σ. The latter can be rewritten $\det(q^{(2)}) = \epsilon^{AB}\epsilon^{CD}q_{AC}q_{BD}/2$ where the indices $A\ldots D$ go from 1 to 2. This can be further rewritten as

$$\det(q^{(2)}) = \frac{\epsilon^{3ab}\epsilon^{3cd}q_{ac}q_{bc}}{2} .$$

The standard formula for an inverse matrix is

$$q^{ab} = \frac{\epsilon^{acd}\epsilon^{bef}q_{ce}q_{df}}{3!\det(q)}$$

Note the similarity between this and the expression for

$\det(q^{(2)})$. But in Ashtekar variables we have $\tilde{E}_i^a \tilde{E}^{bi} = \det(q)q^{ab}$. Therefore

$$A_\Sigma = \int_\Sigma dx^1 dx^2 \sqrt{\tilde{E}_i^3 \tilde{E}^{3i}} .$$

According to the rules of canonical quantization we should promote the triads $\hat{E}_i^3$ to quantum operators,

$$\hat{E}_i^3 \sim \frac{\delta}{\delta A_3^i} .$$

It turns out that the area A_Σ can be promoted to a well defined quantum operator despite the fact that we are dealing with product of two functional derivatives and worse we have a square-root to contend with as well.[13] Putting $N = 2J$, we talk of being in the J-th representation. We note that $\sum_i T^i T^i = J(J+1)1$. This quantity is important in the final formula for the area spectrum. We simply state the result below,

$$\hat{A}_\Sigma W_\gamma[A] = 8\pi\ell_{\text{Planck}}^2 \beta \sum_I \sqrt{j_I(j_I+1)} W_\gamma[A]$$

where the sum is over all edges I of the Wilson loop that pierce the surface Σ.

The formula for the volume of a region R is given by

$$V = \int_R d^3x \sqrt{\det(q)} = \frac{1}{6}\int_R dx^3 \sqrt{\epsilon_{abc}\epsilon^{ijk}\tilde{E}_i^a \tilde{E}_j^b \tilde{E}_k^c} .$$

The quantization of the volume proceeds the same way as with the area. As we take the derivative, and each time we do so we bring down the tangent vector $\dot{\gamma}^a$, when the volume operator acts on non-intersecting Wilson loops the result vanishes. Quantum states with non-zero volume must therefore involve intersections. Given that the anti-symmetric summation is taken over in the formula for the volume we would need at least intersections with three non-coplanar lines. Actually it turns out that one needs at least four-valent vertices for the volume operator to be non-vanishing.

We now consider Wilson loops with intersections. We assume the real representation where the gauge group is $SU(2)$. Wilson loops are an over complete basis as there are identities relating different Wilson loops. These come about from the fact that Wilson loops are based on matrices (the holonomy) and these matrices satisfy identities. Given any two $SU(2)$ matrices $\mathbb{A}$ and $\mathbb{B}$ it is easy to check that,

$$\text{Tr}(\mathbb{A})\text{Tr}(\mathbb{B}) = \text{Tr}(\mathbb{AB}) + \text{Tr}(\mathbb{AB}^{-1}) .$$

This implies that given two loops γ and η that intersect, we will have,

$$W_\gamma[A]W_\eta[A] = W_{\gamma\circ\eta}[A] + W_{\gamma\circ\eta^{-1}}[A]$$

where by η^{-1} we mean the loop η traversed in the opposite direction and $\gamma\circ\eta$ means the loop obtained by going around the loop γ and then along η. See figure below. Given that the matrices are unitary one has that $W_\gamma[A] = W_{\gamma^{-1}}[A]$. Also given the cyclic property of the matrix traces (i.e.

$Tr(AB) = Tr(BA)$) one has that $W_{\gamma \circ \eta}[A] = W_{\eta \circ \gamma}[A]$. These identities can be combined with each other into further identities of increasing complexity adding more loops. These identities are the so-called Mandelstam identities. Spin networks certain are linear combinations of intersecting Wilson loops designed to address the over completeness introduced by the Mandelstam identities (for trivalent intersections they eliminate the over-completeness entirely) and actually constitute a basis for all gauge invariant functions.

Graphical representation of the simplest non-trivial Mandestam identity relating different Wilson loops.

As mentioned above the holonomy tells you how to propagate test spin half particles. A spin network state assigns an amplitude to a set of spin half particles tracing out a path in space, merging and splitting. These are described by spin networks γ : the edges are labelled by spins together with `intertwiners' at the vertices which are prescription for how to sum over different ways the spins are rerouted. The sum over rerouting are chosen as such to make the form of the intertwiner invariant under Gauss gauge transformations.

3.4.5   Real variables, modern analysis and LQG

Main article: Hamiltonian constraint of LQG

Let us go into more detail about the technical difficulties associated with using Ashtekar's variables:

With Ashtekar's variables one uses a complex connection and so the relevant gauge group as actually $SL(2, \mathbb{C})$ and not $SU(2)$. As $SL(2, \mathbb{C})$ is non-compact it creates serious problems for the rigorous construction of the necessary mathematical machinery. The group $SU(2)$ is on the other hand is compact and the relevant constructions needed have been developed.

As mentioned above, because Ashtekar's variables are complex it results in complex general relativity. To recover the real theory one has to impose what are known as the reality conditions. These require that the densitized triad be real and that the real part of the Ashtekar connection equals the compatible spin connection (the compatibility condition being $\nabla_a e_b^i = 0$) determined by the desitized triad. The expression for compatible connection Γ_a^i is rather complicated and as such non-polynomial formula enters through the back door.

Before we state the next difficulty we should give a definition; a tensor density of weight W transforms like an ordinary tensor, except that in additional the W th power of the Jacobian,

$$J = \left| \frac{\partial x^a}{\partial x'^b} \right|$$

appears as a factor, i.e.

$$T'^{a\ldots}_{b\ldots} = J^W \frac{\partial x'^a}{\partial x^c} \cdots \frac{\partial x^d}{\partial x'^b} T^{c\ldots}_{d\ldots} .$$

It turns out that it is impossible, on general grounds, to construct a UV-finite, diffeomorphism non-violating operator corresponding to $\sqrt{\det(q)}H$. The reason is that the rescaled Hamiltonian constraint is a scalar density of weight two while it can be shown that only scalar densities of weight one have a chance to result in a well defined operator. Thus, one is forced to work with the original unrescaled, density one-valued, Hamiltonian constraint. However, this is non-polynomial and the whole virtue of the complex variables is questioned. In fact, all the solutions constructed for Ashtekar's Hamiltonian constraint only vanished for finite regularization (physics), however, this violates spatial diffeomorphism invariance.

Without the implementation and solution of the Hamiltonian constraint no progress can be made and no reliable predictions are possible!

To overcome the first problem one works with the configuration variable

$$A_a^i = \Gamma_a^i + \beta K_a^i$$

where β is real (as pointed out by Barbero, who introduced real variables some time after Ashtekar's variables[14][15]). The Guass law and the spatial diffeomorphism constraints are the same. In real Ashtekar variables the Hamiltonian is

$$H = \frac{\epsilon_{ijk} F_{ab}^k \tilde{E}_i^a \tilde{E}_j^b}{\sqrt{\det(q)}} + 2 \frac{\beta^2 + 1}{\beta^2} \frac{(\tilde{E}_i^a \tilde{E}_j^b - \tilde{E}_j^a \tilde{E}_i^b)}{\sqrt{\det(q)}} (A_a^i - \Gamma_a^i)(A_b^j - \Gamma_b^j) = H_E + H' .$$

The complicated relationship between Γ_a^i and the desitized triads causes serious problems upon quantization. It is with the choice $\beta = \pm i$ that the second more complicated term is made to vanish. However, as mentioned above Γ_a^i reappears in the reality conditions. Also we still have the problem of the $1/\sqrt{\det(q)}$ factor.

Thiemann was able to make it work for real β . First he could simplify the troublesome $1/\sqrt{\det(q)}$ by using the identity

$$\{A_c^k, V\} = \frac{\epsilon_{abc} \epsilon^{ijk} \tilde{E}_i^a \tilde{E}_j^b}{\sqrt{\det(q)}}$$

where V is the volume. The A_c^k and V can be promoted to well defined operators in the loop representation and the Poisson bracket is replaced by a commutator upon quantization; this takes care of the first term. It turns out that a similar trick can be used to treat the second term. One introduces the quantity

$$K = \int d^3x K_a^i \tilde{E}_i^a$$

and notes that

$$K_a^i = \{A_a^i, K\}.$$

We are then able to write

$$A_a^i - \Gamma_a^i = \beta K_a^i = \beta\{A_a^i, K\}.$$

The reason the quantity K is easier to work with at the time of quantization is that it can be written as

$$K = -\{V, \int d^3x H_E\}$$

where we have used that the integrated densitized trace of the extrinsic curvature, K, is the "time derivative of the volume".

In the long history of canonical quantum gravity formulating the Hamiltonian constraint as a quantum operator (Wheeler–DeWitt equation) in a mathematically rigorous manner has been a formidable problem. It was in the loop representation that a mathematically well defined Hamiltonian constraint was finally formulated in 1996.[9] We leave more details of its construction to the article Hamiltonian constraint of LQG. This together with the quantum versions of the Gauss law and spatial diffeomorphism constrains written in the loop representation are the central equations of LQG (modern canonical quantum General relativity).

Finding the states that are annihilated by these constraints (the physical states), and finding the corresponding physical inner product, and observables is the main goal of the technical side of LQG.

A very important aspect of the Hamiltonian operator is that it only acts at vertices (a consequence of this is that Thiemann's Hamiltonian operator, like Ashtekar's operator, annihilates non-intersecting loops except now it is not just formal and has rigorous mathematical meaning). More precisely, its action is non-zero on at least vertices of valence three and greater and results in a linear combination of new spin networks where the original graph has been modified by the addition of lines at each vertex together and a change in the labels of the adjacent links of the vertex.

3.4.6 Implementation and solution the quantum constraints

Main articles: spectrum, dual space, Rigged Hilbert space and quantum configuration space

We solve, at least approximately, all the quantum constraint equations and for the physical inner product to make physical predictions.

Before we move on to the constraints of LQG, lets us consider certain cases. We start with a kinematic Hilbert space $\mathcal{H}_{\text{Kin}}$ as so is equipped with an inner product—the kinematic inner product $\langle \phi, \psi \rangle_{\text{Kin}}$.

i) Say we have constraints $\hat{C}_I$ whose zero eigenvalues lie in their discrete spectrum. Solutions of the first constraint, $\hat{C}_1$, correspond to a subspace of the kinematic Hilbert space, $\mathcal{H}_1 \subset \mathcal{H}_{\text{Kin}}$. There will be a projection operator P_1 mapping $\mathcal{H}_{\text{Kin}}$ onto $\mathcal{H}_1$. The kinematic inner product structure is easily employed to provide the inner product structure after solving this first constraint; the new inner product $\langle \phi, \psi \rangle_1$ is simply

$$\langle \phi, \psi \rangle_1 = \langle P\phi, P\psi \rangle_{\text{Kin}}$$

They are based on the same inner product and are states normalizable with respect to it.

ii) The zero point is not contained in the point spectrum of all the $\hat{C}_I$, there is then no non-trivial solution $\Psi \in \mathcal{H}_{\text{Kin}}$ to the system of quantum constraint equations $\hat{C}_I \Psi = 0$ for all I.

For example the zero eigenvalue of the operator

$$\hat{C} = \left(i\frac{d}{dx} - k \right)$$

on $L_2(\mathbb{R}, dx)$ lies in the continuous spectrum $\mathbb{R}$ but the formal "eigenstate" $\exp(-ikx)$ is not normalizable in the kinematic inner product,

$$\int_{-\infty}^{\infty} dx \psi^*(x)\psi(x) = \int_{-\infty}^{\infty} dx e^{ikx}e^{-ikx} = \int_{-\infty}^{\infty} dx = \infty$$

and so does not belong to the kinematic Hilbert space $\mathcal{H}_{\text{Kin}}$. In these cases we take a dense subset $\mathcal{S}$ of $\mathcal{H}_{\text{Kin}}$ (intuitively this means either any point in $\mathcal{S}$ is either in $\mathcal{H}_{\text{Kin}}$ or arbitrarily close to a point in $\mathcal{H}_{\text{Kin}}$) with very good convergence properties and consider its dual space $\mathcal{S}'$ (intuitively these map elements of $\mathcal{S}$ onto finite complex numbers in a linear manner), then $\mathcal{S} \subset \mathcal{H}_{\text{Kin}} \subset \mathcal{S}'$ (as $\mathcal{S}'$ contains distributional functions). The constraint operator is then implemented on this larger dual space, which contains distributional functions, under the adjoint action on the operator. One looks for solutions on this larger space. This comes at the price that the solutions must be given a new Hilbert space inner product with respect to which they are normalizable (see article on rigged Hilbert space). In this case we have a generalized projection operator on the new space of states. We cannot use the above formula for the new inner product as it diverges, instead the new inner product is given by the simply modification of the above,

$$\langle \phi, \psi \rangle_1 = \langle P\phi, \psi \rangle_{\text{Kin}}.$$

The generalized projector P is known as a rigging map.

Implementation and solution the quantum constraints of LQG.

Let us move to LQG, additional complications will arise from that one cannot define an operator for the quantum spatial diffeomorphism constraint as the infinitesimal generator of finite diffeomorphism transformations and the fact the constraint algebra is not a Lie algebra due to the bracket between two Hamiltonian constraints.

Implementation and solution the Gauss constraint:

One does not actually need to promote the Gauss constraint to an operator since we can work directly with Gauss-gauge-invariant functions (that is, one solves the constraint classically and quantizes only the phase space reduced with respect to the Gauss constraint). The Gauss law is solved by the use of spin network states. They provide a basis for the Kinematic Hilbert space $\mathcal{H}_{\text{Kin}}$.

Implementation of the quantum spatial diffeomorphism constraint:

It turns out that one cannot define an operator for the quantum spatial diffeomorphism constraint as the infinitesimal generator of finite diffeomorphism transformations, represented on $\mathcal{H}_{\text{Kin}}$. The representation of finite diffeomorphisms is a family of unitary operators $\hat{U}_\varphi$ acting on a spin-network state ψ_γ by

$$\hat{U}_\varphi \psi_\gamma := \psi_{\varphi \circ \gamma}$$

for any spatial diffeomorphism φ on Σ. To understand why one cant define an operator for the quantum spatial diffeomorphism constraint consider what is called a 1-parameter subgroup φ_t in the group of spatial diffeomorphisms, this is then represented as a 1-parameter unitary group $\hat{U}_{\varphi_t}$ on $\mathcal{H}_{\text{Kin}}$. However, $\hat{U}_{\varphi_t}$ is not weakly continuous since the subspace $\psi_{\varphi_t \circ \gamma}$ belongs to and the subspace ψ_γ belongs to are orthogonal to each other no matter how small the parameter t is. So one always has

$$|< \psi_\gamma |\hat{U}_{\varphi_t}|\psi_\gamma >_{Kin} - < \psi_\gamma |\psi_\gamma >_{Kin} | =< \psi_\gamma |\psi_\gamma >_{Kin} \neq 0,$$

even in the limit when t goes to zero. Therefore, the infinitesimal generator of $\hat{U}_{\varphi_t}$ does not exist.

Solution of the spatial diffeomorphism constraint.

The spatial diffeomorphism constraint has been solved. The induced inner product $< \cdot, \cdot >_{Diff}$ on $\mathcal{H}_{\text{Diff}}$ (we do not pursue the details) has a very simple description in terms of spin network states: given two spin networks s and s', with associated spin network states ψ_s and $\psi_{s'}$, the inner product is 1 if s and s' are related to each other by a spatial diffeomorphism and zero otherwise.

We have provided a description of the implemented and complete solution of the kinematic constraints, the Gauss and spatial diffeomorphisms constraints which will be the same for any background-independent gauge field theory. The feature that distinguishes such different theories is the Hamiltonian constraint which is the only one that depends on the Lagrangian of the classical theory.

Problem arising from the Hamiltonian constraint.

Details of the implementation the quantum Hamiltonian constraint and solutions are treated in a different article Hamiltonian constraint of LQG. However, in this article we introduce an approximation scheme for the formal solution of the Hamiltonian constraint operator given in the section below on spinfoams. Here we just mention issues that arises with the Hamiltonian constraint.

The Hamiltonian constraint maps diffeomorphism invariant states onto non-diffeomorphism invaiant states as so does not preserve the diffeomorphism Hilbert space $\mathcal{H}_{\text{Diff}}$. This is an unavoidable consequence of the operator algebra, in particular the commutator:

$$[\hat{C}(\vec{N}), \hat{H}(M)] \propto \hat{H}(\mathcal{L}_{\vec{N}} M)$$

as can be seen by applying this to $\psi_s \in \mathcal{H}_{Diff}$,

$$(\vec{C}(\vec{N})\hat{H}(M) - \hat{H}(M)\vec{C}(\vec{N}))\psi_s \propto \hat{H}(\mathcal{L}_{\vec{N}} M)\psi_s$$

and using $\vec{C}(\vec{N})\psi_s = 0$ to obtain

$$\vec{C}(\vec{N})[\hat{H}(M)\psi_s \propto \hat{H}(\mathcal{L}_{\vec{N}} M)\psi_s \neq 0$$

and so $\hat{H}(M)\psi_s$ is not in $\mathcal{H}_{Diff}$.

This means that you can't just solve the spatial diffeomorphism constraint and then the Hamiltonian constraint. This problem can be circumvented by the introduction of the master constraint, with its trivial operator algebra, one is then able in principle to construct the physical inner product from $\mathcal{H}_{\text{Diff}}$.

3.5   Spin foams

Main articles: spin network, spin foam, BF model and Barrett–Crane model

In loop quantum gravity (LQG), a spin network represents a "quantum state" of the gravitational field on a 3-dimensional hypersurface. The set of all possible spin networks (or, more accurately, "s-knots" - that is, equivalence classes of spin networks under diffeomorphisms) is countable; it constitutes a basis of LQG Hilbert space.

In physics, a spin foam is a topological structure made out of two-dimensional faces that represents one of the configurations that must be summed to obtain a Feynman's path integral (functional integration) description of quantum gravity. It is closely related to loop quantum gravity.

3.5.1 Spin foam derived from the Hamiltonian constraint operator

The Hamiltonian constraint generates 'time' evolution. Solving the Hamiltonian constraint should tell us how quantum states evolve in 'time' from an initial spin network state to a final spin network state. One approach to solving the Hamiltonian constraint starts with what is called the Dirac delta function. This is a rather singular function of the real line, denoted $\delta(x)$, that is zero everywhere except at $x = 0$ but whose integral is finite and nonzero. It can be represented as a Fourier integral,

$$\delta(x) = \int e^{ikx} dk \ .$$

One can employ the idea of the delta function to impose the condition that the Hamiltonian constraint should vanish. It is obvious that

$$\prod_{x \in \Sigma} \delta(\hat{H}(x))$$

is non-zero only when $\hat{H}(x) = 0$ for all x in Σ. Using this we can 'project' out solutions to the Hamiltonian constraint. With analogy to the Fourier integral given above, this (generalized) projector can formally be written as

$$\int [dN] e^{i \int d^3x N(x) \hat{H}(x)} \ .$$

Interestingly, this is formally spatially diffeomorphism-invariant. As such it can be applied at the spatially diffeomorphism-invariant level. Using this the physical inner product is formally given by

$$\left\langle \int [dN] e^{i \int d^3x N(x) \hat{H}(x)} s_{\text{int}} s_{\text{fin}} \right\rangle_{\text{Diff}}$$

where s_{int} are the initial spin network and s_{fin} is the final spin network.

The exponential can be expanded

$$\left\langle \int [dN] (1 \quad + \quad i \int d^3x N(x) \hat{H}(x) \quad + \right.$$
$$\frac{i^2}{2!} [\int d^3x N(x) \hat{H}(x)][\int d^3x' N(x') \hat{H}(x')] \quad +$$
$$\left. \dots) s_{\text{int}}, s_{\text{fin}} \right\rangle_{\text{Diff}}$$

and each time a Hamiltonian operator acts it does so by adding a new edge at the vertex. The summation over different sequences of actions of $\hat{H}$ can be visualized as a summation over different histories of 'interaction vertices' in the 'time' evolution sending the initial spin network to the final spin network. This then naturally gives rise to the two-complex (a combinatorial set of faces that join along edges, which in turn join on vertices) underlying the spin foam description; we evolve forward an initial spin network sweeping out a surface, the action of the Hamiltonian constraint operator is to produce a new planar surface starting at the vertex. We are able to use the action of the Hamiltonian constraint on the vertex of a spin network state to associate

an amplitude to each "interaction" (in analogy to Feynman diagrams). See figure below. This opens up a way of trying to directly link canonical LQG to a path integral description. Now just as a spin networks describe quantum space, each configuration contributing to these path integrals, or sums over history, describe 'quantum space-time'. Because of their resemblance to soap foams and the way they are labeled John Baez gave these 'quantum space-times' the name 'spin foams'.

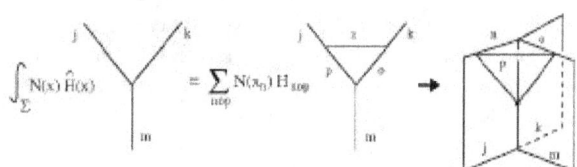

The action of the Hamiltonian constraint translated to the path integral or so-called spin foam description. A single node splits into three nodes, creating a spin foam vertex. $N(x_n)$ is the value of N at the vertex and H_{nop} are the matrix elements of the Hamiltonian constraint $\hat{H}$.

There are however severe difficulties with this particular approach, for example the Hamiltonian operator is not self-adjoint, in fact it is not even a normal operator (i.e. the operator does not commute with its adjoint) and so the spectral theorem cannot be used to define the exponential in general. The most serious problem is that the $\hat{H}(x)$'s are not mutually commuting, it can then be shown the formal quantity $\int [dN] e^{i \int d^3x N(x) \hat{H}(x)}$ cannot even define a (generalized) projector. The master constraint (see below) does not suffer from these problems and as such offers a way of connecting the canonical theory to the path integral formulation.

3.5.2 Spin foams from BF theory

It turns out there are alternative routes to formulating the path integral, however their connection to the Hamiltonian formalism is less clear. One way is to start with the BF theory. This is a simpler theory to general relativity. It has no local degrees of freedom and as such depends only on topological aspects of the fields. BF theory is what is known as a topological field theory. Surprisingly, it turns out that general relativity can be obtained from BF theory by imposing a constraint.[16] BF theory involves a field B_{ab}^{IJ} and if one chooses the field B to be the (anti-symmetric) product of two tetrads

$$B_{ab}^{IJ} = \frac{1}{2}(E_a^I E_b^J - E_b^I E_a^J)$$

(tetrads are like triads but in four spacetime dimensions), one recovers general relativity. The condition that the B field be given by the product of two tetrads is called the simplicity constraint. The spin foam dynamics of the topological field theory is well understood. Given the spin foam 'interaction' amplitudes for this simple theory, one then tries to implement the simplicity conditions to obtain a path integral for general relativity. The non-trivial task of constructing a spin foam model is then reduced to the question of how this simplicity constraint should be imposed in the quantum theory. The first attempt at this was the famous Barrett–Crane model.[17] However this model was shown to be problematic, for example there did not seem to be enough degrees of freedom to ensure the correct classical limit.[18] It has been argued that the simplicity constraint was imposed too strongly at the quantum level and should only be imposed in the sense of expectation values just as with the Lorenz gauge condition $\partial_\mu \hat{A}^\mu$ in the Gupta–Bleuler formalism of quantum electrodynamics. New models have now been put forward, sometimes motivated by imposing the simplicity conditions in a weaker sense.

Another difficulty here is that spin foams are defined on a discretization of spacetime. While this presents no problems for a topological field theory as it has no local degrees of freedom, it presents problems for GR. This is known as the problem triangularization dependence.

3.5.3 Modern formulation of spin foams

Just as imposing the classical simplicity constraint recovers general relativity from BF theory, one expects an appropriate quantum simplicity constraint will recover quantum gravity from quantum BF theory.

Much progress has been made with regard to this issue by Engle, Pereira, and Rovelli[19] and Freidal and Krasnov[20] in defining spin foam interaction amplitudes with much better behaviour.

An attempt to make contact between EPRL-FK spin foam and the canonical formulation of LQG has been made.[21]

3.5.4 Spin foam derived from the master constraint operator

See below.

3.6 The semi-classical limit

3.6.1 What is the semiclassical limit?

Main articles: Correspondence principle and classical limit

The **classical limit** or **correspondence limit** is the ability of a physical theory to approximate or "recover" classical mechanics when considered over special values of its parameters.[22] The classical limit is used with physical theories that predict non-classical behavior.

In physics, the **correspondence principle** states that the behavior of systems described by the theory of quantum mechanics (or by the old quantum theory) reproduces classical physics in the limit of large quantum numbers. In other words, it says that for large orbits and for large energies, quantum calculations must agree with classical calculations.[23]

The principle was formulated by Niels Bohr in 1920,[24] though he had previously made use of it as early as 1913 in developing his model of the atom.[25]

There are two basic requirements in establishing the semiclassical limit of any quantum theory:

i) reproduction of the Poisson brackets (of the diffeomorphism constraints in the case of general relativity). This is extremely important because, as noted above, the Poisson bracket algebra formed between the (smeared) constraints themselves completely determines the classical theory. This is analogous to establishing Ehrenfest's theorem;

ii) the specification of a complete set of classical observables whose corresponding operators (see complete set of commuting observables for the quantum mechanical definition of a complete set of observables) when acted on by appropriate semi-classical states reproduce the same classical variables with small quantum corrections (a subtle point is that states that are semi-classical for one class of observables may not be semi-classical for a different class of observables[26]).

This may be easily done, for example, in ordinary quantum mechanics for a particle but in general relativity this becomes a highly non-trivial problem as we will see below.

3.6.2 Why might LQG not have general relativity as its semiclassical limit?

Any candidate theory of quantum gravity must be able to reproduce Einstein's theory of general relativity as a classical limit of a quantum theory. This is not guaranteed because of a feature of quantum field theories which is that they have different sectors, these are analogous to the different phases that come about in the thermodynamical limit of statistical systems. Just as different phases are physically different, so are different sectors of a quantum field theory. It may turn out that LQG belongs to an unphysical sector - one in which you do not recover general relativity in the semi classical limit (in fact there might not be any physical sector at all).

Moreover, the physical Hilbert space H_{phys} must contain enough semi-classical states to guarantee that the quantum theory one obtains can return to the classical theory when $\hbar \to 0$.

Theorems establishing the uniqueness of the loop representation as defined by Ashtekar et al. (i.e. a certain concrete realization of a Hilbert space and associated operators reproducing the correct loop algebra - the realization that everybody was using) have been given by two groups (Lewandowski, Okolow, Sahlmann and Thiemann;[27] and Christian Fleischhack[28]). Before this result was established it was not known whether there could be other examples of Hilbert spaces with operators invoking the same loop algebra, other realizations, not equivalent to the one that had been used so far. These uniqueness theorems imply no others exist and so if LQG does not have the correct semiclassical limit then this would mean the end of the loop representation of quantum gravity altogether.

3.6.3 Difficulties checking the semiclassical limit of LQG

There are difficulties in trying to establish LQG gives Einstein's theory of general relativity in the semi classical limit. There are a number of particular difficulties in establishing the semi-classical limit

1. There is no operator corresponding to infinitesimal spacial diffeomorphisms (it is not surprising that the theory has no generator of infinitesimal spatial 'translations' as it predicts spatial geometry has a discrete nature, compare to the situation in condensed matter). Instead it must be approximated by finite spatial diffeomorphisms and so the Poisson bracket structure of the classical theory is not exactly reproduced. This problem can be circumvented with the introduction of the so-called master constraint (see below)[29]

2. There is the problem of reconciling the discrete combinatorial nature of the quantum states with the continuous nature of the fields of the classical theory.

3. There are serious difficulties arising from the structure of the Poisson brackets involving the spatial diffeomorphism and Hamiltonian constraints. In particular, the algebra of (smeared) Hamiltonian constraints does not close, it is proportional to a sum over infinitesimal spatial diffeomorphisms (which, as we have just noted, does not exist in the quantum theory) where the coefficients of proportionality are not constants but have non-trivial phase space dependence - as such it does not form a Lie algebra. However, the situation is much improved by the introduction of the master constraint.[29]

4. The semi-classical machinery developed so far is only appropriate to non-graph-changing operators, however, Thiemann's Hamiltonian constraint is a graph-changing operator - the new graph it generates has degrees of freedom upon which the coherent state does not depend and so their quantum fluctuations are not suppressed. There is also the restriction, so far, that these coherent states are only defined at the Kinematic level, and now one has to lift them to the level of $\mathcal{H}_{Diff}$ and $\mathcal{H}_{Phys}$. It can be shown that Thiemann's Hamiltonian constraint is required to be graph changing in order to resolve problem 3 in some sense. The master constraint algebra however is trivial and so the requirement that it be graph changing can be lifted and indeed non-graph changing master constraint operators have been defined.

5. Formulating observables for classical general relativity is a formidable problem by itself because of its non-linear nature and space-time diffeomorphism invariance. In fact a systematic approximation scheme to calculate observables has only been recently developed.[30][31]

Difficulties in trying to examine the semi classical limit of the theory should not be confused with it having the wrong semi classical limit.

3.6.4 Progress in demonstrating LQG has the correct semiclassical limit

Much details here to be written up...

Concerning issue number 2 above one can consider so-called weave states. Ordinary measurements of geometric quantities are macroscopic, and planckian discreteness is smoothed out. The fabric of a T-shirt is analogous. At a distance it is a smooth curved two-dimensional surface. But a closer inspection we see that it is actually composed of thousands of one-dimensional linked threads. The image of space given in LQG is similar, consider a very large spin network formed by a very large number of nodes and links, each of Planck scale. But probed at a macroscopic scale, it appears as a three-dimensional continuous metric geometry.

As far as the editor knows problem 4 of having semi-classical machinery for non-graph changing operators is as the moment still out of reach.

To make contact with familiar low energy physics it is mandatory to have to develop approximation schemes both for the physical inner product and for Dirac observables.

The spin foam models have been intensively studied can be viewed as avenues toward approximation schemes for the physical inner product.

Markopoulou et al. adopted the idea of noiseless subsystems in an attempt to solve the problem of the low energy limit in background independent quantum gravity theories[32][33][34] The idea has even led to the intriguing possibility of matter of the standard model being identified with emergent degrees of freedom from some versions of LQG (see section below: *LQG and related research programs*).

As Wightman emphasized in the 1950s, in Minkowski QFTs the $n-$ point functions

$$W(x_1, \ldots, x_n) = \langle 0| \phi(x_n) \ldots \phi(x_1) |0 \rangle \,.$$

completely determine the theory. In particular, one can calculate the scattering amplitudes from these quantities. As explained below in the section on the *Background independent scattering amplitudes*, in the background-independent context, the $n-$ point functions refer to a state and in gravity that state can naturally encode information about a specific geometry which can then appear in the expressions of these quantities. To leading order LQG calculations have been shown to agree in an appropriate sense with the $n-$ point functions calculated in the effective low energy quantum general relativity.

3.7 Improved dynamics and the master constraint

Main articles: Hamiltonian (quantum mechanics), Hamiltonian constraint of LQG and Friedrichs extension

3.7.1 The master constraint

Thiemann's master constraint should not be confused with the master equation which has to do with random processes. The Master Constraint Programme for Loop Quantum Gravity (LQG) was proposed as a classically equivalent way to impose the infinite number of Hamiltonian constraint equations

$$H(x) = 0$$

(x being a continuous index) in terms of a single master constraint,

$$M = \int d^3x \frac{[H(x)]^2}{\sqrt{\det(q(x))}} \,.$$

which involves the square of the constraints in question. Note that $H(x)$ were infinitely many whereas the master constraint is only one. It is clear that if M vanishes then so do the infinitely many $H(x)$'s. Conversely, if all the $H(x)$

's vanish then so does M, therefore they are equivalent. The master constraint M involves an appropriate averaging over all space and so is invariant under spatial diffeomorphisms (it is invariant under spatial "shifts" as it is a summation over all such spatial "shifts" of a quantity that transforms as a scalar). Hence its Poisson bracket with the (smeared) spacial diffeomorphism constraint, $C(\vec{N})$, is simple:

$$\{M, C(\vec{N})\} = 0 \,.$$

(it is $su(2)$ invariant as well). Also, obviously as any quantity Poisson commutes with itself, and the master constraint being a single constraint, it satisfies

$$\{M, M\} = 0 \,.$$

We also have the usual algebra between spatial diffeomorphisms. This represents a dramatic simplification of the Poisson bracket structure, and raises new hope in understanding the dynamics and establishing the semi-classical limit.[35]

An initial objection to the use of the master constraint was that on first sight it did not seem to encode information about the observables; because the Mater constraint is quadratic in the constraint, when you compute its Poisson bracket with any quantity, the result is proportional to the constraint, therefore it always vanishes when the constraints are imposed and as such does not select out particular phase space functions. However, it was realized that the condition

$$\{\{M, O\}, O\}_{M=0} = 0$$

is equivalent to O being a Dirac observable. So the master constraint does capture information about the observables. Because of its significance this is known as the Master equation.[35]

That the master constraint Poisson algebra is an honest Lie algebra opens up the possibility of using a certain method, known as group averaging, in order to construct solutions of the infinite number of Hamiltonian constraints, a physical inner product thereon and Dirac observables via what is known as refined algebraic quantization RAQ[36]

3.7.2 The quantum master constraint

Define the quantum master constraint (regularisation issues aside) as

$$\hat{M} := \int d^3x \left(\widehat{\frac{H}{\det(q(x))^{1/4}}} \right)^\dagger (x) \left(\widehat{\frac{H}{\det(q(x))^{1/4}}} \right)(x) \,.$$

Obviously,

$$\left(\widehat{\frac{H}{\det(q(x))^{1/4}}} \right)(x)\Psi = 0$$

for all x implies $\hat{M}\Psi = 0$. Conversely, if $\hat{M}\Psi = 0$ then

$$0 = <\Psi, \hat{M}\Psi> = \int d^3x \left\| \left(\widehat{\frac{H}{\det(q(x))^{1/4}}} \right)(x)\Psi \right\|^2 \quad Eq\ 4$$

implies

$$\left(\widehat{\frac{H}{\det(q(x))^{1/4}}} \right)(x)\Psi = 0 .$$

What is done first is, we are able to compute the matrix elements of the would-be operator $\hat{M}$, that is, we compute the quadratic form Q_M . It turns out that as Q_M is a graph changing, diffeomorphism invariant quadratic form it cannot exist on the kinematic Hilbert space H_{Kin} , and must be defined on H_{Diff} . The fact that the master constraint operator $\hat{M}$ is densely defined on H_{Diff} , it is obvious that $\hat{M}$ is a positive and symmetric operator in H_{Diff} . Therefore, the quadratic form Q_M associated with $\hat{M}$ is closable. The closure of Q_M is the quadratic form of a unique self-adjoint operator $\overline{M}$, called the Friedrichs extension of $\hat{M}$. We relabel $\overline{M}$ as $\hat{M}$ for simplicity. (Note that the presence of an inner product, viz Eq 4, means there are no superfluous solutions i.e. there are no Ψ such that $\left(\widehat{\frac{H}{\det(q(x))^{1/4}}} \right)(x)\Psi \neq 0$ but for which $\hat{M}\Psi = 0$).

It is also possible to construct a quadratic form Q_{M_E} for what is called the extended master constraint (discussed below) on H_{Kin} which also involves the weighted integral of the square of the spatial diffeomorphism constraint (this is possible because Q_{M_E} is not graph changing).

The spectrum of the master constraint may not contain zero due to normal or factor ordering effects which are finite but similar in nature to the infinite vacuum energies of background-dependent quantum field theories. In this case it turns out to be physically correct to replace $\hat{M}$ with $\hat{M}' := \hat{M} - min(spec(\hat{M}))\hat{1}$ provided that the "normal ordering constant" vanishes in the classical limit, that is, $\lim_{\hbar\to 0} min(spec(\hat{M})) = 0$, so that $\hat{M}'$ is a valid quantisation of M .

3.7.3   Testing the master constraint

The constraints in their primitive form are rather singular, this was the reason for integrating them over test functions to obtain smeared constraints. However, it would appear that the equation for the master constraint, given above, is even more singular involving the product of two primitive constraints (although integrated over space). Squaring the constraint is dangerous as it could lead to worsened ultraviolent behaviour of the corresponding operator and hence the master constraint programme must be approached with due care.

In doing so the master constraint programme has been satisfactorily tested in a number of model systems with non-trivial constraint algebras, free and interacting field theories.[37][38][39][40][41] The master constraint for LQG was established as a genuine positive self-adjoint operator and the physical Hilbert space of LQG was shown to be non-empty,[42] an obvious consistency test LQG must pass to be a viable theory of quantum General relativity.

3.7.4   Applications of the master constraint

The master constraint has been employed in attempts to approximate the physical inner product and define more rigorous path integrals.[43][44][45][46]

The Consistent Discretizations approach to LQG,[47][48] is an application of the master constraint program to construct the physical Hilbert space of the canonical theory.

3.7.5   Spin foam from the master constraint

It turns out that the master constraint is easily generalized to incorporate the other constraints. It is then referred to as the extended master constraint, denoted M_E . We can define the extended master constraint which imposes both the Hamiltonian constraint and spatial diffeomorphism constraint as a single operator,

$$M_E = \int_\Sigma d^3x \frac{H(x)^2 - q^{ab} V_a(x) V_b(x)}{\sqrt{\det(q)}} .$$

Setting this single constraint to zero is equivalent to $H(x) = 0$ and $V_a(x) = 0$ for all x in Σ . This constraint implements the spatial diffeomorphism and Hamiltonian constraint at the same time on the Kinematic Hilbert space. The physical inner product is then defined as

$$\langle\phi, \psi\rangle_{Phys} = \lim_{T\to\infty} \left\langle \phi, \int_{-T}^{T} dt\, e^{it\hat{M}_E} \psi \right\rangle$$

(as $\delta(\hat{M}_E) = \lim_{T\to\infty} \int_{-T}^{T} dt\, e^{it\hat{M}_E}$). A spin foam representation of this expression is obtained by splitting the t-parameter in discrete steps and writing

$$e^{it\hat{M}_E} = \lim_{n\to\infty} [e^{it\hat{M}_E/n}]^n = \lim_{n\to\infty} [1 + it\hat{M}_E/n]^n.$$

The spin foam description then follows from the application of $[1 + it\hat{M}_E/n]$ on a spin network resulting in a linear combination of new spin networks whose graph and labels have been modified. Obviously an approximation is made by truncating the value of n to some finite integer. An advantage of the extended master constraint is that we are working at the kinematic level and so far it is only here we have access semi-classical coherent states. Moreover, one can find none graph changing versions of this master constraint operator, which are the only type of operators appropriate for these coherent states.

3.7.6 Algebraic quantum gravity

The master constraint programme has evolved into a fully combinatorial treatment of gravity known as Algebraic Quantum Gravity (AQG).[49] The non-graph changing master constraint operator is adapted in the framework of algebraic quantum gravity. While AQG is inspired by LQG, it differs drastically from it because in AQG there is fundamentally no topology or differential structure - it is background independent in a more generalized sense and could possibly have something to say about topology change. In this new formulation of quantum gravity AQG semiclassical states always control the fluctuations of all present degrees of freedom. This makes the AQG semiclassical analysis superior over that of LQG, and progress has been made in establishing it has the correct semiclassical limit and providing contact with familiar low energy physics.[50][51] See Thiemann's book for details.

3.8 Physical applications of LQG

3.8.1 Black hole entropy

Main articles: Black hole thermodynamics, Isolated horizon and Immirzi parameter

The Immirzi parameter (also known as the Barbero-Immirzi parameter) is a numerical coefficient appearing in loop quantum gravity. It may take real or imaginary values.

An artist depiction of two black holes merging, a process in which the laws of thermodynamics are upheld.

Black hole thermodynamics is the area of study that seeks to reconcile the laws of thermodynamics with the existence of black hole event horizons. The no hair conjecture of general relativity states that a black hole is characterized only by its mass, its charge, and its angular momentum; hence, it has

no entropy. It appears, then, that one can violate the second law of thermodynamics by dropping an object with nonzero entropy into a black hole.[52] Work by Stephen Hawking and Jacob Bekenstein showed that one can preserve the second law of thermodynamics by assigning to each black hole a *black-hole entropy*

$$S_{BH} = \frac{k_B A}{4\ell_P^2},$$

where A is the area of the hole's event horizon, k_B is the Boltzmann constant, and $\ell_P = \sqrt{G\hbar/c^3}$ is the Planck length.[53] The fact that the black hole entropy is also the maximal entropy that can be obtained by the Bekenstein bound (wherein the Bekenstein bound becomes an equality) was the main observation that led to the holographic principle.[52]

An oversight in the application of the no-hair theorem is the assumption that the relevant degrees of freedom accounting for the entropy of the black hole must be classical in nature; what if they were purely quantum mechanical instead and had non-zero entropy? Actually, this is what is realized in the LQG derivation of black hole entropy, and can be seen as a consequence of its background-independence – the classical black hole spacetime comes about from the semi-classical limit of the quantum state of the gravitational field, but there are many quantum states that have the same semiclassical limit. Specifically, in LQG[54] it is possible to associate a quantum geometrical interpretation to the microstates: These are the quantum geometries of the horizon which are consistent with the area, A, of the black hole and the topology of the horizon (i.e. spherical). LQG offers a geometric explanation of the finiteness of the entropy and of the proportionality of the area of the horizon.[55][56] These calculations have been generalized to rotating black holes.[57]

It is possible to derive, from the covariant formulation of full quantum theory (Spinfoam) the correct relation between energy and area (1st law), the Unruh temperature and the distribution that yields Hawking entropy.[58] The calculation makes use of the notion of dynamical horizon and is done for non-extremal black holes.

A recent success of the theory in this direction is the computation of the entropy of all non singular black holes directly from theory and independent of Immirzi parameter.[59] The result is the expected formula $S = A/4$, where S is the entropy and A the area of the black hole, derived by Bekenstein and Hawking on heuristic grounds. This is the only known derivation of this formula from a fundamental theory, for the case of generic non singular black holes. Older attempts at this calculation had difficulties. The problem was that although Loop quantum gravity predicted that the entropy of a black hole is proportional to the area of the

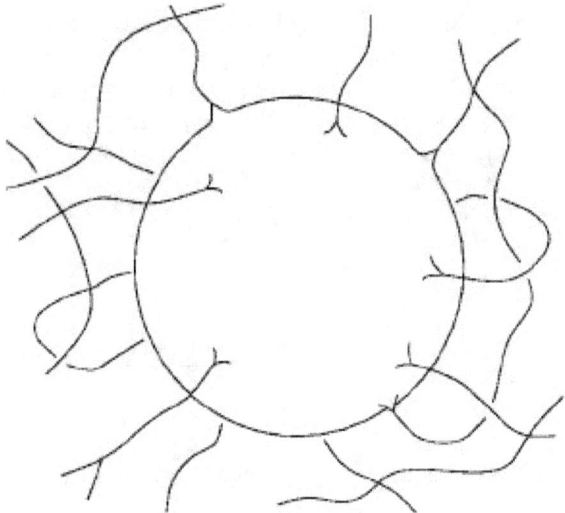

Representation of quantum geometries of the horizon. Polymer excitations in the bulk puncture the horizon, endowing it with quantized area. Intrinsically the horizon is flat except at punctures where it acquires a quantized deficit angle or quantized amount of curvature. These deficit angles add up to 4π.

event horizon, the result depended on a crucial free parameter in the theory, the above-mentioned Immirzi parameter. However, there is no known computation of the Immirzi parameter, so it had to be fixed by demanding agreement with Bekenstein and Hawking's calculation of the black hole entropy.

3.8.2 Loop quantum cosmology

Main articles: loop quantum cosmology, Big bounce and inflation (cosmology)

The popular and technical literature makes extensive references to LQG-related topic of loop quantum cosmology. LQC was mainly developed by Martin Bojowald, it was popularized Loop quantum cosmology in *Scientific American* for predicting a Big Bounce prior to the Big Bang. Loop quantum cosmology (LQC) is a symmetry-reduced model of classical general relativity quantized using methods that mimic those of loop quantum gravity (LQG) that predicts a "quantum bridge" between contracting and expanding cosmological branches.

Achievements of LQC have been the resolution of the big bang singularity, the prediction of a Big Bounce, and a natural mechanism for inflation (cosmology).

LQC models share features of LQG and so is a useful toy model. However, the results obtained are subject to the usual restriction that a truncated classical theory, then quantized, might not display the true behaviour of the full the-

ory due to artificial suppression of degrees of freedom that might have large quantum fluctuations in the full theory. It has been argued that singularity avoidance in LQC are by mechanisms only available in these restrictive models and that singularity avoidance in the full theory can still be obtained but by a more subtle feature of LQG.[60][61]

3.8.3 Loop quantum gravity phenomenology

Quantum gravity effects are notoriously difficult to measure because the Planck length is so incredibly small. However recently physicists have started to consider the possibility of measuring quantum gravity effects mostly from astrophysical observations and gravitational wave detectors.The energy of those fluctuations at scales this small cause space-perturbations which are visible at higher scales.

3.8.4 Background independent scattering amplitudes

Loop quantum gravity is formulated in a background-independent language. No spacetime is assumed a priori, but rather it is built up by the states of theory themselves - however scattering amplitudes are derived from n-point functions (Correlation function (quantum field theory)) and these, formulated in conventional quantum field theory, are functions of points of a background space-time. The relation between the background-independent formalism and the conventional formalism of quantum field theory on a given spacetime is far from obvious, and it is far from obvious how to recover low-energy quantities from the full background-independent theory. One would like to derive the n-point functions of the theory from the background-independent formalism, in order to compare them with the standard perturbative expansion of quantum general relativity and therefore check that loop quantum gravity yields the correct low-energy limit.

A strategy for addressing this problem has been suggested;[62] the idea is to study the boundary amplitude, namely a path integral over a finite space-time region, seen as a function of the boundary value of the field.[63] In conventional quantum field theory, this boundary amplitude is well–defined[64][65] and codes the physical information of the theory; it does so in quantum gravity as well, but in a fully background–independent manner.[66] A generally covariant definition of n-point functions can then be based on the idea that the distance between physical points –arguments of the n-point function is determined by the state of the gravitational field on the boundary of the spacetime region considered.

Progress has been made in calculating background independent scattering amplitudes this way with the use of spin foams. This is a way to extract physical information from the theory. Claims to have reproduced the correct behaviour for graviton scattering amplitudes and to have recovered classical gravity have been made. "We have calculated Newton's law starting from a world with no space and no time." - Carlo Rovelli.

3.9 Gravitons, string theory, supersymmetry, extra dimensions in LQG

Main articles: graviton, string theory, supersymmetry, Kaluza–Klein theory and supergravity

Some quantum theories of gravity posit a spin-2 quantum field that is quantized, giving rise to gravitons. In string theory one generally starts with quantized excitations on top of a classically fixed background. This theory is thus described as background dependent. Particles like photons as well as changes in the spacetime geometry (gravitons) are both described as excitations on the string worldsheet. The background dependence of string theory can have important physical consequences, such as determining the number of quark generations. In contrast, loop quantum gravity, like general relativity, is manifestly background independent, eliminating the background required in string theory. Loop quantum gravity, like string theory, also aims to overcome the nonrenormalizable divergences of quantum field theories.

LQG never introduces a background and excitations living on this background, so LQG does not use gravitons as building blocks. Instead one expects that one may recover a kind of semiclassical limit or weak field limit where something like "gravitons" will show up again. In contrast, gravitons play a key role in string theory where they are among the first (massless) level of excitations of a superstring.

LQG differs from string theory in that it is formulated in 3 and 4 dimensions and without supersymmetry or Kaluza–Klein extra dimensions, while the latter requires both to be true. There is no experimental evidence to date that confirms string theory's predictions of supersymmetry and Kaluza–Klein extra dimensions. In a 2003 paper A dialog on quantum gravity,[67] Carlo Rovelli regards the fact LQG is formulated in 4 dimensions and without supersymmetry as a strength of the theory as it represents the most parsimonious explanation, consistent with current experimental results, over its rival string/M-theory. Proponents of string theory will often point to the fact that, among other things, it demonstrably reproduces the established theories of general relativity and quantum field theory in the appropriate limits, which Loop Quantum Gravity has struggled to do. In that sense string theory's connection to established physics may be considered more reliable and less speculative, at the mathematical level. Peter Woit in Not Even Wrong and Lee Smolin in The Trouble with Physics regard string/M-theory to be in conflict with current known experimental results.

Since LQG has been formulated in 4 dimensions (with and without supersymmetry), and M-theory requires supersymmetry and 11 dimensions, a direct comparison between the two has not been possible. It is possible to extend mainstream LQG formalism to higher-dimensional supergravity, general relativity with supersymmetry and Kaluza–Klein extra dimensions should experimental evidence establish their existence. It would therefore be desirable to have higher-dimensional Supergravity loop quantizations at one's disposal in order to compare these approaches. In fact a series of recent papers have been published attempting just this.[68][69][70][71][72][73][74][75] Most recently, Thiemann (and alumni) have made progress toward calculating black hole entropy for supergravity in higher dimensions. It will be interesting to compare these results to the corresponding super string calculations.[76][77]

As of April 2013 LHC has failed to find evidence of supersymmetry or Kaluza–Klein extra dimensions, which has encouraged LQG researchers. Shaposhnikov in his paper "Is there a new physics between electroweak and Planck scales?" has proposed the neutrino minimal standard model,[78] which claims the most parsimonious theory is a standard model extended with neutrinos, plus gravity, and that extra dimensions, GUT physics, and supersymmetry, string/M-theory physics are unrealized in nature, and that any theory of quantum gravity must be four dimensional, like loop quantum gravity.

3.10 LQG and related research programs

Main articles: noncommutative geometry, twistor theory, entropic gravity, Sundance Bilson-Thompson, Asymptotic safety in quantum gravity, Causal dynamical triangulation, group field theory and consistent discretizations

Several research groups have attempted to combine LQG with other research programs: Johannes Aastrup, Jesper M. Grimstrup et al. research combines noncommutative geometry with loop quantum gravity,[79] Laurent Freidel, Simone Speziale, et al., spinors and twistor theory with loop quantum gravity,[80] and Lee Smolin et al. with Verlinde entropic gravity and loop gravity.[81] Stephon Alexander, Antonino Marciano and Lee Smolin have attempted to explain the origins of weak force chirality in terms of Ashketar's variables, which describe gravity as chiral,[82] and LQG with Yang–Mills theory fields[83] in four dimensions. Sundance Bilson-Thompson, Hackett et al.,[84][85] has attempted to introduce standard model via LQG"s degrees of freedom as an emergent property (by employing the idea noiseless subsystems a useful notion introduced in more general situation for constrained systems by Fotini Markopoulou-Kalamara et al.[86]) LQG has also drawn philosophical comparisons with causal dynamical triangulation[87] and asymptotically safe gravity,[88] and the spinfoam with group field theory and AdS/CFT correspondence.[89] Smolin and Wen have suggested combining LQG with String-net liquid, tensors, and Smolin and Fotini Markopoulou-Kalamara Quantum Graphity. There is the consistent discretizations approach. In addition to what has already mentioned above, Pullin and Gambini provide a framework to connect the path integral and canonical approaches to quantum gravity. They may help reconcile the spin foam and canonical loop representation approaches. Recent research by Chris Duston and Matilde Marcolli introduces topology change via topspin networks.[90]

3.11 Problems and comparisons with alternative approaches

Main article: List of unsolved problems in physics

Some of the major unsolved problems in physics are theoretical, meaning that existing theories seem incapable of explaining a certain observed phenomenon or experimental result. The others are experimental, meaning that there is a difficulty in creating an experiment to test a proposed theory or investigate a phenomenon in greater detail.

Can quantum mechanics and general relativity be realized as a fully consistent theory (perhaps as a quantum field theory)? Is spacetime fundamentally continuous or discrete? Would a consistent theory involve a force mediated by a hypothetical graviton, or be a product of a discrete structure of spacetime itself (as in loop quantum gravity)? Are there deviations from the predictions of general relativity at very small or very large scales or in other extreme circumstances that flow from a quantum gravity theory?

The theory of LQG is one possible solution to the problem of quantum gravity, as is string theory. There are substantial differences however. For example, string theory also addresses unification, the understanding of all known forces and particles as manifestations of a single entity, by postulating extra dimensions and so-far unobserved additional particles and symmetries. Contrary to this, LQG is based

only on quantum theory and general relativity and its scope is limited to understanding the quantum aspects of the gravitational interaction. On the other hand, the consequences of LQG are radical, because they fundamentally change the nature of space and time and provide a tentative but detailed physical and mathematical picture of quantum spacetime.

Presently, no semiclassical limit recovering general relativity has been shown to exist. This means it remains unproven that LQG's description of spacetime at the Planck scale has the right continuum limit (described by general relativity with possible quantum corrections). Specifically, the dynamics of the theory is encoded in the Hamiltonian constraint, but there is no candidate Hamiltonian.[91] Other technical problems include finding off-shell closure of the constraint algebra and physical inner product vector space, coupling to matter fields of Quantum field theory, fate of the renormalization of the graviton in perturbation theory that lead to ultraviolet divergence beyond 2-loops (see One-loop Feynman diagram in Feynman diagram).[91]

While there has been a recent proposal relating to observation of naked singularities,[92] and doubly special relativity as a part of a program called loop quantum cosmology, there is no experimental observation for which loop quantum gravity makes a prediction not made by the Standard Model or general relativity (a problem that plagues all current theories of quantum gravity). Because of the above-mentioned lack of a semiclassical limit, LQG has not yet even reproduced the predictions made by general relativity.

An alternative criticism is that general relativity may be an effective field theory, and therefore quantization ignores the fundamental degrees of freedom.

3.12 See also

3.13 Notes

[1] Rovelli, Carlo (August 2008). "Loop Quantum Gravity" (PDF). *CERN*. Retrieved 14 September 2014.

[2] Rovelli, C. (2011). "Zakopane lectures on loop gravity". arXiv:1102.3660 [gr-qc].

[3] Muxin, H. (2011). "Cosmological constant in loop quantum gravity vertex amplitude". *Physical Review D* **84** (6): 064010. arXiv:1105.2212. Bibcode:2011PhRvD..84f4010H. doi:10.1103/PhysRevD.84.064010.

[4] Fairbairn, W. J.; Meusburger, C. (2011). "q-Deformation of Lorentzian spin foam models". arXiv:1112.2511 [gr-qc].

[5] Rovelli, C. (2004). *Quantum Gravity*. Cambridge Monographs on Mathematical Physics. p. 71. ISBN 978-0-521-83733-0.

[6] Kauffman, S.; Smolin, L. (7 April 1997). "A Possible Solution For The Problem Of Time In Quantum Cosmology". *Edge.org*. Retrieved 2014-08-20.

[7] Smolin, L. (2006). "The Case for Background Independence". In Rickles, D.; French, S.; Saatsi, J. T. *The Structural Foundations of Quantum Gravity*. Clarendon Press. pp. 196*ff*. arXiv:hep-th/0507235. ISBN 978-0-19-926969-3.

[8] Rovelli, C. (2004). *Quantum Gravity*. Cambridge Monographs on Mathematical Physics. p. 13ff. ISBN 978-0-521-83733-0.

[9] Thiemann, T. (1996). "Anomaly-free formulation of nonperturbative, four-dimensional Lorentzian quantum gravity". *Physics Letters B* **380**: 257–264. arXiv:gr-qc/9606088. Bibcode:1996PhLB..380..257T. doi:10.1016/0370-2693(96)00532-1.

[10] Baez, J.; de Muniain, J. P. (1994). *Gauge Fields, Knots and Quantum Gravity*. Series on Knots and Everything. Vol. 4. World Scientific. Part III, chapter 4. ISBN 978-981-02-1729-7.

[11] Thiemann, T. (2003). "Lectures on Loop Quantum Gravity". *Lecture Notes in Physics* **631**: 41–135. arXiv:gr-qc/0210094. Bibcode:2003LNP...631...41T. doi:10.1007/978-3-540-45230-0_3.

[12] Rovelli, C.; Smolin, L. (1988). "Knot Theory and Quantum Gravity". *Physical Review Letters* **61** (10): 1155–1958. Bibcode:1988PhRvL..61.1155R. doi:10.1103/PhysRevLett.61.1155.

[13] Gambini, R.; Pullin, J. (2011). *A First Course in Loop Quantum Gravity*. Oxford University Press. Section 8.2. ISBN 978-0-19-959075-9.

[14] Fernando, J.; Barbero, G. (1995). "Reality Conditions and Ashtekar Variables: A Different Perspective". *Physical Review D* **51**: 5498–5506. arXiv:gr-qc/9410013. Bibcode:1995PhRvD..51.5498B. doi:10.1103/PhysRevD.51.5498.

[15] Fernando, J.; Barbero, G. (1995). "Real Ashtekar Variables for Lorentzian Signature Space-times". *Physical Review D* **51**: 5507–5520. arXiv:gr-qc/9410014. Bibcode:1995PhRvD..51.5507B. doi:10.1103/PhysRevD.51.5507.

[16] Bojowald, M.; Alejandro, P. "Spin Foam Quantization and Anomalies". arXiv:gr-qc/0303026 [gr-qc].

[17] Barrett, J.; Crane, L. (2000). "A Lorentzian signature model for quantum general relativity". *Classical and Quantum Gravity* **17**: 3101–3118. arXiv:gr-qc/9904025. Bibcode:2000CQGra..17.3101B. doi:10.1088/0264-9381/17/16/302..

[18] Rovelli, C.; Alesci, E. (2007). "The complete LQG propagator I. Difficulties with the Barrett–Crane vertex". *Physical Review D* **76**: 104012. arXiv:hep-th/0703074. Bibcode:2007PhRvD..76b4012B. doi:10.1103/PhysRevD.76.024012.

[19] Engle, J.; Pereira, R.; Rovelli, C. (2009). "Loop-Quantum-Gravity Vertex Amplitude". *Physical Review Letters* **99**: 161301. arXiv:0705.2388. Bibcode:2007PhRvL..99p1301E. doi:10.1103/physrevlett.99.161301.

[20] Freidal, L.; Krasnov, K. (2008). "A new spin foam model for 4D gravity". *Classical and Quantum Gravity* **25**: 125018. arXiv:0708.1595. Bibcode:2008CQGra..25l5018F. doi:10.1088/0264-9381/25/12/125018.

[21] Alesci, E.; Thiemann, T.; Zipfel, A. (2011). "Linking covariant and canonical LQG: new solutions to the Euclidean Scalar Constraint". arXiv:1109.1290.

[22] Bohm, D. (1989). *Quantum Theory*. Dover Publications. ISBN 978-0-486-65969-5.

[23] Tipler, P.; Llewellyn, R. (2008). *Modern Physics* (5th ed.). W. H. Freeman and Co. pp. 160–161. ISBN 978-0-7167-7550-8.

[24] Bohr, N. (1920). "Über die Serienspektra der Element". *Zeitschrift für Physik* **2** (5): 423–478. Bibcode:1920ZPhy....2..423B. doi:10.1007/BF01329978. (English translation in Bohr 1976, pp. 241–282)

[25] Jammer, M. (1989). *The Conceptual Development of Quantum Mechanics* (2nd ed.). Tomash Publishers. Section 3.2. ISBN 978-0-88318-617-6.

[26] Ashtekar, A.; Bombelli, L.; Corichi, A. (2005). "Semiclassical States for Constrained Systems". *Physical Review D* **72**: 025008. arXiv:hep-ph/0504114. Bibcode:2005PhRvD..72a5008C. doi:10.1103/PhysRevD.72.015008.

[27] Lewandowski, J.; Okołów, A.; Sahlmann, H.; Thiemann, T. (2005). "Uniqueness of Diffeomorphism Invariant States on Holonomy-Flux Algebras". *Communications in Mathematical Physics* **267**: 703–733. arXiv:gr-qc/0504147. Bibcode:2006CMaPh.267..703L. doi:10.1007/s00220-006-0100-7.

[28] Fleischhack, C. (2006). "Irreducibility of the Weyl algebra in loop quantum gravity". *Physical Review Letters* **97**: 061302. Bibcode:2006PhRvL..97f1302F. doi:10.1103/physrevlett.97.061302.

[29] Thiemann, T. (2008). *Modern Canonical General Relativity*. Cambridge Monographs on Mathematical Physics. Cambridge University Press. Section 10.6. ISBN 978-0-521-74187-3.

[30] "Partial and Complete Observables for Hamiltonian Constrained Systems". *General Relativity and Gravitation* **39**: 1891–1927. 2007. arXiv:gr-qc/0411013. Bibcode:2007GReGr..39.1891D. doi:10.1007/s10714-007-0495-2.

[31] "Partial and Complete Observables for Canonical General Relativity". *Classical and Quantum Gravity* **23**: 6155–6184. arXiv:gr-qc/0507106. Bibcode:2006CQGra..23.6155D. doi:10.1088/0264-9381/23/22/006.

[32] Dreyer, O.; Markopoulou, f.; Smolin, L. (2006). "Symmetry and entropy of black hole horizons". *Nuclear Physics B* **774**: 1–13. arXiv:hep-th/0409056. Bibcode:2006NuPhB.744....1D. doi:10.1016/j.nuclphysb.2006.02.045.

[33] Kribs, D. W.; Markopoulou, F. "Geometry from quantum particles". arXiv:gr-qc/0510052.

[34] Markopoulou, F.; Poulin, D. "Noiseless subsystems and the low energy limit of spin foam models" (unpublished).

[35] *The Phoenix Project: Master Constraint Programme for Loop Quantum Gravity*, Class.Quant.Grav.23:2211-2248,2006 or http://fr.arxiv.org/pdf/gr-qc/0305080

[36] *Modern Canonical Quantum General Relativity* by Thomas Thiemann

[37] *Testing the Master Constraint Programme for Loop Quantum Gravity I. General Framework*, Bianca Dittrich, Thomas Thiemann, Class.Quant.Grav. 23 (2006) 1025-1066.

[38] *Testing the Master Constraint Programme for Loop Quantum Gravity II. Finite Dimensional Systems*, Bianca Dittrich, Thomas Thiemann, Class.Quant.Grav. 23 (2006) 1067-1088.

[39] *Testing the Master Constraint Programme for Loop Quantum Gravity III. SL(2,R) Models*, Bianca Dittrich, Thomas Thiemann, Class.Quant.Grav. 23 (2006) 1089-1120.

[40] *Testing the Master Constraint Programme for Loop Quantum Gravity IV. Free Field Theories*, Bianca Dittrich, Thomas Thiemann, Class.Quant.Grav. 23 (2006) 1121-1142.

[41] *Testing the Master Constraint Programme for Loop Quantum Gravity V. Interacting Field Theories*, Bianca Dittrich, Thomas Thiemann, Class.Quant.Grav. 23 (2006) 1143-1162.

[42] *Quantum Spin Dynamics VIII. The Master Constraint*, Thomas Thiemann, Class.Quant.Grav. 23 (2006) 2249-2266.

[43] *Approximating the physical inner product of Loop Quantum Cosmology*, Benjamin Bahr, Thomas Thiemann, Class.Quant.Grav.24:2109-2138,2007.

[44] *On the Relation between Operator Constraint --, Master Constraint --, Reduced Phase Space --, and Path Integral Quantisation*, Muxin Han, Thomas Thiemann, Class.Quant.Grav.27:225019,2010.

[45] *On the Relation between Rigging Inner Product and Master Constraint Direct Integral Decomposition*, Muxin Han, Thomas Thiemann, J.Math.Phys.51:092501,2010.

[46] *A Path-integral for the Master Constraint of Loop Quantum Gravity*, Muxin Han, Class.Quant.Grav.27:215009,2010

[47] *Emergent diffeomorphism invariance in a discrete loop quantum gravity model*, Rodolfo Gambini, Jorge Pullin, Class.Quant.Grav.26:035002,2009

[48] Section 10.2.2 *A First Course in Loop quantum Gravity*, Rodolfo Gambinni, Jorge Pullin, Oxford University Press, first published 2011.

[49] *Algebraic Quantum Gravity (AQG) I. Conceptual Setup*, K. Giesel, T. Thiemann, Class.Quant.Grav.24:2465-2498,2007.

[50] *Algebraic Quantum Gravity (AQG) II. Semiclassical Analysis*, K. Giesel, T. Thiemann, Class.Quant.Grav.24:2499-2564,2007.

[51] *Algebraic Quantum Gravity (AQG) III. Semiclassical Perturbation Theory*, K. Giesel, T. Thiemann, Class.Quant.Grav.24:2565-2588,2007.

[52] Bousso, Raphael (2002). "The Holographic Principle". *Reviews of Modern Physics* **74** (3): 825–874. arXiv:hep-th/0203101. Bibcode:2002RvMP...74..825B. doi:10.1103/RevModPhys.74.825.

[53] Majumdar, Parthasarathi (1998). "Black Hole Entropy and Quantum Gravity" **73**. p. 147. arXiv:gr-qc/9807045. Bibcode:1999InJPB..73..147M.

[54] See List of loop quantum gravity researchers

[55] Rovelli, Carlo (1996). "Black Hole Entropy from Loop Quantum Gravity". *Physical Review Letters* **77** (16): 3288–3291. arXiv:gr-qc/9603063. Bibcode:1996PhRvL..77.3288R. doi:10.1103/PhysRevLett.77.3288.

[56] Ashtekar, Abhay; Baez, John; Corichi, Alejandro; Krasnov, Kirill (1998). "Quantum Geometry and Black Hole Entropy". *Physical Review Letters* **80** (5): 904–907. arXiv:gr-qc/9710007. Bibcode:1998PhRvL..80..904A. doi:10.1103/PhysRevLett.80.904.

[57] *Quantum horizons and black hole entropy: Inclusion of distortion and rotation*, Abhay Ashtekar, Jonathan Engle, Chris Van Den Broeck, Class.Quant.Grav.22:L27-L34, 2005.

[58] Bianchi, Eugenio (2012). "Entropy of Non-Extremal Black Holes from Loop Gravity". arXiv:1204.5122.

[59] http://inspirehep.net/record/940357?ln=en. http://inspirehep.net/record/1111991.

[60] *On (Cosmological) Singularity Avoidance in Loop Quantum Gravity*, Johannes Brunnemann, Thomas Thiemann, Class.Quant.Grav. 23 (2006) 1395-1428.

[61] *Unboundedness of Triad -- Like Operators in Loop Quantum Gravity*, Johannes Brunnemann, Thomas Thiemann, Class.Quant.Grav. 23 (2006) 1429-1484.

[62] L. Modesto, C. Rovelli:*Particle scattering in loop quantum gravity*, Phys Rev Lett 95 (2005) 191301

[63] R Oeckl, *A 'general boundary' formulation for quantum mechanics and quantum gravity*, Phys Lett B575 (2003) 318-324 : *Schrodinger's cat and the clock: lessons for quantum gravity*, Class Quant Grav 20 (2003) 5371-5380l

[64] F. Conrady, C. Rovelli *Generalized Schrodinger equation in Euclidean field theory"*, Int J Mod Phys A 19, (2004) 1-32.

[65] L. Doplicher, *Generalized Tomonaga-Schwinger equation from the Hadamard formula*, Phys Rev D70 (2004) 064037

[66] F. Conrady, L. Doplicher, R. Oeckl, C. Rovelli, M. Testa, *Minkowski vacuum in background independent quantum gravity*, Phys Rev D69 (2004) 064019.

[67] http://arxiv.org/abs/arXiv:hep-th/0310077

[68] *New Variables for Classical and Quantum Gravity in all Dimensions I. Hamiltonian Analysis*, Norbert Bodendorfer, Thomas Thiemann, Andreas Thurn, Class. Quantum Grav. 30 (2013) 045001

[69] *New Variables for Classical and Quantum Gravity in all Dimensions II. Lagrangian Analysis*, Norbert Bodendorfer, Thomas Thiemann, Andreas Thurn, Quantum Grav. 30 (2013) 045002

[70] *New Variables for Classical and Quantum Gravity in all Dimensions III. Quantum Theory*, Norbert Bodendorfer, Thomas Thiemann, Andreas Thurn, Class. Quantum Grav. 30 (2013) 045003

[71] *New Variables for Classical and Quantum Gravity in all Dimensions IV. Matter Coupling*, Norbert Bodendorfer, Thomas Thiemann, Andreas Thurn, Class. Quantum Grav. 30 (2013) 045004

[72] *On the Implementation of the Canonical Quantum Simplicity Constraint*, Norbert Bodendorfer, Thomas Thiemann, Andreas Thurn, Class. Quantum Grav. 30 (2013) 045005

[73] *Towards Loop Quantum Supergravity (LQSG) I. Rarita-Schwinger Sector*, Norbert Bodendorfer, Thomas Thiemann, Andreas Thurn, Class. Quantum Grav. 30 (2013) 045006

[74] *Towards Loop Quantum Supergravity (LQSG) II. p-Form Sector*, Norbert Bodendorfer, Thomas Thiemann, Andreas Thurn, Class. Quantum Grav. 30 (2013) 045007

[75] *Towards Loop Quantum Supergravity (LQSG)*, Norbert Bodendorfer, Thomas Thiemann, Andreas Thurn, Phys. Lett. B 711: 205-211 (2012)

[76] *New Variables for Classical and Quantum Gravity in all Dimensions V. Isolated Horizon Boundary Degrees of Freedom*, Norbert Bodendorfer, Thomas Thiemann, Andreas Thurn, http://uk.arxiv.org/pdf/1304.2679.

[77] *Black hole entropy from loop quantum gravity in higher dimensions*, Norbert Bodendorfer http://uk.arxiv.org/pdf/1307.5029

[78] http://arxiv.org/abs/0708.3550

[79] http://arxiv.org/abs/1203.6164

[80] http://arxiv.org/abs/1006.0199

[81] http://arxiv.org/abs/1001.3668

[82] http://arxiv.org/abs/1212.5246

[83] http://arxiv.org/abs/1105.3480

[84] *Quantum gravity and the standard model*, Sundance O. Bilson-Thompson, Fotini Markopoulou, Lee Smolin, Class.Quant.Grav.24:3975-3994,2007.

[85] For a precise review and outlook of this research see: *Emergent Braided Matter of Quantum Geometry*, Sundance Bilson-Thompson, Jonathan Hackett, Louis Kauffman, Yidun Wan, SIGMA 8 (2012), 014, 43 pages.

[86] *Constrained Mechanics and Noiseless Subsystems*, Tomasz Konopka, Fotini Markopoulou, arXiv:gr-qc/0601028.

[87] http://www.perimeterinstitute.ca/people/renate-loll

[88] wwnpqft.inln.cnrs.fr/pdf/Bianchi.pdf

[89] http://arxiv.org/abs/0804.0632

[90] http://arxiv.org/abs/1308.2934

[91] Nicolai, Hermann; Peeters, Kasper; Zamaklar, Marija (2005). "Loop quantum gravity: an outside view". *Classical and Quantum Gravity* **22** (19): R193–R247. arXiv:hep-th/0501114. Bibcode:2005CQGra..22R.193N. doi:10.1088/0264-9381/22/19/R01.

[92] Goswami; Joshi, Pankaj S.; Singh, Parampreet; et al. (2006). "Quantum evaporation of a naked singularity". *Physical Review Letters* **96** (3): 31302. arXiv:gr-qc/0506129. Bibcode:2006PhRvL..96c1302G. doi:10.1103/PhysRevLett.96.031302.

3.14 References

- Topical Reviews

  - Rovelli, Carlo (2011). "Zakopane lectures on loop gravity". arXiv:1102.3660.

  - Rovelli, Carlo (1998). "Loop Quantum Gravity". *Living Reviews in Relativity* **1**. Retrieved 2008-03-13.

  - Thiemann, Thomas (2003). "Lectures on Loop Quantum Gravity". *Lectures Notes in Physics*. Lecture Notes in Physics **631**: 41–135. arXiv:gr-qc/0210094. Bibcode:2003LNP...631...41T. doi:10.1007/978-3-540-45230-0_3. ISBN 978-3-540-40810-9.

  - Ashtekar, Abhay; Lewandowski, Jerzy (2004). "Background Independent Quantum Gravity: A Status Report". *Classical and Quantum Gravity* **21** (15): R53–R152. arXiv:gr-qc/0404018. Bibcode:2004CQGra..21R..53A. doi:10.1088/0264-9381/21/15/R01.

- Carlo Rovelli and Marcus Gaul, *Loop Quantum Gravity and the Meaning of Diffeomorphism Invariance*, e-print available as gr-qc/9910079.

- Lee Smolin, *The case for background independence*, e-print available as hep-th/0507235.

- Alejandro Corichi, *Loop Quantum Geometry: A primer*, e-print available as .

- Alejandro Perez, *Introduction to loop quantum gravity and spin foams*, e-print available as .

- Hermann Nicolai and Kasper Peeters *Loop and spin foam quantum gravity: A Brief guide for beginners.*, e-print available as .

- Popular books:

  - Lee Smolin, *Three Roads to Quantum Gravity*

  - Carlo Rovelli, *Che cos'è il tempo? Che cos'è lo spazio?*, Di Renzo Editore, Roma, 2004. French translation: *Qu'est ce que le temps? Qu'est ce que l'espace?*, Bernard Gilson ed, Brussel, 2006. English translation: *What is Time? What is space?*, Di Renzo Editore, Roma, 2006.

  - Julian Barbour, *The End of Time: The Next Revolution in Our Understanding of the Universe*

  - Musser, George (2008). "The Complete Idiot's Guide to String Theory". *The Physics Teacher* (Indianapolis: Alpha) **47** (2): 368. Bibcode:2009PhTea..47Q.128H. doi:10.1119/1.3072469. ISBN 978-1-59257-702-6. – Focuses on string theory but has an extended discussion of loop gravity as well.

- Magazine articles:

  - Lee Smolin, "Atoms of Space and Time", *Scientific American*, January 2004

  - Martin Bojowald, "Following the Bouncing Universe", *Scientific American*, October 2008

- Easier introductory, expository or critical works:

  - Abhay Ashtekar, *Gravity and the quantum*, e-print available as gr-qc/0410054 (2004)

  - John C. Baez and Javier Perez de Muniain, *Gauge Fields, Knots and Quantum Gravity*, World Scientific (1994)

  - Carlo Rovelli, *A Dialog on Quantum Gravity*, e-print available as hep-th/0310077 (2003)

  - Rodolfo Gambini and Jorge Pullin, *A First Course in Loop Quantum Gravity*, Oxford (2011)

  - Carlo Rovelli and Francesca Vidotto, *Covariant Loop Quantum Gravity*, Cambridge (2014); draft available online

- More advanced introductory/expository works:

  - Carlo Rovelli, *Quantum Gravity*, Cambridge University Press (2004); draft available online

  - Thomas Thiemann, *Introduction to modern canonical quantum general relativity*, e-print available as gr-qc/0110034

  - Thomas Thiemann, *Introduction to Modern Canonical Quantum General Relativity*, Cambridge University Press (2007)

  - Abhay Ashtekar, *New Perspectives in Canonical Gravity*, Bibliopolis (1988).

  - Abhay Ashtekar, *Lectures on Non-Perturbative Canonical Gravity*, World Scientific (1991)

  - Rodolfo Gambini and Jorge Pullin, *Loops, Knots, Gauge Theories and Quantum Gravity*, Cambridge University Press (1996)

  - Hermann Nicolai, Kasper Peeters, Marija Zamaklar, *Loop quantum gravity: an outside view*, e-print available as hep-th/0501114

  - H. Nicolai and K. Peeters, *Loop and Spin Foam Quantum Gravity: A Brief Guide for Beginners*, e-print available as hep-th/0601129

  - T. Thiemann The LQG – String: Loop Quantum Gravity Quantization of String Theory (2004)

- Conference proceedings:

  - John C. Baez (ed.), *Knots and Quantum Gravity*

- Fundamental research papers:

  - Ashtekar, Abhay (1986). "New variables for classical and quantum gravity". *Physical Review Letters* **57** (18): 2244–2247. Bibcode:1986PhRvL..57.2244A. doi:10.1103/PhysRevLett.57.2244. PMID 10033673

  - Ashtekar, Abhay (1987). "New Hamiltonian formulation of general relativity". *Physical Review D* **36** (6): 1587–1602. Bibcode:1987PhRvD..36.1587A. doi:10.1103/PhysRevD.36.1587

  - Roger Penrose, *Angular momentum: an approach to combinatorial space-time* in *Quantum Theory and Beyond*, ed. Ted Bastin, Cambridge University Press, 1971

  - Rovelli, Carlo; Smolin, Lee (1988). "Knot theory and quantum gravity". *Physical Review Letters* **61** (10): 1155–1158. Bibcode:1988PhRvL..61.1155R. doi:10.1103/PhysRevLett.61.1155.

- Rovelli, Carlo; Smolin, Lee (1990). "Loop space representation of quantum general relativity". *Nuclear Physics* **B331**: 80–152.

- Carlo Rovelli and Lee Smolin, *Discreteness of area and volume in quantum gravity*, Nucl. Phys., **B442** (1995) 593-622, e-print available as gr-qc/9411005

- Kuchař, Karel (1973). "Canonical Quantization of Gravity". In Israel, Werner. *Relativity, Astrophysics and Cosmology*. D. Reidel. pp. 237–288. ISBN 90-277-0369-8.

- Thiemann, Thomas (2006). "Loop Quantum Gravity: An Inside View". *Approaches to Fundamental Physics*. Lecture Notes in Physics **721**: 185–263. arXiv:hep-th/0608210. Bibcode:2007LNP...721..185T. doi:10.1007/978-3-540-71117-9_10. ISBN 978-3-540-71115-5.

3.15 External links

- "Loop Quantum Gravity" by Carlo Rovelli Physics World, November 2003

- Quantum Foam and Loop Quantum Gravity

- Abhay Ashtekar: Semi-Popular Articles . Some excellent popular articles suitable for beginners about space, time, GR, and LQG.

- Loop Quantum Gravity: Lee Smolin.

- Loop Quantum Gravity on arxiv.org

- A list of LQG references catered to fresh graduates

- Loop Quantum Gravity Lectures Online by Lee Smolin

- Spin networks, spin foams and loop quantum gravity

- Wired magazine, News: *Moving Beyond String Theory*

- April 2006 Scientific American Special Issue, *A Matter of Time*, has Lee Smolin LQG Article *Atoms of Space and Time*

- September 2006, The Economist, article *Looping the loop*

- Gamma-ray Large Area Space Telescope: http://glast.gsfc.nasa.gov/

- Zeno meets modern science. Article from Acta Physica Polonica B by Z.K. Silagadze.

- Did pre-big bang universe leave its mark on the sky? - According to a model based on "loop quantum gravity" theory, a parent universe that existed before ours may have left an imprint (*New Scientist*, 10 April 2008)

Chapter 4

Causal fermion system

The theory of **causal fermion systems** is an approach to describe fundamental physics. It gives quantum mechanics, general relativity and quantum field theory as limiting cases[1][2][3][4][5] and is therefore a candidate for a unified physical theory.

Instead of introducing physical objects on a preexisting space-time manifold, the general concept is to derive space-time as well as all the objects therein as secondary objects from the structures of an underlying causal fermion system. This concept also makes it possible to generalize notions of differential geometry to the non-smooth setting.[6][7] In particular, one can describe situations when space-time no longer has a manifold structure on the microscopic scale (like a space-time lattice or other discrete or continuous structures on the Planck scale). As a result, the theory of causal fermion systems is a proposal for quantum geometry and an approach to quantum gravity.

Causal fermion systems were introduced by Felix Finster and collaborators.

4.1   Motivation and physical concept

The physical starting point is the fact that the Dirac equation in Minkowski space has solutions of negative energy which are usually associated to the Dirac sea. Taking the concept seriously that the states of the Dirac sea form an integral part of the physical system, one finds that many structures (like the causal and metric structures as well as the bosonic fields) can be recovered from the wave functions of the sea states. This leads to the idea that the wave functions of all occupied states (including the sea states) should be regarded as the basic physical objects, and that all structures in space-time arise as a result of the collective interaction of the sea states with each other and with the additional particles and "holes" in the sea. Implementing this picture mathematically leads to the framework of causal fermion systems.

More precisely, the correspondence between the above physical situation and the mathematical framework is ob-

tained as follows. All occupied states span a Hilbert space of wave functions in Minkowski space $\hat{M}$. The observable information on the distribution of the wave functions in space-time is encoded in the *local correlation operators* $F(x), x \in \hat{M}$, which in an orthonormal basis (ψ_i) have the matrix representation

$$\left(F(x) \right)^i_j = -\overline{\psi_i(x)}\psi_j(x)$$

(where $\overline{\psi}$ is the adjoint spinor). In order to make the wave functions into the basic physical objects, one considers the set $\{ F(x) \,|\, x \in \hat{M} \}$ as a set of linear operators on an *abstract* Hilbert space. The structures of Minkowski space are all disregarded, except for the volume measure $d^4 x$, which is transformed to a corresponding measure on the linear operators (the *"universal measure"*). The resulting structures, namely a Hilbert space together with a measure on the linear operators thereon, are the basic ingredients of a causal fermion system.

The above construction can also be carried out in more general space-times. Moreover, taking the abstract definition as the starting point, causal fermion systems allow for the description of generalized "quantum space-times." The physical picture is that one causal fermion system describes a space-time together with all structures and objects therein (like the causal and the metric structures, wave functions and quantum fields). In order to single out the physically admissible causal fermion systems, one must formulate physical equations. In analogy to the Lagrangian formulation of classical field theory, the physical equations for causal fermion systems are formulated via a variational principle, the so-called *causal action principle*. Since one works with different basic objects, the causal action principle has a novel mathematical structure where one minimizes a positive action under variations of the universal measure. The connection to conventional physical equations is obtained in a certain limiting case (the continuum limit) in which the interaction can be described effectively by gauge fields coupled to particles and antiparticles, whereas the Dirac sea is no longer apparent.

4.2   General mathematical setting

In this section the mathematical framework of causal fermion systems is introduced.

4.2.1   Definition of a causal fermion system

A **causal fermion system** of spin dimension $n \in \mathbb{N}$ is a triple $(\mathcal{H}, \mathcal{F}, \rho)$ where

- $(\mathcal{H}, \langle . | . \rangle_{\mathcal{H}})$ is a complex Hilbert space.

- $\mathcal{F}$ is the set of all self-adjoint linear operators of finite rank on $\mathcal{H}$ which (counting multiplicities) have at most n positive and at most n negative eigenvalues.

- ρ is a measure on $\mathcal{F}$.

The measure ρ is referred to as the **universal measure**.

As will be outlined below, this definition is rich enough to encode analogs of the mathematical structures needed to formulate physical theories. In particular, a causal fermion system gives rise to a space-time together with additional structures that generalize objects like spinors, the metric and curvature. Moreover, it comprises quantum objects like wave functions and a fermionic Fock state.[8]

4.2.2   The causal action principle

Inspired by the Langrangian formulation of classical field theory, the dynamics on a causal fermion system is described by a variational principle defined as follows.

Given a Hilbert space $(\mathcal{H}, \langle . | . \rangle_{\mathcal{H}})$ and the spin dimension n, the set $\mathcal{F}$ is defined as above. Then for any $x, y \in \mathcal{F}$, the product xy is an operator of rank at most $2n$. It is not necessarily self-adjoint because in general $(xy)^* = yx \neq xy$. We denote the non-trivial eigenvalues of the operator xy (counting algebraic multiplicities) by

$$\lambda_1^{xy}, \ldots, \lambda_{2n}^{xy} \in \mathbb{C}.$$

Moreover, the **spectral weight** $|.|$ is defined by

$$|xy| = \sum_{i=1}^{2n} |\lambda_i^{xy}| \quad \text{and} \quad |(xy)^2| = \sum_{i=1}^{2n} |\lambda_i^{xy}|^2.$$

The **Lagrangian** is introduced by

$$\mathcal{L}(x, y) = |(xy)^2| - \frac{1}{2n}|xy|^2 = \frac{1}{4n}\sum_{i,j=1}^{2n} \left(|\lambda_i^{xy}| - |\lambda_j^{xy}|\right)^2$$

The **causal action** is defined by

$$S = \iint_{\mathcal{F} \times \mathcal{F}} \mathcal{L}(x, y)\, d\rho(x)\, d\rho(y).$$

The **causal action principle** is to minimize S under variations of ρ within the class of (positive) Borel measures under the following constraints:

- Boundedness constraint: $\iint_{\mathcal{F} \times \mathcal{F}} |xy|^2\, d\rho(x)\, d\rho(y) \leq C$ for some positive constant C.

- Trace constraint: $\int_{\mathcal{F}} \mathrm{tr}(x)\, d\rho(x)$ is kept fixed.

- The total volume $\rho(\mathcal{F})$ is preserved.

Here on $\mathcal{F} \subset L(\mathcal{H})$ one considers the topology induced by the sup -norm on the bounded linear operators on $\mathcal{H}$.

The constraints prevent trivial minimizers and ensure existence, provided that $\mathcal{H}$ is finite-dimensional.[9] This variational principle also makes sense in the case that the total volume $\rho(\mathcal{F})$ is infinite if one considers variations $\delta\rho$ of bounded variation with $(\delta\rho)(\mathcal{F}) = 0$.

4.3   Inherent structures

In contemporary physical theories, the word space-time refers to a Lorentzian manifold (M, g). This means that space-time is a set of points enriched by topological and geometric structures. In the context of causal fermion systems, space-time does not need to have a manifold structure. Instead, space-time M is a set of operators on a Hilbert space (a subset of $\mathcal{F}$). This implies additional inherent structures that correspond to and generalize usual objects on a space-time manifold.

For a causal fermion system $(\mathcal{H}, \mathcal{F}, \rho)$, we define **space-time** M as the support of the universal measure,

$$M := \mathrm{supp}\,\rho \subset \mathcal{F}.$$

With the topology induced by $\mathcal{F}$, space-time M is a topological space.

4.3.1   Causal structure

For $x, y \in M$, we denote the non-trivial eigenvalues of the operator xy (counting algebraic multiplicities) by $\lambda_1^{xy}, \ldots, \lambda_{2n}^{xy} \in \mathbb{C}$. The points x and y are defined to be **spacelike** separated if all the λ_j^{xy} have the same absolute value. They are **timelike** separated if the λ_j^{xy} do not all have the same absolute value and are all real. In all other cases, the points x and y are **lightlike** separated.

This notion of causality fits together with the "causality" of the above causal action in the sense that if two space-time points $x, y \in M$ are space-like separated, then the Lagrangian $\mathcal{L}(x, y)$ vanishes. This corresponds to the physical notion of causality that spatially separated space-time points do not interact. This causal structure is the reason for the notion "causal" in causal fermion system and causal action.

Let π_x the orthogonal projection on the subspace $S_x := x(\mathcal{H}) \subset \mathcal{H}$. Then the sign of the functional

$$i\mathrm{Tr}\left(x\, y\, \pi_x\, \pi_y - y\, x\, \pi_y\, \pi_x\right)$$

distinguishes the **future** from the **past**. In contrast to the structure of a partially ordered set, the relation "lies in the future of" is in general not transitive. But it is transitive on the macroscopic scale in typical examples.[6][7]

4.3.2   Spinors and wave functions

For every $x \in M$ the **spin space** is defined by $S_x = x(\mathcal{H})$: it is a subspace of $\mathcal{H}$ of dimension at most $2n$. The **spin scalar product** $\prec .|. \succ_x$ defined by

$$\prec u | v \succ_x = -\langle u | x u \rangle_{\mathcal{H}} \qquad \text{all for} u, v \in S_x$$

is an indefinite inner product on S_x of signature (p, q) with $p, q \leq n$.

A **wave function** ψ is a mapping

$$\psi : M \to \mathcal{H} \qquad \text{with} \qquad \psi(x) \in S_x \qquad \text{all for} x \in M .$$

On wave functions for which the norm $\|.\|$ defined by

$$\|\psi\|^2 = \int_M \left(\psi(x)\,\big|\,|x|\,\psi(x)\right)_{\mathcal{H}} d\rho(x)$$

is finite (where $|x| = \sqrt{x^2}$ is the absolute value of the symmetric operator x), one can define the inner product

$$<\psi|\phi> = \int_M \prec \psi(x)|\phi(x) \succ_x d\rho(x) .$$

Together with the topology induced by the norm $\|.\|$, one obtains a Krein space $(\mathcal{K}, <.|.>)$.

To any vector $u \in \mathcal{H}$ we can associate the wave function

$$\psi^u(x) := \pi_x u$$

(where $\pi_x : \mathcal{H} \to S_x$ is again the orthogonal projection to the spin space). This gives rise to a distinguished family of wave functions, referred to as the wave functions of the **occupied states**.

4.3.3   The fermionic projector

The **kernel of the fermionic projector** $P(x, y)$ is defined by

$$P(x, y) = \pi_x y|_{S_y} : S_y \to S_x$$

(where $\pi_x : \mathcal{H} \to S_x$ is again the orthogonal projection on the spin space, and $|_{S_y}$ denotes the restriction to S_y). The **fermionic projector** P is the operator

$$P : \mathcal{K} \to \mathcal{K}, \qquad (P\psi)(x) = \int_M P(x, y)\, \psi(y)\, d\rho(y),$$

which has the dense domain of definition given by all vectors $\psi \in \mathcal{K}$ satisfying the conditions

$$\phi := \int_M x\, \psi(x)\, d\rho(x) \in \mathcal{H} \quad \text{and} \quad \|\phi\| < \infty .$$

As a consequence of the causal action principle, the kernel of the fermionic projector has additional normalization properties[10] which justify the name projector.

4.3.4   Connection and curvature

Being an operator from one spin space to another, the kernel of the fermionic projector gives relations between different space-time points. This fact can be used to introduce a **spin connection**

$$D_{x,y} : S_y \to S_x \qquad \text{unitary} .$$

The basic idea is to take a polar decomposition of $P(x, y)$. The construction becomes more involved by the fact that the spin connection should induce a corresponding **metric connection**

$$\nabla_{x,y} : T_y \to T_x \qquad \text{isometric} .$$

where the tangent space T_x is a specific subspace of the linear operators on S_x endowed with a Lorentzian metric. The **spin curvature** is defined as the holonomy of the spin connection,

$$\Re(x, y, z) = D_{x,y}\, D_{y,z}\, D_{z,x} \;:\; S_x \to S_x \,.$$

Similarly, the metric connection gives rise to **metric curvature**. These geometric structures give rise to a proposal for a quantum geometry.[6]

4.3.5 A fermionic Fock state

If $\mathcal{H}$ has finite dimension f, choosing an orthonormal basis $u_1, \ldots, u_f$ of $\mathcal{H}$ and taking the wedge product of the corresponding wave functions

$$\left(\psi^{u_1} \wedge \cdots \wedge \psi^{u_f} \right)(x_1, \ldots, x_f)$$

gives a state of an f-particle fermionic Fock space. Due to the total anti-symmetrization, this state depends on the choice of the basis of $\mathcal{H}$ only by a phase factor.[11] This correspondence explains why the vectors in the particle space are to be interpreted as fermions. It also motivates the name causal **fermion** system.

4.4 Underlying physical principles

Causal fermion systems incorporate several physical principles in a specific way:

- A **local gauge principle**: In order to represent the wave functions in components, one chooses bases of the spin spaces. Denoting the signature of the spin scalar product at x by $(\mathfrak{p}_x, \mathfrak{q}_x)$, a pseudo-orthonormal basis $(\mathfrak{e}_\alpha(x))_{\alpha=1,\ldots,\mathfrak{p}_x+\mathfrak{q}_x}$ of S_x is given by

$$\prec\mathfrak{e}_\alpha|\mathfrak{e}_\beta\succ = s_\alpha\, \delta_{\alpha\beta} \quad \text{with} \quad s_1, \ldots, s_{\mathfrak{p}_x} = 1, \; s_{\mathfrak{p}_x+1}, \ldots, s_{\mathfrak{p}_x+\mathfrak{q}_x} = -1 \,.$$

  Then a wave function ψ can be represented with component functions,

$$\psi(x) = \sum_{\alpha=1}^{\mathfrak{p}_x+\mathfrak{q}_x} \psi^\alpha(x)\,\mathfrak{e}_\alpha(x) \,.$$

  The freedom of choosing the bases $(\mathfrak{e}_\alpha(x))$ independently at every space-time point corresponds to local unitary transformations of the wave functions,

$$\psi^\alpha(x) \to \sum_{\beta=1}^{\mathfrak{p}_x+\mathfrak{q}_x} U(x)^\alpha_\beta\, \psi^\beta(x) \quad \text{with} \quad U(x) \in \mathrm{U}(\mathfrak{p}_x, \mathfrak{q}_x) \,.$$

  These transformations have the interpretation as local gauge transformations. The gauge group is determined to be the isometry group of the spin scalar product. The causal action is gauge invariant in the sense that it does not depend on the choice of spinor bases.

- The **equivalence principle**: For an explicit description of space-time one must work with local coordinates. The freedom in choosing such coordinates generalizes the freedom in choosing general reference frames in a space-time manifold. Therefore, the equivalence principle of general relativity is respected. The causal action is generally covariant in the sense that it does not depend on the choice of coordinates.

- The **Pauli exclusion principle**: The fermionic Fock state associated to the causal fermion system makes it possible to describe the many-particle state by a totally antisymmetric wave function. This gives agreement with the Pauli exclusion principle.

- The principle of **causality** is incorporated by the form of the causal action in the sense that space-time points with spacelike separation do not interact.

4.5 Limiting cases

Causal fermion systems have mathematically sound limiting cases that give a connection to conventional physical structures.

4.5.1 Lorentzian spin geometry of globally hyperbolic space-times

Starting on any globally hyperbolic Lorentzian spin manifold (M, g) with spinor bundle SM, one gets into the framework of causal fermion systems by choosing $(\mathcal{H}, \langle.|.\rangle_\mathcal{H})$ as a subspace of the solution space of the Dirac equation. Defining the so-called **local correlation operator** $F(p)$ for $p \in M$ by

$$\langle \psi | F(p)\phi \rangle_\mathcal{H} = -\prec\psi|\phi\succ_p$$

(where $\prec\!\psi|\phi\!\succ_p$ is the inner product on the fibre $S_p\hat{M}$) and introducing the universal measure as the push-forward of the volume measure on $\hat{M}$.

$$\rho = F_* d\mu ,$$

one obtains a causal fermion system. For the local correlation operators to be well-defined, $\mathcal{H}$ must consist of continuous sections, typically making it necessary to introduce a regularization on the microscopic scale ε . In the limit $\varepsilon \searrow 0$, all the intrinsic structures on the causal fermion system (like the causal structure, connection and curvature) go over to the corresponding structures on the Lorentzian spin manifold.[6] Thus the geometry of space-time is encoded completely in the corresponding causal fermion systems.

4.5.2   Quantum mechanics and classical field equations

The Euler-Lagrange equations corresponding to the causal action principle have a well-defined limit if the space-times $M := \mathrm{supp}\,\rho$ of the causal fermion systems go over to Minkowski space. More specifically, one considers a sequence of causal fermion systems (for example with $\mathcal{H}$ finite-dimensional in order to ensure the existence of the fermionick Fock state as well as of minimizers of the causal action), such that the corresponding wave functions go over to a configuration of interacting Dirac seas involving additional particle states or "holes" in the seas. This procedure, referred to as the continuum limit, gives effective equations having the structure of the Dirac equation coupled to classical field equations. For example, for a simplified model involving three elementary fermionic particles in spin dimension two, one obtains an interaction via a classical axial gauge field A [3] described by the coupled Dirac- and Yang-Mills equations

$$(i\partial\!\!\!/ + \gamma^5\!A\!\!\!/ - m)\psi = 0$$
$$C_0(\partial_j^k A^j - \Box A^k) - C_2 A^k = 12\pi^2 \bar{\psi}\gamma^5\gamma^k\psi .$$

Taking the non-relativistic limit of the Dirac equation, one obtains the Pauli equation or the Schrödinger equation, giving the correspondence to quantum mechanics. Here C_0 and C_2 depend on the regularization and determine the coupling constant as well as the rest mass.

Likewise, for a system involving neutrinos in spin dimension 4, one gets effectively a massive $SU(2)$ gauge field coupled to the left-handed component of the Dirac spinors.[4] The fermion configuration of the standard model can be described in spin dimension 16.[1]

4.5.3   The Einstein field equations

For the just-mentioned system involving neutrinos,[4] the continuum limit also yields the Einstein field equations coupled to the Dirac spinors,

$$R_{jk} - \frac{1}{2} R\, g_{jk} + \Lambda\, g_{jk} = \kappa\, T_{jk}[\Psi, A] ,$$

up to corrections of higher order in the curvature tensor. Here the cosmological constant Λ is undetermined, and T_{jk} denotes the energy-momentum tensor of the spinors and the $SU(2)$ gauge field. The gravitation constant κ depends on the regularization length.

4.5.4   Quantum field theory in Minkowski space

Starting from the coupled system of equations obtained in the continuum limit and expanding in powers of the coupling constant, one obtains integrals which correspond to Feynman diagrams on the tree level. Fermionic loop diagrams arise due to the interaction with the sea states, whereas bosonic loop diagrams appear when taking averages over the microscopic (in generally non-smooth) space-time structure of a causal fermion system (method of microscopic mixing).[5] The detailed analysis and comparison with standard quantum field theory is work in progress.

4.6   References

[1] F. Finster, *The Principle of the Fermionic Projector*, hep-th/0001048, hep-th/0202059, hep-th/0210121, AMS/IP Studies in Advanced Mathematics, vol. **35**, American Mathematical Society, Providence, RI, 2006.

[2] F. Finster, *A formulation of quantum field theory realizing a sea of interacting Dirac particles*, arXiv:0911.2102 [hep-th], Lett. Math. Phys. **97** (2011), no. 2, 165–183.

[3] F. Finster, *An action principle for an interacting fermion system and its analysis in the continuum limit*, arXiv:0908.1542 [math-ph] (2009).

[4] F. Finster, *The continuum limit of a fermion system involving neutrinos: Weak and gravitational interactions*, arXiv:1211.3351 [math-ph] (2012).

[5] F. Finster, *Perturbative quantum field theory in the framework of the fermionic projector*, arXiv:1310.4121 [math-ph], J. Math. Phys. **55** (2014), no. 4, 042301.

[6] F. Finster and A. Grotz, *A Lorentzian quantum geometry*, arXiv:1107.2026 [math-ph], Adv. Theor. Math. Phys. **16** (2012), no. 4, 1197–1290.

[7] F. Finster and N. Kamran, *Spinors on singular spaces and the topology of causal fermion systems*, arXiv:1403.7885 [math-ph] (2014).

[8] F. Finster, A. Grotz, and D. Schiefeneder, *Causal fermion systems: A quantum space-time emerging from an action principle*, arXiv:1102.2585 [math-ph], Quantum Field Theory and Gravity (F. Finster, O. Müller, M. Nardmann, J. Tolksdorf, and E. Zeidler, eds.), Birkhäuser Verlag, Basel, 2012, pp. 157–182.

[9] F. Finster, *Causal variational principles on measure spaces*, arXiv:0811.2666 [math-ph], J. Reine Angew. Math. **646** (2010), 141–194.

[10] F. Finster and J. Kleiner, *Noether-like theorems for causal variational principles*, arXiv:1506.09076 [math-ph] (2015).

[11] F. Finster, *Entanglement and second quantization in the framework of the fermionic projector*, arXiv:0911.0076 [math-ph], J. Phys. A: Math. Theor. **43** (2010), 395302.

4.7 Further reading

- For a non-technical introduction see Finster, Kleiner: Causal fermion systems as a candidate for a unified physical theory, arXiv:1502.03587 [math-ph] 2015, Online.

- Talk "Causal fermion systems as an approach to quantum theory" at Conference Quantum Mathematical Physics - A bridge between Mathematics and Physics, Regensburg, September 2014, Video online.

- The Principle of the Fermionic Projector, AMS/IP Studies in Advanced Mathematics Series **35**, American Mathematical Society, Providence, RI; International Press, Cambridge, MA, 2006, ISBN 978-0-8218-3974-4, Online.

- Finster, Grotz, Schiefeneder: Causal fermion systems: A quantum space-time emerging from an action principle, in "Quantum Field Theory and Gravity", Birkhäuser, 2012, Online.

- A formulation of quantum field theory realizing a sea of interacting Dirac particles, Letters in Mathematical Physics **97**, Springer, 2011, 165–183, Online.

Chapter 5

Grand Unified Theory

For the album, see Grand Unification (album).

A **Grand Unified Theory (GUT)** is a model in particle physics in which at high energy, the three gauge interactions of the Standard Model which define the electromagnetic, weak, and strong interactions or forces, are merged into one single force. This unified interaction is characterized by one larger gauge symmetry and thus several force carriers, but one unified coupling constant. If Grand Unification is realized in nature, there is the possibility of a grand unification epoch in the early universe in which the fundamental forces are not yet distinct.

Models that do not unify all interactions using one simple Lie group as the gauge symmetry, but do so using semisimple groups, can exhibit similar properties and are sometimes referred to as Grand Unified Theories as well.

Unifying gravity with the other three interactions would provide a theory of everything (TOE), rather than a GUT. Nevertheless, GUTs are often seen as an intermediate step towards a TOE.

The novel particles predicted by GUT models are expected to have masses around the GUT scale—just a few orders of magnitude below the Planck scale—and so will be well beyond the reach of any foreseen particle collider experiments. Therefore, the particles predicted by GUT models will be unable to be observed directly and instead the effects of grand unification might be detected through indirect observations such as proton decay, electric dipole moments of elementary particles, or the properties of neutrinos.[1] Some grand unified theories predict the existence of magnetic monopoles.

As of 2012, all GUT models which aim to be completely realistic are quite complicated, even compared to the Standard Model, because they need to introduce additional fields and interactions, or even additional dimensions of space. The main reason for this complexity lies in the difficulty of reproducing the observed fermion masses and mixing angles. Due to this difficulty, and due to the lack of any

observed effect of grand unification so far, there is no generally accepted GUT model.

5.1 History

Historically, the first true GUT which was based on the simple Lie group SU(5), was proposed by Howard Georgi and Sheldon Glashow in 1974.[2] The Georgi–Glashow model was preceded by the semisimple Lie algebra Pati–Salam model by Abdus Salam and Jogesh Pati,[3] who pioneered the idea to unify gauge interactions.

The acronym GUT was first coined in 1978 by CERN researchers John Ellis, Andrzej Buras, Mary K. Gaillard, and Dimitri Nanopoulos, however in the final version of their paper[4] they opted for the less anatomical *GUM* (Grand Unification Mass). Nanopoulos later that year was the first to use[5] the acronym in a paper.[6]

5.2 Motivation

The fact that the electric charges of electrons and protons seem to cancel each other exactly to extreme precision is essential for the existence of the macroscopic world as we know it, but this important property of elementary particles is not explained in the Standard Model of particle physics. While the description of strong and weak interactions within the Standard Model is based on gauge symmetries governed by the simple symmetry groups SU(3) and SU(2) which allow only discrete charges, the remaining component, the weak hypercharge interaction is described by an abelian symmetry U(1) which in principle allows for arbitrary charge assignments.[note 1] The observed charge quantization, namely the fact that all known elementary particles carry electric charges which appear to be exact multiples of 1/3 of the "elementary" charge, has led to the idea that hypercharge interactions and possibly the strong and weak interactions might be embedded in one Grand Unified interaction described by a single, larger simple symmetry group containing the Standard Model. This would automatically predict the quantized nature and values of all elementary particle charges. Since this also results in a prediction for the relative strengths of the fundamental interactions which we observe, in particular the weak mixing angle, Grand Unification ideally reduces the number of independent input parameters, but is also constrained by observations.

Grand Unification is reminiscent of the unification of electric and magnetic forces by Maxwell's theory of electromagnetism in the 19th century, but its physical implications and mathematical structure are qualitatively different.

5.3 Unification of matter particles

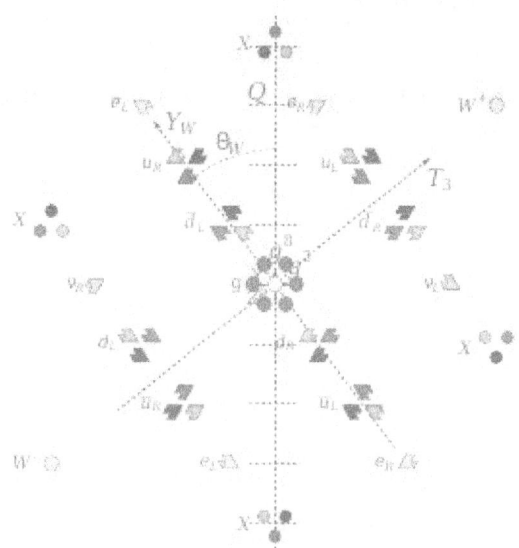

The pattern of weak isospins, weak hypercharges, and strong charges for particles in the SU(5) model, rotated by the predicted weak mixing angle, showing electric charge roughly along the vertical. In addition to Standard Model particles, the theory includes twelve colored X bosons, responsible for proton decay.

Schematic representation of fermions and bosons in SU(5) GUT showing 5 + 10 split in the multiplets. Neutral bosons (photon, Z-boson, and neutral gluons) are not shown but occupy the diagonal entries of the matrix in complex superpositions.

For an elementary introduction to how Lie algebras are related to particle physics, see the article Particle physics and representation theory.

5.3.1 SU(5)

Main article: SU(5) (physics)
SU(5) is the simplest GUT. The smallest simple Lie group which contains the standard model, and upon which the first Grand Unified Theory was based, is

$$SU(5) \supset SU(3) \times SU(2) \times U(1)$$

Such group symmetries allow the reinterpretation of several known particles as different states of a single particle field.

However, it is not obvious that the simplest possible choices for the extended "Grand Unified" symmetry should yield the correct inventory of elementary particles. The fact that all currently known (2009) matter particles fit nicely into three copies of the smallest group representations of SU(5) and immediately carry the correct observed charges, is one of the first and most important reasons why people believe that a Grand Unified Theory might actually be realized in nature.

The two smallest irreducible representations of SU(5) are **5** and **10**. In the standard assignment, the **5** contains the charge conjugates of the right-handed down-type quark color triplet and a left-handed lepton isospin doublet, while the **10** contains the six up-type quark components, the left-handed down-type quark color triplet, and the right-handed electron. This scheme has to be replicated for each of the three known generations of matter. It is notable that the theory is anomaly free with this matter content.

The hypothetical right-handed neutrinos are a singlet of SU(5), which makes that its mass is not forbidden by any symmetry so it doesn't need a spontaneous symmetry breaking which explains why its mass would be heavy. (see seesaw mechanism).

5.3.2   SO(10)

Main article: SO(10) (physics)

The next simple Lie group which contains the standard

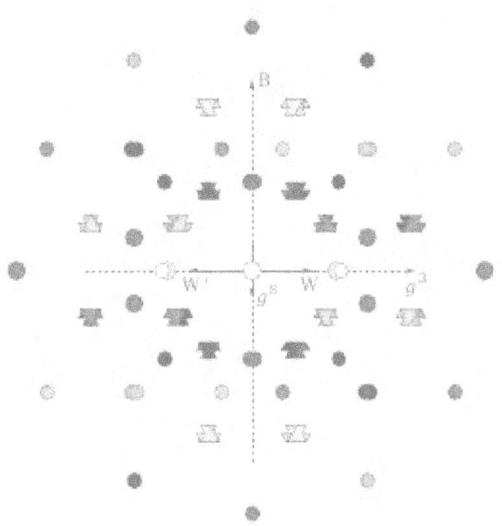

The pattern of weak isospin, W, weaker isospin, W', strong g3 and g8, and baryon minus lepton, B, charges for particles in the SO(10) Grand Unified Theory, rotated to show the embedding in E_6.

model is

$$SO(10) \supset SU(5) \supset SU(3) \times SU(2) \times U(1)$$

Here, the unification of matter is even more complete, since the irreducible spinor representation **16** contains both the **5** and **10** of SU(5) and a right-handed neutrino, and thus the complete particle content of one generation of the extended standard model with neutrino masses. This is already the largest simple group which achieves the unification of matter in a scheme involving only the already known matter particles (apart from the Higgs sector).

Since different standard model fermions are grouped together in larger representations, GUTs specifically predict relations among the fermion masses, such as between the electron and the down quark, the muon and the strange quark, and the tau lepton and the bottom quark for SU(5) and SO(10). Some of these mass relations hold approximately, but most don't (see Georgi-Jarlskog mass relation).

The boson matrix for SO(10) is found by taking the 15 × 15 matrix from the **10 + 5** representation of SU(5) and adding an extra row and column for the right handed neutrino. The bosons are found by adding a partner to each of the 20 charged bosons (2 right-handed W bosons, 6 massive charged gluons and 12 X/Y type bosons) and adding

an extra heavy neutral Z-boson to make 5 neutral bosons in total. The boson matrix will have a boson or its new partner in each row and column. These pairs combine to create the familiar 16D Dirac spinor matrices of SO(10).

5.3.3   SU(8)

Assuming 4 generations of fermions instead of 3 makes a total of **64** types of particles. These can be put into **64 = 8 + 56** representations of SU(8). This can be divided into SU(5) × SU(3)F × U(1) which is the SU(5) theory together with some heavy bosons which act on the generation number.

5.3.4   O(16)

Again assuming 4 generations of fermions, the **128** particles and anti-particles can be put into a single spinor representation of O(16).

5.3.5   Symplectic groups and quaternion representations

Symplectic gauge groups could also be considered. For example Sp(8) (which is called Sp(4) in the article symplectic group) has a representation in terms of 4 × 4 quaternion unitary matrices which has a **16** dimensional real representation and so might be considered as a candidate for a gauge group. Sp(8) has 32 charged bosons and 4 neutral bosons. Its subgroups include SU(4) so can at least contain the gluons and photon of SU(3) × U(1). Although it's probably not possible to have weak bosons acting on chiral fermions in this representation. A quaternion representation of the fermions might be:

$$\begin{bmatrix} e + i\overline{e} + jv + k\overline{v} \\ u_r + i\overline{u_r} + jd_r + k\overline{d_r} \\ u_g + i\overline{u_g} + jd_g + k\overline{d_g} \\ u_b + i\overline{u_b} + jd_b + k\overline{d_b} \end{bmatrix}_L$$

A further complication with quaternion representations of fermions is that there are two types of multiplication: left multiplication and right multiplication which must be taken into account. It turns out that including left and right-handed 4 × 4 quaternion matrices is equivalent to including a single right-multiplication by a unit quaternion which adds an extra SU(2) and so has an extra neutral boson and two more charged bosons. Thus the group of left and right handed 4 × 4 quaternion matrcies is Sp(8) × SU(2) which does include the standard model bosons:

$$SU(4,H)L \times HR$$

$$= Sp(8) \times SU(2) \supset SU(4) \times SU(2) \supset SU(3) \times SU(2) \times U(1)$$

If ψ is a quaternion valued spinor, A_μ^{ab} is quaternion hermitian 4×4 matrix coming from Sp(8) and B_μ is a pure imaginary quaternion (both of which are 4-vector bosons) then the interaction term is:

$$\overline{\psi^a} \gamma_\mu \left(A_\mu^{ab} \psi^b + \psi^a B_\mu \right)$$

5.3.6 E8 and octonion representations

It can be noted that a generation of 16 fermions can be put into the form of an octonion with each element of the octonion being an 8-vector. If the 3 generations are then put in a 3x3 hermitian matrix with certain additions for the diagonal elements then these matrices form an exceptional (grassman-) Jordan algebra, which has the symmetry group of one of the exceptional Lie groups (F_4, E_6, E_7 or E_8) depending on the details.

$$\psi = \begin{bmatrix} a & e & \mu \\ \overline{e} & b & \tau \\ \overline{\mu} & \overline{\tau} & c \end{bmatrix}$$

$$[\psi_A, \psi_B] \subset J_3(O)$$

Because they are fermions the anti-commutators of the Jordan algebra become commutators. It is known that E_6 has subgroup O(10) and so is big enough to include the Standard Model. An E_8 gauge group, for example, would have 8 neutral bosons, 120 charged bosons and 120 charged antibosons. To account for the 248 fermions in the lowest multiplet of E_8, these would either have to include anti-particles (and so have baryogenesis), have new undiscovered particles, or have gravity-like (spin connection) bosons affecting elements of the particles spin direction. Each of these poses theoretical problems.

5.3.7 Beyond Lie groups

Other structures have been suggested including Lie 3-algebras and Lie superalgebras. Neither of these fit with Yang–Mills theory. In particular Lie superalgebras would introduce bosons with the wrong statistics. Supersymmetry however does fit with Yang–Mills. For example N=4 Super Yang Mills Theory requires an SU(N) gauge group.

5.4 Unification of forces and the role of supersymmetry

The unification of forces is possible due to the energy scale dependence of force coupling parameters in quantum field theory called renormalization group running, which allows parameters with vastly different values at usual energies to converge to a single value at a much higher energy scale.[7]

The renormalization group running of the three gauge couplings in the Standard Model has been found to nearly, but not quite, meet at the same point if the hypercharge is normalized so that it is consistent with SU(5) or SO(10) GUTs, which are precisely the GUT groups which lead to a simple fermion unification. This is a significant result, as other Lie groups lead to different normalizations. However, if the supersymmetric extension MSSM is used instead of the Standard Model, the match becomes much more accurate. In this case, the coupling constants of the strong and electroweak interactions meet at the grand unification energy, also known as the GUT scale:

$$\Lambda_{GUT} \approx 10^{16} \text{ GeV}$$

It is commonly believed that this matching is unlikely to be a coincidence, and is often quoted as one of the main motivations to further investigate supersymmetric theories despite the fact that no supersymmetric partner particles have been experimentally observed (May 2015). Also, most model builders simply assume supersymmetry because it solves the hierarchy problem — i.e., it stabilizes the electroweak Higgs mass against radiative corrections.

5.5 Neutrino masses

Since Majorana masses of the right-handed neutrino are forbidden by SO(10) symmetry, SO(10) GUTs predict the Majorana masses of right-handed neutrinos to be close to the GUT scale where the symmetry is spontaneously broken in those models. In supersymmetric GUTs, this scale tends to be larger than would be desirable to obtain realistic masses of the light, mostly left-handed neutrinos (see neutrino oscillation) via the seesaw mechanism.

5.6 Proposed theories

Several such theories have been proposed, but none is currently universally accepted. An even more ambitious theory that includes *all* fundamental forces, including gravitation, is termed a theory of everything. Some common mainstream GUT models are:

Not quite GUTs:

Note: These models refer to Lie algebras not to Lie groups. The Lie group could be [SU(4) × SU(2) × SU(2)]/$\mathbf{Z}_2$, just to take a random example.

The most promising candidate is SO(10). (Minimal) SO(10) does not contain any exotic fermions (i.e. additional fermions besides the Standard Model fermions and the right-handed neutrino), and it unifies each generation into a single irreducible representation. A number of other GUT models are based upon subgroups of SO(10). They are the minimal left-right model, SU(5), flipped SU(5) and the Pati–Salam model. The GUT group E_6 contains SO(10), but models based upon it are significantly more complicated. The primary reason for studying E_6 models comes from $E_8 \times E_8$ heterotic string theory.

GUT models generically predict the existence of topological defects such as monopoles, cosmic strings, domain walls, and others. But none have been observed. Their absence is known as the monopole problem in cosmology. Most GUT models also predict proton decay, although not the Pati–Salam model; current experiments still haven't detected proton decay. This experimental limit on the proton's lifetime pretty much rules out minimal SU(5).

- Proton Decay. These graphics refer to the X bosons and Higgs bosons.

- Dimension 6 proton decay mediated by the X boson in SU(5) GUT

- Dimension 6 proton decay mediated by the X boson in flipped SU(5) GUT

- Dimension 6 proton decay mediated by the triplet Higgs and the anti-triplet Higgs in SU(5) GUT

Some GUT theories like SU(5) and SO(10) suffer from what is called the doublet-triplet problem. These theories predict that for each electroweak Higgs doublet, there is a corresponding colored Higgs triplet field with a very small mass (many orders of magnitude smaller than the GUT scale here). In theory, unifying quarks with leptons, the Higgs doublet would also be unified with a Higgs triplet. Such triplets have not been observed. They would also cause extremely rapid proton decay (far below current experimental limits) and prevent the gauge coupling strengths from running together in the renormalization group.

Most GUT models require a threefold replication of the matter fields. As such, they do not explain why there are three generations of fermions. Most GUT models also fail to explain the little hierarchy between the fermion masses for different generations.

5.7   Ingredients

A GUT model basically consists of a gauge group which is a compact Lie group, a connection form for that Lie group, a Yang–Mills action for that connection given by an invariant symmetric bilinear form over its Lie algebra (which is specified by a coupling constant for each factor), a Higgs sector consisting of a number of scalar fields taking on values within real/complex representations of the Lie group and chiral Weyl fermions taking on values within a complex rep of the Lie group. The Lie group contains the Standard Model group and the Higgs fields acquire VEVs leading to a spontaneous symmetry breaking to the Standard Model. The Weyl fermions represent matter.

5.8   Current status

As of 2012, there is still no hard evidence that nature is described by a Grand Unified Theory. The discovery of neutrino oscillations indicates that the Standard Model is incomplete and has led to renewed interest toward certain GUT such as SO(10). One of the few possible experimental tests of certain GUT is proton decay and also fermion masses. There are a few more special tests for supersymmetric GUT.

The gauge coupling strengths of QCD, the weak interaction and hypercharge seem to meet at a common length scale called the GUT scale and equal approximately to 10^{16} GeV, which is slightly suggestive. This interesting numerical observation is called the **gauge coupling unification**, and it works particularly well if one assumes the existence of superpartners of the Standard Model particles. Still it is possible to achieve the same by postulating, for instance, that ordinary (non supersymmetric) SO(10) models break with an intermediate gauge scale, such as the one of Pati–Salam group

5.9   See also

- Paradigm shift
- Classical unified field theories
- X and Y bosons
- B – L quantum number

5.10   Notes

[1] There are however certain constraints on the choice of particle charges from theoretical consistency, in particular

anomaly cancellation.

5.11 References

[1] Ross, G. (1984). *Grand Unified Theories*. Westview Press. ISBN 978-0-8053-6968-7.

[2] Georgi, H.; Glashow, S.L. (1974). "Unity of All Elementary Particle Forces". *Physical Review Letters* **32**: 438–441. Bibcode:1974PhRvL..32..438G. doi:10.1103/PhysRevLett.32.438.

[3] Pati, J.; Salam, A. (1974). "Lepton Number as the Fourth Color". *Physical Review D* **10**: 275–289. Bibcode:1974PhRvD..10..275P. doi:10.1103/PhysRevD.10.275.

[4] Buras, A.J.; Ellis, J.; Gaillard, M.K.; Nanopoulos, D.V. (1978). "Aspects of the grand unification of strong, weak and electromagnetic interactions" (PDF). *Nuclear Physics B* **135** (1): 66–92. Bibcode:1978NuPhB.135...66B. doi:10.1016/0550-3213(78)90214-6. Retrieved 2011-03-21.

[5] Nanopoulos, D.V. (1979). "Protons Are Not Forever". *Orbis Scientiae* **1**: 91. Harvard Preprint HUTP-78/A062.

[6] Ellis, J. (2002). "Physics gets physical". *Nature* **415** (6875): 957. Bibcode:2002Natur.415..957E. doi:10.1038/415957b.

[7] Ross, G. (1984). *Grand Unified Theories*. Westview Press. ISBN 978-0-8053-6968-7.

5.12 Further reading

- Stephen Hawking, A Brief History of Time, includes a brief popular overview.

- Grand unification Paul Langacker Scholarpedia 7(10):11419. doi:10.4249/scholarpedia.11419

5.13 External links

- The Algebra of Grand Unified Theories

Chapter 6

Quantum field theory

"Relativistic quantum field theory" redirects here. For other uses, see Relativity.

In theoretical physics, **quantum field theory** (**QFT**) is a theoretical framework for constructing quantum mechanical models of subatomic particles in particle physics and quasiparticles in condensed matter physics. A QFT treats particles as excited states of an underlying physical field, so these are called field quanta.

In quantum field theory, quantum mechanical interactions between particles are described by interaction terms between the corresponding underlying fields.

6.1 Definition

Quantum electrodynamics (QED) has one electron field and one photon field; quantum chromodynamics (QCD) has one field for each type of quark; and, in condensed matter, there is an atomic displacement field that gives rise to phonon particles. Edward Witten describes QFT as "by far" the most difficult theory in modern physics.[1]

6.1.1 Dynamics

See also: Relativistic dynamics

Ordinary quantum mechanical systems have a fixed number of particles, with each particle having a finite number of degrees of freedom. In contrast, the excited states of a QFT can represent any number of particles. This makes quantum field theories especially useful for describing systems where the particle count/number may change over time, a crucial feature of relativistic dynamics.

6.1.2 States

QFT interaction terms are similar in spirit to those between charges with electric and magnetic fields in Maxwell's equations. However, unlike the classical fields of Maxwell's theory, fields in QFT generally exist in quantum superpositions of states and are subject to the laws of quantum mechanics.

Because the fields are continuous quantities over space, there exist excited states with arbitrarily large numbers of particles in them, providing QFT systems with an effectively infinite number of degrees of freedom. Infinite degrees of freedom can easily lead to divergences of calculated quantities (i.e., the quantities become infinite). Techniques such as renormalization of QFT parameters or discretization of spacetime, as in lattice QCD, are often used to avoid such infinities so as to yield physically meaningful results.

6.1.3 Fields and radiation

The gravitational field and the electromagnetic field are the only two fundamental fields in nature that have infinite range and a corresponding classical low-energy limit, which greatly diminishes and hides their "particle-like" excitations. Albert Einstein in 1905, attributed "particle-like" and discrete exchanges of momenta and energy, characteristic of "field quanta", to the electromagnetic field. Originally, his principal motivation was to explain the thermodynamics of radiation. Although the photoelectric effect and Compton scattering strongly suggest the existence of the photon, it might alternately be explained by a mere quantization of emission; more definitive evidence of the quantum nature of radiation is now taken up into modern quantum optics as in the antibunching effect.[2]

6.2 Theories

There is currently no complete quantum theory of the remaining fundamental force, gravity. Many of the proposed theories to describe gravity as a QFT postulate the existence of a graviton particle that mediates the gravitational force. Presumably, the as yet unknown correct quantum field-theoretic treatment of the gravitational field will behave like Einstein's general theory of relativity in the low-energy limit. Quantum field theory of the fundamental forces itself has been postulated to be the low-energy effective field theory limit of a more fundamental theory such as superstring theory.

Most theories in standard particle physics are formulated as **relativistic quantum field theories**, such as QED, QCD, and the Standard Model. QED, the quantum field-theoretic description of the electromagnetic field, approximately reproduces Maxwell's theory of electrodynamics in the low-energy limit, with small non-linear corrections to the Maxwell equations required due to virtual electron–positron pairs.

In the perturbative approach to quantum field theory, the full field interaction terms are approximated as a perturbative expansion in the number of particles involved. Each term in the expansion can be thought of as forces between particles being mediated by other particles. In QED, the electromagnetic force between two electrons is caused by an exchange of photons. Similarly, intermediate vector bosons mediate the weak force and gluons mediate the strong force in QCD. The notion of a force-mediating particle comes from perturbation theory, and does not make sense in the context of non-perturbative approaches to QFT, such as with bound states.

6.3 History

Main article: History of quantum field theory

6.3.1 Foundations

The early development of the field involved Dirac, Fock, Pauli, Heisenberg and Bogolyubov. This phase of development culminated with the construction of the theory of quantum electrodynamics in the 1950s.

6.3.2 Gauge theory

Gauge theory was formulated and quantized, leading to the **unification of forces** embodied in the standard model of particle physics. This effort started in the 1950s with the work of Yang and Mills, was carried on by Martinus Veltman and a host of others during the 1960s and completed by the 1970s through the work of Gerard 't Hooft, Frank Wilczek, David Gross and David Politzer.

6.3.3 Grand synthesis

Parallel developments in the understanding of phase transitions in condensed matter physics led to the study of the renormalization group. This in turn led to the **grand synthesis** of theoretical physics, which unified theories of particle and condensed matter physics through quantum field theory. This involved the work of Michael Fisher and Leo Kadanoff in the 1970s, which led to the seminal reformulation of quantum field theory by Kenneth G. Wilson in 1975.

6.4 Principles

6.4.1 Classical and quantum fields

Main article: Classical field theory

A classical field is a function defined over some region of space and time.[3] Two physical phenomena which are described by classical fields are Newtonian gravitation, described by Newtonian gravitational field $g(x, t)$, and classical electromagnetism, described by the electric and magnetic fields $E(x, t)$ and $B(x, t)$. Because such fields can in principle take on distinct values at each point in space, they are said to have infinite degrees of freedom.[3]

Classical field theory does not, however, account for the quantum-mechanical aspects of such physical phenomena. For instance, it is known from quantum mechanics that certain aspects of electromagnetism involve discrete particles—photons—rather than continuous fields. The business of *quantum* field theory is to write down a field that is, like a classical field, a function defined over space and time, but which also accommodates the observations of quantum mechanics. This is a *quantum field*.

It is not immediately clear *how* to write down such a quantum field, since quantum mechanics has a structure very unlike a field theory. In its most general formulation, quantum mechanics is a theory of abstract operators (observables) acting on an abstract state space (Hilbert space), where the observables represent physically observable quantities and the state space represents the possible states of the system under study.[4] For instance, the fundamental observables associated with the motion of a single quantum mechanical particle are the position and momentum operators $\hat{x}$ and $\hat{p}$.

Field theory, in contrast, treats x as a way to index the field rather than as an operator.[5]

There are two common ways of developing a quantum field: the path integral formalism and canonical quantization.[6] The latter of these is pursued in this article.

Lagrangian formalism

Quantum field theory frequently makes use of the Lagrangian formalism from classical field theory. This formalism is analogous to the Lagrangian formalism used in classical mechanics to solve for the motion of a particle under the influence of a field. In classical field theory, one writes down a Lagrangian density, $\mathcal{L}$, involving a field, $\varphi(\mathbf{x},t)$, and possibly its first derivatives ($\partial\varphi/\partial t$ and $\nabla\varphi$), and then applies a field-theoretic form of the Euler–Lagrange equation. Writing coordinates $(t, \mathbf{x}) = (x^0, x^1, x^2, x^3) = x^\mu$, this form of the Euler–Lagrange equation is[3]

$$\frac{\partial}{\partial x^\mu}\left[\frac{\partial\mathcal{L}}{\partial(\partial\phi/\partial x^\mu)}\right] - \frac{\partial\mathcal{L}}{\partial\phi} = 0,$$

where a sum over μ is performed according to the rules of Einstein notation.

By solving this equation, one arrives at the "equations of motion" of the field.[3] For example, if one begins with the Lagrangian density

$$\mathcal{L}(\phi, \nabla\phi) = -\rho(t, \mathbf{x})\,\phi(t, \mathbf{x}) - \frac{1}{8\pi G}|\nabla\phi|^2,$$

and then applies the Euler–Lagrange equation, one obtains the equation of motion

$$4\pi G\rho(t, \mathbf{x}) = \nabla^2\phi.$$

This equation is Newton's law of universal gravitation, expressed in differential form in terms of the gravitational potential $\varphi(t, \mathbf{x})$ and the mass density $\rho(t, \mathbf{x})$. Despite the nomenclature, the "field" under study is the gravitational potential, φ, rather than the gravitational field, $\mathbf{g}$. Similarly, when classical field theory is used to study electromagnetism, the "field" of interest is the electromagnetic four-potential $(V/c, \mathbf{A})$, rather than the electric and magnetic fields $\mathbf{E}$ and $\mathbf{B}$.

Quantum field theory uses this same Lagrangian procedure to determine the equations of motion for quantum fields. These equations of motion are then supplemented by commutation relations derived from the canonical quantization procedure described below, thereby incorporating quantum mechanical effects into the behavior of the field.

6.4.2 Single- and many-particle quantum mechanics

Main articles: Quantum mechanics and First quantization

In quantum mechanics, a particle (such as an electron or proton) is described by a complex wavefunction, $\psi(x, t)$, whose time-evolution is governed by the Schrödinger equation:

$$-\frac{\hbar^2}{2m}\frac{\partial^2}{\partial x^2}\psi(x,t) + V(x)\psi(x,t) = i\hbar\frac{\partial}{\partial t}\psi(x,t).$$

Here m is the particle's mass and $V(x)$ is the applied potential. Physical information about the behavior of the particle is extracted from the wavefunction by constructing expected values for various quantities; for example, the expected value of the particle's position is given by integrating $\psi^*(x)\,x\,\psi(x)$ over all space, and the expected value of the particle's momentum is found by integrating $-i\hbar\psi^*(x)\mathrm{d}\psi/\mathrm{d}x$. The quantity $\psi^*(x)\psi(x)$ is itself in the Copenhagen interpretation of quantum mechanics interpreted as a probability density function. This treatment of quantum mechanics, where a particle's wavefunction evolves against a classical background potential $V(x)$, is sometimes called *first quantization*.

This description of quantum mechanics can be extended to describe the behavior of multiple particles, so long as the number and the type of particles remain fixed. The particles are described by a wavefunction $\psi(x_1, x_2, ..., xN, t)$, which is governed by an extended version of the Schrödinger equation.

Often one is interested in the case where N particles are all of the same type (for example, the 18 electrons orbiting a neutral argon nucleus). As described in the article on identical particles, this implies that the state of the entire system must be either symmetric (bosons) or antisymmetric (fermions) when the coordinates of its constituent particles are exchanged. This is achieved by using a Slater determinant as the wavefunction of a fermionic system (and a Slater permanent for a bosonic system), which is equivalent to an element of the symmetric or antisymmetric subspace of a tensor product.

For example, the general quantum state of a system of N bosons is written as

$$|\phi_1 \cdots \phi_N\rangle = \sqrt{\frac{\prod_j N_j!}{N!}}\sum_{p\in S_N}|\phi_{p(1)}\rangle \otimes \cdots \otimes |\phi_{p(N)}\rangle,$$

where $|\phi_i\rangle$ are the single-particle states, Nj is the number of particles occupying state j, and the sum is taken over all possible permutations p acting on N elements. In general, this is a sum of $N!$ (N factorial) distinct terms. $\sqrt{\frac{\prod_j N_j!}{N!}}$ is a normalizing factor.

There are several shortcomings to the above description of quantum mechanics, which are addressed by quantum field theory. First, it is unclear how to extend quantum mechanics to include the effects of special relativity.[7] Attempted replacements for the Schrödinger equation, such as the Klein–Gordon equation or the Dirac equation, have many unsatisfactory qualities; for instance, they possess energy eigenvalues that extend to $-\infty$, so that there seems to be no easy definition of a ground state. It turns out that such inconsistencies arise from relativistic wavefunctions not having a well-defined probabilistic interpretation in position space, as probability conservation is not a relativistically covariant concept. The second shortcoming, related to the first, is that in quantum mechanics there is no mechanism to describe particle creation and annihilation;[8] this is crucial for describing phenomena such as pair production, which result from the conversion between mass and energy according to the relativistic relation $E = mc^2$.

6.4.3 Second quantization

Main article: Second quantization

In this section, we will describe a method for constructing a quantum field theory called **second quantization**. This basically involves choosing a way to index the quantum mechanical degrees of freedom in the space of multiple identical-particle states. It is based on the Hamiltonian formulation of quantum mechanics.

Several other approaches exist, such as the Feynman path integral,[9] which uses a Lagrangian formulation. For an overview of some of these approaches, see the article on quantization.

Bosons

For simplicity, we will first discuss second quantization for bosons, which form perfectly symmetric quantum states. Let us denote the mutually orthogonal single-particle states which are possible in the system by $|\phi_1\rangle$, $|\phi_2\rangle$, $|\phi_3\rangle$, and so on. For example, the 3-particle state with one particle in state $|\phi_1\rangle$ and two in state $|\phi_2\rangle$ is

$$\frac{1}{\sqrt{3}} \left[|\phi_1\rangle|\phi_2\rangle|\phi_2\rangle + |\phi_2\rangle|\phi_1\rangle|\phi_2\rangle + |\phi_2\rangle|\phi_2\rangle|\phi_1\rangle \right].$$

The first step in second quantization is to express such quantum states in terms of **occupation numbers**, by listing the number of particles occupying each of the single-particle states $|\phi_1\rangle$, $|\phi_2\rangle$, etc. This is simply another way of labelling the states. For instance, the above 3-particle state is denoted as

$$|1, 2, 0, 0, 0, \ldots\rangle.$$

An N-particle state belongs to a space of states describing systems of N particles. The next step is to combine the individual N-particle state spaces into an extended state space, known as Fock space, which can describe systems of any number of particles. This is composed of the state space of a system with no particles (the so-called vacuum state, written as $|0\rangle$), plus the state space of a 1-particle system, plus the state space of a 2-particle system, and so forth. States describing a definite number of particles are known as Fock states: a general element of Fock space will be a linear combination of Fock states. There is a one-to-one correspondence between the occupation number representation and valid boson states in the Fock space.

At this point, the quantum mechanical system has become a quantum field in the sense we described above. The field's elementary degrees of freedom are the occupation numbers, and each occupation number is indexed by a number j indicating which of the single-particle states $|\phi_1\rangle$, $|\phi_2\rangle$, ..., $|\phi_j\rangle$, ... it refers to:

$$|N_1, N_2, N_3, \ldots, N_j, \ldots\rangle.$$

The properties of this quantum field can be explored by defining creation and annihilation operators, which add and subtract particles. They are analogous to ladder operators in the quantum harmonic oscillator problem, which added and subtracted energy quanta. However, these operators literally create and annihilate particles of a given quantum state. The bosonic annihilation operator a_2 and creation operator $a_2^\dagger$ are easily defined in the occupation number representation as having the following effects:

$$a_2|N_1, N_2, N_3, \ldots\rangle = \sqrt{N_2}\,|N_1, (N_2 - 1), N_3, \ldots\rangle,$$

$$a_2^\dagger|N_1, N_2, N_3, \ldots\rangle = \sqrt{N_2 + 1}\,|N_1, (N_2+1), N_3, \ldots\rangle.$$

It can be shown that these are operators in the usual quantum mechanical sense, i.e. linear operators acting on the Fock space. Furthermore, they are indeed Hermitian conjugates, which justifies the way we have written them. They can be shown to obey the commutation relation

$$[a_i, a_j] = 0 \quad , \quad \left[a_i^\dagger, a_j^\dagger\right] = 0 \quad , \quad \left[a_i, a_j^\dagger\right] = \delta_{ij}.$$

where δ stands for the Kronecker delta. These are precisely the relations obeyed by the ladder operators for an infinite set of independent quantum harmonic oscillators, one for each single-particle state. Adding or removing bosons from each state is therefore analogous to exciting or de-exciting a quantum of energy in a harmonic oscillator.

Applying an annihilation operator a_k followed by its corresponding creation operator $a_k^\dagger$ returns the number N_k of particles in the kth single-particle eigenstate:

$$a_k^\dagger a_k | \ldots, N_k, \ldots \rangle = N_k | \ldots, N_k, \ldots \rangle.$$

The combination of operators $a_k^\dagger a_k$ is known as the number operator for the kth eigenstate.

The Hamiltonian operator of the quantum field (which, through the Schrödinger equation, determines its dynamics) can be written in terms of creation and annihilation operators. For instance, for a field of free (non-interacting) bosons, the total energy of the field is found by summing the energies of the bosons in each energy eigenstate. If the kth single-particle energy eigenstate has energy E_k and there are N_k bosons in this state, then the total energy of these bosons is $E_k N_k$. The energy in the *entire* field is then a sum over k :

$$E_{\text{tot}} = \sum_k E_k N_k$$

This can be turned into the Hamiltonian operator of the field by replacing N_k with the corresponding number operator, $a_k^\dagger a_k$. This yields

$$H = \sum_k E_k \, a_k^\dagger a_k.$$

Fermions

It turns out that a different definition of creation and annihilation must be used for describing fermions. According to the Pauli exclusion principle, fermions cannot share quantum states, so their occupation numbers N_i can only take on the value 0 or 1. The fermionic annihilation operators c and creation operators $c^\dagger$ are defined by their actions on a Fock state thus

$$c_j | N_1, N_2, \ldots, N_j = 0, \ldots \rangle = 0$$

$$c_j | N_1, N_2, \ldots, N_j = 1, \ldots \rangle = (-1)^{(N_1 + \cdots + N_{j-1})} | N_1, N_2, \ldots, N_j$$

$$c_j^\dagger | N_1, N_2, \ldots, N_j = 0, \ldots \rangle = (-1)^{(N_1 + \cdots + N_{j-1})} | N_1, N_2, \ldots, N_j$$

$$c_j^\dagger | N_1, N_2, \ldots, N_j = 1, \ldots \rangle = 0.$$

These obey an anticommutation relation:

$$\{c_i, c_j\} = 0 \quad , \quad \{c_i^\dagger, c_j^\dagger\} = 0 \quad , \quad \{c_i, c_j^\dagger\} = \delta_{ij}.$$

One may notice from this that applying a fermionic creation operator twice gives zero, so it is impossible for the particles to share single-particle states, in accordance with the exclusion principle.

Field operators

We have previously mentioned that there can be more than one way of indexing the degrees of freedom in a quantum field. Second quantization indexes the field by enumerating the single-particle quantum states. However, as we have discussed, it is more natural to think about a "field", such as the electromagnetic field, as a set of degrees of freedom indexed by position.

To this end, we can define *field operators* that create or destroy a particle at a particular point in space. In particle physics, these operators turn out to be more convenient to work with, because they make it easier to formulate theories that satisfy the demands of relativity.

Single-particle states are usually enumerated in terms of their momenta (as in the particle in a box problem.) We can construct field operators by applying the Fourier transform to the creation and annihilation operators for these states. For example, the bosonic field annihilation operator $\phi(\mathbf{r})$ is

$$\phi(\mathbf{r}) \overset{\text{def}}{=} \sum_j e^{i\mathbf{k}_j \cdot \mathbf{r}} a_j.$$

The bosonic field operators obey the commutation relation

$$[\phi(\mathbf{r}), \phi(\mathbf{r}')] = 0 \quad , \quad [\phi^\dagger(\mathbf{r}), \phi^\dagger(\mathbf{r}')] = 0 \quad ,$$
$$[\phi(\mathbf{r}), \phi^\dagger(\mathbf{r}')] = \delta^3(\mathbf{r} - \mathbf{r}')$$

where $\delta(x)$ stands for the Dirac delta function. As before, the fermionic relations are the same, with the commutators replaced by anticommutators.

The field operator is not the same thing as a single-particle wavefunction. The former is an operator acting on the Fock space, and the latter is a quantum-mechanical amplitude for finding a particle in some position. However, they are closely related, and are indeed commonly denoted with the same symbol. If we have a Hamiltonian with a space representation, say

$$H = -\frac{\bar{h}^2}{2m} \sum_{=1, \{\ldots\}}^{0, \ldots \rangle} \nabla_i^2 + \sum_{i<j} U(|\mathbf{r}_i - \mathbf{r}_j|)$$

where the indices i and j run over all particles, then the field theory Hamiltonian (in the non-relativistic limit and for negligible self-interactions) is

$$H = -\frac{\hbar^2}{2m} \int d^3r \, \phi^\dagger(\mathbf{r}) \nabla^2 \phi(\mathbf{r}) +$$
$$\frac{1}{2} \int d^3r \int d^3r' \, \phi^\dagger(\mathbf{r}) \phi^\dagger(\mathbf{r}') U(|\mathbf{r}-\mathbf{r}'|) \phi(\mathbf{r}') \phi(\mathbf{r}).$$

This looks remarkably like an expression for the expectation value of the energy, with ϕ playing the role of the wavefunction. This relationship between the field operators and wavefunctions makes it very easy to formulate field theories starting from space-projected Hamiltonians.

6.4.4 Dynamics

Once the Hamiltonian operator is obtained as part of the canonical quantization process, the time dependence of the state is described with the Schrödinger equation, just as with other quantum theories. Alternatively, the Heisenberg picture can be used where the time dependence is in the operators rather than in the states.

6.4.5 Implications

Unification of fields and particles

The "second quantization" procedure that we have outlined in the previous section takes a set of single-particle quantum states as a starting point. Sometimes, it is impossible to define such single-particle states, and one must proceed directly to quantum field theory. For example, a quantum theory of the electromagnetic field *must* be a quantum field theory, because it is impossible (for various reasons) to define a wavefunction for a single photon.[10] In such situations, the quantum field theory can be constructed by examining the mechanical properties of the classical field and guessing the corresponding quantum theory. For free (non-interacting) quantum fields, the quantum field theories obtained in this way have the same properties as those obtained using second quantization, such as well-defined creation and annihilation operators obeying commutation or anticommutation relations.

Quantum field theory thus provides a unified framework for describing "field-like" objects (such as the electromagnetic field, whose excitations are photons) and "particle-like" objects (such as electrons, which are treated as excitations of an underlying electron field), so long as one can treat interactions as "perturbations" of free fields. There are still unsolved problems relating to the more general case of interacting fields that may or may not be adequately described by perturbation theory. For more on this topic, see Haag's theorem.

Physical meaning of particle indistinguishability

The second quantization procedure relies crucially on the particles being identical. We would not have been able to construct a quantum field theory from a distinguishable many-particle system, because there would have been no way of separating and indexing the degrees of freedom.

Many physicists prefer to take the converse interpretation, which is that *quantum field theory explains what identical particles are*. In ordinary quantum mechanics, there is not much theoretical motivation for using symmetric (bosonic) or antisymmetric (fermionic) states, and the need for such states is simply regarded as an empirical fact. From the point of view of quantum field theory, particles are identical if and only if they are excitations of the same underlying quantum field. Thus, the question "why are all electrons identical?" arises from mistakenly regarding individual electrons as fundamental objects, when in fact it is only the electron field that is fundamental.

Particle conservation and non-conservation

During second quantization, we started with a Hamiltonian and state space describing a fixed number of particles (N), and ended with a Hamiltonian and state space for an arbitrary number of particles. Of course, in many common situations N is an important and perfectly well-defined quantity, e.g. if we are describing a gas of atoms sealed in a box. From the point of view of quantum field theory, such situations are described by quantum states that are eigenstates of the number operator $\hat{N}$, which measures the total number of particles present. As with any quantum mechanical observable, $\hat{N}$ is conserved if it commutes with the Hamiltonian. In that case, the quantum state is trapped in the N-particle subspace of the total Fock space, and the situation could equally well be described by ordinary N-particle quantum mechanics. (Strictly speaking, this is only true in the noninteracting case or in the low energy density limit of renormalized quantum field theories)

For example, we can see that the free-boson Hamiltonian described above conserves particle number. Whenever the Hamiltonian operates on a state, each particle destroyed by an annihilation operator a_k is immediately put back by the creation operator $a_k^\dagger$.

On the other hand, it is possible, and indeed common, to encounter quantum states that are *not* eigenstates of $\hat{N}$, which do not have well-defined particle numbers. Such states are difficult or impossible to handle using ordinary quantum mechanics, but they can be easily described in quantum field theory as quantum superpositions of states having different values of N. For example, suppose we have a bosonic field whose particles can be created or destroyed by interactions with a fermionic field. The Hamiltonian of the combined system would be given by the Hamiltonians of the free boson and free fermion fields, plus a "potential energy" term such as

$$H_I = \sum_{k,q} V_q (a_q + a_{-q}^\dagger) c_{k+q}^\dagger c_k,$$

where $a_k^\dagger$ and a_k denotes the bosonic creation and annihilation operators, $c_k^\dagger$ and c_k denotes the fermionic creation and annihilation operators, and V_q is a parameter that describes the strength of the interaction. This "interaction term" describes processes in which a fermion in state k either absorbs or emits a boson, thereby being kicked into a different eigenstate $k + q$. (In fact, this type of Hamiltonian is used to describe interaction between conduction electrons and phonons in metals. The interaction between electrons and photons is treated in a similar way, but is a little more complicated because the role of spin must be taken into account.) One thing to notice here is that even if we start out with a fixed number of bosons, we will typically end up with a superposition of states with different numbers of bosons at later times. The number of fermions, however, is conserved in this case.

In condensed matter physics, states with ill-defined particle numbers are particularly important for describing the various superfluids. Many of the defining characteristics of a superfluid arise from the notion that its quantum state is a superposition of states with different particle numbers. In addition, the concept of a coherent state (used to model the laser and the BCS ground state) refers to a state with an ill-defined particle number but a well-defined phase.

6.4.6   Axiomatic approaches

The preceding description of quantum field theory follows the spirit in which most physicists approach the subject. However, it is not mathematically rigorous. Over the past several decades, there have been many attempts to put quantum field theory on a firm mathematical footing by formulating a set of axioms for it. These attempts fall into two broad classes.

The first class of axioms, first proposed during the 1950s, include the Wightman, Osterwalder–Schrader, and Haag–Kastler systems. They attempted to formalize the physicists' notion of an "operator-valued field" within the context of functional analysis, and enjoyed limited success. It was possible to prove that any quantum field theory satisfying these axioms satisfied certain general theorems, such as the spin-statistics theorem and the CPT theorem. Unfortunately, it proved extraordinarily difficult to show that any

realistic field theory, including the Standard Model, satisfied these axioms. Most of the theories that could be treated with these analytic axioms were physically trivial, being restricted to low-dimensions and lacking interesting dynamics. The construction of theories satisfying one of these sets of axioms falls in the field of constructive quantum field theory. Important work was done in this area in the 1970s by Segal, Glimm, Jaffe and others.

During the 1980s, a second set of axioms based on geometric ideas was proposed. This line of investigation, which restricts its attention to a particular class of quantum field theories known as topological quantum field theories, is associated most closely with Michael Atiyah and Graeme Segal, and was notably expanded upon by Edward Witten, Richard Borcherds, and Maxim Kontsevich. However, most of the physically relevant quantum field theories, such as the Standard Model, are not topological quantum field theories; the quantum field theory of the fractional quantum Hall effect is a notable exception. The main impact of axiomatic topological quantum field theory has been on mathematics, with important applications in representation theory, algebraic topology, and differential geometry.

Finding the proper axioms for quantum field theory is still an open and difficult problem in mathematics. One of the Millennium Prize Problems—proving the existence of a mass gap in Yang–Mills theory—is linked to this issue.

6.5   Associated phenomena

In the previous part of the article, we described the most general features of quantum field theories. Some of the quantum field theories studied in various fields of theoretical physics involve additional special ideas, such as renormalizability, gauge symmetry, and supersymmetry. These are described in the following sections.

6.5.1   Renormalization

Main article: Renormalization

Early in the history of quantum field theory, it was found that many seemingly innocuous calculations, such as the perturbative shift in the energy of an electron due to the presence of the electromagnetic field, give infinite results. The reason is that the perturbation theory for the shift in an energy involves a sum over all other energy levels, and there are infinitely many levels at short distances that each give a finite contribution which results in a divergent series.

Many of these problems are related to failures in classical electrodynamics that were identified but unsolved in the

19th century, and they basically stem from the fact that many of the supposedly "intrinsic" properties of an electron are tied to the electromagnetic field that it carries around with it. The energy carried by a single electron—its self energy—is not simply the bare value, but also includes the energy contained in its electromagnetic field, its attendant cloud of photons. The energy in a field of a spherical source diverges in both classical and quantum mechanics, but as discovered by Weisskopf with help from Furry, in quantum mechanics the divergence is much milder, going only as the logarithm of the radius of the sphere.

The solution to the problem, presciently suggested by Stueckelberg, independently by Bethe after the crucial experiment by Lamb, implemented at one loop by Schwinger, and systematically extended to all loops by Feynman and Dyson, with converging work by Tomonaga in isolated postwar Japan, comes from recognizing that all the infinities in the interactions of photons and electrons can be isolated into redefining a finite number of quantities in the equations by replacing them with the observed values: specifically the electron's mass and charge: this is called renormalization. The technique of renormalization recognizes that the problem is essentially purely mathematical, that extremely short distances are at fault. In order to define a theory on a continuum, first place a cutoff on the fields, by postulating that quanta cannot have energies above some extremely high value. This has the effect of replacing continuous space by a structure where very short wavelengths do not exist, as on a lattice. Lattices break rotational symmetry, and one of the crucial contributions made by Feynman, Pauli and Villars, and modernized by 't Hooft and Veltman, is a symmetry-preserving cutoff for perturbation theory (this process is called regularization). There is no known symmetrical cutoff outside of perturbation theory, so for rigorous or numerical work people often use an actual lattice.

On a lattice, every quantity is finite but depends on the spacing. When taking the limit of zero spacing, we make sure that the physically observable quantities like the observed electron mass stay fixed, which means that the constants in the Lagrangian defining the theory depend on the spacing. Hopefully, by allowing the constants to vary with the lattice spacing, all the results at long distances become insensitive to the lattice, defining a continuum limit.

The renormalization procedure only works for a certain class of quantum field theories, called **renormalizable quantum field theories**. A theory is **perturbatively renormalizable** when the constants in the Lagrangian only diverge at worst as logarithms of the lattice spacing for very short spacings. The continuum limit is then well defined in perturbation theory, and even if it is not fully well defined non-perturbatively, the problems only show up at distance scales that are exponentially small in the inverse coupling for weak couplings. The Standard Model of particle physics

is perturbatively renormalizable, and so are its component theories (quantum electrodynamics/electroweak theory and quantum chromodynamics). Of the three components, quantum electrodynamics is believed to not have a continuum limit, while the asymptotically free SU(2) and SU(3) weak hypercharge and strong color interactions are nonperturbatively well defined.

The renormalization group describes how renormalizable theories emerge as the long distance low-energy effective field theory for any given high-energy theory. Because of this, renormalizable theories are insensitive to the precise nature of the underlying high-energy short-distance phenomena. This is a blessing because it allows physicists to formulate low energy theories without knowing the details of high energy phenomenon. It is also a curse, because once a renormalizable theory like the standard model is found to work, it gives very few clues to higher energy processes. The only way high energy processes can be seen in the standard model is when they allow otherwise forbidden events, or if they predict quantitative relations between the coupling constants.

6.5.2   Haag's theorem

See also: Haag's theorem

From a mathematically rigorous perspective, there exists no interaction picture in a Lorentz-covariant quantum field theory. This implies that the perturbative approach of Feynman diagrams in QFT is not strictly justified, despite producing vastly precise predictions validated by experiment. This is called Haag's theorem, but most particle physicists relying on QFT largely shrug it off.

6.5.3   Gauge freedom

A gauge theory is a theory that admits a symmetry with a local parameter. For example, in every quantum theory the global phase of the wave function is arbitrary and does not represent something physical. Consequently, the theory is invariant under a global change of phases (adding a constant to the phase of all wave functions, everywhere); this is a global symmetry. In quantum electrodynamics, the theory is also invariant under a *local* change of phase, that is – one may shift the phase of all wave functions so that the shift may be different at every point in space-time. This is a *local* symmetry. However, in order for a well-defined derivative operator to exist, one must introduce a new field, the gauge field, which also transforms in order for the local change of variables (the phase in our example) not to affect the derivative. In quantum electrodynamics this gauge field is the electromagnetic field. The change of local gauge of variables is termed gauge transformation. It is worth noting that by Noether's theorem, for every such symmetry there exists an associated conserved current. The aforementioned symmetry of the wavefunction under global phase changes implies the conservation of electric charge.

In quantum field theory the excitations of fields represent particles. The particle associated with excitations of the gauge field is the gauge boson, which is the photon in the case of quantum electrodynamics.

The degrees of freedom in quantum field theory are local fluctuations of the fields. The existence of a gauge symmetry reduces the number of degrees of freedom, simply because some fluctuations of the fields can be transformed to zero by gauge transformations, so they are equivalent to having no fluctuations at all, and they therefore have no physical meaning. Such fluctuations are usually called "non-physical degrees of freedom" or *gauge artifacts*; usually some of them have a negative norm, making them inadequate for a consistent theory. Therefore, if a classical field theory has a gauge symmetry, then its quantized version (i.e. the corresponding quantum field theory) will have this symmetry as well. In other words, a gauge symmetry cannot have a quantum anomaly. If a gauge symmetry is anomalous (i.e. not kept in the quantum theory) then the theory is non-consistent: for example, in quantum electrodynamics, had there been a gauge anomaly, this would require the appearance of photons with longitudinal polarization and polarization in the time direction, the latter having a negative norm, rendering the theory inconsistent; another possibility would be for these photons to appear only in intermediate processes but not in the final products of any interaction, making the theory non-unitary and again inconsistent (see optical theorem).

In general, the gauge transformations of a theory consist of several different transformations, which may not be commutative. These transformations are together described by a mathematical object known as a gauge group. Infinitesimal gauge transformations are the gauge group generators. Therefore, the number of gauge bosons is the group dimension (i.e. number of generators forming a basis).

All the fundamental interactions in nature are described by gauge theories. These are:

- Quantum chromodynamics, whose gauge group is $SU(3)$. The gauge bosons are eight gluons.

- The electroweak theory, whose gauge group is $U(1) \times SU(2)$, (a direct product of $U(1)$ and $SU(2)$).

- Gravity, whose classical theory is general relativity, admits the equivalence principle, which is a form

of gauge symmetry. However, it is explicitly non-renormalizable.

6.5.4   Multivalued gauge transformations

The gauge transformations which leave the theory invariant involve, by definition, only single-valued gauge functions $\Lambda(x_i)$ which satisfy the Schwarz integrability criterion

$$\partial_{x_i x_j} \Lambda = \partial_{x_j x_i} \Lambda.$$

An interesting extension of gauge transformations arises if the gauge functions $\Lambda(x_i)$ are allowed to be multivalued functions which violate the integrability criterion. These are capable of changing the physical field strengths and are therefore not proper symmetry transformations. Nevertheless, the transformed field equations describe correctly the physical laws in the presence of the newly generated field strengths. See the textbook by H. Kleinert cited below for the applications to phenomena in physics.

6.5.5   Supersymmetry

Main article: Supersymmetry

Supersymmetry assumes that every fundamental fermion has a superpartner that is a boson and vice versa. It was introduced in order to solve the so-called Hierarchy Problem, that is, to explain why particles not protected by any symmetry (like the Higgs boson) do not receive radiative corrections to its mass driving it to the larger scales (GUT, Planck...). It was soon realized that supersymmetry has other interesting properties: its gauged version is an extension of general relativity (Supergravity), and it is a key ingredient for the consistency of string theory.

The way supersymmetry protects the hierarchies is the following: since for every particle there is a superpartner with the same mass, any loop in a radiative correction is cancelled by the loop corresponding to its superpartner, rendering the theory UV finite.

Since no superpartners have yet been observed, if supersymmetry exists it must be broken (through a so-called soft term, which breaks supersymmetry without ruining its helpful features). The simplest models of this breaking require that the energy of the superpartners not be too high; in these cases, supersymmetry is expected to be observed by experiments at the Large Hadron Collider. The Higgs particle has been detected at the LHC, and no such superparticles have been discovered.

6.6 See also

- Abraham–Lorentz force
- Basic concepts of quantum mechanics
- Common integrals in quantum field theory
- Einstein–Maxwell–Dirac equations
- Form factor (quantum field theory)
- Green–Kubo relations
- Green's function (many-body theory)
- Invariance mechanics
- List of quantum field theories
- Quantum electrodynamics
- Quantum field theory in curved spacetime
- Quantum flavordynamics
- Quantum hydrodynamics
- Quantum triviality
- Relation between Schrödinger's equation and the path integral formulation of quantum mechanics
- Relationship between string theory and quantum field theory
- Schwinger–Dyson equation
- Static forces and virtual-particle exchange
- Symmetry in quantum mechanics
- Theoretical and experimental justification for the Schrödinger equation
- Ward–Takahashi identity
- Wheeler–Feynman absorber theory
- Wigner's classification
- Wigner's theorem

6.7 Notes

6.8 References

[1] "Beautiful Minds, Vol. 20: Ed Witten". la Repubblica. 2010. Retrieved 22 June 2012. See here.

[2] J. J. Thorn et al. (2004). Observing the quantum behavior of light in an undergraduate laboratory. . J. J. Thorn, M. S. Neel, V. W. Donato, G. S. Bergreen, R. E. Davies, and M. Beck. American Association of Physics Teachers, 2004.DOI: 10.1119/1.1737397.

[3] David Tong, *Lectures on Quantum Field Theory*, chapter 1.

[4] Srednicki, Mark. *Quantum Field Theory* (1st ed.). p. 19.

[5] Srednicki, Mark. *Quantum Field Theory* (1st ed.). pp. 25–6.

[6] Zee, Anthony. *Quantum Field Theory in a Nutshell* (2nd ed.). p. 61.

[7] David Tong, *Lectures on Quantum Field Theory*, Introduction.

[8] Zee, Anthony. *Quantum Field Theory in a Nutshell* (2nd ed.). p. 3.

[9] Abraham Pais, *Inward Bound: Of Matter and Forces in the Physical World* ISBN 0-19-851997-4. Pais recounts how his astonishment at the rapidity with which Feynman could calculate using his method. Feynman's method is now part of the standard methods for physicists.

[10] Newton, T.D.; Wigner, E.P. (1949). "Localized states for elementary systems". *Reviews of Modern Physics* **21** (3): 400–406. Bibcode:1949RvMP...21..400N. doi:10.1103/RevModPhys.21.400.

6.9 Further reading

General readers

- Feynman, R.P. (2001) [1964]. *The Character of Physical Law*. MIT Press. ISBN 0-262-56003-8.
- Feynman, R.P. (2006) [1985]. *QED: The Strange Theory of Light and Matter*. Princeton University Press. ISBN 0-691-12575-9.
- Gribbin, J. (1998). *Q is for Quantum: Particle Physics from A to Z*. Weidenfeld & Nicolson. ISBN 0-297-81752-3.
- Schumm, Bruce A. (2004) *Deep Down Things*. Johns Hopkins Univ. Press. Chpt. 4.

Introductory texts

- McMahon, D. (2008).   *Quantum Field Theory*. McGraw-Hill. ISBN 978-0-07-154382-8.

- Bogoliubov, N.; Shirkov, D. (1982). *Quantum Fields*. Benjamin-Cummings. ISBN 0-8053-0983-7.

- Frampton, P.H. (2000). *Gauge Field Theories. Frontiers in Physics (2nd ed.)*. Wiley.

- Greiner, W; Müller, B. (2000). *Gauge Theory of Weak Interactions*. Springer. ISBN 3-540-67672-4.

- Itzykson, C.; Zuber, J.-B. (1980). *Quantum Field Theory*. McGraw-Hill. ISBN 0-07-032071-3.

- Kane, G.L. (1987).   *Modern Elementary Particle Physics*. Perseus Books. ISBN 0-201-11749-5.

- Kleinert, H.; Schulte-Frohlinde, Verena (2001). *Critical Properties of φ^4-Theories*. World Scientific. ISBN 981-02-4658-7.

- Kleinert, H. (2008). *Multivalued Fields in Condensed Matter, Electrodynamics, and Gravitation* (PDF). World Scientific. ISBN 978-981-279-170-2.

- Loudon, R (1983).   *The Quantum Theory of Light*. Oxford University Press. ISBN 0-19-851155-8.

- Mandl, F.; Shaw, G. (1993). *Quantum Field Theory*. John Wiley & Sons. ISBN 978-0-471-94186-6.

- Peskin, M.; Schroeder, D. (1995). *An Introduction to Quantum Field Theory*. Westview Press. ISBN 0-201-50397-2.

- Ryder, L.H. (1985).   *Quantum Field Theory*. Cambridge University Press. ISBN 0-521-33859-X.

- Schwartz, M.D. (2014). *Quantum Field Theory and the Standard Model*. Cambridge University Press. ISBN 978-1107034730.

- Srednicki, Mark (2007) *Quantum Field Theory*. Cambridge Univ. Press.

- Ynduráin, F.J. (1996).   *Relativistic Quantum Mechanics and Introduction to Field Theory* (1st ed.). Springer. ISBN 978-3-540-60453-2.

- Zee, A. (2003). *Quantum Field Theory in a Nutshell*. Princeton University Press. ISBN 0-691-01019-6.

Advanced texts

- Brown, Lowell S. (1994).   *Quantum Field Theory*. Cambridge University Press. ISBN 978-0-521-46946-3.

- Bogoliubov, N.; Logunov, A.A.; Oksak, A.I.; Todorov, I.T. (1990). *General Principles of Quantum Field Theory*. Kluwer Academic Publishers. ISBN 978-0-7923-0540-8.

- Weinberg, S. (1995). *The Quantum Theory of Fields 1–3*. Cambridge University Press.

Articles:

- Gerard 't Hooft (2007) "The Conceptual Basis of Quantum Field Theory" in Butterfield, J., and John Earman, eds., *Philosophy of Physics, Part A*. Elsevier: 661–730.

- Frank Wilczek (1999) "Quantum field theory", *Reviews of Modern Physics* 71: S83–S95.   Also doi=10.1103/Rev. Mod. Phys. 71.

6.10   External links

- Hazewinkel, Michiel, ed. (2001), "Quantum field theory", *Encyclopedia of Mathematics*, Springer, ISBN 978-1-55608-010-4

- Stanford Encyclopedia of Philosophy: "Quantum Field Theory", by Meinard Kuhlmann.

- Siegel, Warren, 2005. *Fields*. A free text, also available from arXiv:hep-th/9912205.

- Quantum Field Theory by P. J. Mulders

Chapter 7

Special relativity

For history and motivation, see History of special relativity.

In physics, **special relativity** (**SR**, also known as the **special theory of relativity** or **STR**) is the generally accepted and experimentally well confirmed physical theory regarding the relationship between space and time. In Einstein's original pedagogical treatment, it is based on two postulates: (1) that the laws of physics are invariant (i.e. identical) in all inertial systems (non-accelerating frames of reference); and (2) that the speed of light in a vacuum is the same for all observers, regardless of the motion of the light source. It was originally proposed in 1905 by Albert Einstein in the paper "On the Electrodynamics of Moving Bodies".[1] The inconsistency of Newtonian mechanics with Maxwell's equations of electromagnetism and the inability to discover Earth's motion through a luminiferous aether led to the development of special relativity, which corrects mechanics to handle situations involving motions nearing the speed of light. As of today, special relativity is the most accurate model of motion at any speed. Even so, Newtonian mechanics is still useful (due to its simplicity and high accuracy) as an approximation at small velocities relative to the speed of light.

Special relativity implies a wide range of consequences, which have been experimentally verified,[2] including length contraction, time dilation, relativistic mass, mass–energy equivalence, a universal speed limit, and relativity of simultaneity. It has replaced the conventional notion of an absolute universal time with the notion of a time that is dependent on reference frame and spatial position. Rather than an invariant time interval between two events, there is an invariant spacetime interval. Combined with other laws of physics, the two postulates of special relativity predict the equivalence of mass and energy, as expressed in the mass–energy equivalence formula $E = mc^2$, where c is the speed of light in vacuum.[3][4]

A defining feature of special relativity is the replacement of the Galilean transformations of Newtonian mechanics with the Lorentz transformations. Time and space cannot be defined separately from each other. Rather space and time are interwoven into a single continuum known as spacetime. Events that occur at the same time for one observer could occur at different times for another.

The theory is "special" in that it only applies in the special case where the curvature of spacetime due to gravity is negligible.[5][6] In order to include gravity, Einstein formulated general relativity in 1915. (Special relativity, contrary to some outdated descriptions, is capable of handling accelerated frames of reference.[7])

As Galilean relativity is now considered an approximation of special relativity that is valid for low speeds, special relativity is considered an approximation of general relativity that is valid for weak gravitational fields, i.e. at a sufficiently small scale and in conditions of free fall. Whereas general relativity incorporates noneuclidean geometry in order to represent gravitational effects as the geometric curvature of spacetime, special relativity is restricted to the flat spacetime known as Minkowski space. A locally Lorentz-invariant frame that abides by special relativity can be defined at sufficiently small scales, even in curved spacetime.

Galileo Galilei had already postulated that there is no absolute and well-defined state of rest (no privileged reference frames), a principle now called Galileo's principle of relativity. Einstein extended this principle so that it accounted for the constant speed of light,[8] a phenomenon that had been recently observed in the Michelson–Morley experiment. He also postulated that it holds for all the laws of physics, including both the laws of mechanics and of electrodynamics.[9]

7.1 Postulates

Einstein discerned two fundamental propositions that seemed to be the most assured, regardless of the exact validity of the (then) known laws of either mechanics or electrodynamics. These propositions were the constancy of the speed of light and the independence of physical laws (espe-

Following Einstein's original presentation of special relativity in 1905, many different sets of postulates have been proposed in various alternative derivations.[12] However, the most common set of postulates remains those employed by Einstein in his original paper. A more mathematical statement of the Principle of Relativity made later by Einstein, which introduces the concept of simplicity not mentioned above is:

> *Special principle of relativity*: If a system of coordinates K is chosen so that, in relation to it, physical laws hold good in their simplest form, the *same* laws hold good in relation to any other system of coordinates K' moving in uniform translation relatively to K.[13]

Henri Poincaré provided the mathematical framework for relativity theory by proving that Lorentz transformations are a subset of his Poincaré group of symmetry transformations. Einstein later derived these transformations from his axioms.

Many of Einstein's papers present derivations of the Lorentz transformation based upon these two principles.[14]

Einstein consistently based the derivation of Lorentz invariance (the essential core of special relativity) on just the two basic principles of relativity and light-speed invariance. He wrote:

> The insight fundamental for the special theory of relativity is this: The assumptions relativity and light speed invariance are compatible if relations of a new type ("Lorentz transformation") are postulated for the conversion of coordinates and times of events... The universal principle of the special theory of relativity is contained in the postulate: The laws of physics are invariant with respect to Lorentz transformations (for the transition from one inertial system to any other arbitrarily chosen inertial system). This is a restricting principle for natural laws...[10]

Thus many modern treatments of special relativity base it on the single postulate of universal Lorentz covariance, or, equivalently, on the single postulate of Minkowski spacetime.[15][16]

From the principle of relativity alone without assuming the constancy of the speed of light (i.e. using the isotropy of space and the symmetry implied by the principle of special relativity) one can show that the spacetime transformations between inertial frames are either Euclidean, Galilean, or Lorentzian. In the Lorentzian case, one can then obtain relativistic interval conservation and a certain finite limiting

Albert Einstein around 1905, the year his "Annus Mirabilis papers" – which included Zur Elektrodynamik bewegter Körper, *the paper founding special relativity – were published.*

cially the constancy of the speed of light) from the choice of inertial system. In his initial presentation of special relativity in 1905 he expressed these postulates as:[1]

- The Principle of Relativity – The laws by which the states of physical systems undergo change are not affected, whether these changes of state be referred to the one or the other of two systems in uniform translatory motion relative to each other.[1]

- The Principle of Invariant Light Speed – "... light is always propagated in empty space with a definite velocity [speed] c which is independent of the state of motion of the emitting body" (from the preface).[1] That is, light in vacuum propagates with the speed c (a fixed constant, independent of direction) in at least one system of inertial coordinates (the "stationary system"), regardless of the state of motion of the light source.

The derivation of special relativity depends not only on these two explicit postulates, but also on several tacit assumptions (made in almost all theories of physics), including the isotropy and homogeneity of space and the independence of measuring rods and clocks from their past history.[11]

speed. Experiments suggest that this speed is the speed of light in vacuum.[17][18]

The constancy of the speed of light was motivated by Maxwell's theory of electromagnetism and the lack of evidence for the luminiferous ether. There is conflicting evidence on the extent to which Einstein was influenced by the null result of the Michelson–Morley experiment.[19][20] In any case, the null result of the Michelson–Morley experiment helped the notion of the constancy of the speed of light gain widespread and rapid acceptance.

7.2 Lack of an absolute reference frame

The principle of relativity, which states that there is no preferred inertial reference frame, dates back to Galileo, and was incorporated into Newtonian physics. However, in the late 19th century, the existence of electromagnetic waves led physicists to suggest that the universe was filled with a substance that they called "aether", which would act as the medium through which these waves, or vibrations travelled. The aether was thought to constitute an absolute reference frame against which speeds could be measured, and could be considered fixed and motionless. Aether supposedly possessed some wonderful properties: it was sufficiently elastic to support electromagnetic waves, and those waves could interact with matter, yet it offered no resistance to bodies passing through it. The results of various experiments, including the Michelson–Morley experiment, led to the theory of special relativity, by showing that there was no aether.[21] Einstein's solution was to discard the notion of an aether and the absolute state of rest. In relativity, any reference frame moving with uniform motion will observe the same laws of physics. In particular, the speed of light in vacuum is always measured to be c, even when measured by multiple systems that are moving at different (but constant) velocities.

7.3 Reference frames, coordinates and the Lorentz transformation

Main article: Lorentz transformation

Reference frames play a crucial role in relativity theory. The term reference frame as used here is an observational perspective in space which is not undergoing any change in motion (acceleration), from which a position can be measured along 3 spatial axes. In addition, a reference frame has the ability to determine measurements of the time of events using a 'clock' (any reference device with uniform periodicity).

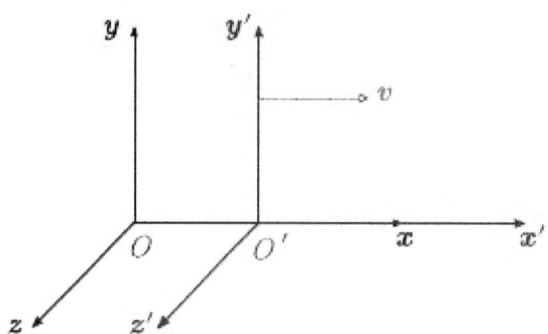

The primed system is in motion relative to the unprimed system with constant velocity v only along the x-axis, from the perspective of an observer stationary in the unprimed system. By the principle of relativity, an observer stationary in the primed system will view a likewise construction except that the velocity they record will be −v. The changing of the speed of propagation of interaction from infinite in non-relativistic mechanics to a finite value will require a modification of the transformation equations mapping events in one frame to another.

An event is an occurrence that can be assigned a single unique time and location in space relative to a reference frame: it is a "point" in spacetime. Since the speed of light is constant in relativity in each and every reference frame, pulses of light can be used to unambiguously measure distances and refer back the times that events occurred to the clock, even though light takes time to reach the clock after the event has transpired.

For example, the explosion of a firecracker may be considered to be an "event". We can completely specify an event by its four spacetime coordinates: The time of occurrence and its 3-dimensional spatial location define a reference point. Let's call this reference frame S.

In relativity theory we often want to calculate the position of a point from a different reference point.

Suppose we have a second reference frame S', whose spatial axes and clock exactly coincide with that of S at time zero, but it is moving at a constant velocity v with respect to S along the x-axis.

Since there is no absolute reference frame in relativity theory, a concept of 'moving' doesn't strictly exist, as everything is always moving with respect to some other reference frame. Instead, any two frames that move at the same speed in the same direction are said to be *comoving*. Therefore, S and S' are not *comoving*.

Define the event to have spacetime coordinates (t,x,y,z) in system S and (t',x',y',z') in S'. Then the Lorentz transformation specifies that these coordinates are related in the following way:

$$t' = \gamma \left(t - vx/c^2\right)$$
$$x' = \gamma \left(x - vt\right)$$
$$y' = y$$
$$z' = z,$$

where

$$\gamma = \frac{1}{\sqrt{1 - \frac{v^2}{c^2}}}$$

is the Lorentz factor and c is the speed of light in vacuum, and the velocity v of S' is parallel to the x-axis. The y and z coordinates are unaffected; only the x and t coordinates are transformed. These Lorentz transformations form a one-parameter group of linear mappings, that parameter being called rapidity.

There is nothing special about the x-axis, the transformation can apply to the y or z axes, or indeed in any direction, which can be done by directions parallel to the motion (which are warped by the γ factor) and perpendicular; see main article for details.

A quantity invariant under Lorentz transformations is known as a Lorentz scalar.

Writing the Lorentz transformation and its inverse in terms of coordinate differences, where for instance one event has coordinates (x_1, t_1) and (x'_1, t'_1), another event has coordinates (x_2, t_2) and (x'_2, t'_2), and the differences are defined as

$$\Delta x' = x'_2 - x'_1, \quad \Delta x = x_2 - x_1,$$
$$\Delta t' = t'_2 - t'_1, \quad \Delta t = t_2 - t_1,$$

we get

$$\Delta x' = \gamma \left(\Delta x - v\,\Delta t\right), \quad \Delta x = \gamma \left(\Delta x' + v\,\Delta t'\right),$$
$$\Delta t' = \gamma \left(\Delta t - \frac{v\,\Delta x}{c^2}\right), \quad \Delta t = \gamma \left(\Delta t' + \frac{v\,\Delta x'}{c^2}\right).$$

These effects are not merely appearances; they are explicitly related to our way of measuring *time intervals* between events which occur at the same place in a given coordinate system (called "co-local" events). These time intervals will be *different* in another coordinate system moving with respect to the first, unless the events are also simultaneous. Similarly, these effects also relate to our measured distances between separated but simultaneous events in a given coordinate system of choice. If these events are not co-local, but are separated by distance (space), they will *not* occur at the

same *spatial distance* from each other when seen from another moving coordinate system. However, the spacetime interval will be the same for all observers. The underlying reality remains the same. Only our perspective changes.

7.4 Consequences derived from the Lorentz transformation

See also: Twin paradox and Relativistic mechanics

The consequences of special relativity can be derived from the Lorentz transformation equations.[22] These transformations, and hence special relativity, lead to different physical predictions than those of Newtonian mechanics when relative velocities become comparable to the speed of light. The speed of light is so much larger than anything humans encounter that some of the effects predicted by relativity are initially counterintuitive.

7.4.1 Relativity of simultaneity

See also: Relativity of simultaneity and Ladder paradox

Two events happening in two different locations that oc-

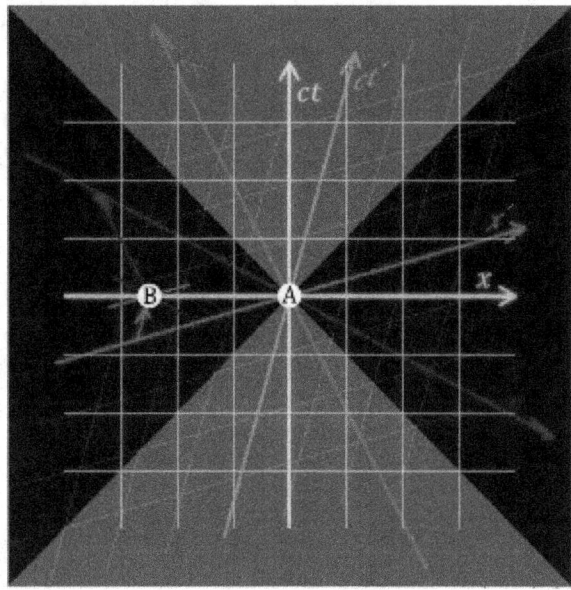

Event B is simultaneous with A in the green reference frame, but it occurs before A in the blue frame, and occurs after A in the red frame.

cur simultaneously in the reference frame of one inertial observer, may occur non-simultaneously in the reference frame of another inertial observer (lack of absolute simultaneity).

From the first equation of the Lorentz transformation in terms of coordinate differences

$$\Delta t' = \gamma \left(\Delta t - \frac{v \Delta x}{c^2} \right)$$

it is clear that two events that are simultaneous in frame S (satisfying $\Delta t = 0$), are not necessarily simultaneous in another inertial frame S' (satisfying $\Delta t' = 0$). Only if these events are additionally co-local in frame S (satisfying $\Delta x = 0$), will they be simultaneous in another frame S'.

7.4.2 Time dilation

See also: Time dilation

The time lapse between two events is not invariant from one observer to another, but is dependent on the relative speeds of the observers' reference frames (e.g., the twin paradox which concerns a twin who flies off in a spaceship traveling near the speed of light and returns to discover that his or her twin sibling has aged much more).

Suppose a clock is at rest in the unprimed system S. The location of the clock on two different ticks is then characterized by $\Delta x = 0$. To find the relation between the times between these ticks as measured in both systems, the first equation can be used to find:

$$\Delta t' = \gamma \, \Delta t \text{ for events satisfying } \Delta x = 0 \;.$$

This shows that the time ($\Delta t'$) between the two ticks as seen in the frame in which the clock is moving (S'), is *longer* than the time (Δt) between these ticks as measured in the rest frame of the clock (S). Time dilation explains a number of physical phenomena; for example, the decay rate of muons produced by cosmic rays impinging on the Earth's atmosphere.[23]

7.4.3 Length contraction

See also: Lorentz contraction

The dimensions (e.g., length) of an object as measured by one observer may be smaller than the results of measurements of the same object made by another observer (e.g., the ladder paradox involves a long ladder traveling near the speed of light and being contained within a smaller garage).

Similarly, suppose a measuring rod is at rest and aligned along the x-axis in the unprimed system S. In this system, the length of this rod is written as Δx. To measure the length

of this rod in the system S', in which the clock is moving, the distances x' to the end points of the rod must be measured simultaneously in that system S'. In other words, the measurement is characterized by $\Delta t' = 0$, which can be combined with the fourth equation to find the relation between the lengths Δx and $\Delta x'$:

$$\Delta x' = \frac{\Delta x}{\gamma} \text{ for events satisfying } \Delta t' = 0 \;.$$

This shows that the length ($\Delta x'$) of the rod as measured in the frame in which it is moving (S'), is *shorter* than its length (Δx) in its own rest frame (S).

7.4.4 Composition of velocities

See also: Velocity-addition formula

Velocities (speeds) do not simply add. If the observer in S measures an object moving along the x axis at velocity u, then the observer in the S' system, a frame of reference moving at velocity v in the x direction with respect to S, will measure the object moving with velocity u' where (from the Lorentz transformations above):

$$u' = \frac{dx'}{dt'} = \frac{\gamma \, (dx - v dt)}{\gamma \, (dt - v dx/c^2)} = \frac{(dx/dt) - v}{1 - (v/c^2)(dx/dt)}$$
$$= \frac{u - v}{1 - uv/c^2} \;.$$

The other frame S will measure:

$$u = \frac{dx}{dt} = \frac{\gamma \, (dx' + v dt')}{\gamma \, (dt' + v dx'/c^2)} = \frac{(dx'/dt') + v}{1 + (v/c^2)(dx'/dt')}$$
$$= \frac{u' + v}{1 + u'v/c^2} \;.$$

Notice that if the object were moving at the speed of light in the S system (i.e. $u = c$), then it would also be moving at the speed of light in the S' system. Also, if both u and v are small with respect to the speed of light, we will recover the intuitive Galilean transformation of velocities

$$u' \approx u - v \;.$$

The usual example given is that of a train (frame S' above) traveling due east with a velocity v with respect to the tracks (frame S). A child inside the train throws a baseball due east with a velocity u' with respect to the train. In nonrelativistic physics, an observer at rest on the tracks will measure the velocity of the baseball (due east) as $u = u' + v$, while in special relativity this is no longer true; instead the velocity of the baseball (due east) is given by the second equation: $u = (u' + v)/(1 + u'v/c^2)$. Again, there is nothing special about the x or east directions. This formalism applies to any direction by considering parallel and perpendicular motion to the direction of relative velocity v. see main article for details.

7.5   Other consequences

7.5.1   Thomas rotation

See also: Thomas rotation

The orientation of an object (i.e. the alignment of its axes with the observer's axes) may be different for different observers. Unlike other relativistic effects, this effect becomes quite significant at fairly low velocities as can be seen in the spin of moving particles.

7.5.2   Equivalence of mass and energy

Main article: Mass–energy equivalence

As an object's speed approaches the speed of light from an observer's point of view, its relativistic mass increases thereby making it more and more difficult to accelerate it from within the observer's frame of reference.

The energy content of an object at rest with mass m equals mc^2. Conservation of energy implies that, in any reaction, a decrease of the sum of the masses of particles must be accompanied by an increase in kinetic energies of the particles after the reaction. Similarly, the mass of an object can be increased by taking in kinetic energies.

In addition to the papers referenced above—which give derivations of the Lorentz transformation and describe the foundations of special relativity—Einstein also wrote at least four papers giving heuristic arguments for the equivalence (and transmutability) of mass and energy, for $E = mc^2$.

Mass–energy equivalence is a consequence of special relativity. The energy and momentum, which are separate in Newtonian mechanics, form a four-vector in relativity, and this relates the time component (the energy) to the space components (the momentum) in a nontrivial way. For an object at rest, the energy–momentum four-vector is $(E, 0, 0, 0)$: it has a time component which is the energy, and three space components which are zero. By changing frames with a Lorentz transformation in the x direction with a small value of the velocity v, the energy momentum four-vector becomes $(E, Ev/c^2, 0, 0)$. The momentum is equal to the energy multiplied by the velocity divided by c^2. As such, the Newtonian mass of an object, which is the ratio of the momentum to the velocity for slow velocities, is equal to E/c^2.

The energy and momentum are properties of matter and radiation, and it is impossible to deduce that they form a four-vector just from the two basic postulates of special relativity by themselves, because these don't talk about matter or radiation, they only talk about space and time. The derivation therefore requires some additional physical reasoning. In his 1905 paper, Einstein used the additional principles that Newtonian mechanics should hold for slow velocities, so that there is one energy scalar and one three-vector momentum at slow velocities, and that the conservation law for energy and momentum is exactly true in relativity. Furthermore, he assumed that the energy of light is transformed by the same Doppler-shift factor as its frequency, which he had previously shown to be true based on Maxwell's equations.[1] The first of Einstein's papers on this subject was "Does the Inertia of a Body Depend upon its Energy Content?" in 1905.[24] Although Einstein's argument in this paper is nearly universally accepted by physicists as correct, even self-evident, many authors over the years have suggested that it is wrong.[25] Other authors suggest that the argument was merely inconclusive because it relied on some implicit assumptions.[26]

Einstein acknowledged the controversy over his derivation in his 1907 survey paper on special relativity. There he notes that it is problematic to rely on Maxwell's equations for the heuristic mass–energy argument. The argument in his 1905 paper can be carried out with the emission of any massless particles, but the Maxwell equations are implicitly used to make it obvious that the emission of light in particular can be achieved only by doing work. To emit electromagnetic waves, all you have to do is shake a charged particle, and this is clearly doing work, so that the emission is of energy.[27][28]

7.5.3   How far can one travel from the Earth?

See also: Space travel using constant acceleration

Since one can not travel faster than light, one might conclude that a human can never travel farther from Earth than 40 light years if the traveler is active between the age of 20 and 60. One would easily think that a traveler would never be able to reach more than the very few solar systems which exist within the limit of 20–40 light years from the earth. But that would be a mistaken conclusion. Because of time dilation, a hypothetical spaceship can travel thousands of light years during the pilot's 40 active years. If a spaceship could be built that accelerates at a constant 1 g, it will after a little less than a year be traveling at almost the speed of light as seen from Earth. Time dilation will increase his life span as seen from the reference system of the Earth, but his lifespan measured by a clock traveling with him will not thereby change. During his journey, people on Earth will experience more time than he does. A 5-year round trip for him will take 6½ Earth years and cover a distance of over 6 light-years. A 20-year round trip for him (5 years accelerating, 5 decelerating, twice each) will land him back on Earth having traveled for 335 Earth years and a distance of 331 light years.[29] A full 40-year trip at 1 *g* will appear on Earth to last 58,000 years and cover a distance of 55,000 light years. A 40-year trip at 1.1 *g* will take 148,000 Earth years and cover about 140,000 light years. A one-way 28 year (14 years accelerating, 14 decelerating as measured with the cosmonaut's clock) trip at 1 *g* acceleration could reach 2,000,000 light-years to the Andromeda Galaxy.[30] This same time dilation is why a muon traveling close to *c* is observed to travel much further than *c* times its half-life (when at rest).[31]

7.6 Causality and prohibition of motion faster than light

See also: Causality (physics) and Tachyonic antitelephone
In diagram 2 the interval AB is 'time-like'; i.e., there is a

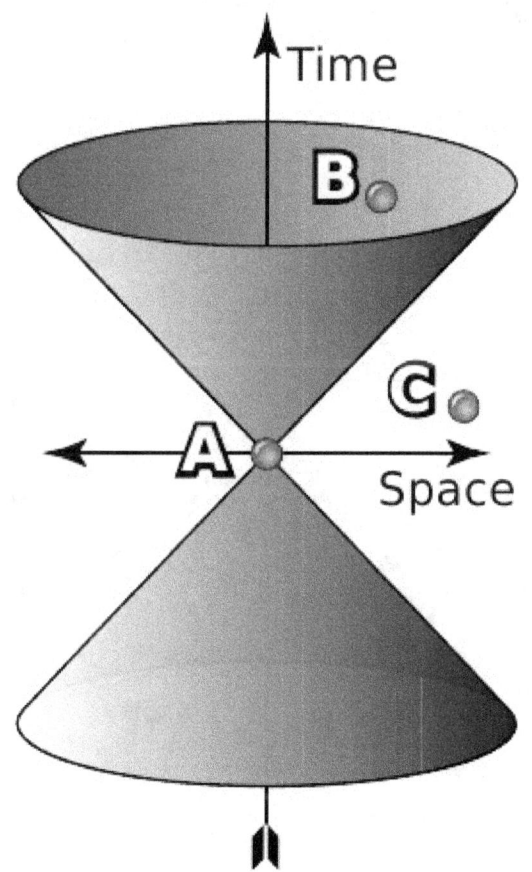

Diagram 2. Light cone

frame of reference in which events A and B occur at the

same location in space, separated only by occurring at different times. If A precedes B in that frame, then A precedes B in all frames. It is hypothetically possible for matter (or information) to travel from A to B, so there can be a causal relationship (with A the cause and B the effect).

The interval AC in the diagram is 'space-like'; i.e., there is a frame of reference in which events A and C occur simultaneously, separated only in space. There are also frames in which A precedes C (as shown) and frames in which C precedes A. If it were possible for a cause-and-effect relationship to exist between events A and C, then paradoxes of causality would result. For example, if A was the cause, and C the effect, then there would be frames of reference in which the effect preceded the cause. Although this in itself won't give rise to a paradox, one can show[32][33] that faster than light signals can be sent back into one's own past. A causal paradox can then be constructed by sending the signal if and only if no signal was received previously.

Therefore, if causality is to be preserved, one of the consequences of special relativity is that no information signal or material object can travel faster than light in vacuum. However, some "things" can still move faster than light. For example, the location where the beam of a search light hits the bottom of a cloud can move faster than light when the search light is turned rapidly.[34]

Even without considerations of causality, there are other strong reasons why faster-than-light travel is forbidden by special relativity. For example, if a constant force is applied to an object for a limitless amount of time, then integrating $F = dp/dt$ gives a momentum that grows without bound, but this is simply because $p = m\gamma v$ approaches infinity as v approaches *c*. To an observer who is not accelerating, it appears as though the object's inertia is increasing, so as to produce a smaller acceleration in response to the same force. This behavior is observed in particle accelerators, where each charged particle is accelerated by the electromagnetic force.

7.7 Geometry of spacetime

Main article: Minkowski space

7.7.1 Comparison between flat Euclidean space and Minkowski space

See also: line element
Special relativity uses a 'flat' 4-dimensional Minkowski space – an example of a spacetime. Minkowski spacetime appears to be very similar to the standard 3-dimensional

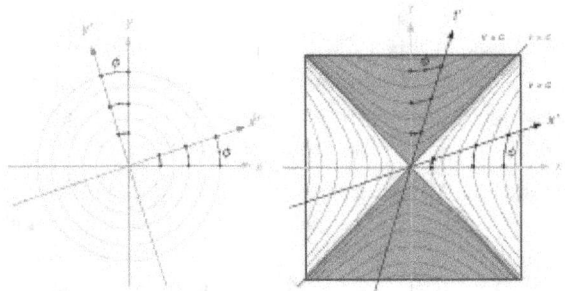

*Orthogonality and rotation of coordinate systems compared between **left:** Euclidean space through circular angle φ, **right:** in Minkowski spacetime through hyperbolic angle φ (red lines labelled c denote the worldlines of a light signal, a vector is orthogonal to itself if it lies on this line).[35]*

Euclidean space, but there is a crucial difference with respect to time.

In 3D space, the differential of distance (line element) ds is defined by

$$ds^2 = d\mathbf{x} \cdot d\mathbf{x} = dx_1^2 + dx_2^2 + dx_3^2,$$

where $d\mathbf{x} = (dx_1, dx_2, dx_3)$ are the differentials of the three spatial dimensions. In Minkowski geometry, there is an extra dimension with coordinate X^0 derived from time, such that the distance differential fulfills

$$ds^2 = -dX_0^2 + dX_1^2 + dX_2^2 + dX_3^2,$$

where $d\mathbf{X} = (dX_0, dX_1, dX_2, dX_3)$ are the differentials of the four spacetime dimensions. This suggests a deep theoretical insight: special relativity is simply a rotational symmetry of our spacetime, analogous to the rotational symmetry of Euclidean space (see image right).[36] Just as Euclidean space uses a Euclidean metric, so spacetime uses a Minkowski metric. Basically, special relativity can be stated as the *invariance of any spacetime interval* (that is the 4D distance between any two events) when viewed from *any inertial reference frame*. All equations and effects of special relativity can be derived from this rotational symmetry (the Poincaré group) of Minkowski spacetime.

The actual form of ds above depends on the metric and on the choices for the X^0 coordinate. To make the time coordinate look like the space coordinates, it can be treated as imaginary: $X_0 = ict$ (this is called a Wick rotation). According to Misner, Thorne and Wheeler (1971, §2.3), ultimately the deeper understanding of both special and general relativity will come from the study of the Minkowski metric (described below) and to take $X^0 = ct$, rather than a "disguised" Euclidean metric using ict as the time coordinate.

Some authors use $X^0 = t$, with factors of c elsewhere to compensate; for instance, spatial coordinates are divided by c or factors of $c^{\pm 2}$ are included in the metric tensor.[37] These numerous conventions can be superseded by using natural units where $c = 1$. Then space and time have equivalent units, and no factors of c appear anywhere.

7.7.2 3D spacetime

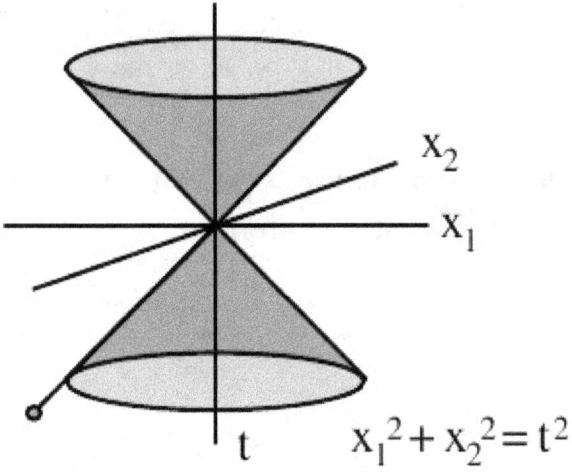

Three-dimensional dual-cone.

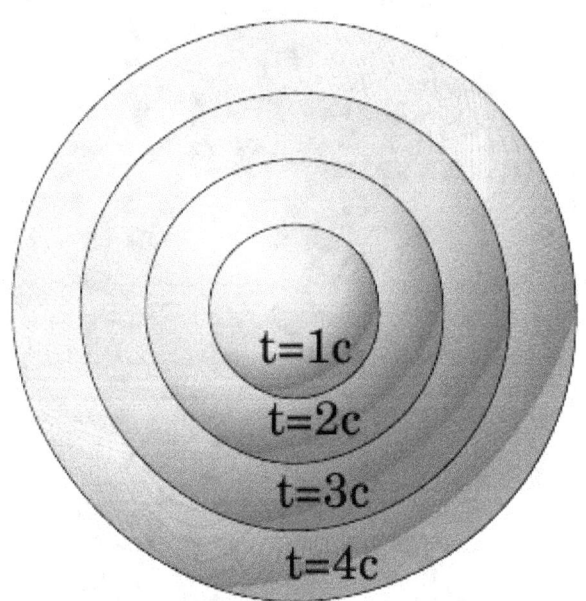

Null spherical space.

If we reduce the spatial dimensions to 2, so that we can represent the physics in a 3D space

$$ds^2 = dx_1^2 + dx_2^2 - c^2 dt^2,$$

we see that the null geodesics lie along a dual-cone (see image right) defined by the equation:

$$ds^2 = 0 = dx_1^2 + dx_2^2 - c^2 dt^2$$

or simply

$$dx_1^2 + dx_2^2 = c^2 dt^2,$$

which is the equation of a circle of radius $c\, dt$.

7.7.3 4D spacetime

If we extend this to three spatial dimensions, the null geodesics are the 4-dimensional cone:

$$ds^2 = 0 = dx_1^2 + dx_2^2 + dx_3^2 - c^2 dt^2$$

so

$$dx_1^2 + dx_2^2 + dx_3^2 = c^2 dt^2.$$

This null dual-cone represents the "line of sight" of a point in space. That is, when we look at the stars and say "The light from that star which I am receiving is X years old", we are looking down this line of sight: a null geodesic. We are looking at an event a distance $d = \sqrt{x_1^2 + x_2^2 + x_3^2}$ away and a time d/c in the past. For this reason the null dual cone is also known as the 'light cone'. (The point in the lower left of the picture above right represents the star, the origin represents the observer, and the line represents the null geodesic "line of sight".)

The cone in the $-t$ region is the information that the point is 'receiving', while the cone in the $+t$ section is the information that the point is 'sending'.

The geometry of Minkowski space can be depicted using Minkowski diagrams, which are useful also in understanding many of the thought-experiments in special relativity.

Note that, in 4d spacetime, the concept of the center of mass becomes more complicated, see center of mass (relativistic).

7.8 Physics in spacetime

7.8.1 Transformations of physical quantities between reference frames

Above, the Lorentz transformation for the time coordinate and three space coordinates illustrates that they are intertwined. This is true more generally: certain pairs of "time-like" and "spacelike" quantities naturally combine on equal footing under the same Lorentz transformation.

The Lorentz transformation in standard configuration above, i.e. for a boost in the x direction, can be recast into matrix form as follows:

$$\begin{pmatrix} ct' \\ x' \\ y' \\ z' \end{pmatrix} = \begin{pmatrix} \gamma & -\beta\gamma & 0 & 0 \\ -\beta\gamma & \gamma & 0 & 0 \\ 0 & 0 & 1 & 0 \\ 0 & 0 & 0 & 1 \end{pmatrix} \begin{pmatrix} ct \\ x \\ y \\ z \end{pmatrix} = \begin{pmatrix} \gamma ct - \gamma\beta x \\ \gamma x - \beta\gamma ct \\ y \\ z \end{pmatrix}.$$

In Newtonian mechanics, quantities which have magnitude and direction are mathematically described as 3d vectors in Euclidean space, and in general they are parametrized by time. In special relativity, this notion is extended by adding the appropriate timelike quantity to a spacelike vector quantity, and we have 4d vectors, or "four vectors", in Minkowski spacetime. The components of vectors are written using tensor index notation, as this has numerous advantages. The notation makes it clear the equations are manifestly covariant under the Poincaré group, thus bypassing the tedious calculations to check this fact. In constructing such equations, we often find that equations previously thought to be unrelated are, in fact, closely connected being part of the same tensor equation. Recognizing other physical quantities as tensors simplifies their transformation laws. Throughout, upper indices (superscripts) are contravariant indices rather than exponents except when they indicate a square (this is should be clear from the context), and lower indices (subscripts) are covariant indices. For simplicity and consistency with the earlier equations, Cartesian coordinates will be used.

The simplest example of a four-vector is the position of an event in spacetime, which constitutes a timelike component ct and spacelike component $\mathbf{x} = (x, y, z)$, in a contravariant position four vector with components:

$$X^\nu = (X^0, X^1, X^2, X^3) = (ct, x, y, z) = (ct, \mathbf{x}).$$

where we define $X^0 = ct$ so that the time coordinate has the same dimension of distance as the other spatial dimensions; so that space and time are treated equally.[38][39][40] Now the transformation of the contravariant components of the position 4-vector can be compactly written as:

$$X^{\mu'} = \Lambda^{\mu'}{}_\nu X^\nu$$

where there is an implied summation on ν from 0 to 3, and $\Lambda^{\mu'}{}_{\nu}$ is a matrix.

More generally, all contravariant components of a four-vector T^{ν} transform from one frame to another frame by a Lorentz transformation:

$$T^{\mu'} = \Lambda^{\mu'}{}_{\nu} T^{\nu}$$

Examples of other 4-vectors include the four-velocity U^{μ}, defined as the derivative of the position 4-vector with respect to proper time:

$$U^{\mu} = \frac{dX^{\mu}}{d\tau} = \gamma(v)(c, v_x, v_y, v_z) = \gamma(v)(c, \mathbf{v}).$$

where the Lorentz factor is:

$$\gamma(v) = \frac{1}{\sqrt{1 - (v/c)^2}}, \quad v^2 = v_x^2 + v_y^2 + v_z^2.$$

The relativistic energy $E = \gamma(v)mc^2$ and relativistic momentum $\mathbf{p} = \gamma(v)m\mathbf{v}$ of an object are respectively the timelike and spacelike components of a contravariant four momentum vector:

$$P^{\mu} = mU^{\mu}$$

$$= m\gamma(v)(c, v_x, v_y, v_z) = (E/c, p_x, p_y, p_z) = (E/c, \mathbf{p}).$$

where m is the invariant mass.

The four-acceleration is the proper time derivative of 4-velocity:

$$A^{\mu} = \frac{dU^{\mu}}{d\tau}.$$

The transformation rules for *three*-dimensional velocities and accelerations are very awkward; even above in standard configuration the velocity equations are quite complicated owing to their non-linearity. On the other hand, the transformation of *four*-velocity and *four*-acceleration are simpler by means of the Lorentz transformation matrix.

The four-gradient of a scalar field φ transforms covariantly rather than contravariantly:

$$\left(\frac{1}{c} \frac{\partial \phi}{\partial t'} \quad \frac{\partial \phi}{\partial x'} \quad \frac{\partial \phi}{\partial y'} \quad \frac{\partial \phi}{\partial z'} \right)$$

$$= \left(\frac{1}{c} \frac{\partial \phi}{\partial t} \quad \frac{\partial \phi}{\partial x} \quad \frac{\partial \phi}{\partial y} \quad \frac{\partial \phi}{\partial z} \right) \begin{pmatrix} \gamma & -\beta\gamma & 0 & 0 \\ -\beta\gamma & \gamma & 0 & 0 \\ 0 & 0 & 1 & 0 \\ 0 & 0 & 0 & 1 \end{pmatrix}.$$

that is:

$$(\partial_{\mu'}\phi) = \Lambda_{\mu'}{}^{\nu}(\partial_{\nu}\phi), \quad \partial_{\mu} \equiv \frac{\partial}{\partial x^{\mu}}.$$

only in Cartesian coordinates. It's the covariant derivative which transforms in manifest covariance, in Cartesian coordinates this happens to reduce to the partial derivatives, but not in other coordinates.

More generally, the *covariant* components of a 4-vector transform according to the *inverse* Lorentz transformation:

$$\Lambda_{\mu'}{}^{\nu} T^{\mu'} = T^{\nu}$$

where $\Lambda_{\mu'}{}^{\nu}$ is the reciprocal matrix of $\Lambda^{\mu'}{}_{\nu}$.

The postulates of special relativity constrain the exact form the Lorentz transformation matrices take.

More generally, most physical quantities are best described as (components of) tensors. So to transform from one frame to another, we use the well-known tensor transformation law[41]

$$T^{\alpha'\beta'\cdots\zeta'}_{\theta'\iota'\cdots\kappa'} = \Lambda^{\alpha'}{}_{\mu}\Lambda^{\beta'}{}_{\nu}\cdots\Lambda^{\zeta'}{}_{\rho}\Lambda_{\theta'}{}^{\sigma}\Lambda_{\iota'}{}^{\upsilon}\cdots\Lambda_{\kappa'}{}^{\phi}T^{\mu\nu\cdots\rho}_{\sigma\upsilon\cdots\phi}$$

where $\Lambda_{\chi'}{}^{\psi}$ is the reciprocal matrix of $\Lambda^{\chi'}{}_{\nu}$. All tensors transform by this rule.

An example of a four dimensional second order antisymmetric tensor is the relativistic angular momentum, which has six components: three are the classical angular momentum, and the other three are related to the boost of the center of mass of the system. The derivative of the relativistic angular momentum with respect to proper time is the relativistic torque, also second order antisymmetric tensor.

The electromagnetic field tensor is another second order antisymmetric tensor field, with six components: three for the electric field and another three for the magnetic field. There is also the stress–energy tensor for the electromagnetic field, namely the electromagnetic stress–energy tensor.

7.8.2   Metric

The metric tensor allows one to define the inner product of two vectors, which in turn allows one to assign a magnitude to the vector. Given the four-dimensional nature of spacetime the Minkowski metric η has components (valid in any inertial reference frame) which can be arranged in a 4 × 4 matrix:

$$\eta_{\alpha\beta} = \begin{pmatrix} -1 & 0 & 0 & 0 \\ 0 & 1 & 0 & 0 \\ 0 & 0 & 1 & 0 \\ 0 & 0 & 0 & 1 \end{pmatrix}$$

which is equal to its reciprocal, $\eta^{\alpha\beta}$, in those frames. Throughout we use the signs as above, different authors use different conventions – see Minkowski metric alternative signs.

The Poincaré group is the most general group of transformations which preserves the Minkowski metric:

$$\eta_{\alpha\beta} = \eta_{\mu'\nu'}\Lambda^{\mu'}{}_{\alpha}\Lambda^{\nu'}{}_{\beta}$$

and this is the physical symmetry underlying special relativity.

The metric can be used for raising and lowering indices on vectors and tensors. Invariants can be constructed using the metric, the inner product of a 4-vector T with another 4-vector S is:

$$T^{\alpha}S_{\alpha} = T^{\alpha}\eta_{\alpha\beta}S^{\beta} = T_{\alpha}\eta^{\alpha\beta}S_{\beta} = \text{scalar invariant}$$

Invariant means that it takes the same value in all inertial frames, because it is a scalar (0 rank tensor), and so no Λ appears in its trivial transformation. The magnitude of the 4-vector T is the positive square root of the inner product with itself:

$$|\mathbf{T}| = \sqrt{T^{\alpha}T_{\alpha}}$$

One can extend this idea to tensors of higher order, for a second order tensor we can form the invariants:

$$T^{\alpha}{}_{\alpha}\,,T^{\alpha}{}_{\beta}T^{\beta}{}_{\alpha}\,,T^{\alpha}{}_{\beta}T^{\beta}{}_{\gamma}T^{\gamma}{}_{\alpha} = \text{scalars invariant}\,,$$

similarly for higher order tensors. Invariant expressions, particularly inner products of 4-vectors with themselves, provide equations that are useful for calculations, because one doesn't need to perform Lorentz transformations to determine the invariants.

7.8.3 Relativistic kinematics and invariance

The coordinate differentials transform also contravariantly:

$$dX^{\mu'} = \Lambda^{\mu'}{}_{\nu}dX^{\nu}$$

so the squared length of the differential of the position four-vector dX^{μ} constructed using

$$dX^2 = dX^{\mu}dX_{\mu} = \eta_{\mu\nu}dX^{\mu}dX^{\nu}$$
$$= -(cdt)^2 + (dx)^2 + (dy)^2 + (dz)^2$$

is an invariant. Notice that when the line element dX^2 is negative that $\sqrt{-dX^2}$ is the differential of proper time, while when dX^2 is positive, $\sqrt{dX^2}$ is differential of the proper distance.

The 4-velocity U^{μ} has an invariant form:

$$\mathbf{U}^2 = \eta_{\nu\mu}U^{\nu}U^{\mu} = -c^2\,,$$

which means all velocity four-vectors have a magnitude of c. This is an expression of the fact that there is no such thing as being at coordinate rest in relativity: at the least, you are always moving forward through time. Differentiating the above equation by τ produces:

$$2\eta_{\mu\nu}A^{\mu}U^{\nu} = 0.$$

So in special relativity, the acceleration four-vector and the velocity four-vector are orthogonal.

7.8.4 Relativistic dynamics and invariance

The invariant magnitude of the momentum 4-vector generates the energy–momentum relation:

$$\mathbf{P}^2 = \eta^{\mu\nu}P_{\mu}P_{\nu} = -(E/c)^2 + p^2.$$

We can work out what this invariant is by first arguing that, since it is a scalar, it doesn't matter in which reference frame we calculate it, and then by transforming to a frame where the total momentum is zero.

$$\mathbf{P}^2 = -(E_{\text{rest}}/c)^2 = -(mc)^2.$$

We see that the rest energy is an independent invariant. A rest energy can be calculated even for particles and systems in motion, by translating to a frame in which momentum is zero.

The rest energy is related to the mass according to the celebrated equation discussed above:

$$E_{\text{rest}} = mc^2.$$

Note that the mass of systems measured in their center of momentum frame (where total momentum is zero) is given by the total energy of the system in this frame. It may not be equal to the sum of individual system masses measured in other frames.

To use Newton's third law of motion, both forces must be defined as the rate of change of momentum with respect to the same time coordinate. That is, it requires the 3D force defined above. Unfortunately, there is no tensor in 4D which contains the components of the 3D force vector among its components.

If a particle is not traveling at c, one can transform the 3D force from the particle's co-moving reference frame into the observer's reference frame. This yields a 4-vector called the four-force. It is the rate of change of the above energy momentum four-vector with respect to proper time. The covariant version of the four-force is:

$$F_\nu = \frac{dP_\nu}{d\tau} = m A_\nu$$

In the rest frame of the object, the time component of the four force is zero unless the "invariant mass" of the object is changing (this requires a non-closed system in which energy/mass is being directly added or removed from the object) in which case it is the negative of that rate of change of mass, times c. In general, though, the components of the four force are not equal to the components of the three-force, because the three force is defined by the rate of change of momentum with respect to coordinate time, i.e. dp/dt while the four force is defined by the rate of change of momentum with respect to proper time, i.e. $dp/d\tau$.

In a continuous medium, the 3D *density of force* combines with the *density of power* to form a covariant 4-vector. The spatial part is the result of dividing the force on a small cell (in 3-space) by the volume of that cell. The time component is $-1/c$ times the power transferred to that cell divided by the volume of the cell. This will be used below in the section on electromagnetism.

7.9   Relativity and unifying electromagnetism

Main articles: Classical electromagnetism and special relativity and Covariant formulation of classical electromagnetism

Theoretical investigation in classical electromagnetism led to the discovery of wave propagation. Equations generalizing the electromagnetic effects found that finite propagation speed of the **E** and **B** fields required certain behaviors on charged particles. The general study of moving charges forms the Liénard–Wiechert potential, which is a step towards special relativity.

The Lorentz transformation of the electric field of a moving charge into a non-moving observer's reference frame results in the appearance of a mathematical term commonly called the magnetic field. Conversely, the *magnetic* field generated by a moving charge disappears and becomes a purely *electrostatic* field in a comoving frame of reference. Maxwell's equations are thus simply an empirical fit to special relativistic effects in a classical model of the Universe. As electric and magnetic fields are reference frame dependent and thus intertwined, one speaks of *electromagnetic* fields. Special relativity provides the transformation rules for how an electromagnetic field in one inertial frame appears in another inertial frame.

Maxwell's equations in the 3D form are already consistent with the physical content of special relativity, although they are easier to manipulate in a manifestly covariant form, i.e. in the language of tensor calculus.[42] See main links for more detail.

7.10   Status

Main articles: Tests of special relativity and Criticism of relativity theory

Special relativity in its Minkowski spacetime is accurate only when the absolute value of the gravitational potential is much less than c^2 in the region of interest.[43] In a strong gravitational field, one must use general relativity. General relativity becomes special relativity at the limit of weak field. At very small scales, such as at the Planck length and below, quantum effects must be taken into consideration resulting in quantum gravity. However, at macroscopic scales and in the absence of strong gravitational fields, special relativity is experimentally tested to extremely high degree of accuracy (10^{-20})[44] and thus accepted by the physics community. Experimental results which appear to contradict it are not reproducible and are thus widely believed to be due to experimental errors.

Special relativity is mathematically self-consistent, and it is an organic part of all modern physical theories, most notably quantum field theory, string theory, and general relativity (in the limiting case of negligible gravitational fields).

Newtonian mechanics mathematically follows from special relativity at small velocities (compared to the speed of light) – thus Newtonian mechanics can be considered as a special relativity of slow moving bodies. See classical mechanics for a more detailed discussion.

Several experiments predating Einstein's 1905 paper are now interpreted as evidence for relativity. Of these it is known Einstein was aware of the Fizeau experiment before 1905,[45] and historians have concluded that Einstein was at least aware of the Michelson–Morley experiment as early as 1899 despite claims he made in his later years that it played no role in his development of the theory.[20]

- The Fizeau experiment (1851, repeated by Michelson and Morley in 1886) measured the speed of light in moving media, with results that are consistent with relativistic addition of colinear velocities.

- The famous Michelson–Morley experiment (1881, 1887) gave further support to the postulate that detecting an absolute reference velocity was not achievable. It should be stated here that, contrary to many alternative claims, it said little about the invariance of the speed of light with respect to the source and observer's velocity, as both source and observer were travelling together at the same velocity at all times.

- The Trouton–Noble experiment (1903) showed that the torque on a capacitor is independent of position and inertial reference frame.

- The Experiments of Rayleigh and Brace (1902, 1904) showed that length contraction doesn't lead to birefringence for a co-moving observer, in accordance with the relativity principle.

Particle accelerators routinely accelerate and measure the properties of particles moving at near the speed of light, where their behavior is completely consistent with relativity theory and inconsistent with the earlier Newtonian mechanics. These machines would simply not work if they were not engineered according to relativistic principles. In addition, a considerable number of modern experiments have been conducted to test special relativity. Some examples:

- Tests of relativistic energy and momentum – testing the limiting speed of particles

- Ives–Stilwell experiment – testing relativistic Doppler effect and time dilation

- Time dilation of moving particles – relativistic effects on a fast-moving particle's half-life

- Kennedy–Thorndike experiment – time dilation in accordance with Lorentz transformations

- Hughes–Drever experiment – testing isotropy of space and mass

- Modern searches for Lorentz violation – various modern tests

- Experiments to test emission theory demonstrated that the speed of light is independent of the speed of the emitter.

- Experiments to test the aether drag hypothesis – no "aether flow obstruction".

7.11 Theories of relativity and quantum mechanics

Special relativity can be combined with quantum mechanics to form relativistic quantum mechanics. It is an unsolved problem in physics how *general* relativity and quantum mechanics can be unified; quantum gravity and a "theory of everything", which require such a unification, are active and ongoing areas in theoretical research.

The early Bohr–Sommerfeld atomic model explained the fine structure of alkali metal atoms using both special relativity and the preliminary knowledge on quantum mechanics of the time.[46]

In 1928, Paul Dirac constructed an influential relativistic wave equation, now known as the Dirac equation in his honour,[47] that is fully compatible both with special relativity and with the final version of quantum theory existing after 1926. This equation explained not only the intrinsic angular momentum of the electrons called *spin*, it also led to the prediction of the antiparticle of the electron (the positron),[47][48] and fine structure could only be fully explained with special relativity. It was the first foundation of *relativistic quantum mechanics*. In non-relativistic quantum mechanics, spin is phenomenological and cannot be explained.

On the other hand, the existence of antiparticles leads to the conclusion that relativistic quantum mechanics is not enough for a more accurate and complete theory of particle interactions. Instead, a theory of particles interpreted as quantized fields, called *quantum field theory*, becomes necessary; in which particles can be created and destroyed throughout space and time.

7.12 See also

People: Hendrik Lorentz | Henri Poincaré | Albert Einstein | Max Planck | Hermann Minkowski | Max von Laue | Arnold Sommerfeld | Max Born | Gustav Herglotz | Richard C. Tolman

Relativity: Theory of relativity | History of special relativity | Principle of relativity | General relativity | Frame of reference | Inertial frame of reference | Lorentz transformations | Bondi k-calculus | Einstein synchronisation | Rietdijk–Putnam argument | Special relativity (alternative formulations) | Criticism of relativity theory | Relativity priority dispute

Physics: Newtonian Mechanics | spacetime | speed of light | simultaneity | center of mass (relativistic) | physical cosmology | Doppler effect | relativistic Euler equations | Aether drag hypothesis | Lorentz ether theory | Moving magnet and conductor problem | Shape waves | Relativistic heat conduction | Relativistic disk | Thomas precession | Born rigidity | Born coordinates

Mathematics: Derivations of the Lorentz transformations | Minkowski space | four-vector | world line | light cone | Lorentz group | Poincaré group | geometry | tensors | split-complex number | Relativity in the APS formalism

Philosophy: actualism | conventionalism | formalism

Paradoxes: Twin paradox | Ehrenfest paradox | Ladder paradox | Bell's spaceship paradox | Velocity composition paradox

7.13　References

[1] Albert Einstein (1905) *"Zur Elektrodynamik bewegter Körper"*, *Annalen der Physik* 17: 891; English translation On the Electrodynamics of Moving Bodies by George Barker Jeffery and Wilfrid Perrett (1923); Another English translation On the Electrodynamics of Moving Bodies by Megh Nad Saha (1920).

[2] Tom Roberts and Siegmar Schleif (October 2007). "What is the experimental basis of Special Relativity?". *Usenet Physics FAQ*. Retrieved 2008-09-17.

[3] Albert Einstein (2001). *Relativity: The Special and the General Theory* (Reprint of 1920 translation by Robert W. Lawson ed.). Routledge. p. 48. ISBN 0-415-25384-5.

[4] Richard Phillips Feynman (1998). *Six Not-so-easy Pieces: Einstein's relativity, symmetry, and space–time* (Reprint of 1995 ed.). Basic Books. p. 68. ISBN 0-201-32842-9.

[5] Sean Carroll, Lecture Notes on General Relativity, ch. 1, "Special relativity and flat spacetime," http://ned.ipac.caltech.edu/level5/March01/Carroll3/Carroll1.html

[6] Wald, General Relativity, p. 60: "...the special theory of relativity asserts that spacetime is the manifold $\mathbb{R}^4$ with a flat metric of Lorentz signature defined on it. Conversely, the entire content of special relativity ... is contained in this statement ..."

[7] Rindler, W., 1969, Essential Relativity: Special, General, and Cosmological

[8] Edwin F. Taylor and John Archibald Wheeler (1992). *Spacetime Physics: Introduction to Special Relativity*. W. H. Freeman. ISBN 0-7167-2327-1.

[9] Wolfgang Rindler (1977). *Essential Relativity*. Birkhäuser. p. §1.11 p. 7. ISBN 3-540-07970-X.

[10] Einstein, Autobiographical Notes, 1949.

[11] Einstein, "Fundamental Ideas and Methods of the Theory of Relativity", 1920

[12] For a survey of such derivations, see Lucas and Hodgson, Spacetime and Electromagnetism, 1990

[13] Einstein, A., Lorentz, H. A., Minkowski, H., & Weyl, H. (1952). *The Principle of Relativity: a collection of original memoirs on the special and general theory of relativity.* Courier Dover Publications. p. 111. ISBN 0-486-60081-5.

[14] Einstein, On the Relativity Principle and the Conclusions Drawn from It, 1907; "The Principle of Relativity and Its Consequences in Modern Physics", 1910; "The Theory of Relativity", 1911; Manuscript on the Special Theory of Relativity, 1912; Theory of Relativity, 1913; Einstein, Relativity, the Special and General Theory, 1916; The Principle Ideas of the Theory of Relativity, 1916; What Is The Theory of Relativity?, 1919; The Principle of Relativity (Princeton Lectures), 1921; Physics and Reality, 1936; The Theory of Relativity, 1949.

[15] Das, A. (1993) *The Special Theory of Relativity, A Mathematical Exposition*, Springer, ISBN 0-387-94042-1.

[16] Schutz, J. (1997) Independent Axioms for Minkowski Spacetime, Addison Wesley Longman Limited, ISBN 0-582-31760-6.

[17] Yaakov Friedman (2004). *Physical Applications of Homogeneous Balls*. Progress in Mathematical Physics **40**. pp. 1–21. ISBN 0-8176-3339-1.

[18] David Morin (2007) *Introduction to Classical Mechanics*, Cambridge University Press, Cambridge, chapter 11, Appendix I. ISBN 1-139-46837-5.

[19] Michael Polanyi (1974) *Personal Knowledge: Towards a Post-Critical Philosophy*, ISBN 0-226-67288-3, footnote page 10–11: Einstein reports, via Dr N Balzas in response to Polanyi's query, that "The Michelson–Morley experiment had no role in the foundation of the theory." and "..the theory of relativity was not founded to explain its outcome at all."

[20] Jeroen van Dongen (2009). "On the role of the Michelson–Morley experiment: Einstein in Chicago" (PDF). *Eprint arXiv:0908.1545* **0908**: 1545. arXiv:0908.1545. Bibcode:2009arXiv0908.1545V.

[21] Staley, Richard (2009). "Albert Michelson, the Velocity of Light, and the Ether Drift", *Einstein's generation. The origins of the relativity revolution*, Chicago: University of Chicago Press. ISBN 0-226-77057-5

[22] Robert Resnick (1968). *Introduction to special relativity*. Wiley. pp. 62–63.

[23] Daniel Kleppner and David Kolenkow (1973). *An Introduction to Mechanics*. pp. 468–70. ISBN 0-07-035048-5.

[24] Does the inertia of a body depend upon its energy content? A. Einstein, *Annalen der Physik*. **18**:639, 1905 (English translation by W. Perrett and G.B. Jeffery)

[25] Max Jammer (1997). *Concepts of Mass in Classical and Modern Physics*. Courier Dover Publications. pp. 177–178. ISBN 0-486-29998-8.

[26] John J. Stachel (2002). *Einstein from B to Z*. Springer. p. 221. ISBN 0-8176-4143-2.

[27] *On the Inertia of Energy Required by the Relativity Principle*, A. Einstein, Annalen der Physik 23 (1907): 371–384

[28] In a letter to Carl Seelig in 1955, Einstein wrote "I had already previously found that Maxwell's theory did not account for the micro-structure of radiation and could therefore have no general validity.", Einstein letter to Carl Seelig, 1955.

[29] Philip Gibbs and Don Koks. "The Relativistic Rocket". Retrieved 30 August 2012.

[30] Philip Gibbs and Don Koks. "The Relativistic Rocket". Retrieved 13 October 2013.

[31] The special theory of relativity shows that time and space are affected by motion. Library.thinkquest.org. Retrieved on 2013-04-24.

[32] R. C. Tolman, *The theory of the Relativity of Motion*. (Berkeley 1917), p. 54

[33] G. A. Benford, D. L. Book, and W. A. Newcomb (1970). "The Tachyonic Antitelephone". *Physical Review D* **2** (2): 263. Bibcode:1970PhRvD...2..263B. doi:10.1103/PhysRevD.2.263.

[34] Wesley C. Salmon (2006). *Four Decades of Scientific Explanation*. University of Pittsburgh. p. 107. ISBN 0-8229-5926-7., Section 3.7 page 107

[35] J.A. Wheeler, C. Misner, K.S. Thorne (1973). *Gravitation*. W.H. Freeman & Co. p. 58. ISBN 0-7167-0344-0.

[36] J.R. Forshaw, A.G. Smith (2009). *Dynamics and Relativity*. Wiley. p. 247. ISBN 978-0-470-01460-8.

[37] R. Penrose (2007). *The Road to Reality*. Vintage books. ISBN 0-679-77631-1.

[38] Jean-Bernard Zuber & Claude Itzykson, *Quantum Field Theory*, pg 5, ISBN 0-07-032071-3

[39] Charles W. Misner, Kip S. Thorne & John A. Wheeler, *Gravitation*, pg 51, ISBN 0-7167-0344-0

[40] George Sterman, *An Introduction to Quantum Field Theory*, pg 4 , ISBN 0-521-31132-2

[41] Sean M. Carroll (2004). *Spacetime and Geometry: An Introduction to General Relativity*. Addison Wesley. p. 22. ISBN 0-8053-8732-3.

[42] E. J. Post (1962). *Formal Structure of Electromagnetics: General Covariance and Electromagnetics*. Dover Publications Inc. ISBN 0-486-65427-3.

[43] Øyvind Grøn and Sigbjørn Hervik (2007). *Einstein's general theory of relativity: with modern applications in cosmology*. Springer. p. 195. ISBN 0-387-69199-5., Extract of page 195 (with units where c=1)

[44] The number of works is vast, see as example: Sidney Coleman, Sheldon L. Glashow (1997). "Cosmic Ray and Neutrino Tests of Special Relativity". *Phys. Lett.* **B405** (3–4): 249–252. arXiv:hep-ph/9703240. Bibcode:1997PhLB..405..249C. doi:10.1016/S0370-2693(97)00638-2. An overview can be found on this page

[45] John D. Norton, John D. (2004). "Einstein's Investigations of Galilean Covariant Electrodynamics prior to 1905". *Archive for History of Exact Sciences* **59**: 45–105. Bibcode:2004AHES...59...45N. doi:10.1007/s00407-004-0085-6.

[46] R. Resnick, R. Eisberg (1985). *Quantum Physics of Atoms, Molecules, Solids, Nuclei and Particles* (2nd ed.). John Wiley & Sons. pp. 114–116. ISBN 978-0-471-87373-0.

[47] P.A.M. Dirac (1930). "A Theory of Electrons and Protons". *Proceedings of the Royal Society* **A126** (801): 360. Bibcode:1930RSPSA.126..360D. doi:10.1098/rspa.1930.0013. JSTOR 95359.

[48] C.D. Anderson (1933). "The Positive Electron". *Phys. Rev.* **43** (6): 491–494. Bibcode:1933PhRv...43..491A. doi:10.1103/PhysRev.43.491.

7.13.1   Textbooks

- Einstein, Albert (1920). Relativity: The Special and General Theory.

- Einstein, Albert (1996). *The Meaning of Relativity*. Fine Communications. ISBN 1-56731-136-9

- Logunov, Anatoly A. (2005) Henri Poincaré and the Relativity Theory (transl. from Russian by G. Pontocorvo and V. O. Soleviev, edited by V. A. Petrov) Nauka, Moscow.

- Charles Misner, Kip Thorne, and John Archibald Wheeler (1971) *Gravitation*. W. H. Freeman & Co. ISBN 0-7167-0334-3

- Post, E.J., 1997 (1962) *Formal Structure of Electromagnetics: General Covariance and Electromagnetics*. Dover Publications.

- Wolfgang Rindler (1991). Introduction to Special Relativity (2nd ed.), Oxford University Press. ISBN 978-0-19-853952-0; ISBN 0-19-853952-5

- Harvey R. Brown (2005). Physical relativity: space–time structure from a dynamical perspective, Oxford University Press, ISBN 0-19-927583-1; ISBN 978-0-19-927583-0

- Qadir, Asghar (1989). *Relativity: An Introduction to the Special Theory*. Singapore: World Scientific Publications. p. 128. ISBN 9971-5-0612-2.

- Silberstein, Ludwik (1914) The Theory of Relativity.

- Lawrence Sklar (1977). *Space, Time and Spacetime*. University of California Press. ISBN 0-520-03174-1.

- Lawrence Sklar (1992). *Philosophy of Physics*. Westview Press. ISBN 0-8133-0625-6.

- Taylor, Edwin, and John Archibald Wheeler (1992) *Spacetime Physics* (2nd ed.). W.H. Freeman & Co. ISBN 0-7167-2327-1

- Tipler, Paul, and Llewellyn, Ralph (2002). *Modern Physics* (4th ed.). W. H. Freeman & Co. ISBN 0-7167-4345-0

7.13.2 Journal articles

- Alvager; Farley, F. J. M.; Kjellman, J.; Wallin, L.; et al. (1964). "Test of the Second Postulate of Special Relativity in the GeV region". *Physics Letters* **12** (3): 260. Bibcode:1964PhL.....12..260A. doi:10.1016/0031-9163(64)91095-9.

- Darrigol, Olivier (2004). "The Mystery of the Poincaré–Einstein Connection". *Isis* **95** (4): 614–26. doi:10.1086/430652. PMID 16011297.

- Wolf, Peter; Petit, Gerard (1997). "Satellite test of Special Relativity using the Global Positioning System". *Physical Review A* **56** (6): 4405–09. Bibcode:1997PhRvA..56.4405W. doi:10.1103/PhysRevA.56.4405.

- Special Relativity Scholarpedia

7.14   External links

7.14.1   Original works

- *Zur Elektrodynamik bewegter Körper* Einstein's original work in German, Annalen der Physik, Bern 1905

- *On the Electrodynamics of Moving Bodies* English Translation as published in the 1923 book *The Principle of Relativity*.

7.14.2   Special relativity for a general audience (no mathematical knowledge required)

- Einstein Light An award-winning, non-technical introduction (film clips and demonstrations) supported by dozens of pages of further explanations and animations, at levels with or without mathematics.

- Einstein Online Introduction to relativity theory, from the Max Planck Institute for Gravitational Physics.

- Audio: Cain/Gay (2006) – Astronomy Cast. Einstein's Theory of Special Relativity

7.14.3   Special relativity explained (using simple or more advanced mathematics)

- Greg Egan's *Foundations*.

- The Hogg Notes on Special Relativity A good introduction to special relativity at the undergraduate level, using calculus.

- Relativity Calculator: Special Relativity – An algebraic and integral calculus derivation for $E = mc^2$.

- MathPages – Reflections on Relativity A complete online book on relativity with an extensive bibliography.

- Relativity An introduction to special relativity at the undergraduate level, without calculus.

-

- *Relativity: the Special and General Theory* at Project Gutenberg, by Albert Einstein

- Special Relativity Lecture Notes is a standard introduction to special relativity containing illustrative explanations based on drawings and spacetime diagrams from Virginia Polytechnic Institute and State University.

- Understanding Special Relativity The theory of special relativity in an easily understandable way.

- An Introduction to the Special Theory of Relativity (1964) by Robert Katz, "an introduction ... that is accessible to any student who has had an introduction to general physics and some slight acquaintance with the calculus" (130 pp; pdf format).

- Lecture Notes on Special Relativity by J D Cresser Department of Physics Macquarie University.

- SpecialRelativity.net - An overview with visualizations and minimal mathematics.

7.14.4 Visualization

- Raytracing Special Relativity Software visualizing several scenarios under the influence of special relativity.

- Real Time Relativity The Australian National University. Relativistic visual effects experienced through an interactive program.

- Spacetime travel A variety of visualizations of relativistic effects, from relativistic motion to black holes.

- Through Einstein's Eyes The Australian National University. Relativistic visual effects explained with movies and images.

- Warp Special Relativity Simulator A computer program to show the effects of traveling close to the speed of light.

- Animation clip on YouTube visualizing the Lorentz transformation.

- Original interactive FLASH Animations from John de Pillis illustrating Lorentz and Galilean frames, Train and Tunnel Paradox, the Twin Paradox, Wave Propagation, Clock Synchronization, etc.

- Relativistic Optics at the ANU

- lightspeed An OpenGL-based program developed to illustrate the effects of special relativity on the appearance of moving objects.

- Animation showing the stars near Earth, as seen from a spacecraft accelerating rapidly to light speed.

Chapter 8

General relativity

For the book by Robert Wald, see General Relativity (book).

For a more accessible and less technical introduction to this topic, see Introduction to general relativity.

General relativity, also known as the **general theory of**

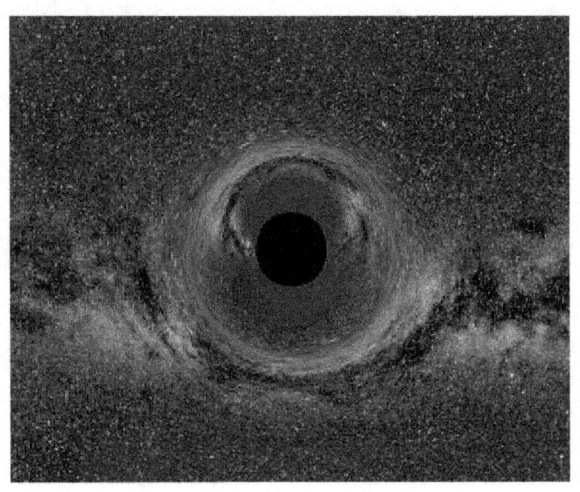

A simulated black hole of 10 solar masses within the Milky Way, seen from a distance of 600 kilometers.

relativity, is the geometric theory of gravitation published by Albert Einstein in 1915[1] and the current description of gravitation in modern physics. General relativity generalizes special relativity and Newton's law of universal gravitation, providing a unified description of gravity as a geometric property of space and time, or spacetime. In particular, the curvature of spacetime is directly related to the energy and momentum of whatever matter and radiation are present. The relation is specified by the Einstein field equations, a system of partial differential equations.

Some predictions of general relativity differ significantly from those of classical physics, especially concerning the passage of time, the geometry of space, the motion of bodies in free fall, and the propagation of light. Examples of such differences include gravitational time dilation, gravitational lensing, the gravitational redshift of light, and the gravitational time delay. The predictions of general relativity have been confirmed in all observations and experiments to date. Although general relativity is not the only relativistic theory of gravity, it is the simplest theory that is consistent with experimental data. However, unanswered questions remain, the most fundamental being how general relativity can be reconciled with the laws of quantum physics to produce a complete and self-consistent theory of quantum gravity.

Einstein's theory has important astrophysical implications. For example, it implies the existence of black holes—regions of space in which space and time are distorted in such a way that nothing, not even light, can escape—as an end-state for massive stars. There is ample evidence that the intense radiation emitted by certain kinds of astronomical objects is due to black holes; for example, microquasars and active galactic nuclei result from the presence of stellar black holes and black holes of a much more massive type, respectively. The bending of light by gravity can lead to the phenomenon of gravitational lensing, in which multiple images of the same distant astronomical object are visible in the sky. General relativity also predicts the existence of gravitational waves, which have since been observed indirectly; a direct measurement is the aim of projects such as LIGO and NASA/ESA Laser Interferometer Space Antenna and various pulsar timing arrays. In addition, general relativity is the basis of current cosmological models of a consistently expanding universe.

8.1 History

Main articles: History of general relativity and Classical theories of gravitation

Soon after publishing the special theory of relativity in 1905, Einstein started thinking about how to incorporate gravity into his new relativistic framework. In 1907, beginning with a simple thought experiment involving an observer in free fall, he embarked on what would be an eight-year search for a relativistic theory of gravity. After nu-

Albert Einstein developed the theories of special and general relativity. Picture from 1921.

merous detours and false starts, his work culminated in the presentation to the Prussian Academy of Science in November 1915 of what are now known as the Einstein field equations. These equations specify how the geometry of space and time is influenced by whatever matter and radiation are present, and form the core of Einstein's general theory of relativity.[2]

The Einstein field equations are nonlinear and very difficult to solve. Einstein used approximation methods in working out initial predictions of the theory. But as early as 1916, the astrophysicist Karl Schwarzschild found the first non-trivial exact solution to the Einstein field equations, the so-called Schwarzschild metric. This solution laid the groundwork for the description of the final stages of gravitational collapse, and the objects known today as black holes. In the same year, the first steps towards generalizing Schwarzschild's solution to electrically charged objects were taken, which eventually resulted in the Reissner–Nordström solution, now associated with electrically charged black holes.[3] In 1917, Einstein applied his theory to the universe as a whole, initiating the field of relativistic cosmology. In line with contemporary thinking, he assumed a static universe, adding a new parameter to his original field equations—the cosmological constant—

to match that observational presumption.[4] By 1929, however, the work of Hubble and others had shown that our universe is expanding. This is readily described by the expanding cosmological solutions found by Friedmann in 1922, which do not require a cosmological constant. Lemaître used these solutions to formulate the earliest version of the Big Bang models, in which our universe has evolved from an extremely hot and dense earlier state.[5] Einstein later declared the cosmological constant the biggest blunder of his life.[6]

During that period, general relativity remained something of a curiosity among physical theories. It was clearly superior to Newtonian gravity, being consistent with special relativity and accounting for several effects unexplained by the Newtonian theory. Einstein himself had shown in 1915 how his theory explained the anomalous perihelion advance of the planet Mercury without any arbitrary parameters ("fudge factors").[7] Similarly, a 1919 expedition led by Eddington confirmed general relativity's prediction for the deflection of starlight by the Sun during the total solar eclipse of May 29, 1919,[8] making Einstein instantly famous.[9] Yet the theory entered the mainstream of theoretical physics and astrophysics only with the developments between approximately 1960 and 1975, now known as the golden age of general relativity.[10] Physicists began to understand the concept of a black hole, and to identify quasars as one of these objects' astrophysical manifestations.[11] Ever more precise solar system tests confirmed the theory's predictive power,[12] and relativistic cosmology, too, became amenable to direct observational tests.[13]

8.2 From classical mechanics to general relativity

General relativity can be understood by examining its similarities with and departures from classical physics. The first step is the realization that classical mechanics and Newton's law of gravity admit a geometric description. The combination of this description with the laws of special relativity results in a heuristic derivation of general relativity.[14]

8.2.1 Geometry of Newtonian gravity

At the base of classical mechanics is the notion that a body's motion can be described as a combination of free (or inertial) motion, and deviations from this free motion. Such deviations are caused by external forces acting on a body in accordance with Newton's second law of motion, which states that the net force acting on a body is equal to that body's (inertial) mass multiplied by its acceleration.[15]

*According to general relativity, objects in a gravitational field be-
have similarly to objects within an accelerating enclosure. For example, an observer will see a ball fall the same way in a rocket
(left) as it does on Earth (right), provided that the acceleration of
the rocket is equal to 9.8 m/s² (the acceleration due to gravity at the
surface of the Earth).*

The preferred inertial motions are related to the geometry
of space and time: in the standard reference frames of classical mechanics, objects in free motion move along straight
lines at constant speed. In modern parlance, their paths are
geodesics, straight world lines in curved spacetime.[16]

Conversely, one might expect that inertial motions, once
identified by observing the actual motions of bodies
and making allowances for the external forces (such as
electromagnetism or friction), can be used to define the geometry of space, as well as a time coordinate. However,
there is an ambiguity once gravity comes into play. According to Newton's law of gravity, and independently verified
by experiments such as that of Eötvös and its successors
(see Eötvös experiment), there is a universality of free fall
(also known as the weak equivalence principle, or the universal equality of inertial and passive-gravitational mass):
the trajectory of a test body in free fall depends only on
its position and initial speed, but not on any of its material
properties.[17] A simplified version of this is embodied in
Einstein's elevator experiment, illustrated in the figure on
the right: for an observer in a small enclosed room, it is
impossible to decide, by mapping the trajectory of bodies
such as a dropped ball, whether the room is at rest in a gravitational field, or in free space aboard a rocket that is accelerating at a rate equal to that of the gravitational field.[18]

Given the universality of free fall, there is no observable
distinction between inertial motion and motion under the
influence of the gravitational force. This suggests the definition of a new class of inertial motion, namely that of objects in free fall under the influence of gravity. This new
class of preferred motions, too, defines a geometry of space
and time—in mathematical terms, it is the geodesic motion
associated with a specific connection which depends on the

gradient of the gravitational potential. Space, in this construction, still has the ordinary Euclidean geometry. However, space*time* as a whole is more complicated. As can be
shown using simple thought experiments following the free-
fall trajectories of different test particles, the result of trans-
porting spacetime vectors that can denote a particle's velocity (time-like vectors) will vary with the particle's trajectory; mathematically speaking, the Newtonian connection
is not integrable. From this, one can deduce that spacetime
is curved. The result is a geometric formulation of Newtonian gravity using only covariant concepts, i.e. a description which is valid in any desired coordinate system.[19] In
this geometric description, tidal effects—the relative acceleration of bodies in free fall—are related to the derivative
of the connection, showing how the modified geometry is
caused by the presence of mass.[20]

8.2.2   Relativistic generalization

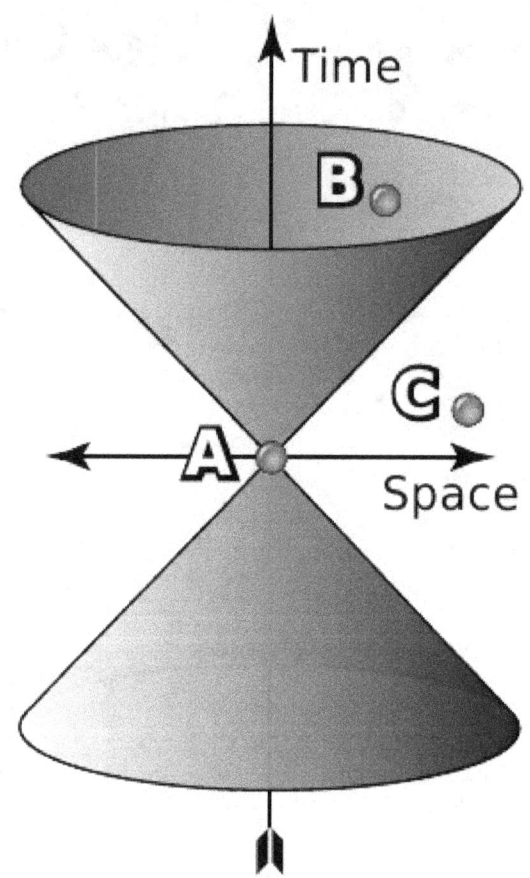

Light cone

As intriguing as geometric Newtonian gravity may be,
its basis, classical mechanics, is merely a limiting case
of (special) relativistic mechanics.[21] In the language of

symmetry: where gravity can be neglected, physics is Lorentz invariant as in special relativity rather than Galilei invariant as in classical mechanics. (The defining symmetry of special relativity is the Poincaré group, which includes translations and rotations.) The differences between the two become significant when dealing with speeds approaching the speed of light, and with high-energy phenomena.[22]

With Lorentz symmetry, additional structures come into play. They are defined by the set of light cones (see image). The light-cones define a causal structure: for each event A, there is a set of events that can, in principle, either influence or be influenced by A via signals or interactions that do not need to travel faster than light (such as event B in the image), and a set of events for which such an influence is impossible (such as event C in the image). These sets are observer-independent.[23] In conjunction with the world-lines of freely falling particles, the light-cones can be used to reconstruct the space–time's semi-Riemannian metric, at least up to a positive scalar factor. In mathematical terms, this defines a conformal structure,[24] or much better, a conformal geometry, as it is difficult to understand how space or time or space-time can have a structure.

Special relativity is defined in the absence of gravity, so for practical applications, it is a suitable model whenever gravity can be neglected. Bringing gravity into play, and assuming the universality of free fall, an analogous reasoning as in the previous section applies: there are no global inertial frames. Instead there are approximate inertial frames moving alongside freely falling particles. Translated into the language of spacetime: the straight time-like lines that define a gravity-free inertial frame are deformed to lines that are curved relative to each other, suggesting that the inclusion of gravity necessitates a change in spacetime geometry.[25]

A priori, it is not clear whether the new local frames in free fall coincide with the reference frames in which the laws of special relativity hold—that theory is based on the propagation of light, and thus on electromagnetism, which could have a different set of preferred frames. But using different assumptions about the special-relativistic frames (such as their being earth-fixed, or in free fall), one can derive different predictions for the gravitational redshift, that is, the way in which the frequency of light shifts as the light propagates through a gravitational field (cf. below). The actual measurements show that free-falling frames are the ones in which light propagates as it does in special relativity.[26] The generalization of this statement, namely that the laws of special relativity hold to good approximation in freely falling (and non-rotating) reference frames, is known as the Einstein equivalence principle, a crucial guiding principle for generalizing special-relativistic physics to include gravity.[27]

The same experimental data shows that time as measured by clocks in a gravitational field—proper time, to give the technical term—does not follow the rules of special relativity. In the language of spacetime geometry, it is not measured by the Minkowski metric. As in the Newtonian case, this is suggestive of a more general geometry. At small scales, all reference frames that are in free fall are equivalent, and approximately Minkowskian. Consequently, we are now dealing with a curved generalization of Minkowski space. The metric tensor that defines the geometry—in particular, how lengths and angles are measured—is not the Minkowski metric of special relativity, it is a generalization known as a semi- or pseudo-Riemannian metric. Furthermore, each Riemannian metric is naturally associated with one particular kind of connection, the Levi-Civita connection, and this is, in fact, the connection that satisfies the equivalence principle and makes space locally Minkowskian (that is, in suitable locally inertial coordinates, the metric is Minkowskian, and its first partial derivatives and the connection coefficients vanish).[28]

8.2.3 Einstein's equations

Main articles: Einstein field equations and Mathematics of general relativity

Having formulated the relativistic, geometric version of the effects of gravity, the question of gravity's source remains. In Newtonian gravity, the source is mass. In special relativity, mass turns out to be part of a more general quantity called the energy–momentum tensor, which includes both energy and momentum densities as well as stress (that is, pressure and shear).[29] Using the equivalence principle, this tensor is readily generalized to curved space-time. Drawing further upon the analogy with geometric Newtonian gravity, it is natural to assume that the field equation for gravity relates this tensor and the Ricci tensor, which describes a particular class of tidal effects: the change in volume for a small cloud of test particles that are initially at rest, and then fall freely. In special relativity, conservation of energy–momentum corresponds to the statement that the energy–momentum tensor is divergence-free. This formula, too, is readily generalized to curved spacetime by replacing partial derivatives with their curved-manifold counterparts, covariant derivatives studied in differential geometry. With this additional condition—the covariant divergence of the energy–momentum tensor, and hence of whatever is on the other side of the equation, is zero— the simplest set of equations are what are called Einstein's (field) equations:

On the left-hand side is the Einstein tensor, a specific divergence-free combination of the Ricci tensor $R_{\mu\nu}$ and the metric. Where $G_{\mu\nu}$ is symmetric. In particular,

$$R = g^{\mu\nu} R_{\mu\nu}$$

is the curvature scalar. The Ricci tensor itself is related to the more general Riemann curvature tensor as

$$R_{\mu\nu} = R^{\alpha}{}_{\mu\alpha\nu}.$$

On the right-hand side, $T_{\mu\nu}$ is the energy–momentum tensor. All tensors are written in abstract index notation.[30] Matching the theory's prediction to observational results for planetary orbits (or, equivalently, assuring that the weak-gravity, low-speed limit is Newtonian mechanics), the proportionality constant can be fixed as $\kappa = 8\pi G/c^4$, with G the gravitational constant and c the speed of light.[31] When there is no matter present, so that the energy–momentum tensor vanishes, the results are the vacuum Einstein equations,

$$R_{\mu\nu} = 0.$$

There are alternatives to general relativity built upon the same premises, which include additional rules and/or constraints, leading to different field equations. Examples are Brans–Dicke theory, teleparallelism, and Einstein–Cartan theory.[32]

8.3   Definition and basic applications

See also: Mathematics of general relativity and Physical theories modified by general relativity

The derivation outlined in the previous section contains all the information needed to define general relativity, describe its key properties, and address a question of crucial importance in physics, namely how the theory can be used for model-building.

8.3.1   Definition and basic properties

General relativity is a metric theory of gravitation. At its core are Einstein's equations, which describe the relation between the geometry of a four-dimensional, pseudo-Riemannian manifold representing spacetime, and the energy–momentum contained in that spacetime.[33] Phenomena that in classical mechanics are ascribed to the action of the force of gravity (such as free-fall, orbital motion,

and spacecraft trajectories), correspond to inertial motion within a curved geometry of spacetime in general relativity; there is no gravitational force deflecting objects from their natural, straight paths. Instead, gravity corresponds to changes in the properties of space and time, which in turn changes the straightest-possible paths that objects will naturally follow.[34] The curvature is, in turn, caused by the energy–momentum of matter. Paraphrasing the relativist John Archibald Wheeler, spacetime tells matter how to move; matter tells spacetime how to curve.[35]

While general relativity replaces the scalar gravitational potential of classical physics by a symmetric rank-two tensor, the latter reduces to the former in certain limiting cases. For weak gravitational fields and slow speed relative to the speed of light, the theory's predictions converge on those of Newton's law of universal gravitation.[36]

As it is constructed using tensors, general relativity exhibits general covariance: its laws—and further laws formulated within the general relativistic framework—take on the same form in all coordinate systems.[37] Furthermore, the theory does not contain any invariant geometric background structures, i.e. it is background independent. It thus satisfies a more stringent general principle of relativity, namely that the laws of physics are the same for all observers.[38] Locally, as expressed in the equivalence principle, spacetime is Minkowskian, and the laws of physics exhibit local Lorentz invariance.[39]

8.3.2   Model-building

The core concept of general-relativistic model-building is that of a solution of Einstein's equations. Given both Einstein's equations and suitable equations for the properties of matter, such a solution consists of a specific semi-Riemannian manifold (usually defined by giving the metric in specific coordinates), and specific matter fields defined on that manifold. Matter and geometry must satisfy Einstein's equations, so in particular, the matter's energy–momentum tensor must be divergence-free. The matter must, of course, also satisfy whatever additional equations were imposed on its properties. In short, such a solution is a model universe that satisfies the laws of general relativity, and possibly additional laws governing whatever matter might be present.[40]

Einstein's equations are nonlinear partial differential equations and, as such, difficult to solve exactly.[41] Nevertheless, a number of exact solutions are known, although only a few have direct physical applications.[42] The best-known exact solutions, and also those most interesting from a physics point of view, are the Schwarzschild solution, the Reissner–Nordström solution and the Kerr metric, each corresponding to a certain type of black hole in an otherwise empty universe,[43] and the Friedmann–Lemaître–

Robertson–Walker and de Sitter universes, each describing an expanding cosmos.[44] Exact solutions of great theoretical interest include the Gödel universe (which opens up the intriguing possibility of time travel in curved spacetimes), the Taub-NUT solution (a model universe that is homogeneous, but anisotropic), and anti-de Sitter space (which has recently come to prominence in the context of what is called the Maldacena conjecture).[45]

Given the difficulty of finding exact solutions, Einstein's field equations are also solved frequently by numerical integration on a computer, or by considering small perturbations of exact solutions. In the field of numerical relativity, powerful computers are employed to simulate the geometry of spacetime and to solve Einstein's equations for interesting situations such as two colliding black holes.[46] In principle, such methods may be applied to any system, given sufficient computer resources, and may address fundamental questions such as naked singularities. Approximate solutions may also be found by perturbation theories such as linearized gravity[47] and its generalization, the post-Newtonian expansion, both of which were developed by Einstein. The latter provides a systematic approach to solving for the geometry of a spacetime that contains a distribution of matter that moves slowly compared with the speed of light. The expansion involves a series of terms; the first terms represent Newtonian gravity, whereas the later terms represent ever smaller corrections to Newton's theory due to general relativity.[48] An extension of this expansion is the parametrized post-Newtonian (PPN) formalism, which allows quantitative comparisons between the predictions of general relativity and alternative theories.[49]

8.4 Consequences of Einstein's theory

General relativity has a number of physical consequences. Some follow directly from the theory's axioms, whereas others have become clear only in the course of many years of research that followed Einstein's initial publication.

8.4.1 Gravitational time dilation and frequency shift

Main article: Gravitational time dilation

Assuming that the equivalence principle holds,[50] gravity influences the passage of time. Light sent down into a gravity well is blueshifted, whereas light sent in the opposite direction (i.e., climbing out of the gravity well) is redshifted; collectively, these two effects are known as the gravitational frequency shift. More generally, processes close to a massive body run more slowly when compared

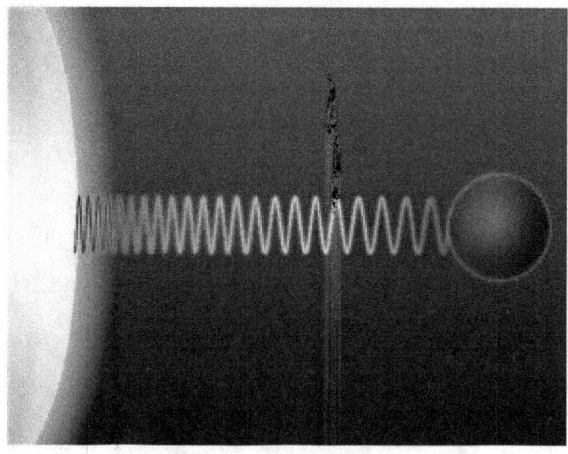

Schematic representation of the gravitational redshift of a light wave escaping from the surface of a massive body

with processes taking place farther away; this effect is known as gravitational time dilation.[51]

Gravitational redshift has been measured in the laboratory[52] and using astronomical observations.[53] Gravitational time dilation in the Earth's gravitational field has been measured numerous times using atomic clocks,[54] while ongoing validation is provided as a side effect of the operation of the Global Positioning System (GPS).[55] Tests in stronger gravitational fields are provided by the observation of binary pulsars.[56] All results are in agreement with general relativity.[57] However, at the current level of accuracy, these observations cannot distinguish between general relativity and other theories in which the equivalence principle is valid.[58]

8.4.2 Light deflection and gravitational time delay

Main articles: Kepler problem in general relativity, Gravitational lens and Shapiro delay

General relativity predicts that the path of light is bent in a gravitational field; light passing a massive body is deflected towards that body. This effect has been confirmed by observing the light of stars or distant quasars being deflected as it passes the Sun.[59]

This and related predictions follow from the fact that light follows what is called a light-like or null geodesic—a generalization of the straight lines along which light travels in classical physics. Such geodesics are the generalization of the invariance of lightspeed in special relativity.[60] As one examines suitable model spacetimes (either the exterior Schwarzschild solution or, for more than a single mass, the post-Newtonian expansion),[61] several effects of gravity on light propagation emerge. Although the bending of light can

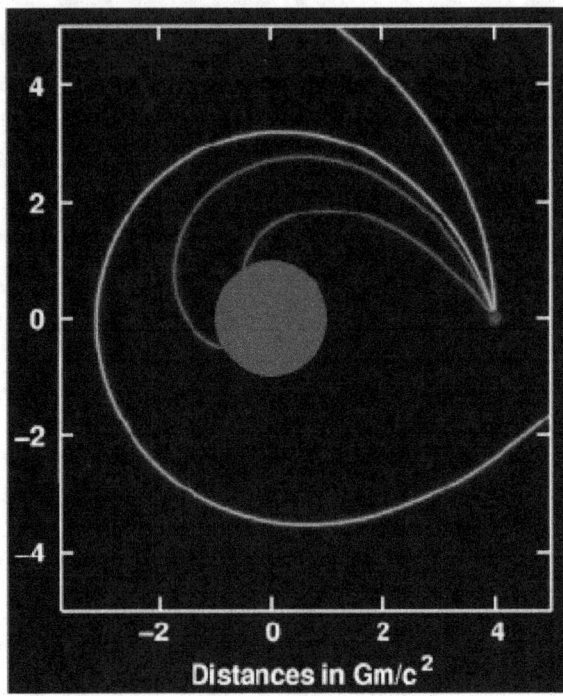

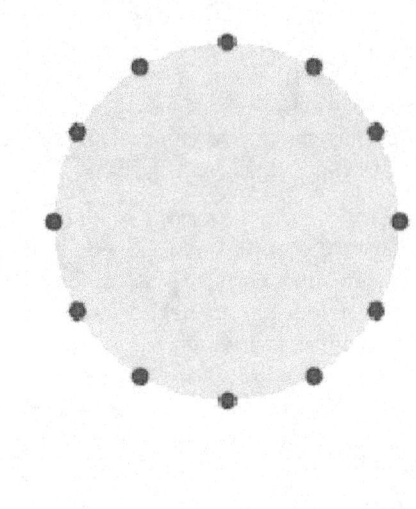

Ring of test particles influenced by gravitational wave

Deflection of light (sent out from the location shown in blue) near a compact body (shown in gray)

also be derived by extending the universality of free fall to light,[62] the angle of deflection resulting from such calculations is only half the value given by general relativity.[63]

Closely related to light deflection is the gravitational time delay (or Shapiro delay), the phenomenon that light signals take longer to move through a gravitational field than they would in the absence of that field. There have been numerous successful tests of this prediction.[64] In the parameterized post-Newtonian formalism (PPN), measurements of both the deflection of light and the gravitational time delay determine a parameter called γ, which encodes the influence of gravity on the geometry of space.[65]

8.4.3 Gravitational waves

Main article: Gravitational wave
One of several analogies between weak-field gravity and electromagnetism is that, analogous to electromagnetic waves, there are gravitational waves: ripples in the metric of spacetime that propagate at the speed of light.[66] The simplest type of such a wave can be visualized by its action on a ring of freely floating particles. A sine wave propagating through such a ring towards the reader distorts the ring in a characteristic, rhythmic fashion (animated image to the right).[67] Since Einstein's equations are non-linear, arbitrarily strong gravitational waves do not obey linear superposition, making their description difficult. However,

for weak fields, a linear approximation can be made. Such linearized gravitational waves are sufficiently accurate to describe the exceedingly weak waves that are expected to arrive here on Earth from far-off cosmic events, which typically result in relative distances increasing and decreasing by 10^{-21} or less. Data analysis methods routinely make use of the fact that these linearized waves can be Fourier decomposed.[68]

Some exact solutions describe gravitational waves without any approximation, e.g., a wave train traveling through empty space[69] or so-called Gowdy universes, varieties of an expanding cosmos filled with gravitational waves.[70] But for gravitational waves produced in astrophysically relevant situations, such as the merger of two black holes, numerical methods are presently the only way to construct appropriate models.[71]

8.4.4 Orbital effects and the relativity of direction

Main article: Kepler problem in general relativity

General relativity differs from classical mechanics in a number of predictions concerning orbiting bodies. It predicts an overall rotation (precession) of planetary orbits, as well as orbital decay caused by the emission of gravitational waves and effects related to the relativity of direction.

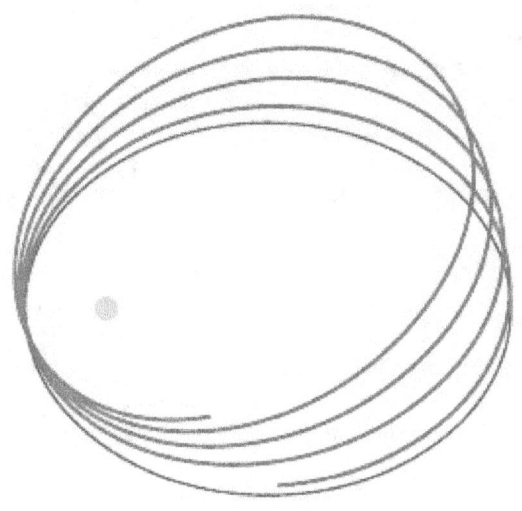

Newtonian (red) vs. Einsteinian orbit (blue) of a lone planet orbiting a star

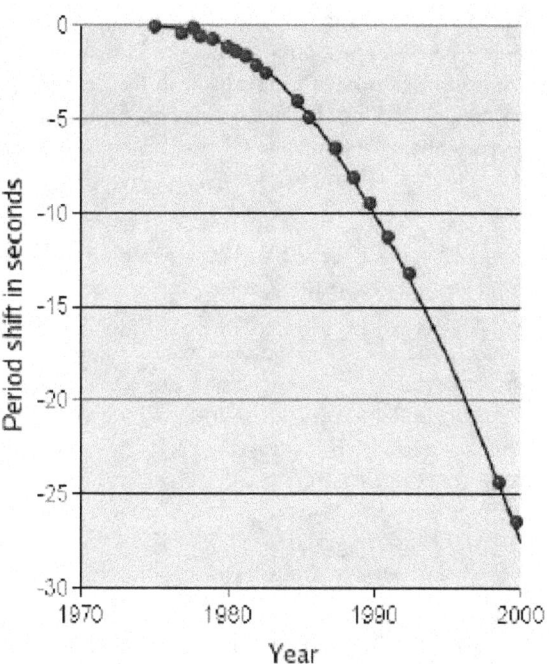

Orbital decay for PSR1913+16: time shift in seconds, tracked over three decades.[78]

Precession of apsides

In general relativity, the apsides of any orbit (the point of the orbiting body's closest approach to the system's center of mass) will precess—the orbit is not an ellipse, but akin to an ellipse that rotates on its focus, resulting in a rose curve-like shape (see image). Einstein first derived this result by using an approximate metric representing the Newtonian limit and treating the orbiting body as a test particle. For him, the fact that his theory gave a straightforward explanation of the anomalous perihelion shift of the planet Mercury, discovered earlier by Urbain Le Verrier in 1859, was important evidence that he had at last identified the correct form of the gravitational field equations.[72]

The effect can also be derived by using either the exact Schwarzschild metric (describing spacetime around a spherical mass)[73] or the much more general post-Newtonian formalism.[74] It is due to the influence of gravity on the geometry of space and to the contribution of self-energy to a body's gravity (encoded in the nonlinearity of Einstein's equations).[75] Relativistic precession has been observed for all planets that allow for accurate precession measurements (Mercury, Venus, and Earth),[76] as well as in binary pulsar systems, where it is larger by five orders of magnitude.[77]

Orbital decay

According to general relativity, a binary system will emit gravitational waves, thereby losing energy. Due to this loss, the distance between the two orbiting bodies decreases, and so does their orbital period. Within the Solar System or for

ordinary double stars, the effect is too small to be observable. This is not the case for a close binary pulsar, a system of two orbiting neutron stars, one of which is a pulsar: from the pulsar, observers on Earth receive a regular series of radio pulses that can serve as a highly accurate clock, which allows precise measurements of the orbital period. Because neutron stars are very compact, significant amounts of energy are emitted in the form of gravitational radiation.[79]

The first observation of a decrease in orbital period due to the emission of gravitational waves was made by Hulse and Taylor, using the binary pulsar PSR1913+16 they had discovered in 1974. This was the first detection of gravitational waves, albeit indirect, for which they were awarded the 1993 Nobel Prize in physics.[80] Since then, several other binary pulsars have been found, in particular the double pulsar PSR J0737-3039, in which both stars are pulsars.[81]

Geodetic precession and frame-dragging

Main articles: Geodetic precession and Frame dragging

Several relativistic effects are directly related to the relativity of direction.[82] One is geodetic precession: the axis direction of a gyroscope in free fall in curved spacetime will change when compared, for instance, with the direction of light received from distant stars—even though such a gyroscope represents the way of keeping a direction as stable as possible ("parallel transport").[83] For the Moon–Earth system, this effect has been measured with the help of lunar laser ranging.[84] More recently, it has been measured for test masses aboard the satellite Gravity Probe B to a precision of better than 0.3%.[85][86]

Near a rotating mass, there are so-called gravitomagnetic or frame-dragging effects. A distant observer will determine that objects close to the mass get "dragged around". This is most extreme for rotating black holes where, for any object entering a zone known as the ergosphere, rotation is inevitable.[87] Such effects can again be tested through their influence on the orientation of gyroscopes in free fall.[88] Somewhat controversial tests have been performed using the LAGEOS satellites, confirming the relativistic prediction.[89] Also the Mars Global Surveyor probe around Mars has been used.[90][91]

8.5 Astrophysical applications

8.5.1 Gravitational lensing

Main article: Gravitational lensing

The deflection of light by gravity is responsible for a new

Einstein cross: four images of the same astronomical object, produced by a gravitational lens

class of astronomical phenomena. If a massive object is situated between the astronomer and a distant target object with appropriate mass and relative distances, the astronomer will see multiple distorted images of the target. Such effects are known as gravitational lensing.[92] Depending on the configuration, scale, and mass distribution, there can be two or more images, a bright ring known as an Einstein ring, or partial rings called arcs.[93] The earliest example was discovered in 1979;[94] since then, more than a hundred gravitational lenses have been observed.[95] Even if the multiple images are too close to each other to be resolved, the effect can still be measured, e.g., as an overall brightening of the target object; a number of such "microlensing events" have been observed.[96]

Gravitational lensing has developed into a tool of observational astronomy. It is used to detect the presence and distribution of dark matter, provide a "natural telescope" for observing distant galaxies, and to obtain an independent estimate of the Hubble constant. Statistical evaluations of lensing data provide valuable insight into the structural evolution of galaxies.[97]

8.5.2 Gravitational wave astronomy

Main articles: Gravitational wave and Gravitational wave astronomy

Observations of binary pulsars provide strong indirect ev-

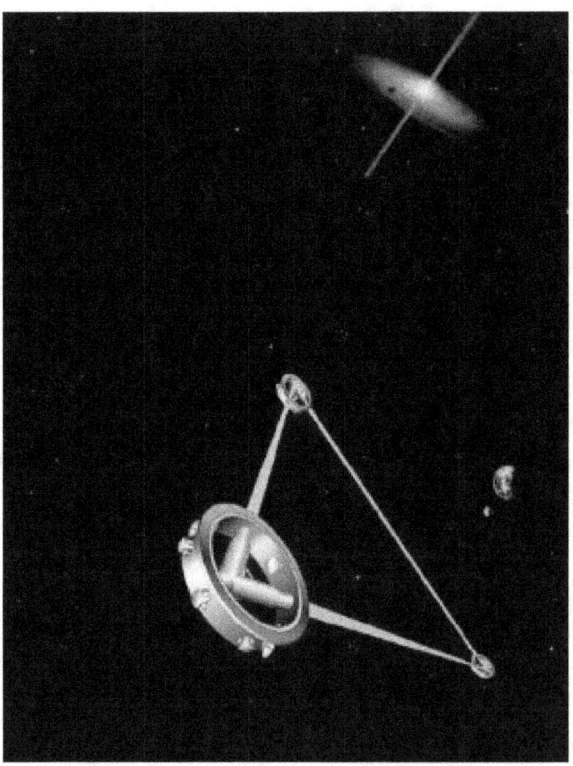

Artist's impression of the space-borne gravitational wave detector LISA

idence for the existence of gravitational waves (see Orbital decay, above). However, gravitational waves reaching us from the depths of the cosmos have not been detected directly. Such detection is a major goal of current relativity-related research.[98] Several land-based gravitational wave detectors are currently in operation, most notably the interferometric detectors GEO 600, LIGO (two detectors), TAMA 300 and VIRGO.[99] Various pulsar timing arrays are using millisecond pulsars to detect gravitational waves in the 10^{-9} to 10^{-6} Hertz frequency range, which originate from binary supermassive blackholes.[100] European space-based detector, eLISA / NGO, is currently under development.[101] with a precursor mission (LISA Pathfinder) due for launch in 2015.[102]

Observations of gravitational waves promise to complement observations in the electromagnetic spectrum.[103] They are expected to yield information about black holes and other dense objects such as neutron stars and white dwarfs, about certain kinds of supernova implosions, and about processes in the very early universe, including the signature of certain types of hypothetical cosmic string.[104]

8.5.3 Black holes and other compact objects

Main article: Black hole

Whenever the ratio of an object's mass to its radius becomes sufficiently large, general relativity predicts the formation of a black hole, a region of space from which nothing, not even light, can escape. In the currently accepted models of stellar evolution, neutron stars of around 1.4 solar masses, and stellar black holes with a few to a few dozen solar masses, are thought to be the final state for the evolution of massive stars.[105] Usually a galaxy has one supermassive black hole with a few million to a few billion solar masses in its center,[106] and its presence is thought to have played an important role in the formation of the galaxy and larger cosmic structures.[107]

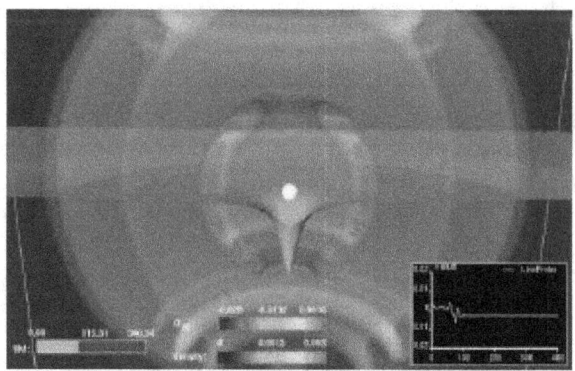

Simulation based on the equations of general relativity: a star collapsing to form a black hole while emitting gravitational waves

Astronomically, the most important property of compact objects is that they provide a supremely efficient mechanism for converting gravitational energy into electromag-

netic radiation.[108] Accretion, the falling of dust or gaseous matter onto stellar or supermassive black holes, is thought to be responsible for some spectacularly luminous astronomical objects, notably diverse kinds of active galactic nuclei on galactic scales and stellar-size objects such as microquasars.[109] In particular, accretion can lead to relativistic jets, focused beams of highly energetic particles that are being flung into space at almost light speed.[110] General relativity plays a central role in modelling all these phenomena,[111] and observations provide strong evidence for the existence of black holes with the properties predicted by the theory.[112]

Black holes are also sought-after targets in the search for gravitational waves (cf. Gravitational waves, above). Merging black hole binaries should lead to some of the strongest gravitational wave signals reaching detectors here on Earth, and the phase directly before the merger ("chirp") could be used as a "standard candle" to deduce the distance to the merger events–and hence serve as a probe of cosmic expansion at large distances.[113] The gravitational waves produced as a stellar black hole plunges into a supermassive one should provide direct information about the supermassive black hole's geometry.[114]

8.5.4 Cosmology

This blue horseshoe is a distant galaxy that has been magnified and warped into a nearly complete ring by the strong gravitational pull of the massive foreground luminous red galaxy.

Main article: Physical cosmology

The current models of cosmology are based on Einstein's field equations, which include the cosmological constant Λ since it has important influence on the large-scale dynamics of the cosmos,

$$R_{\mu\nu} - \frac{1}{2}R\,g_{\mu\nu} + \Lambda\,g_{\mu\nu} = \frac{8\pi G}{c^4}\,T_{\mu\nu}$$

where $g_{\mu\nu}$ is the spacetime metric.[115] Isotropic and homogeneous solutions of these enhanced equations, the Friedmann–Lemaître–Robertson–Walker solutions,[116] allow physicists to model a universe that has evolved over the past 14 billion years from a hot, early Big Bang phase.[117] Once a small number of parameters (for example the universe's mean matter density) have been fixed by astronomical observation,[118] further observational data can be used to put the models to the test.[119] Predictions, all successful, include the initial abundance of chemical elements formed in a period of primordial nucleosynthesis,[120] the large-scale structure of the universe,[121] and the existence and properties of a "thermal echo" from the early cosmos, the cosmic background radiation.[122]

Astronomical observations of the cosmological expansion rate allow the total amount of matter in the universe to be estimated, although the nature of that matter remains mysterious in part. About 90% of all matter appears to be so-called dark matter, which has mass (or, equivalently, gravitational influence), but does not interact electromagnetically and, hence, cannot be observed directly.[123] There is no generally accepted description of this new kind of matter, within the framework of known particle physics[124] or otherwise.[125] Observational evidence from redshift surveys of distant supernovae and measurements of the cosmic background radiation also show that the evolution of our universe is significantly influenced by a cosmological constant resulting in an acceleration of cosmic expansion or, equivalently, by a form of energy with an unusual equation of state, known as dark energy, the nature of which remains unclear.[126]

A so-called inflationary phase,[127] an additional phase of strongly accelerated expansion at cosmic times of around 10^{-33} seconds, was hypothesized in 1980 to account for several puzzling observations that were unexplained by classical cosmological models, such as the nearly perfect homogeneity of the cosmic background radiation.[128] Recent measurements of the cosmic background radiation have resulted in the first evidence for this scenario.[129] However, there is a bewildering variety of possible inflationary scenarios, which cannot be restricted by current observations.[130] An even larger question is the physics of the earliest universe, prior to the inflationary phase and close to where the classical models predict the big bang singularity. An authoritative answer would require a complete theory of quantum gravity, which has not yet been developed[131] (cf. the section on quantum gravity, below).

8.5.5 Time travel

Kurt Gödel showed that solutions to Einstein's equations exist that contain closed timelike curves (CTCs), which allow for loops in time. The solutions require extreme physical conditions unlikely ever to occur in practice, and it remains an open question whether further laws of physics will eliminate them completely. Since then other—similarly impractical—GR solutions containing CTCs have been found, such as the Tipler cylinder and traversable wormholes.

8.6 Advanced concepts

8.6.1 Causal structure and global geometry

Main article: Causal structure
In general relativity, no material body can catch up with

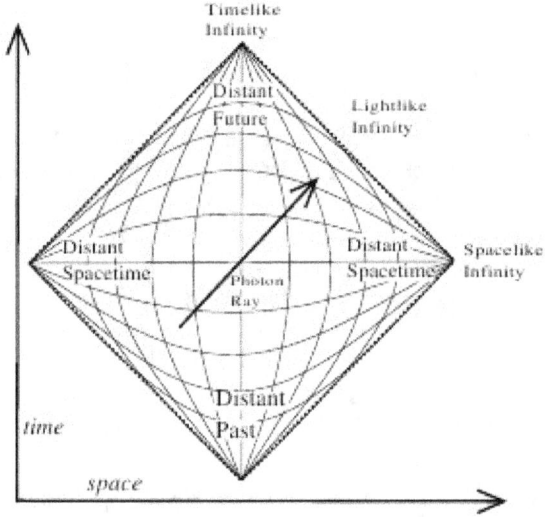

Penrose–Carter diagram of an infinite Minkowski universe

or overtake a light pulse. No influence from an event A can reach any other location X before light sent out at A to X. In consequence, an exploration of all light worldlines (null geodesics) yields key information about the spacetime's causal structure. This structure can be displayed using Penrose–Carter diagrams in which infinitely large regions of space and infinite time intervals are shrunk ("compactified") so as to fit onto a finite map, while light still travels along diagonals as in standard spacetime diagrams.[132]

Aware of the importance of causal structure, Roger Penrose and others developed what is known as global geometry. In global geometry, the object of study is not one particular solution (or family of solutions) to Einstein's equations.

Rather, relations that hold true for all geodesics, such as the Raychaudhuri equation, and additional non-specific assumptions about the nature of matter (usually in the form of so-called energy conditions) are used to derive general results.[133]

8.6.2 Horizons

Main articles: Horizon (general relativity), No hair theorem and Black hole mechanics

Using global geometry, some spacetimes can be shown to contain boundaries called horizons, which demarcate one region from the rest of spacetime. The best-known examples are black holes: if mass is compressed into a sufficiently compact region of space (as specified in the hoop conjecture, the relevant length scale is the Schwarzschild radius[134]), no light from inside can escape to the outside. Since no object can overtake a light pulse, all interior matter is imprisoned as well. Passage from the exterior to the interior is still possible, showing that the boundary, the black hole's *horizon*, is not a physical barrier.[135]

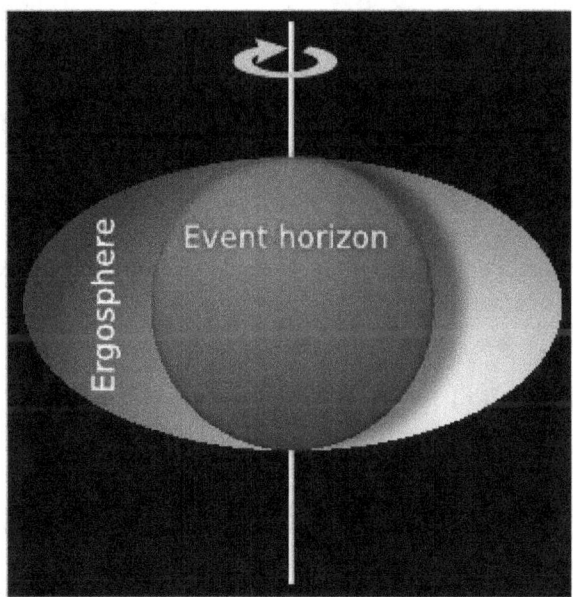

The ergosphere of a rotating black hole, which plays a key role when it comes to extracting energy from such a black hole

Early studies of black holes relied on explicit solutions of Einstein's equations, notably the spherically symmetric Schwarzschild solution (used to describe a static black hole) and the axisymmetric Kerr solution (used to describe a rotating, stationary black hole, and introducing interesting features such as the ergosphere). Using global geometry, later studies have revealed more general properties of black holes. In the long run, they are rather simple objects charac-

terized by eleven parameters specifying energy, linear momentum, angular momentum, location at a specified time and electric charge. This is stated by the black hole uniqueness theorems: "black holes have no hair", that is, no distinguishing marks like the hairstyles of humans. Irrespective of the complexity of a gravitating object collapsing to form a black hole, the object that results (having emitted gravitational waves) is very simple.[136]

Even more remarkably, there is a general set of laws known as black hole mechanics, which is analogous to the laws of thermodynamics. For instance, by the second law of black hole mechanics, the area of the event horizon of a general black hole will never decrease with time, analogous to the entropy of a thermodynamic system. This limits the energy that can be extracted by classical means from a rotating black hole (e.g. by the Penrose process).[137] There is strong evidence that the laws of black hole mechanics are, in fact, a subset of the laws of thermodynamics, and that the black hole area is proportional to its entropy.[138] This leads to a modification of the original laws of black hole mechanics: for instance, as the second law of black hole mechanics becomes part of the second law of thermodynamics, it is possible for black hole area to decrease—as long as other processes ensure that, overall, entropy increases. As thermodynamical objects with non-zero temperature, black holes should emit thermal radiation. Semi-classical calculations indicate that indeed they do, with the surface gravity playing the role of temperature in Planck's law. This radiation is known as Hawking radiation (cf. the quantum theory section, below).[139]

There are other types of horizons. In an expanding universe, an observer may find that some regions of the past cannot be observed ("particle horizon"), and some regions of the future cannot be influenced (event horizon).[140] Even in flat Minkowski space, when described by an accelerated observer (Rindler space), there will be horizons associated with a semi-classical radiation known as Unruh radiation.[141]

8.6.3 Singularities

Main article: Spacetime singularity

Another general feature of general relativity is the appearance of spacetime boundaries known as singularities. Spacetime can be explored by following up on timelike and lightlike geodesics—all possible ways that light and particles in free fall can travel. But some solutions of Einstein's equations have "ragged edges"—regions known as spacetime singularities, where the paths of light and falling particles come to an abrupt end, and geometry becomes ill-defined. In the more interesting cases, these are "curvature singularities", where geometrical quantities characterizing spacetime curvature, such as the Ricci scalar, take on infinite values.[142] Well-known examples of spacetimes with future singularities—where worldlines end—are the Schwarzschild solution, which describes a singularity inside an eternal static black hole,[143] or the Kerr solution with its ring-shaped singularity inside an eternal rotating black hole.[144] The Friedmann–Lemaître–Robertson–Walker solutions and other spacetimes describing universes have past singularities on which worldlines begin, namely Big Bang singularities, and some have future singularities (Big Crunch) as well.[145]

Given that these examples are all highly symmetric—and thus simplified—it is tempting to conclude that the occurrence of singularities is an artifact of idealization.[146] The famous singularity theorems, proved using the methods of global geometry, say otherwise: singularities are a generic feature of general relativity, and unavoidable once the collapse of an object with realistic matter properties has proceeded beyond a certain stage[147] and also at the beginning of a wide class of expanding universes.[148] However, the theorems say little about the properties of singularities, and much of current research is devoted to characterizing these entities' generic structure (hypothesized e.g. by the so-called BKL conjecture).[149] The cosmic censorship hypothesis states that all realistic future singularities (no perfect symmetries, matter with realistic properties) are safely hidden away behind a horizon, and thus invisible to all distant observers. While no formal proof yet exists, numerical simulations offer supporting evidence of its validity.[150]

8.6.4   Evolution equations

Main article: Initial value formulation (general relativity)

Each solution of Einstein's equation encompasses the whole history of a universe — it is not just some snapshot of how things are, but a whole, possibly matter-filled, spacetime. It describes the state of matter and geometry everywhere and at every moment in that particular universe. Due to its general covariance, Einstein's theory is not sufficient by itself to determine the time evolution of the metric tensor. It must be combined with a coordinate condition, which is analogous to gauge fixing in other field theories.[151]

To understand Einstein's equations as partial differential equations, it is helpful to formulate them in a way that describes the evolution of the universe over time. This is done in so-called "3+1" formulations, where spacetime is split into three space dimensions and one time dimension. The best-known example is the ADM formalism.[152] These decompositions show that the spacetime evolution equations of general relativity are well-behaved: solutions always exist, and are uniquely defined, once suitable initial conditions have been specified.[153] Such formulations of Einstein's field equations are the basis of numerical relativity.[154]

8.6.5   Global and quasi-local quantities

Main article: Mass in general relativity

The notion of evolution equations is intimately tied in with another aspect of general relativistic physics. In Einstein's theory, it turns out to be impossible to find a general definition for a seemingly simple property such as a system's total mass (or energy). The main reason is that the gravitational field—like any physical field—must be ascribed a certain energy, but that it proves to be fundamentally impossible to localize that energy.[155]

Nevertheless, there are possibilities to define a system's total mass, either using a hypothetical "infinitely distant observer" (ADM mass)[156] or suitable symmetries (Komar mass).[157] If one excludes from the system's total mass the energy being carried away to infinity by gravitational waves, the result is the so-called Bondi mass at null infinity.[158] Just as in classical physics, it can be shown that these masses are positive.[159] Corresponding global definitions exist for momentum and angular momentum.[160] There have also been a number of attempts to define *quasi-local* quantities, such as the mass of an isolated system formulated using only quantities defined within a finite region of space containing that system. The hope is to obtain a quantity useful for general statements about isolated systems, such as a more precise formulation of the hoop conjecture.[161]

8.7   Relationship with quantum theory

If general relativity were considered to be one of the two pillars of modern physics, then quantum theory, the basis of understanding matter from elementary particles to solid state physics, would be the other.[162] However, how to reconcile quantum theory with general relativity is still an open question.

8.7.1   Quantum field theory in curved spacetime

Main article: Quantum field theory in curved spacetime

Ordinary quantum field theories, which form the basis of

modern elementary particle physics, are defined in flat Minkowski space, which is an excellent approximation when it comes to describing the behavior of microscopic particles in weak gravitational fields like those found on Earth.[163] In order to describe situations in which gravity is strong enough to influence (quantum) matter, yet not strong enough to require quantization itself, physicists have formulated quantum field theories in curved spacetime. These theories rely on general relativity to describe a curved background spacetime, and define a generalized quantum field theory to describe the behavior of quantum matter within that spacetime.[164] Using this formalism, it can be shown that black holes emit a blackbody spectrum of particles known as Hawking radiation, leading to the possibility that they evaporate over time.[165] As briefly mentioned above, this radiation plays an important role for the thermodynamics of black holes.[166]

8.7.2 Quantum gravity

Main article: Quantum gravity
See also: String theory, Canonical general relativity, Loop quantum gravity, Causal Dynamical Triangulations and Causal sets

The demand for consistency between a quantum description of matter and a geometric description of spacetime,[167] as well as the appearance of singularities (where curvature length scales become microscopic), indicate the need for a full theory of quantum gravity: for an adequate description of the interior of black holes, and of the very early universe, a theory is required in which gravity and the associated geometry of spacetime are described in the language of quantum physics.[168] Despite major efforts, no complete and consistent theory of quantum gravity is currently known, even though a number of promising candidates exist.[169]

Attempts to generalize ordinary quantum field theories, used in elementary particle physics to describe fundamental interactions, so as to include gravity have led to serious problems. At low energies, this approach proves successful, in that it results in an acceptable effective (quantum) field theory of gravity.[170] At very high energies, however, the result are models devoid of all predictive power ("non-renormalizability").[171]

One attempt to overcome these limitations is string theory, a quantum theory not of point particles, but of minute one-dimensional extended objects.[172] The theory promises to be a unified description of all particles and interactions, including gravity;[173] the price to pay is unusual features such as six extra dimensions of space in addition to the usual three.[174] In what is called the second superstring revolution, it was conjectured that both string theory and a

Projection of a Calabi–Yau manifold, one of the ways of compactifying the extra dimensions posited by string theory

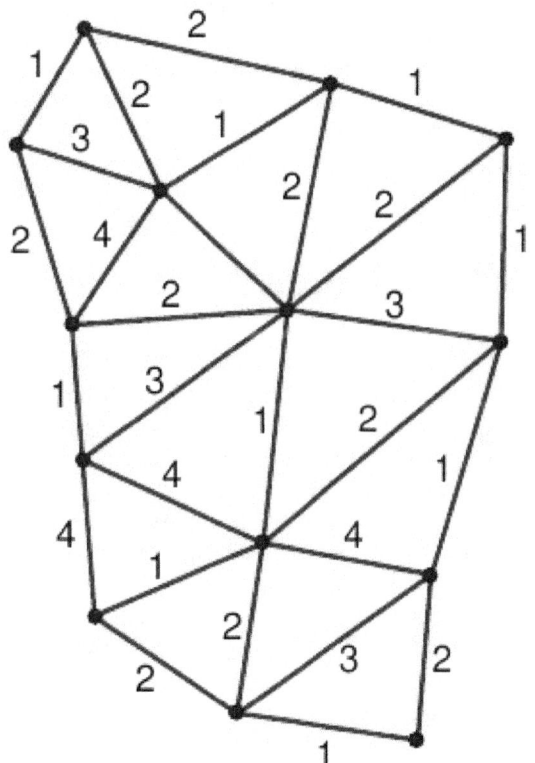

Simple spin network of the type used in loop quantum gravity

unification of general relativity and supersymmetry known as supergravity[175] form part of a hypothesized eleven-dimensional model known as M-theory, which would constitute a uniquely defined and consistent theory of quantum gravity.[176]

Another approach starts with the canonical quantization procedures of quantum theory. Using the initial-value-formulation of general relativity (cf. evolution equations above), the result is the Wheeler–deWitt equation (an analogue of the Schrödinger equation) which, regrettably, turns out to be ill-defined.[177] However, with the introduction of what are now known as Ashtekar variables,[178] this leads to a promising model known as loop quantum gravity. Space is represented by a web-like structure called a spin network, evolving over time in discrete steps.[179]

Depending on which features of general relativity and quantum theory are accepted unchanged, and on what level changes are introduced,[180] there are numerous other attempts to arrive at a viable theory of quantum gravity, some examples being dynamical triangulations,[181] causal sets,[182] twistor models[183] or the path-integral based models of quantum cosmology.[184]

All candidate theories still have major formal and conceptual problems to overcome. They also face the common problem that, as yet, there is no way to put quantum gravity predictions to experimental tests (and thus to decide between the candidates where their predictions vary), although there is hope for this to change as future data from cosmological observations and particle physics experiments becomes available.[185]

8.8   Current status

General relativity has emerged as a highly successful model of gravitation and cosmology, which has so far passed many unambiguous observational and experimental tests. However, there are strong indications the theory is incomplete.[186] The problem of quantum gravity and the question of the reality of spacetime singularities remain open.[187] Observational data that is taken as evidence for dark energy and dark matter could indicate the need for new physics.[188] Even taken as is, general relativity is rich with possibilities for further exploration. Mathematical relativists seek to understand the nature of singularities and the fundamental properties of Einstein's equations,[189] and increasingly powerful computer simulations (such as those describing merging black holes) are run.[190] The race for the first direct detection of gravitational waves continues,[191] in the hope of creating opportunities to test the theory's validity for much stronger gravitational fields than has been possible to date.[192] Almost a hundred years after its publication, general relativity remains a highly active area of research.[193]

8.9   See also

- Center of mass (relativistic)
- Contributors to general relativity
- Derivations of the Lorentz transformations
- Ehrenfest paradox
- Einstein–Hilbert action
- Introduction to mathematics of general relativity
- Relativity priority dispute
- Ricci calculus
- Tests of general relativity
- Timeline of gravitational physics and relativity
- Two-body problem in general relativity

8.10   Notes

[1] O'Connor, J.J. and E.F. Robertson (1996), "General relativity". *Mathematical Physics index*, School of Mathematics and Statistics, University of St. Andrews, Scotland, May, 1996. Retrieved 2015-02-04.

[2] Pais 1982, ch. 9 to 15, Janssen 2005; an up-to-date collection of current research, including reprints of many of the original articles, is Renn 2007; an accessible overview can be found in Renn 2005, pp. 110ff. An early key article is Einstein 1907, cf. Pais 1982, ch. 9. The publication featuring the field equations is Einstein 1915, cf. Pais 1982, ch. 11–15

[3] Schwarzschild 1916a, Schwarzschild 1916b and Reissner 1916 (later complemented in Nordström 1918)

[4] Einstein 1917, cf. Pais 1982, ch. 15e

[5] Hubble's original article is Hubble 1929; an accessible overview is given in Singh 2004, ch. 2–4

[6] As reported in Gamow 1970. Einstein's condemnation would prove to be premature, cf. the section Cosmology, below

[7] Pais 1982, pp. 253–254

[8] Kennefick 2005, Kennefick 2007

[9] Pais 1982, ch. 16

[10] Thorne, Kip (2003). "Warping spacetime". *The future of theoretical physics and cosmology: celebrating Stephen Hawking's 60th birthday*. Cambridge University Press. p. 74. ISBN 0-521-82081-2., Extract of page 74

[11] Israel 1987, ch. 7.8–7.10, Thorne 1994, ch. 3–9

[12] Sections Orbital effects and the relativity of direction, Gravitational time dilation and frequency shift and Light deflection and gravitational time delay, and references therein

[13] Section Cosmology and references therein; the historical development is in Overbye 1999

[14] The following exposition re-traces that of Ehlers 1973, sec. 1

[15] Arnold 1989, ch. 1

[16] Ehlers 1973, pp. 5f

[17] Will 1993, sec. 2.4, Will 2006, sec. 2

[18] Wheeler 1990, ch. 2

[19] Ehlers 1973, sec. 1.2, Havas 1964, Künzle 1972. The simple thought experiment in question was first described in Heckmann & Schücking 1959

[20] Ehlers 1973, pp. 10f

[21] Good introductions are, in order of increasing presupposed knowledge of mathematics, Giulini 2005, Mermin 2005, and Rindler 1991; for accounts of precision experiments, cf. part IV of Ehlers & Lämmerzahl 2006

[22] An in-depth comparison between the two symmetry groups can be found in Giulini 2006a

[23] Rindler 1991, sec. 22, Synge 1972, ch. 1 and 2

[24] Ehlers 1973, sec. 2.3

[25] Ehlers 1973, sec. 1.4, Schutz 1985, sec. 5.1

[26] Ehlers 1973, pp. 17ff; a derivation can be found in Mermin 2005, ch. 12. For the experimental evidence, cf. the section Gravitational time dilation and frequency shift, below

[27] Rindler 2001, sec. 1.13; for an elementary account, see Wheeler 1990, ch. 2; there are, however, some differences between the modern version and Einstein's original concept used in the historical derivation of general relativity, cf. Norton 1985

[28] Ehlers 1973, sec. 1.4 for the experimental evidence, see once more section Gravitational time dilation and frequency shift. Choosing a different connection with non-zero torsion leads to a modified theory known as Einstein–Cartan theory

[29] Ehlers 1973, p. 16, Kenyon 1990, sec. 7.2, Weinberg 1972, sec. 2.8

[30] Ehlers 1973, pp. 19–22; for similar derivations, see sections 1 and 2 of ch. 7 in Weinberg 1972. The Einstein tensor is the only divergence-free tensor that is a function of the metric coefficients, their first and second derivatives at most, and allows the spacetime of special relativity as a solution in the absence of sources of gravity, cf. Lovelock 1972. The tensors on both side are of second rank, that is, they can each

be thought of as 4×4 matrices, each of which contains ten independent terms; hence, the above represents ten coupled equations. The fact that, as a consequence of geometric relations known as Bianchi identities, the Einstein tensor satisfies a further four identities reduces these to six independent equations, e.g. Schutz 1985, sec. 8.3

[31] Kenyon 1990, sec. 7.4

[32] Brans & Dicke 1961, Weinberg 1972, sec. 3 in ch. 7, Goenner 2004, sec. 7.2, and Trautman 2006, respectively

[33] Wald 1984, ch. 4, Weinberg 1972, ch. 7 or, in fact, any other textbook on general relativity

[34] At least approximately, cf. Poisson 2004

[35] Wheeler 1990, p. xi

[36] Wald 1984, sec. 4.4

[37] Wald 1984, sec. 4.1

[38] For the (conceptual and historical) difficulties in defining a general principle of relativity and separating it from the notion of general covariance, see Giulini 2006b

[39] section 5 in ch. 12 of Weinberg 1972

[40] Introductory chapters of Stephani et al. 2003

[41] A review showing Einstein's equation in the broader context of other PDEs with physical significance is Geroch 1996

[42] For background information and a list of solutions, cf. Stephani et al. 2003; a more recent review can be found in MacCallum 2006

[43] Chandrasekhar 1983, ch. 3,5,6

[44] Narlikar 1993, ch. 4, sec. 3.3

[45] Brief descriptions of these and further interesting solutions can be found in Hawking & Ellis 1973, ch. 5

[46] Lehner 2002

[47] For instance Wald 1984, sec. 4.4

[48] Will 1993, sec. 4.1 and 4.2

[49] Will 2006, sec. 3.2, Will 1993, ch. 4

[50] Rindler 2001, pp. 24–26 vs. pp. 236–237 and Ohanian & Ruffini 1994, pp. 164–172. Einstein derived these effects using the equivalence principle as early as 1907, cf. Einstein 1907 and the description in Pais 1982, pp. 196–198

[51] Rindler 2001, pp. 24–26; Misner, Thorne & Wheeler 1973, § 38.5

[52] Pound–Rebka experiment, see Pound & Rebka 1959, Pound & Rebka 1960; Pound & Snider 1964; a list of further experiments is given in Ohanian & Ruffini 1994, table 4.1 on p. 186

[53] Greenstein, Oke & Shipman 1971; the most recent and most accurate Sirius B measurements are published in Barstow, Bond et al. 2005.

[54] Starting with the Hafele–Keating experiment, Hafele & Keating 1972a and Hafele & Keating 1972b, and culminating in the Gravity Probe A experiment; an overview of experiments can be found in Ohanian & Ruffini 1994, table 4.1 on p. 186

[55] GPS is continually tested by comparing atomic clocks on the ground and aboard orbiting satellites; for an account of relativistic effects, see Ashby 2002 and Ashby 2003

[56] Stairs 2003 and Kramer 2004

[57] General overviews can be found in section 2.1. of Will 2006; Will 2003, pp. 32–36; Ohanian & Ruffini 1994, sec. 4.2

[58] Ohanian & Ruffini 1994, pp. 164–172

[59] Cf. Kennefick 2005 for the classic early measurements by the Eddington expeditions; for an overview of more recent measurements, see Ohanian & Ruffini 1994, ch. 4.3. For the most precise direct modern observations using quasars, cf. Shapiro et al. 2004

[60] This is not an independent axiom; it can be derived from Einstein's equations and the Maxwell Lagrangian using a WKB approximation, cf. Ehlers 1973, sec. 5

[61] Blanchet 2006, sec. 1.3

[62] Rindler 2001, sec. 1.16; for the historical examples, Israel 1987, pp. 202–204; in fact, Einstein published one such derivation as Einstein 1907. Such calculations tacitly assume that the geometry of space is Euclidean, cf. Ehlers & Rindler 1997

[63] From the standpoint of Einstein's theory, these derivations take into account the effect of gravity on time, but not its consequences for the warping of space, cf. Rindler 2001, sec. 11.11

[64] For the Sun's gravitational field using radar signals reflected from planets such as Venus and Mercury, cf. Shapiro 1964, Weinberg 1972, ch. 8, sec. 7; for signals actively sent back by space probes (transponder measurements), cf. Bertotti, Iess & Tortora 2003; for an overview, see Ohanian & Ruffini 1994, table 4.4 on p. 200; for more recent measurements using signals received from a pulsar that is part of a binary system, the gravitational field causing the time delay being that of the other pulsar, cf. Stairs 2003, sec. 4.4

[65] Will 1993, sec. 7.1 and 7.2

[66] These have been indirectly observed through the loss of energy in binary pulsar systems such as the Hulse–Taylor binary, the subject of the 1993 Nobel Prize in physics. A number of projects are underway to attempt to observe directly the effects of gravitational waves. For an overview, see Misner, Thorne & Wheeler 1973, part VIII. Unlike electromagnetic waves, the dominant contribution for gravitational waves is not the dipole, but the quadrupole; see Schutz 2001

[67] Most advanced textbooks on general relativity contain a description of these properties, e.g. Schutz 1985, ch. 9

[68] For example Jaranowski & Królak 2005

[69] Rindler 2001, ch. 13

[70] Gowdy 1971, Gowdy 1974

[71] See Lehner 2002 for a brief introduction to the methods of numerical relativity, and Seidel 1998 for the connection with gravitational wave astronomy

[72] Schutz 2003, pp. 48–49, Pais 1982, pp. 253–254

[73] Rindler 2001, sec. 11.9

[74] Will 1993, pp. 177–181

[75] In consequence, in the parameterized post-Newtonian formalism (PPN), measurements of this effect determine a linear combination of the terms β and γ, cf. Will 2006, sec. 3.5 and Will 1993, sec. 7.3

[76] The most precise measurements are VLBI measurements of planetary positions; see Will 1993, ch. 5, Will 2006, sec. 3.5, Anderson et al. 1992; for an overview, Ohanian & Ruffini 1994, pp. 406–407

[77] Kramer et al. 2006

[78] A figure that includes error bars is fig. 7 in Will 2006, sec. 5.1

[79] Stairs 2003, Schutz 2003, pp. 317–321, Bartusiak 2000, pp. 70–86

[80] Weisberg & Taylor 2003; for the pulsar discovery, see Hulse & Taylor 1975; for the initial evidence for gravitational radiation, see Taylor 1994

[81] Kramer 2004

[82] Penrose 2004, §14.5, Misner, Thorne & Wheeler 1973, §11.4

[83] Weinberg 1972, sec. 9.6, Ohanian & Ruffini 1994, sec. 7.8

[84] Bertotti, Ciufolini & Bender 1987, Nordtvedt 2003

[85] Kahn 2007

[86] A mission description can be found in Everitt et al. 2001; a first post-flight evaluation is given in Everitt, Parkinson & Kahn 2007; further updates will be available on the mission website Kahn 1996–2012.

[87] Townsend 1997, sec. 4.2.1, Ohanian & Ruffini 1994, pp. 469–471

[88] Ohanian & Ruffini 1994, sec. 4.7, Weinberg 1972, sec. 9.7; for a more recent review, see Schäfer 2004

[89] Ciufolini & Pavlis 2004, Ciufolini, Pavlis & Peron 2006, Iorio 2009

[90] Iorio L. (August 2006), "COMMENTS, REPLIES AND NOTES: A note on the evidence of the gravitomagnetic field of Mars", *Classical Quantum Gravity* **23** (17): 5451–5454, arXiv:gr-qc/0606092, Bibcode:2006CQGra..23.5451I, doi:10.1088/0264-9381/23/17/N01

[91] Iorio L. (June 2010), "On the Lense–Thirring test with the Mars Global Surveyor in the gravitational field of Mars", *Central European Journal of Physics* **8** (3): 509–513, arXiv:gr-qc/0701146, Bibcode:2010CEJPh...8..509I, doi:10.2478/s11534-009-0117-6

[92] For overviews of gravitational lensing and its applications, see Ehlers, Falco & Schneider 1992 and Wambsganss 1998

[93] For a simple derivation, see Schutz 2003, ch. 23; cf. Narayan & Bartelmann 1997, sec. 3

[94] Walsh, Carswell & Weymann 1979

[95] Images of all the known lenses can be found on the pages of the CASTLES project, Kochanek et al. 2007

[96] Roulet & Mollerach 1997

[97] Narayan & Bartelmann 1997, sec. 3.7

[98] Barish 2005, Bartusiak 2000, Blair & McNamara 1997

[99] Hough & Rowan 2000

[100] Hobbs, George; Archibald, A.; Arzoumanian, Z.; Backer, D.; Bailes, M.; Bhat, N. D. R.; Burgay, M.; Burke-Spolaor, S.; et al. (2010), "The international pulsar timing array project: using pulsars as a gravitational wave detector", *Classical and Quantum Gravity* **27** (8): 084013, arXiv:0911.5206, Bibcode:2010CQGra..27h4013H, doi:10.1088/0264-9381/27/8/084013

[101] Danzmann & Rüdiger 2003

[102] "LISA pathfinder overview". ESA. Retrieved 2012-04-23.

[103] Thorne 1995

[104] Cutler & Thorne 2002

[105] Miller 2002, lectures 19 and 21

[106] Celotti, Miller & Sciama 1999, sec. 3

[107] Springel et al. 2005 and the accompanying summary Gnedin 2005

[108] Blandford 1987, sec. 8.2.4

[109] For the basic mechanism, see Carroll & Ostlie 1996, sec. 17.2; for more about the different types of astronomical objects associated with this, cf. Robson 1996

[110] For a review, see Begelman, Blandford & Rees 1984. To a distant observer, some of these jets even appear to move faster than light; this, however, can be explained as an optical illusion that does not violate the tenets of relativity, see Rees 1966

[111] For stellar end states, cf. Oppenheimer & Snyder 1939 or, for more recent numerical work, Font 2003, sec. 4.1; for supernovae, there are still major problems to be solved, cf. Buras et al. 2003; for simulating accretion and the formation of jets, cf. Font 2003, sec. 4.2. Also, relativistic lensing effects are thought to play a role for the signals received from X-ray pulsars, cf. Kraus 1998

[112] The evidence includes limits on compactness from the observation of accretion-driven phenomena ("Eddington luminosity"), see Celotti, Miller & Sciama 1999, observations of stellar dynamics in the center of our own Milky Way galaxy, cf. Schödel et al. 2003, and indications that at least some of the compact objects in question appear to have no solid surface, which can be deduced from the examination of X-ray bursts for which the central compact object is either a neutron star or a black hole; cf. Remillard et al. 2006 for an overview, Narayan 2006, sec. 5. Observations of the "shadow" of the Milky Way galaxy's central black hole horizon are eagerly sought for, cf. Falcke, Melia & Agol 2000

[113] Dalal et al. 2006

[114] Barack & Cutler 2004

[115] Originally Einstein 1917; cf. Pais 1982, pp. 285–288

[116] Carroll 2001, ch. 2

[117] Bergström & Goobar 2003, ch. 9–11; use of these models is justified by the fact that, at large scales of around hundred million light-years and more, our own universe indeed appears to be isotropic and homogeneous, cf. Peebles et al. 1991

[118] E.g. with WMAP data, see Spergel et al. 2003

[119] These tests involve the separate observations detailed further on, see, e.g., fig. 2 in Bridle et al. 2003

[120] Peebles 1966; for a recent account of predictions, see Coc, Vangioni-Flam et al. 2004; an accessible account can be found in Weiss 2006; compare with the observations in Olive & Skillman 2004, Bania, Rood & Balser 2002, O'Meara et al. 2001, and Charbonnel & Primas 2005

[121] Lahav & Suto 2004, Bertschinger 1998, Springel et al. 2005

[122] Alpher & Herman 1948, for a pedagogical introduction, see Bergström & Goobar 2003, ch. 11; for the initial detection, see Penzias & Wilson 1965 and, for precision measurements by satellite observatories, Mather et al. 1994 (COBE) and Bennett et al. 2003 (WMAP). Future measurements could also reveal evidence about gravitational waves in the early universe; this additional information is contained in the background radiation's polarization, cf. Kamionkowski, Kosowsky & Stebbins 1997 and Seljak & Zaldarriaga 1997

[123] Evidence for this comes from the determination of cosmological parameters and additional observations involving the dynamics of galaxies and galaxy clusters cf. Peebles 1993, ch. 18, evidence from gravitational lensing, cf. Peacock 1999, sec. 4.6, and simulations of large-scale structure formation, see Springel et al. 2005

[124] Peacock 1999, ch. 12, Peskin 2007; in particular, observations indicate that all but a negligible portion of that matter is not in the form of the usual elementary particles ("non-baryonic matter"), cf. Peacock 1999, ch. 12

[125] Namely, some physicists have questioned whether or not the evidence for dark matter is, in fact, evidence for deviations from the Einsteinian (and the Newtonian) description of gravity cf. the overview in Mannheim 2006, sec. 9

[126] Carroll 2001; an accessible overview is given in Caldwell 2004. Here, too, scientists have argued that the evidence indicates not a new form of energy, but the need for modifications in our cosmological models, cf. Mannheim 2006, sec. 10; aforementioned modifications need not be modifications of general relativity, they could, for example, be modifications in the way we treat the inhomogeneities in the universe, cf. Buchert 2007

[127] A good introduction is Linde 1990; for a more recent review, see Linde 2005

[128] More precisely, these are the flatness problem, the horizon problem, and the monopole problem; a pedagogical introduction can be found in Narlikar 1993, sec. 6.4, see also Börner 1993, sec. 9.1

[129] Spergel et al. 2007, sec. 5,6

[130] More concretely, the potential function that is crucial to determining the dynamics of the inflaton is simply postulated, but not derived from an underlying physical theory

[131] Brandenberger 2007, sec. 2

[132] Frauendiener 2004, Wald 1984, sec. 11.1, Hawking & Ellis 1973, sec. 6.8, 6.9

[133] Wald 1984, sec. 9.2–9.4 and Hawking & Ellis 1973, ch. 6

[134] Thorne 1972; for more recent numerical studies, see Berger 2002, sec. 2.1

[135] Israel 1987. A more exact mathematical description distinguishes several kinds of horizon, notably event horizons and apparent horizons cf. Hawking & Ellis 1973, pp. 312–320 or Wald 1984, sec. 12.2; there are also more intuitive definitions for isolated systems that do not require knowledge of spacetime properties at infinity, cf. Ashtekar & Krishnan 2004

[136] For first steps, cf. Israel 1971; see Hawking & Ellis 1973, sec. 9.3 or Heusler 1996, ch. 9 and 10 for a derivation, and Heusler 1998 as well as Beig & Chruściel 2006 as overviews of more recent results

[137] The laws of black hole mechanics were first described in Bardeen, Carter & Hawking 1973; a more pedagogical presentation can be found in Carter 1979; for a more recent review, see Wald 2001, ch. 2. A thorough, book-length introduction including an introduction to the necessary mathematics Poisson 2004. For the Penrose process, see Penrose 1969

[138] Bekenstein 1973, Bekenstein 1974

[139] The fact that black holes radiate, quantum mechanically, was first derived in Hawking 1975; a more thorough derivation can be found in Wald 1975. A review is given in Wald 2001, ch. 3

[140] Narlikar 1993, sec. 4.4.4, 4.4.5

[141] Horizons: cf. Rindler 2001, sec. 12.4. Unruh effect: Unruh 1976, cf. Wald 2001, ch. 3

[142] Hawking & Ellis 1973, sec. 8.1, Wald 1984, sec. 9.1

[143] Townsend 1997, ch. 2; a more extensive treatment of this solution can be found in Chandrasekhar 1983, ch. 3

[144] Townsend 1997, ch. 4; for a more extensive treatment, cf. Chandrasekhar 1983, ch. 6

[145] Ellis & Van Elst 1999; a closer look at the singularity itself is taken in Börner 1993, sec. 1.2

[146] Here one should remind to the well-known fact that the important "quasi-optical" singularities of the so-called eikonal approximations of many wave-equations, namely the "caustics", are resolved into finite peaks beyond that approximation.

[147] Namely when there are trapped null surfaces, cf. Penrose 1965

[148] Hawking 1966

[149] The conjecture was made in Belinskii, Khalatnikov & Lifschitz 1971; for a more recent review, see Berger 2002. An accessible exposition is given by Garfinkle 2007

[150] The restriction to future singularities naturally excludes initial singularities such as the big bang singularity, which in principle be visible to observers at later cosmic time. The cosmic censorship conjecture was first presented in Penrose 1969; a textbook-level account is given in Wald 1984, pp. 302–305. For numerical results, see the review Berger 2002, sec. 2.1

[151] Hawking & Ellis 1973, sec. 7.1

[152] Arnowitt, Deser & Misner 1962; for a pedagogical introduction, see Misner, Thorne & Wheeler 1973, §21.4–§21.7

[153] Fourès-Bruhat 1952 and Bruhat 1962; for a pedagogical introduction, see Wald 1984, ch. 10; an online review can be found in Reula 1998

[154] Gourgoulhon 2007; for a review of the basics of numerical relativity, including the problems arising from the peculiarities of Einstein's equations, see Lehner 2001

[155] Misner, Thorne & Wheeler 1973, §20.4

[156] Arnowitt, Deser & Misner 1962

[157] Komar 1959; for a pedagogical introduction, see Wald 1984, sec. 11.2; although defined in a totally different way, it can be shown to be equivalent to the ADM mass for stationary spacetimes, cf. Ashtekar & Magnon-Ashtekar 1979

[158] For a pedagogical introduction, see Wald 1984, sec. 11.2

[159] Wald 1984, p. 295 and refs therein; this is important for questions of stability—if there were negative mass states, then flat, empty Minkowski space, which has mass zero, could evolve into these states

[160] Townsend 1997, ch. 5

[161] Such quasi-local mass–energy definitions are the Hawking energy, Geroch energy, or Penrose's quasi-local energy–momentum based on twistor methods; cf. the review article Szabados 2004

[162] An overview of quantum theory can be found in standard textbooks such as Messiah 1999; a more elementary account is given in Hey & Walters 2003

[163] Ramond 1990, Weinberg 1995, Peskin & Schroeder 1995; a more accessible overview is Auyang 1995

[164] Wald 1994, Birrell & Davies 1984

[165] For Hawking radiation Hawking 1975, Wald 1975; an accessible introduction to black hole evaporation can be found in Traschen 2000

[166] Wald 2001, ch. 3

[167] Put simply, matter is the source of spacetime curvature, and once matter has quantum properties, we can expect spacetime to have them as well. Cf. Carlip 2001, sec. 2

[168] Schutz 2003, p. 407

[169] A timeline and overview can be found in Rovelli 2000

[170] Donoghue 1995

[171] In particular, a technique known as renormalization, an integral part of deriving predictions which take into account higher-energy contributions, cf. Weinberg 1996, ch. 17, 18, fails in this case; cf. Goroff & Sagnotti 1985

[172] An accessible introduction at the undergraduate level can be found in Zwiebach 2004; more complete overviews can be found in Polchinski 1998a and Polchinski 1998b

[173] At the energies reached in current experiments, these strings are indistinguishable from point-like particles, but, crucially, different modes of oscillation of one and the same type of fundamental string appear as particles with different (electric and other) charges, e.g. Ibanez 2000. The theory is successful in that one mode will always correspond to a graviton, the messenger particle of gravity, e.g. Green, Schwarz & Witten 1987, sec. 2.3, 5.3

[174] Green, Schwarz & Witten 1987, sec. 4.2

[175] Weinberg 2000, ch. 31

[176] Townsend 1996, Duff 1996

[177] Kuchař 1973, sec. 3

[178] These variables represent geometric gravity using mathematical analogues of electric and magnetic fields; cf. Ashtekar 1986, Ashtekar 1987

[179] For a review, see Thiemann 2006; more extensive accounts can be found in Rovelli 1998, Ashtekar & Lewandowski 2004 as well as in the lecture notes Thiemann 2003

[180] Isham 1994, Sorkin 1997

[181] Loll 1998

[182] Sorkin 2005

[183] Penrose 2004, ch. 33 and refs therein

[184] Hawking 1987

[185] Ashtekar 2007, Schwarz 2007

[186] Maddox 1998, pp. 52–59, 98–122; Penrose 2004, sec. 34.1, ch. 30

[187] section Quantum gravity, above

[188] section Cosmology, above

[189] Friedrich 2005

[190] A review of the various problems and the techniques being developed to overcome them, see Lehner 2002

[191] See Bartusiak 2000 for an account up to that year; up-to-date news can be found on the websites of major detector collaborations such as GEO 600 and LIGO

[192] For the most recent papers on gravitational wave polarizations of inspiralling compact binaries, see Blanchet et al. 2008, and Arun et al. 2007; for a review of work on compact binaries, see Blanchet 2006 and Futamase & Itoh 2006; for a general review of experimental tests of general relativity, see Will 2006

[193] See, e.g., the electronic review journal Living Reviews in Relativity

8.11 References

- Alpher, R. A.; Herman, R. C. (1948), "Evolution of the universe", *Nature* **162** (4124): 774–775, Bibcode:1948Natur.162..774A, doi:10.1038/162774b0

- Anderson, J. D.; Campbell, J. K.; Jurgens, R. F.; Lau, E. L. (1992), "Recent developments in solar-system tests of general relativity", in Sato, H.; Nakamura, T., *Proceedings of the Sixth Marcel Großmann Meeting on General Relativity*, World Scientific, pp. 353–355, ISBN 981-02-0950-9

- Arnold, V. I. (1989), *Mathematical Methods of Classical Mechanics*, Springer, ISBN 3-540-96890-3

- Arnowitt, Richard; Deser, Stanley; Misner, Charles W. (1962), "The dynamics of general relativity", in Witten, Louis, *Gravitation: An Introduction to Current Research*, Wiley, pp. 227–265

- Arun, K.G.; Blanchet, L.; Iyer, B. R.; Qusailah, M. S. S. (2007), "Inspiralling compact binaries in quasi-elliptical orbits: The complete 3PN energy flux", *Physical Review D* **77** (6), arXiv:0711.0302, Bibcode:2008PhRvD..77f4035A, doi:10.1103/PhysRevD.77.064035

- Ashby, Neil (2002), "Relativity and the Global Positioning System" (PDF), *Physics Today* **55** (5): 41–47, Bibcode:2002PhT....55e..41A, doi:10.1063/1.1485583

- Ashby, Neil (2003), "Relativity in the Global Positioning System", *Living Reviews in Relativity* **6**, retrieved 2007-07-06 External link in |work= (help)

- Ashtekar, Abhay (1986), "New variables for classical and quantum gravity", *Phys. Rev. Lett.* **57** (18): 2244–2247, Bibcode:1986PhRvL..57.2244A, doi:10.1103/PhysRevLett.57.2244, PMID 10033673

- Ashtekar, Abhay (1987), "New Hamiltonian formulation of general relativity", *Phys. Rev.* **D36** (6): 1587–1602, Bibcode:1987PhRvD..36.1587A, doi:10.1103/PhysRevD.36.1587

- Ashtekar, Abhay (2007), "LOOP QUANTUM GRAVITY: FOUR RECENT ADVANCES AND A DOZEN FREQUENTLY ASKED QUESTIONS", *The Eleventh Marcel Grossmann Meeting - on Recent Developments in Theoretical and Experimental General Relativity, Gravitation and Relativistic Field Theories - Proceedings of the MG11 Meeting on General Relativity*, p. 126, arXiv:0705.2222, Bibcode:2008mgm..conf..126A, doi:10.1142/9789812834300_0008, ISBN 9789812834263

- Ashtekar, Abhay; Krishnan, Badri (2004), "Isolated and Dynamical Horizons and Their Applications", *Living Rev. Relativity* **7**, arXiv:gr-qc/0407042, Bibcode:2004LRR.....7...10A, doi:10.12942/lrr-2004-10, retrieved 2007-08-28

- Ashtekar, Abhay; Lewandowski, Jerzy (2004), "Background Independent Quantum Gravity: A Status Report", *Class. Quant. Grav.* **21** (15): R53–R152, arXiv:gr-qc/0404018, Bibcode:2004CQGra..21R..53A, doi:10.1088/0264-9381/21/15/R01

- Ashtekar, Abhay; Magnon-Ashtekar, Anne (1979), "On conserved quantities in general relativity", *Journal of Mathematical Physics* **20** (5): 793–800, Bibcode:1979JMP....20..793A, doi:10.1063/1.524151

- Auyang, Sunny Y. (1995), *How is Quantum Field Theory Possible?*, Oxford University Press, ISBN 0-19-509345-3

- Bania, T. M.; Rood, R. T.; Balser, D. S. (2002), "The cosmological density of baryons from observations of 3He+ in the Milky Way", *Nature* **415** (6867): 54–57, Bibcode:2002Natur.415...54B, doi:10.1038/415054a, PMID 11780112

- Barack, Leor; Cutler, Curt (2004), "LISA Capture Sources: Approximate Waveforms, Signal-to-Noise Ratios, and Parameter Estimation Accuracy", *Phys. Rev.* **D69** (8): 082005, arXiv:gr-qc/0310125, Bibcode:2004PhRvD..69h2005B, doi:10.1103/PhysRevD.69.082005

- Bardeen, J. M.; Carter, B.; Hawking, S. W. (1973), "The Four Laws of Black Hole Mechanics", *Comm. Math. Phys.* **31** (2): 161–170, Bibcode:1973CMaPh..31..161B, doi:10.1007/BF01645742

- Barish, Barry (2005), "Towards detection of gravitational waves", in Florides, P.; Nolan, B.; Ottewil, A., *General Relativity and Gravitation. Proceedings of the 17th International Conference*, World Scientific, pp. 24–34, ISBN 981-256-424-1

- Barstow, M; Bond, Howard E.; Holberg, J. B.; Burleigh, M. R.; Hubeny, I.; Koester, D. (2005), "Hubble Space Telescope Spectroscopy of the Balmer lines in Sirius B", *Mon. Not. Roy. Astron. Soc.* **362** (4): 1134–1142, arXiv:astro-ph/0506600, Bibcode:2005MNRAS.362.1134B, doi:10.1111/j.1365-2966.2005.09359.x

- Bartusiak, Marcia (2000), *Einstein's Unfinished Symphony: Listening to the Sounds of Space-Time*, Berkley, ISBN 978-0-425-18620-6

- Begelman, Mitchell C.; Blandford, Roger D.; Rees, Martin J. (1984), "Theory of extragalactic radio sources", *Rev. Mod. Phys.* **56**

(2): 255–351, Bibcode:1984RvMP...56..255B, doi:10.1103/RevModPhys.56.255

- Beig, Robert; Chruściel, Piotr T. (2006), "Stationary black holes", in Françoise, J.-P.; Naber, G.; Tsou, T.S., *Encyclopedia of Mathematical Physics, Volume 2*, Elsevier, p. 2041, arXiv:gr-qc/0502041, Bibcode:2005gr.qc.....2041B, ISBN 0-12-512660-3

- Bekenstein, Jacob D. (1973), "Black Holes and Entropy", *Phys. Rev.* **D7** (8): 2333–2346, Bibcode:1973PhRvD...7.2333B, doi:10.1103/PhysRevD.7.2333

- Bekenstein, Jacob D. (1974), "Generalized Second Law of Thermodynamics in Black-Hole Physics", *Phys. Rev.* **D9** (12): 3292–3300, Bibcode:1974PhRvD...9.3292B, doi:10.1103/PhysRevD.9.3292

- Belinskii, V. A.; Khalatnikov, I. M.; Lifschitz. E. M. (1971), "Oscillatory approach to the singular point in relativistic cosmology", *Advances in Physics* **19** (80): 525–573, Bibcode:1970AdPhy..19..525B, doi:10.1080/00018737000101171; original paper in Russian: Belinsky, V. A.; Lifshits, I. M.; Khalatnikov, E. M. (1970), "Колебательный Режим Приближения К Особой Точке В Релятивистской Космологии", *Uspekhi Fizicheskikh Nauk (Успехи Физических Наук)*, 102(3) (11): 463–500, Bibcode:1970UsFiN.102..463B

- Bennett, C. L.; Halpern, M.; Hinshaw, G.; Jarosik, N.; Kogut, A.; Limon, M.; Meyer, S. S.; Page, L.; et al. (2003), "First Year Wilkinson Microwave Anisotropy Probe (WMAP) Observations: Preliminary Maps and Basic Results", *Astrophys. J. Suppl.* **148** (1): 1–27, arXiv:astro-ph/0302207, Bibcode:2003ApJS..148....1B, doi:10.1086/377253

- Berger, Beverly K. (2002), "Numerical Approaches to Spacetime Singularities", *Living Rev. Relativity* **5**, arXiv:gr-qc/0201056, Bibcode:2002LRR.....5....1B, doi:10.12942/lrr-2002-1, retrieved 2007-08-04

- Bergström, Lars; Goobar, Ariel (2003), *Cosmology and Particle Astrophysics* (2nd ed.), Wiley & Sons, ISBN 3-540-43128-4

- Bertotti, Bruno; Ciufolini, Ignazio; Bender, Peter L. (1987), "New test of general relativity: Measurement of de Sitter geodetic precession rate for lunar perigee", *Physical Review Letters* **58** (11): 1062–1065, Bibcode:1987PhRvL..58.1062B, doi:10.1103/PhysRevLett.58.1062, PMID 10034329

- Bertotti, Bruno; Iess, L.; Tortora, P. (2003), "A test of general relativity using radio links with the Cassini spacecraft", *Nature* **425** (6956): 374–376, Bibcode:2003Natur.425..374B, doi:10.1038/nature01997, PMID 14508481

- Bertschinger, Edmund (1998), "Simulations of structure formation in the universe", *Annu. Rev. Astron. Astrophys.* **36** (1): 599–654, Bibcode:1998ARA&A..36..599B, doi:10.1146/annurev.astro.36.1.599

- Birrell, N. D.; Davies, P. C. (1984), *Quantum Fields in Curved Space*, Cambridge University Press, ISBN 0-521-27858-9

- Blair, David; McNamara, Geoff (1997), *Ripples on a Cosmic Sea. The Search for Gravitational Waves*, Perseus, ISBN 0-7382-0137-5

- Blanchet, L.; Faye, G.; Iyer, B. R.; Sinha, S. (2008), "The third post-Newtonian gravitational wave polarisations and associated spherical harmonic modes for inspiralling compact binaries in quasi-circular orbits", *Classical and Quantum Gravity* **25** (16): 165003, arXiv:0802.1249, Bibcode:2008CQGra..25p5003B, doi:10.1088/0264-9381/25/16/165003

- Blanchet, Luc (2006), "Gravitational Radiation from Post-Newtonian Sources and Inspiralling Compact Binaries", *Living Rev. Relativity* **9**, Bibcode:2006LRR.....9....4B, doi:10.12942/lrr-2006-4, retrieved 2007-08-07

- Blandford, R. D. (1987), "Astrophysical Black Holes", in Hawking, Stephen W.; Israel, Werner, *300 Years of Gravitation*, Cambridge University Press, pp. 277–329, ISBN 0-521-37976-8

- Börner, Gerhard (1993), *The Early Universe. Facts and Fiction*, Springer, ISBN 0-387-56729-1

- Brandenberger, Robert H. (2007), "Conceptual Problems of Inflationary Cosmology and a New Approach to Cosmological Structure Formation", *Inflationary Cosmology*, Lecture Notes in Physics **738**, p. 393, arXiv:hep-th/0701111, Bibcode:2008LNP...738..393B, doi:10.1007/978-3-540-74353-8_11, ISBN 978-3-540-74352-1

- Brans, C. H.; Dicke, R. H. (1961), "Mach's Principle and a Relativistic Theory of Gravitation", *Physical Review* **124** (3): 925–935, Bibcode:1961PhRv..124..925B, doi:10.1103/PhysRev.124.925

- Bridle, Sarah L.; Lahav, Ofer; Ostriker, Jeremiah P.; Steinhardt, Paul J. (2003), "Precision Cosmology? Not Just Yet", *Science* **299** (5612): 1532–1533, arXiv:astro-ph/0303180, Bibcode:2003Sci...299.1532B, doi:10.1126/science.1082158, PMID 12624255

- Bruhat, Yvonne (1962), "The Cauchy Problem", in Witten, Louis, *Gravitation: An Introduction to Current Research*, Wiley, p. 130, ISBN 978-1-114-29166-9

- Buchert, Thomas (2007), "Dark Energy from Structure—A Status Report", *General Relativity and Gravitation* **40** (2–3): 467–527, arXiv:0707.2153, Bibcode:2008GReGr..40..467B, doi:10.1007/s10714-007-0554-8

- Buras, R.; Rampp, M.; Janka, H.-Th.; Kifonidis, K. (2003), "Improved Models of Stellar Core Collapse and Still no Explosions: What is Missing?", *Phys. Rev. Lett.* **90** (24): 241101, arXiv:astro-ph/0303171, Bibcode:2003PhRvL..90x1101B, doi:10.1103/PhysRevLett.90.241101, PMID 12857181

- Caldwell, Robert R. (2004), "Dark Energy", *Physics World* **17** (5): 37–42

- Carlip, Steven (2001), "Quantum Gravity: a Progress Report", *Rept. Prog. Phys.* **64** (8): 885–942, arXiv:gr-qc/0108040, Bibcode:2001RPPh...64..885C, doi:10.1088/0034-4885/64/8/301

- Carroll, Bradley W.; Ostlie, Dale A. (1996), *An Introduction to Modern Astrophysics*, Addison-Wesley, ISBN 0-201-54730-9

- Carroll, Sean M. (2001), "The Cosmological Constant", *Living Rev. Relativity* **4**, arXiv:astro-ph/0004075, Bibcode:2001LRR.....4....1C, doi:10.12942/lrr-2001-1, retrieved 2007-07-21

- Carter, Brandon (1979), "The general theory of the mechanical, electromagnetic and thermodynamic properties of black holes", in Hawking, S. W.; Israel, W., *General Relativity, an Einstein Centenary Survey*, Cambridge University Press, pp. 294–369 and 860–863, ISBN 0-521-29928-4

- Celotti, Annalisa; Miller, John C.; Sciama, Dennis W. (1999), "Astrophysical evidence for the existence of black holes", *Class. Quant. Grav.* **16** (12A): A3–A21, arXiv:astro-ph/9912186, doi:10.1088/0264-9381/16/12A/301

- Chandrasekhar, Subrahmanyan (1983), *The Mathematical Theory of Black Holes*, Oxford University Press, ISBN 0-19-850370-9

- Charbonnel, C.; Primas, F. (2005), "The Lithium Content of the Galactic Halo Stars", *Astronomy & Astrophysics* **442** (3): 961–992, arXiv:astro-ph/0505247, Bibcode:2005A&A...442..961C, doi:10.1051/0004-6361:20042491

- Ciufolini, Ignazio; Pavlis, Erricos C. (2004), "A confirmation of the general relativistic prediction of the Lense-Thirring effect", *Nature* **431** (7011): 958–960, Bibcode:2004Natur.431..958C, doi:10.1038/nature03007, PMID 15496915

- Ciufolini, Ignazio; Pavlis, Erricos C.; Peron, R. (2006), "Determination of frame-dragging using Earth gravity models from CHAMP and GRACE", *New Astron.* **11** (8): 527–550, Bibcode:2006NewA...11..527C, doi:10.1016/j.newast.2006.02.001

- Coc, A.; Vangioni-Flam, Elisabeth; Descouvemont, Pierre; Adahchour, Abderrahim; Angulo, Carmen (2004), "Updated Big Bang Nucleosynthesis confronted to WMAP observations and to the Abundance of Light Elements", *Astrophysical Journal* **600** (2): 544–552, arXiv:astro-ph/0309480, Bibcode:2004ApJ...600..544C, doi:10.1086/380121

- Cutler, Curt; Thorne, Kip S. (2002), "An overview of gravitational wave sources", in Bishop, Nigel; Maharaj, Sunil D., *Proceedings of 16th International Conference on General Relativity and Gravitation (GR16)*, World Scientific, p. 4090, arXiv:gr-qc/0204090, Bibcode:2002gr.qc.....4090C, ISBN 981-238-171-6

- Dalal, Neal; Holz, Daniel E.; Hughes, Scott A.; Jain, Bhuvnesh (2006), "Short GRB and binary black hole standard sirens as a probe of dark energy", *Phys.Rev.* **D74** (6): 063006, arXiv:astro-ph/0601275, Bibcode:2006PhRvD..74f3006D, doi:10.1103/PhysRevD.74.063006

- Danzmann, Karsten; Rüdiger, Albrecht (2003), "LISA Technology—Concepts, Status, Prospects" (PDF), *Class. Quant. Grav.* **20** (10): S1–S9, Bibcode:2003CQGra..20S...1D, doi:10.1088/0264-9381/20/10/301

- Dirac, Paul (1996), *General Theory of Relativity*, Princeton University Press, ISBN 0-691-01146-X

- Donoghue, John F. (1995), "Introduction to the Effective Field Theory Description of Gravity", in Cornet, Fernando, *Effective Theories: Proceedings of the Advanced School, Almunecar, Spain, 26 June–1 July 1995*, Singapore: World Scientific, p. 12024, arXiv:gr-qc/9512024, Bibcode:1995gr.qc....12024D, ISBN 981-02-2908-9

- Duff, Michael (1996), "M-Theory (the Theory Formerly Known as Strings)", *Int. J. Mod. Phys.* **A11** (32): 5623–5641, arXiv:hep-th/9608117, Bibcode:1996IJMPA..11.5623D, doi:10.1142/S0217751X96002583

- Ehlers, Jürgen (1973), "Survey of general relativity theory", in Israel, Werner, *Relativity, Astrophysics and Cosmology*, D. Reidel, pp. 1–125, ISBN 90-277-0369-8

- Ehlers, Jürgen; Falco, Emilio E.; Schneider, Peter (1992), *Gravitational lenses*, Springer, ISBN 3-540-66506-4

- Ehlers, Jürgen; Lämmerzahl, Claus, eds. (2006), *Special Relativity—Will it Survive the Next 101 Years?*, Springer, ISBN 3-540-34522-1

- Ehlers, Jürgen; Rindler, Wolfgang (1997), "Local and Global Light Bending in Einstein's and other Gravitational Theories", *General Relativity and Gravitation* **29** (4): 519–529, Bibcode:1997GReGr..29..519E, doi:10.1023/A:1018843001842

- Einstein, Albert (1907), "Über das Relativitätsprinzip und die aus demselben gezogene Folgerungen" (PDF), *Jahrbuch der Radioaktivität und Elektronik* **4**: 411, retrieved 2008-05-05

- Einstein, Albert (1915), "Die Feldgleichungen der Gravitation", *Sitzungsberichte der Preussischen Akademie der Wissenschaften zu Berlin*: 844–847, retrieved 2006-09-12

- Einstein, Albert (1916), "Die Grundlage der allgemeinen Relativitätstheorie", *Annalen der Physik* **49**: 769–822, Bibcode:1916AnP...354..769E, doi:10.1002/andp.19163540702, archived from the original (PDF) on 2006-08-29, retrieved 2006-09-03

- Einstein, Albert (1917), "Kosmologische Betrachtungen zur allgemeinen Relativitätstheorie", *Sitzungsberichte der Preußischen Akademie der Wissenschaften*: 142

- Ellis, George F R; Van Elst, Henk (1999), Lachièze-Rey, Marc, ed., "Theoretical and Observational Cosmology: Cosmological models (Cargèse lectures 1998)", *Theoretical and observational cosmology : proceedings of the NATO Advanced Study Institute on Theoretical and Observational Cosmology* (Kluwer): 1–116, arXiv:gr-qc/9812046, Bibcode:1999toc..conf....1E, doi:10.1007/978-94-011-4455-1_1, ISBN 978-0-7923-5946-3

- Everitt, C. W. F.; Buchman, S.; DeBra, D. B.; Keiser, G. M. (2001), "Gravity Probe B: Countdown to launch", in Lämmerzahl, C.; Everitt, C. W. F.; Hehl, F. W., *Gyros, Clocks, and Interferometers: Testing Relativistic Gravity in Space (Lecture Notes in Physics 562)*, Springer, pp. 52–82, ISBN 3-540-41236-0

- Everitt, C. W. F.; Parkinson, Bradford; Kahn, Bob (2007), *The Gravity Probe B experiment. Post Flight Analysis—Final Report (Preface and Executive Summary)* (PDF), Project Report: NASA, Stanford University and Lockheed Martin, retrieved 2007-08-05

- Falcke, Heino; Melia, Fulvio; Agol, Eric (2000), "Viewing the Shadow of the Black Hole at the Galactic Center", *Astrophysical Journal* **528** (1): L13–L16, arXiv:astro-ph/9912263, Bibcode:2000ApJ...528L..13F, doi:10.1086/312423, PMID 10587484

- Flanagan, Éanna É.; Hughes, Scott A. (2005), "The basics of gravitational wave theory", *New J.Phys.* **7**: 204, arXiv:gr-qc/0501041, Bibcode:2005NJPh....7..204F, doi:10.1088/1367-2630/7/1/204

- Font, José A. (2003), "Numerical Hydrodynamics in General Relativity", *Living Rev. Relativity* **6**, doi:10.12942/lrr-2003-4, retrieved 2007-08-19

- Fourès-Bruhat, Yvonne (1952), "Théoréme d'existence pour certains systémes d'équations aux derivées partielles non linéaires", *Acta Mathematica* **88** (1): 141–225, Bibcode:1952AcM....88..141F, doi:10.1007/BF02392131

- Frauendiener, Jörg (2004), "Conformal Infinity", *Living Rev. Relativity* **7**, Bibcode:2004LRR.....7....1F, doi:10.12942/lrr-2004-1, retrieved 2007-07-21

- Friedrich, Helmut (2005), "Is general relativity 'essentially understood'?", *Annalen Phys.* **15** (1–2): 84–108, arXiv:gr-qc/0508016, Bibcode:2006AnP...518...84F, doi:10.1002/andp.200510173

- Futamase, T.; Itoh, Y. (2006), "The Post-Newtonian Approximation for Relativistic Compact Binaries", *Living Rev. Relativity* **10**, retrieved 2008-02-29

- Gamow, George (1970), *My World Line*, Viking Press, ISBN 0-670-50376-2

- Garfinkle, David (2007), "Of singularities and breadmaking", *Einstein Online*, retrieved 2007-08-03 External link in |work= (help)

- Geroch, Robert (1996). "Partial Differential Equations of Physics". arXiv:gr-qc/9602055 [gr-qc].

- Giulini, Domenico (2005), *Special Relativity: A First Encounter*, Oxford University Press, ISBN 0-19-856746-4

- Giulini, Domenico (2006a), "Algebraic and Geometric Structures in Special Relativity", in Ehlers, Jürgen; Lämmerzahl, Claus, *Special Relativity—Will it Survive the Next 101 Years?*, Springer, pp. 45–111, arXiv:math-ph/0602018, Bibcode:2006math.ph...2018G, ISBN 3-540-34522-1

- Giulini, Domenico (2006b), Stamatescu, I. O., ed., "An assessment of current paradigms in the physics of fundamental interactions: Some remarks on the notions of general covariance and background independence", *Approaches to Fundamental Physics*, Lecture Notes in Physics (Springer) **721**: 105, arXiv:gr-qc/0603087, Bibcode:2007LNP...721..105G, doi:10.1007/978-3-540-71117-9_6, ISBN 978-3-540-71115-5

- Gnedin, Nickolay Y. (2005), "Digitizing the Universe", *Nature* **435** (7042): 572–573, Bibcode:2005Natur.435..572G, doi:10.1038/435572a, PMID 15931201

- Goenner, Hubert F. M. (2004), "On the History of Unified Field Theories", *Living Rev. Relativity* **7**, Bibcode:2004LRR.....7....2G, doi:10.12942/lrr-2004-2, retrieved 2008-02-28

- Goroff, Marc H.; Sagnotti, Augusto (1985), "Quantum gravity at two loops", *Phys. Lett.* **160B** (1–3): 81–86, Bibcode:1985PhLB..160...81G, doi:10.1016/0370-2693(85)91470-4

- Gourgoulhon, Eric (2007). "3+1 Formalism and Bases of Numerical Relativity". arXiv:gr-qc/0703035 [gr-qc].

- Gowdy, Robert H. (1971), "Gravitational Waves in Closed Universes", *Phys. Rev. Lett.* **27** (12): 826–829, Bibcode:1971PhRvL..27..826G, doi:10.1103/PhysRevLett.27.826

- Gowdy, Robert H. (1974), "Vacuum spacetimes with two-parameter spacelike isometry groups and compact invariant hypersurfaces: Topologies and boundary conditions", *Ann. Phys. (N.Y.)* **83** (1): 203–241, Bibcode:1974AnPhy..83..203G, doi:10.1016/0003-4916(74)90384-4

- Green, M. B.; Schwarz, J. H.; Witten, E. (1987), *Superstring theory. Volume 1: Introduction*, Cambridge University Press, ISBN 0-521-35752-7

- Greenstein, J. L.; Oke, J. B.; Shipman, H. L. (1971), "Effective Temperature, Radius, and Gravitational Redshift of Sirius B", *Astrophysical Journal* **169**: 563, Bibcode:1971ApJ...169..563G, doi:10.1086/151174

- Hafele, J. C.; Keating, R. E. (July 14, 1972). "Around-the-World Atomic Clocks: Predicted Relativistic Time Gains". *Science* **177** (4044): 166–168. Bibcode:1972Sci...177..166H. doi:10.1126/science.177.4044.166. PMID 17779917.

- Hafele, J. C.; Keating, R. E. (July 14, 1972). "Around-the-World Atomic Clocks: Observed Relativistic Time Gains". *Science* **177** (4044): 168–170. Bibcode:1972Sci...177..168H. doi:10.1126/science.177.4044.168. PMID 17779918.

- Havas, P. (1964), "Four-Dimensional Formulation of Newtonian Mechanics and Their Relation to the Special and the General Theory of Relativity", *Rev. Mod. Phys.* **36** (4): 938–965, Bibcode:1964RvMP...36..938H, doi:10.1103/RevModPhys.36.938

- Hawking, Stephen W. (1966), "The occurrence of singularities in cosmology", *Proceedings of the Royal Society* **A294** (1439): 511–521, Bibcode:1966RSPSA.294..511H, doi:10.1098/rspa.1966.0221

- Hawking, S. W. (1975), "Particle Creation by Black Holes", *Communications in Mathematical Physics* **43** (3): 199–220, Bibcode:1975CMaPh..43..199H, doi:10.1007/BF02345020

- Hawking, Stephen W. (1987), "Quantum cosmology", in Hawking, Stephen W.; Israel, Werner, *300 Years of Gravitation*, Cambridge University Press, pp. 631–651, ISBN 0-521-37976-8

- Hawking, Stephen W.; Ellis, George F. R. (1973), *The large scale structure of space-time*, Cambridge University Press, ISBN 0-521-09906-4

- Heckmann, O. H. L.; Schücking, E. (1959), "Newtonsche und Einsteinsche Kosmologie", in Flügge, S., *Encyclopedia of Physics* **53**, p. 489

- Heusler, Markus (1998), "Stationary Black Holes: Uniqueness and Beyond", *Living Rev. Relativity* **1**, doi:10.12942/lrr-1998-6, retrieved 2007-08-04

- Heusler, Markus (1996), *Black Hole Uniqueness Theorems*, Cambridge University Press, ISBN 0-521-56735-1

- Hey, Tony; Walters, Patrick (2003), *The new quantum universe*, Cambridge University Press, ISBN 0-521-56457-3

- Hough, Jim; Rowan, Sheila (2000), "Gravitational Wave Detection by Interferometry (Ground and Space)", *Living Rev. Relativity* **3**, retrieved 2007-07-21

- Hubble, Edwin (1929), "A Relation between Distance and Radial Velocity among Extra-Galactic Nebulae" (PDF), *Proc. Nat. Acad. Sci.* **15** (3): 168–173, Bibcode:1929PNAS...15..168H, doi:10.1073/pnas.15.3.168, PMC 522427, PMID 16577160

- Hulse, Russell A.; Taylor, Joseph H. (1975), "Discovery of a pulsar in a binary system", *Astrophys. J.* **195**: L51–L55, Bibcode:1975ApJ...195L..51H, doi:10.1086/181708

- Ibanez, L. E. (2000), "The second string (phenomenology) revolution", *Class. Quant. Grav.* **17** (5): 1117–1128, arXiv:hep-ph/9911499, Bibcode:2000CQGra..17.1117I, doi:10.1088/0264-9381/17/5/321

- Iorio, L. (2009), "An Assessment of the Systematic Uncertainty in Present and Future Tests of the Lense-Thirring Effect with Satellite Laser Ranging", *Space Sci. Rev.* **148** (1–4): 363, arXiv:0809.1373, Bibcode:2009SSRv..148..363I, doi:10.1007/s11214-008-9478-1

- Isham, Christopher J. (1994), "Prima facie questions in quantum gravity", in Ehlers, Jürgen; Friedrich, Helmut, *Canonical Gravity: From Classical to Quantum*, Springer, ISBN 3-540-58339-4

- Israel, Werner (1971), "Event Horizons and Gravitational Collapse", *General Relativity and Gravitation* **2** (1): 53–59, Bibcode:1971GReGr...2...53I, doi:10.1007/BF02450518

- Israel, Werner (1987), "Dark stars: the evolution of an idea", in Hawking, Stephen W.; Israel, Werner, *300 Years of Gravitation*, Cambridge University Press, pp. 199–276, ISBN 0-521-37976-8

- Janssen, Michel (2005), "Of pots and holes: Einstein's bumpy road to general relativity" (PDF), *Ann. Phys. (Leipzig)* **14** (S1): 58–85, Bibcode:2005AnP...517S..58J, doi:10.1002/andp.200410130

- Jaranowski, Piotr; Królak, Andrzej (2005), "Gravitational-Wave Data Analysis. Formalism and Sample Applications: The Gaussian Case", *Living Rev. Relativity* **8**, doi:10.12942/lrr-2005-3, retrieved 2007-07-30

- Kahn, Bob (1996–2012), *Gravity Probe B Website*, Stanford University, retrieved 2012-04-20

- Kahn, Bob (April 14, 2007), *Was Einstein right? Scientists provide first public peek at Gravity Probe B results (Stanford University Press Release)* (PDF), Stanford University News Service

- Kamionkowski, Marc; Kosowsky, Arthur; Stebbins, Albert (1997), "Statistics of Cosmic Microwave Background Polarization", *Phys. Rev.* **D55** (12): 7368–7388, arXiv:astro-ph/9611125, Bibcode:1997PhRvD..55.7368K, doi:10.1103/PhysRevD.55.7368

- Kennefick, Daniel (2005), "Astronomers Test General Relativity: Light-bending and the Solar Redshift", in Renn, Jürgen. *One hundred authors for Einstein*, Wiley-VCH, pp. 178–181, ISBN 3-527-40574-7

- Kennefick, Daniel (2007), "Not Only Because of Theory: Dyson, Eddington and the Competing Myths of the 1919 Eclipse Expedition", *Proceedings of the 7th Conference on the History of General Relativity, Tenerife, 2005* **0709**, p. 685, arXiv:0709.0685, Bibcode:2007arXiv0709.0685K

- Kenyon, I. R. (1990), *General Relativity*, Oxford University Press, ISBN 0-19-851996-6

- Kochanek, C.S.; Falco, E.E.; Impey, C.; Lehar, J. (2007), *CASTLES Survey Website*, Harvard-Smithsonian Center for Astrophysics, retrieved 2007-08-21

- Komar, Arthur (1959), "Covariant Conservation Laws in General Relativity", *Phys. Rev.* **113** (3): 934–936, Bibcode:1959PhRv..113..934K, doi:10.1103/PhysRev.113.934

- Kramer, Michael (2004), Karshenboim, S. G.; Peik, E., eds., "Astrophysics, Clocks and Fundamental Constants: Millisecond Pulsars as Tools of Fundamental Physics", *Lecture Notes in Physics* (Springer) **648**: 33–54, arXiv:astro-ph/0405178, Bibcode:2004LNP...648...33K, doi:10.1007/978-3-540-40991-5_3, ISBN 978-3-540-21967-5

- Kramer, M.; Stairs, I. H.; Manchester, R. N.; McLaughlin, M. A.; Lyne, A. G.; Ferdman, R. D.; Burgay, M.; Lorimer, D. R.; et al. (2006), "Tests of general relativity from timing the double pulsar", *Science* **314** (5796): 97–102, arXiv:astro-ph/0609417, Bibcode:2006Sci...314...97K, doi:10.1126/science.1132305, PMID 16973838

• Kraus, Ute (1998), "Light Deflection Near Neutron Stars", *Relativistic Astrophysics*, Vieweg, pp. 66–81, ISBN 3-528-06909-0

• Kuchař, Karel (1973), "Canonical Quantization of Gravity", in Israel, Werner, *Relativity, Astrophysics and Cosmology*, D. Reidel, pp. 237–288, ISBN 90-277-0369-8

• Künzle, H. P. (1972), "Galilei and Lorentz Structures on spacetime: comparison of the corresponding geometry and physics", *Ann. Inst. Henri Poincaré a* **17**: 337–362

• Lahav, Ofer; Suto, Yasushi (2004), "Measuring our Universe from Galaxy Redshift Surveys", *Living Rev. Relativity* **7**, arXiv:astro-ph/0310642, Bibcode:2004LRR.....7....8L, doi:10.12942/lrr-2004-8, retrieved 2007-08-19

• Landgraf, M.; Hechler, M.; Kemble, S. (2005), "Mission design for LISA Pathfinder", *Class. Quant. Grav.* **22** (10): S487–S492, arXiv:gr-qc/0411071, Bibcode:2005CQGra..22S.487L, doi:10.1088/0264-9381/22/10/048

• Lehner, Luis (2001), "Numerical Relativity: A review", *Class. Quant. Grav.* **18** (17): R25–R86, arXiv:gr-qc/0106072, Bibcode:2001CQGra..18R..25L, doi:10.1088/0264-9381/18/17/202

• Lehner, Luis (2002), "NUMERICAL RELATIVITY: STATUS AND PROSPECTS", *General Relativity and Gravitation - Proceedings of the 16th International Conference*, p. 210, arXiv:gr-qc/0202055, Bibcode:2002grg..conf..210L, doi:10.1142/9789812776556_0010, ISBN 9789812381712

• Linde, Andrei (1990), *Particle Physics and Inflationary Cosmology*, Harwood, p. 3203, arXiv:hep-th/0503203, Bibcode:2005hep.th....3203L, ISBN 3-7186-0489-2

• Linde, Andrei (2005), "Towards inflation in string theory", *J. Phys. Conf. Ser.* **24**: 151–160, arXiv:hep-th/0503195, Bibcode:2005JPhCS..24..151L, doi:10.1088/1742-6596/24/1/018

• Loll, Renate (1998), "Discrete Approaches to Quantum Gravity in Four Dimensions", *Living Rev. Relativity* **1**, arXiv:gr-qc/9805049, Bibcode:1998LRR.....1...13L, doi:10.12942/lrr-1998-13, retrieved 2008-03-09

• Lovelock, David (1972), "The Four-Dimensionality of Space and the Einstein Tensor", *J. Math. Phys.* **13** (6): 874–876, Bibcode:1972JMP....13..874L, doi:10.1063/1.1666069

• Ludyk, Günter (2013). *Einstein in Matrix Form* (1st ed.). Berlin: Springer. ISBN 9783642357978.

• MacCallum, M. (2006), "Finding and using exact solutions of the Einstein equations", in Mornas, L.; Alonso, J. D., *A Century of Relativity Physics (ERE05, the XXVIII Spanish Relativity Meeting)* **841**, American Institute of Physics, p. 129, arXiv:gr-qc/0601102, Bibcode:2006AIPC..841..129M, doi:10.1063/1.2218172

• Maddox, John (1998), *What Remains To Be Discovered*, Macmillan, ISBN 0-684-82292-X

• Mannheim, Philip D. (2006), "Alternatives to Dark Matter and Dark Energy", *Prog. Part. Nucl. Phys.* **56** (2): 340–445, arXiv:astro-ph/0505266, Bibcode:2006PrPNP..56..340M, doi:10.1016/j.ppnp.2005.08.001

• Mather, J. C.; Cheng, E. S.; Cottingham, D. A.; Eplee, R. E.; Fixsen, D. J.; Hewagama, T.; Isaacman, R. B.; Jensen, K. A.; et al. (1994), "Measurement of the cosmic microwave spectrum by the COBE FIRAS instrument", *Astrophysical Journal* **420**: 439–444, Bibcode:1994ApJ...420..439M, doi:10.1086/173574

• Mermin, N. David (2005), *It's About Time. Understanding Einstein's Relativity*, Princeton University Press, ISBN 0-691-12201-6

• Messiah, Albert (1999), *Quantum Mechanics*, Dover Publications, ISBN 0-486-40924-4

• Miller, Cole (2002), *Stellar Structure and Evolution (Lecture notes for Astronomy 606)*, University of Maryland, retrieved 2007-07-25

• Misner, Charles W.; Thorne, Kip. S.; Wheeler, John A. (1973), *Gravitation*, W. H. Freeman, ISBN 0-7167-0344-0

• Møller, Christian (1952), *The Theory of Relativity* (3rd ed.), Oxford University Press

• Narayan, Ramesh (2006), "Black holes in astrophysics", *New Journal of Physics* **7**: 199, arXiv:gr-qc/0506078, Bibcode:2005NJPh....7..199N, doi:10.1088/1367-2630/7/1/199

• Narayan, Ramesh; Bartelmann, Matthias (1997). "Lectures on Gravitational Lensing". arXiv:astro-ph/9606001 [astro-ph].

- Narlikar, Jayant V. (1993), *Introduction to Cosmology*, Cambridge University Press, ISBN 0-521-41250-1

- Nieto, Michael Martin (2006), "The quest to understand the Pioneer anomaly" (PDF), *EurophysicsNews* **37** (6): 30–34, Bibcode:2006ENews..37...30N, doi:10.1051/epn:2006604

- Nordström, Gunnar (1918), "On the Energy of the Gravitational Field in Einstein's Theory", *Verhandl. Koninkl. Ned. Akad. Wetenschap.*, **26**: 1238–1245

- Nordtvedt, Kenneth (2003). "Lunar Laser Ranging—a comprehensive probe of post-Newtonian gravity". arXiv:gr-qc/0301024 [gr-qc].

- Norton, John D. (1985), "What was Einstein's principle of equivalence?" (PDF), *Studies in History and Philosophy of Science* **16** (3): 203–246, doi:10.1016/0039-3681(85)90002-0, retrieved 2007-06-11

- Ohanian, Hans C.; Ruffini, Remo (1994), *Gravitation and Spacetime*, W. W. Norton & Company, ISBN 0-393-96501-5

- Olive, K. A.; Skillman, E. A. (2004), "A Realistic Determination of the Error on the Primordial Helium Abundance", *Astrophysical Journal* **617** (1): 29–49, arXiv:astro-ph/0405588, Bibcode:2004ApJ...617...29O, doi:10.1086/425170

- O'Meara, John M.; Tytler, David; Kirkman, David; Suzuki, Nao; Prochaska, Jason X.; Lubin, Dan; Wolfe, Arthur M. (2001), "The Deuterium to Hydrogen Abundance Ratio Towards a Fourth QSO: HS0105+1619", *Astrophysical Journal* **552** (2): 718–730, arXiv:astro-ph/0011179, Bibcode:2001ApJ...552..718O, doi:10.1086/320579

- Oppenheimer, J. Robert; Snyder, H. (1939), "On continued gravitational contraction", *Physical Review* **56** (5): 455–459, Bibcode:1939PhRv...56..455O, doi:10.1103/PhysRev.56.455

- Overbye, Dennis (1999), *Lonely Hearts of the Cosmos: the story of the scientific quest for the secret of the Universe*, Back Bay, ISBN 0-316-64896-5

- Pais, Abraham (1982), *'Subtle is the Lord...' The Science and life of Albert Einstein*, Oxford University Press, ISBN 0-19-853907-X

- Peacock, John A. (1999), *Cosmological Physics*, Cambridge University Press, ISBN 0-521-41072-X

- Peebles, P. J. E. (1966), "Primordial Helium abundance and primordial fireball II", *Astrophysical Journal* **146**: 542–552, Bibcode:1966ApJ...146..542P, doi:10.1086/148918

- Peebles, P. J. E. (1993), *Principles of physical cosmology*, Princeton University Press, ISBN 0-691-01933-9

- Peebles, P.J.E.; Schramm, D.N.; Turner, E.L.; Kron, R.G. (1991), "The case for the relativistic hot Big Bang cosmology", *Nature* **352** (6338): 769–776, Bibcode:1991Natur.352..769P, doi:10.1038/352769a0

- Penrose, Roger (1965), "Gravitational collapse and spacetime singularities", *Physical Review Letters* **14** (3): 57–59, Bibcode:1965PhRvL..14...57P, doi:10.1103/PhysRevLett.14.57

- Penrose, Roger (1969), "Gravitational collapse: the role of general relativity", *Rivista del Nuovo Cimento* **1**: 252–276, Bibcode:1969NCimR...1..252P

- Penrose, Roger (2004), *The Road to Reality*, A. A. Knopf, ISBN 0-679-45443-8

- Penzias, A. A.; Wilson, R. W. (1965), "A measurement of excess antenna temperature at 4080 Mc/s", *Astrophysical Journal* **142**: 419–421, Bibcode:1965ApJ...142..419P, doi:10.1086/148307

- Peskin, Michael E.; Schroeder, Daniel V. (1995), *An Introduction to Quantum Field Theory*, Addison-Wesley, ISBN 0-201-50397-2

- Peskin, Michael E. (2007), "Dark Matter and Particle Physics", *Journal of the Physical Society of Japan* **76** (11): 111017, arXiv:0707.1536, Bibcode:2007JPSJ...76k1017P, doi:10.1143/JPSJ.76.111017

- Poisson, Eric (2004), "The Motion of Point Particles in Curved Spacetime", *Living Rev. Relativity* **7**, doi:10.12942/lrr-2004-6, retrieved 2007-06-13

- Poisson, Eric (2004), *A Relativist's Toolkit. The Mathematics of Black-Hole Mechanics*, Cambridge University Press, ISBN 0-521-83091-5

- Polchinski, Joseph (1998a), *String Theory Vol. I: An Introduction to the Bosonic String*, Cambridge University Press, ISBN 0-521-63303-6

- Polchinski, Joseph (1998b), *String Theory Vol. II: Superstring Theory and Beyond*, Cambridge University Press, ISBN 0-521-63304-4

- Pound, R. V.; Rebka, G. A. (1959), "Gravitational Red-Shift in Nuclear Resonance", *Physical Review Letters* **3** (9): 439–441, Bibcode:1959PhRvL...3..439P, doi:10.1103/PhysRevLett.3.439

- Pound, R. V.; Rebka, G. A. (1960), "Apparent weight of photons", *Phys. Rev. Lett.* **4** (7): 337–341, Bibcode:1960PhRvL...4..337P, doi:10.1103/PhysRevLett.4.337

- Pound, R. V.; Snider, J. L. (1964), "Effect of Gravity on Nuclear Resonance", *Phys. Rev. Lett.* **13** (18): 539–540, Bibcode:1964PhRvL..13..539P, doi:10.1103/PhysRevLett.13.539

- Ramond, Pierre (1990), *Field Theory: A Modern Primer*, Addison-Wesley, ISBN 0-201-54611-6

- Rees, Martin (1966), "Appearance of Relativistically Expanding Radio Sources", *Nature* **211** (5048): 468–470, Bibcode:1966Natur.211..468R, doi:10.1038/211468a0

- Reissner, H. (1916), "Über die Eigengravitation des elektrischen Feldes nach der Einsteinschen Theorie", *Annalen der Physik* **355** (9): 106–120, Bibcode:1916AnP...355..106R, doi:10.1002/andp.19163550905

- Remillard, Ronald A.; Lin, Dacheng; Cooper, Randall L.; Narayan, Ramesh (2006), "The Rates of Type I X-Ray Bursts from Transients Observed with RXTE: Evidence for Black Hole Event Horizons", *Astrophysical Journal* **646** (1): 407–419, arXiv:astro-ph/0509758, Bibcode:2006ApJ...646..407R, doi:10.1086/504862

- Renn, Jürgen, ed. (2007), *The Genesis of General Relativity (4 Volumes)*, Dordrecht: Springer, ISBN 1-4020-3999-9

- Renn, Jürgen, ed. (2005), *Albert Einstein—Chief Engineer of the Universe: Einstein's Life and Work in Context*, Berlin: Wiley-VCH, ISBN 3-527-40571-2

- Reula, Oscar A. (1998), "Hyperbolic Methods for Einstein's Equations", *Living Rev. Relativity* **1**. Bibcode:1998LRR.....1....3R, doi:10.12942/lrr-1998-3, retrieved 2007-08-29

- Rindler, Wolfgang (2001), *Relativity. Special, General and Cosmological*, Oxford University Press, ISBN 0-19-850836-0

- Rindler, Wolfgang (1991), *Introduction to Special Relativity*, Clarendon Press, Oxford, ISBN 0-19-853952-5

- Robson, Ian (1996), *Active galactic nuclei*, John Wiley, ISBN 0-471-95853-0

- Roulet, E.; Mollerach, S. (1997), "Microlensing", *Physics Reports* **279** (2): 67–118, arXiv:astro-ph/9603119, Bibcode:1997PhR...279...67R, doi:10.1016/S0370-1573(96)00020-8

- Rovelli, Carlo (2000). "Notes for a brief history of quantum gravity". arXiv:gr-qc/0006061 [gr-qc].

- Rovelli, Carlo (1998), "Loop Quantum Gravity", *Living Rev. Relativity* **1**, doi:10.12942/lrr-1998-1, retrieved 2008-03-13

- Schäfer, Gerhard (2004), "Gravitomagnetic Effects", *General Relativity and Gravitation* **36** (10): 2223–2235, arXiv:gr-qc/0407116, Bibcode:2004GReGr..36.2223S, doi:10.1023/B:GERG.0000046180.97877.32

- Schödel, R.; Ott, T.; Genzel, R.; Eckart, A.; Mouawad, N.; Alexander, T. (2003), "Stellar Dynamics in the Central Arcsecond of Our Galaxy", *Astrophysical Journal* **596** (2): 1015–1034, arXiv:astro-ph/0306214, Bibcode:2003ApJ...596.1015S, doi:10.1086/378122

- Schutz, Bernard F. (1985), *A first course in general relativity*, Cambridge University Press, ISBN 0-521-27703-5

- Schutz, Bernard F. (2001), "Gravitational radiation", in Murdin, Paul, *Encyclopedia of Astronomy and Astrophysics*, Grove's Dictionaries, ISBN 1-56159-268-4

- Schutz, Bernard F. (2003), *Gravity from the ground up*, Cambridge University Press, ISBN 0-521-45506-5

- Schwarz, John H. (2007), "String Theory: Progress and Problems", *Progress of Theoretical Physics Supplement* **170**: 214, arXiv:hep-th/0702219, Bibcode:2007PThPS.170..214S, doi:10.1143/PTPS.170.214

- Schwarzschild, Karl (1916a), "Über das Gravitationsfeld eines Massenpunktes nach der Einsteinschen Theorie", *Sitzungsber. Preuss. Akad. D. Wiss.*: 189–196

- Schwarzschild, Karl (1916b), "Über das Gravitationsfeld eines Kugel aus inkompressibler Flüssigkeit nach der Einsteinschen Theorie", *Sitzungsber. Preuss. Akad. D. Wiss.*: 424–434

- Seidel, Edward (1998), "Numerical Relativity: Towards Simulations of 3D Black Hole Coalescence", in Narlikar, J. V.; Dadhich, N., *Gravitation and Relativity: At the turn of the millennium (Proceedings of the GR-15 Conference, held at IUCAA, Pune, India, December 16–21, 1997)*, IUCAA, p. 6088, arXiv:gr-qc/9806088, Bibcode:1998gr.qc.....6088S, ISBN 81-900378-3-8

- Seljak, Uroš; Zaldarriaga, Matias (1997), "Signature of Gravity Waves in the Polarization of the Microwave Background", *Phys. Rev. Lett.* **78** (11): 2054–2057, arXiv:astro-ph/9609169, Bibcode:1997PhRvL..78.2054S, doi:10.1103/PhysRevLett.78.2054

- Shapiro, S. S.; Davis, J. L.; Lebach, D. E.; Gregory, J. S. (2004), "Measurement of the solar gravitational deflection of radio waves using geodetic very-long-baseline interferometry data, 1979–1999", *Phys. Rev. Lett.* **92** (12): 121101, Bibcode:2004PhRvL..92l1101S, doi:10.1103/PhysRevLett.92.121101, PMID 15089661

- Shapiro, Irwin I. (1964), "Fourth test of general relativity", *Phys. Rev. Lett.* **13** (26): 789–791, Bibcode:1964PhRvL..13..789S, doi:10.1103/PhysRevLett.13.789

- Shapiro, I. I.; Pettengill, Gordon; Ash, Michael; Stone, Melvin; Smith, William; Ingalls, Richard; Brockelman, Richard (1968), "Fourth test of general relativity: preliminary results", *Phys. Rev. Lett.* **20** (22): 1265–1269, Bibcode:1968PhRvL..20.1265S, doi:10.1103/PhysRevLett.20.1265

- Singh, Simon (2004), *Big Bang: The Origin of the Universe*, Fourth Estate, ISBN 0-00-715251-5

- Sorkin, Rafael D. (2005), "Causal Sets: Discrete Gravity", in Gomberoff, Andres; Marolf, Donald, *Lectures on Quantum Gravity*, Springer, p. 9009, arXiv:gr-qc/0309009, Bibcode:2003gr.qc.....9009S, ISBN 0-387-23995-2

- Sorkin, Rafael D. (1997), "Forks in the Road, on the Way to Quantum Gravity", *Int. J. Theor. Phys.* **36** (12): 2759–2781, arXiv:gr-qc/9706002, Bibcode:1997IJTP...36.2759S, doi:10.1007/BF02435709

- Spergel, D. N.; Verde, L.; Peiris, H. V.; Komatsu, E.; Nolta, M. R.; Bennett, C. L.; Halpern, M.; Hinshaw, G.; et al. (2003), "First Year Wilkinson Microwave Anisotropy Probe (WMAP) Observations: Determination of Cosmological Parameters", *Astrophys. J. Suppl.* **148** (1): 175–194, arXiv:astro-ph/0302209, Bibcode:2003ApJS..148..175S, doi:10.1086/377226

- Spergel, D. N.; Bean, R.; Doré, O.; Nolta, M. R.; Bennett, C. L.; Dunkley, J.; Hinshaw, G.; Jarosik, N.; et al. (2007), "Wilkinson Microwave Anisotropy Probe (WMAP) Three Year Results: Implications for Cosmology", *Astrophysical Journal Supplement* **170** (2): 377–408, arXiv:astro-ph/0603449, Bibcode:2007ApJS..170..377S, doi:10.1086/513700

- Springel, Volker; White, Simon D. M.; Jenkins, Adrian; Frenk, Carlos S.; Yoshida, Naoki; Gao, Liang; Navarro, Julio; Thacker, Robert; et al. (2005), "Simulations of the formation, evolution and clustering of galaxies and quasars", *Nature* **435** (7042): 629–636, arXiv:astro-ph/0504097, Bibcode:2005Natur.435..629S, doi:10.1038/nature03597, PMID 15931216

- Stairs, Ingrid H. (2003), "Testing General Relativity with Pulsar Timing", *Living Rev. Relativity* **6**, arXiv:astro-ph/0307536, Bibcode:2003LRR.....6....5S, doi:10.12942/lrr-2003-5, retrieved 2007-07-21

- Stephani, H.; Kramer, D.; MacCallum, M.; Hoenselaers, C.; Herlt, E. (2003), *Exact Solutions of Einstein's Field Equations* (2 ed.), Cambridge University Press, ISBN 0-521-46136-7

- Synge, J. L. (1972), *Relativity: The Special Theory*, North-Holland Publishing Company, ISBN 0-7204-0064-3

- Szabados, László B. (2004), "Quasi-Local Energy-Momentum and Angular Momentum in GR", *Living Rev. Relativity* **7**, doi:10.12942/lrr-2004-4, retrieved 2007-08-23

- Taylor, Joseph H. (1994), "Binary pulsars and relativistic gravity", *Rev. Mod. Phys.* **66** (3): 711–719, Bibcode:1994RvMP...66..711T, doi:10.1103/RevModPhys.66.711

- Thiemann, Thomas (2006), "Approaches to Fundamental Physics: Loop Quantum Gravity: An Inside View", *Lecture Notes in Physics* **721**: 185–263, arXiv:hep-th/0608210, Bibcode:2007LNP...721..185T, doi:10.1007/978-3-540-71117-9_10, ISBN 978-3-540-71115-5

- Thiemann, Thomas (2003), "Lectures on Loop Quantum Gravity", *Lecture Notes in Physics* **631**: 41–135, arXiv:gr-qc/0210094, doi:10.1007/978-3-540-45230-0_3, ISBN 978-3-540-40810-9

- Thorne, Kip S. (1972), "Nonspherical Gravitational Collapse—A Short Review", in Klauder, J., *Magic without Magic*, W. H. Freeman, pp. 231–258

- Thorne, Kip S. (1994), *Black Holes and Time Warps: Einstein's Outrageous Legacy*, W W Norton & Company, ISBN 0-393-31276-3

- Thorne, Kip S. (1995), "Gravitational radiation", *Particle and Nuclear Astrophysics and Cosmology in the Next Millenium*: 160, arXiv:gr-qc/9506086, Bibcode:1995pnac.conf..160T, ISBN 0-521-36853-7

- Townsend, Paul K. (1997). "Black Holes (Lecture notes)". arXiv:gr-qc/9707012 [gr-qc].

- Townsend, Paul K. (1996). "Four Lectures on M-Theory". arXiv:hep-th/9612121 [hep-th].

- Traschen, Jenny (2000), Bytsenko, A.; Williams, F., eds., "An Introduction to Black Hole Evaporation", *Mathematical Methods of Physics (Proceedings of the 1999 Londrina Winter School)* (World Scientific): 180. arXiv:gr-qc/0010055, Bibcode:2000mmp..conf..180T

- Trautman, Andrzej (2006), "Einstein–Cartan theory", in Françoise, J.-P.; Naber, G. L.; Tsou, S. T., *Encyclopedia of Mathematical Physics, Vol. 2*, Elsevier, pp. 189–195, arXiv:gr-qc/0606062, Bibcode:2006gr.qc.....6062T

- Unruh, W. G. (1976), "Notes on Black Hole Evaporation", *Phys. Rev. D* **14** (4): 870–892, Bibcode:1976PhRvD..14..870U, doi:10.1103/PhysRevD.14.870

- Valtonen, M. J.; Lehto, H. J.; Nilsson, K.; Heidt, J.; Takalo, L. O.; Sillanpää, A.; Villforth, C.; Kidger, M.; et al. (2008), "A massive binary black-hole system in OJ 287 and a test of general relativity", *Nature* **452** (7189): 851–853, arXiv:0809.1280, Bibcode:2008Natur.452..851V, doi:10.1038/nature06896, PMID 18421348

- Wald, Robert M. (1975), "On Particle Creation by Black Holes", *Commun. Math. Phys.* **45** (3): 9–34, Bibcode:1975CMaPh..45....9W, doi:10.1007/BF01609863

- Wald, Robert M. (1984), *General Relativity*, University of Chicago Press, ISBN 0-226-87033-2

- Wald, Robert M. (1994), *Quantum field theory in curved spacetime and black hole thermodynamics*, University of Chicago Press, ISBN 0-226-87027-8

- Wald, Robert M. (2001), "The Thermodynamics of Black Holes", *Living Rev. Relativity* **4**, Bibcode:2001LRR.....4....6W, doi:10.12942/lrr-2001-6, retrieved 2007-08-08

- Walsh, D.; Carswell, R. F.; Weymann, R. J. (1979), "0957 + 561 A, B: twin quasistellar objects or gravitational lens?", *Nature* **279** (5712): 381–4, Bibcode:1979Natur.279..381W, doi:10.1038/279381a0, PMID 16068158

- Wambsganss, Joachim (1998), "Gravitational Lensing in Astronomy", *Living Rev. Relativity* **1**, arXiv:astro-ph/9812021, Bibcode:1998LRR.....1...12W, doi:10.12942/lrr-1998-12, retrieved 2007-07-20

- Weinberg, Steven (1972), *Gravitation and Cosmology*, John Wiley, ISBN 0-471-92567-5

- Weinberg, Steven (1995), *The Quantum Theory of Fields I: Foundations*, Cambridge University Press, ISBN 0-521-55001-7

- Weinberg, Steven (1996), *The Quantum Theory of Fields II: Modern Applications*, Cambridge University Press, ISBN 0-521-55002-5

- Weinberg, Steven (2000), *The Quantum Theory of Fields III: Supersymmetry*, Cambridge University Press, ISBN 0-521-66000-9

- Weisberg, Joel M.; Taylor, Joseph H. (2003), "The Relativistic Binary Pulsar B1913+16"", in Bailes, M.; Nice, D. J.; Thorsett, S. E., *Proceedings of "Radio Pulsars," Chania, Crete, August, 2002*, ASP Conference Series

- Weiss, Achim (2006), "Elements of the past: Big Bang Nucleosynthesis and observation", *Einstein Online* (Max Planck Institute for Gravitational Physics), retrieved 2007-02-24 External link in |work= (help)

- Wheeler, John A. (1990), *A Journey Into Gravity and Spacetime*, Scientific American Library, San Francisco: W. H. Freeman, ISBN 0-7167-6034-7

- Will, Clifford M. (1993), *Theory and experiment in gravitational physics*, Cambridge University Press, ISBN 0-521-43973-6

- Will, Clifford M. (2006), "The Confrontation between General Relativity and Experiment", *Living Rev. Relativity* **9**, arXiv:gr-qc/0510072, Bibcode:2006LRR.....9....3W, doi:10.12942/lrr-2006-3, retrieved 2007-06-12

- Zwiebach, Barton (2004), *A First Course in String Theory*, Cambridge University Press, ISBN 0-521-83143-1

8.12   Further reading

Popular books

- Geroch, R (1981), *General Relativity from A to B*, Chicago: University of Chicago Press, ISBN 0-226-28864-1

- Lieber, Lillian (2008), *The Einstein Theory of Relativity: A Trip to the Fourth Dimension*, Philadelphia: Paul Dry Books, Inc., ISBN 978-1-58988-044-3

- Wald, Robert M. (1992), *Space, Time, and Gravity: the Theory of the Big Bang and Black Holes*, Chicago: University of Chicago Press, ISBN 0-226-87029-4

- Wheeler, John; Ford, Kenneth (1998), *Geons, Black Holes, & Quantum Foam: a life in physics*, New York: W. W. Norton, ISBN 0-393-31991-1

Beginning undergraduate textbooks

- Callahan, James J. (2000), *The Geometry of Spacetime: an Introduction to Special and General Relativity*, New York: Springer, ISBN 0-387-98641-3

- Taylor, Edwin F.; Wheeler, John Archibald (2000), *Exploring Black Holes: Introduction to General Relativity*, Addison Wesley, ISBN 0-201-38423-X

Advanced undergraduate textbooks

- B. F. Schutz (2009), *A First Course in General Relativity (Second Edition)*, Cambridge University Press, ISBN 978-0-521-88705-2

- Cheng, Ta-Pei (2005), *Relativity, Gravitation and Cosmology: a Basic Introduction*, Oxford and New York: Oxford University Press, ISBN 0-19-852957-0

- Gron, O.; Hervik, S. (2007), *Einstein's General theory of Relativity*, Springer, ISBN 978-0-387-69199-2

- Hartle, James B. (2003), *Gravity: an Introduction to Einstein's General Relativity*, San Francisco: Addison-Wesley, ISBN 0-8053-8662-9

- Hughston, L. P. & Tod, K. P. (1991), *Introduction to General Relativity*, Cambridge: Cambridge University Press, ISBN 0-521-33943-X

- d'Inverno, Ray (1992), *Introducing Einstein's Relativity*, Oxford: Oxford University Press, ISBN 0-19-859686-3

- Ludyk, Günter (2013). *Einstein in Matrix Form* (1st ed.). Berlin: Springer. ISBN 9783642357978.

Graduate-level textbooks

- Carroll, Sean M. (2004), *Spacetime and Geometry: An Introduction to General Relativity*, San Francisco: Addison-Wesley, ISBN 0-8053-8732-3

- Grøn, Øyvind; Hervik, Sigbjørn (2007), *Einstein's General Theory of Relativity*, New York: Springer, ISBN 978-0-387-69199-2

- Landau, Lev D.; Lifshitz, Evgeny F. (1980), *The Classical Theory of Fields (4th ed.)*, London: Butterworth-Heinemann, ISBN 0-7506-2768-9

- Misner, Charles W.; Thorne, Kip. S.; Wheeler, John A. (1973), *Gravitation*, W. H. Freeman, ISBN 0-7167-0344-0

- Stephani, Hans (1990), *General Relativity: An Introduction to the Theory of the Gravitational Field*, Cambridge: Cambridge University Press, ISBN 0-521-37941-5

- Wald, Robert M. (1984), *General Relativity*, University of Chicago Press, ISBN 0-226-87033-2

8.13 External links

- Einstein Online – Articles on a variety of aspects of relativistic physics for a general audience; hosted by the Max Planck Institute for Gravitational Physics

- NCSA Spacetime Wrinkles – produced by the numerical relativity group at the NCSA, with an elementary introduction to general relativity

- **Courses**

- **Lectures**

- **Tutorials**

- Einstein's General Theory of Relativity on YouTube (lecture by Leonard Susskind recorded September 22, 2008 at Stanford University).

- Series of lectures on General Relativity given in 2006 at the Institut Henri Poincaré (introductory/advanced).

- General Relativity Tutorials by John Baez.

- Brown, Kevin. "Reflections on relativity". *Mathpages.com*. Retrieved May 29, 2005.

- Carroll, Sean M. "Lecture Notes on General Relativity". Retrieved January 5, 2014.

- Moor, Rafi. "Understanding General Relativity". Retrieved July 11, 2006.

- Waner, Stefan. "Introduction to Differential Geometry and General Relativity" (PDF). Retrieved 2015-04-05.

Chapter 9

Spacetime

For other uses of this term, see Spacetime (disambiguation).

In physics, **spacetime** is any mathematical model that combines space and time into a single interwoven continuum. The spacetime of our universe is usually interpreted from a Euclidean space perspective, which regards space as consisting of three dimensions, and time as consisting of one dimension, the "fourth dimension". By combining space and time into a single manifold called Minkowski space, physicists have significantly simplified a large number of physical theories, as well as described in a more uniform way the workings of the universe at both the supergalactic and subatomic levels.

9.1 Explanation

In non-relativistic classical mechanics, the use of Euclidean space instead of spacetime is appropriate, because time is treated as universal with a constant rate of passage that is independent of the state of motion of an observer. In relativistic contexts, time cannot be separated from the three dimensions of space, because the observed rate at which time passes for an object depends on the object's velocity relative to the observer and also on the strength of gravitational fields, which can slow the passage of time for an object as seen by an observer outside the field.

In cosmology, the concept of spacetime combines space and time to a single abstract universe. Mathematically it is a manifold consisting of "events" which are described by some type of coordinate system. Typically **three spatial dimensions** (length, width, height), and one **temporal dimension** (time) are required. Dimensions are independent components of a coordinate grid needed to locate a point in a certain defined "space". For example, on the globe the latitude and longitude are two independent coordinates which together uniquely determine a location. In spacetime, a coordinate grid that spans the 3+1 dimensions locates

events (rather than just points in space), i.e., time is added as another dimension to the coordinate grid. This way the coordinates specify *where* and *when* events occur. However, the unified nature of spacetime and the freedom of coordinate choice it allows imply that to express the temporal coordinate in one coordinate system requires both temporal and spatial coordinates in another coordinate system. Unlike in normal spatial coordinates, there are still restrictions for how measurements can be made spatially and temporally (see Spacetime intervals). These restrictions correspond roughly to a particular mathematical model which differs from Euclidean space in its manifest symmetry.

Until the beginning of the 20th century, time was believed to be independent of motion, progressing at a fixed rate in all reference frames; however, later experiments revealed that time slows at higher speeds of the reference frame relative to another reference frame. Such slowing, called time dilation, is explained in special relativity theory. Many experiments have confirmed time dilation, such as the relativistic decay of muons from cosmic ray showers and the slowing of atomic clocks aboard a Space Shuttle relative to synchronized Earth-bound inertial clocks. The duration of time can therefore vary according to events and reference frames.

When dimensions are understood as mere components of the grid system, rather than physical attributes of space, it is easier to understand the alternate dimensional views as being simply the result of coordinate transformations.

The term *spacetime* has taken on a generalized meaning beyond treating spacetime events with the normal 3+1 dimensions. It is really the combination of space and time. Other proposed spacetime theories include additional dimensions—normally spatial but there exist some speculative theories that include additional temporal dimensions and even some that include dimensions that are neither temporal nor spatial (e.g., superspace). How many dimensions are needed to describe the universe is still an open question. Speculative theories such as string theory predict 10 or 26 dimensions (with M-theory predicting 11 dimensions: 10

spatial and 1 temporal), but the existence of more than four dimensions would only appear to make a difference at the subatomic level.[1]

9.2 Spacetime in literature

Incas regarded space and time as a single concept, referred to as **pacha** (Quechua: *pacha*, Aymara: *pacha*).[2][3] The peoples of the Andes maintain a similar understanding.[4]

Arthur Schopenhauer wrote in §18 of *On the Fourfold Root of the Principle of Sufficient Reason* (1813): "the representation of coexistence is impossible in Time alone; it depends, for its completion, upon the representation of Space; because, in mere Time, all things follow one another, and in mere Space all things are side by side; it is accordingly only by the combination of Time and Space that the representation of coexistence arises".

The idea of a unified spacetime is stated by Edgar Allan Poe in his essay on cosmology titled *Eureka* (1848) that "Space and duration are one". In 1895, in his novel *The Time Machine*, H. G. Wells wrote, "There is no difference between time and any of the three dimensions of space except that our consciousness moves along it", and that "any real body must have extension in four directions: it must have Length, Breadth, Thickness, and Duration".

Marcel Proust, in his novel *Swann's Way* (published 1913), describes the village church of his childhood's Combray as "a building which occupied, so to speak, four dimensions of space—the name of the fourth being Time".

9.2.1 Mathematical concept

In Encyclopedie under the term *dimension* Jean le Rond d'Alembert speculated that duration (time) might be considered a fourth dimension if the idea was not too novel.[5]

Another early venture was by Joseph Louis Lagrange in his *Theory of Analytic Functions* (1797, 1813). He said, "One may view mechanics as a geometry of four dimensions, and mechanical analysis as an extension of geometric analysis".[6]

The ancient idea of the cosmos gradually was described mathematically with differential equations, differential geometry, and abstract algebra. These mathematical articulations blossomed in the nineteenth century as electrical technology stimulated men like Michael Faraday and James Clerk Maxwell to describe the reciprocal relations of electric and magnetic fields. Daniel Siegel phrased Maxwell's role in relativity as follows:

[...] the idea of the propagation of forces

at the velocity of light through the electromagnetic field as described by Maxwell's equations—rather than instantaneously at a distance—formed the necessary basis for relativity theory.[7]

Maxwell used vortex models in his papers on On Physical Lines of Force, but ultimately gave up on any substance but the electromagnetic field. Pierre Duhem wrote:

[Maxwell] was not able to create the theory that he envisaged except by giving up the use of any model, and by extending by means of analogy the abstract system of electrodynamics to displacement currents.[8]

In Siegel's estimation, "this very abstract view of the electromagnetic fields, involving no visualizable picture of what is going on out there in the field, is Maxwell's legacy."[9] Describing the behaviour of electric fields and magnetic fields led Maxwell to view the combination as an electromagnetic field. These fields have a value at every point of spacetime. It is the intermingling of electric and magnetic manifestations, described by Maxwell's equations, that give spacetime its structure. In particular, the rate of motion of an observer determines the electric and magnetic profiles of the electromagnetic field. The propagation of the field is determined by the electromagnetic wave equation, which requires spacetime for description.

Spacetime was described as an affine space with quadratic form in Minkowski space of 1908.[10] In his 1914 textbook *The Theory of Relativity*, Ludwik Silberstein used biquaternions to represent events in Minkowski space. He also exhibited the Lorentz transformations between observers of differing velocities as biquaternion mappings. Biquaternions were described in 1853 by W. R. Hamilton, so while the physical interpretation was new, the mathematics was well known in English literature, making relativity an instance of applied mathematics.

The first inkling of general relativity in spacetime was articulated by W. K. Clifford. Description of the effect of gravitation on space and time was found to be most easily visualized as a "warp" or stretching in the geometrical fabric of space and time, in a smooth and continuous way that changed smoothly from point-to-point along the spacetime fabric. In 1947 James Jeans provided a concise summary of the development of spacetime theory in his book *The Growth of Physical Science*.[11]

9.3   Basic concepts

The basic elements of spacetime are events. In any given spacetime, an event is a unique position at a unique time. Because events are spacetime points, an example of an event in classical relativistic physics is (x, y, z, t), the location of an elementary (point-like) particle at a particular time. A spacetime itself can be viewed as the union of all events in the same way that a line is the union of all of its points, formally organized into a manifold, a space which can be described at small scales using coordinate systems.

A spacetime is independent of any observer.[12] However, in describing physical phenomena (which occur at certain moments of time in a given region of space), each observer chooses a convenient metrical coordinate system. Events are specified by four real numbers in any such coordinate system. The trajectories of elementary (point-like) particles through space and time are thus a continuum of events called the world line of the particle. Extended or composite objects (consisting of many elementary particles) are thus a union of many world lines twisted together by virtue of their interactions through spacetime into a "world-braid".

However, in physics, it is common to treat an extended object as a "particle" or "field" with its own unique (e.g., center of mass) position at any given time, so that the world line of a particle or light beam is the path that this particle or beam takes in the spacetime and represents the history of the particle or beam. The world line of the orbit of the Earth (in such a description) is depicted in two spatial dimensions x and y (the plane of the Earth's orbit) and a time dimension orthogonal to x and y. The orbit of the Earth is an ellipse in space alone, but its world line is a helix in spacetime.[13]

The unification of space and time is exemplified by the common practice of selecting a metric (the measure that specifies the interval between two events in spacetime) such that all four dimensions are measured in terms of units of distance: representing an event as $(x_0, x_1, x_2, x_3) = (ct, x, y, z)$ (in the Lorentz metric) or $(x_1, x_2, x_3, x_4) = (x, y, z, ict)$ (in the original Minkowski metric) where c is the speed of light.[14] The metrical descriptions of Minkowski Space and spacelike, lightlike, and timelike intervals given below follow this convention, as do the conventional formulations of the Lorentz transformation.

9.3.1   Spacetime intervals in flat space

In a Euclidean space, the separation between two points is measured by the distance between the two points. The distance is purely spatial, and is always positive. In spacetime, the displacement four-vector ΔR is given by the space displacement vector Δr and the time difference Δt between the events. The *spacetime interval*, also called *invariant interval*, between the two events, s^2,[15] is defined as:

$$s^2 = \Delta r^2 - c^2 \Delta t^2 \text{ (spacetime interval)},$$

where c is the speed of light. The choice of signs for s^2 above follows the space-like convention $(-+++)$.[16] Spacetime intervals may be classified into three distinct types, based on whether the temporal separation ($c^2 \Delta t^2$) or the spatial separation (Δr^2) of the two events is greater: timelike, light-like or space-like.

Certain types of world lines are called geodesics of the spacetime – straight lines in the case of Minkowski space and their closest equivalent in the curved spacetime of general relativity. In the case of purely time-like paths, geodesics are (locally) the paths of greatest separation (spacetime interval) as measured along the path between two events, whereas in Euclidean space and Riemannian manifolds, geodesics are paths of shortest distance between two points.[17][18] The concept of geodesics becomes central in general relativity, since geodesic motion may be thought of as "pure motion" (inertial motion) in spacetime, that is, free from any external influences.

Time-like interval

$$c^2 \Delta t^2 > \Delta r^2$$
$$s^2 < 0$$

For two events separated by a time-like interval, enough time passes between them that there could be a cause–effect relationship between the two events. For a particle traveling through space at less than the speed of light, any two events which occur to or by the particle must be separated by a time-like interval. Event pairs with time-like separation define a negative spacetime interval ($s^2 < 0$) and may be said to occur in each other's future or past. There exists a reference frame such that the two events are observed to occur in the same spatial location, but there is no reference frame in which the two events can occur at the same time.

The measure of a time-like spacetime interval is described by the proper time interval, $\Delta \tau$:

$$\Delta \tau = \sqrt{\Delta t^2 - \frac{\Delta r^2}{c^2}} \text{ (proper time interval)}.$$

The proper time interval would be measured by an observer with a clock traveling between the two events in an inertial reference frame, when the observer's path intersects each event as that event occurs. (The proper time interval defines a real number, since the interior of the square root is positive.)

Light-like interval

$$c^2 \Delta t^2 = \Delta r^2$$
$$s^2 = 0$$

In a light-like interval, the spatial distance between two events is exactly balanced by the time between the two events. The events define a spacetime interval of zero ($s^2 = 0$). Light-like intervals are also known as "null" intervals.

Events which occur to or are initiated by a photon along its path (i.e., while traveling at c , the speed of light) all have light-like separation. Given one event, all those events which follow at light-like intervals define the propagation of a light cone, and all the events which preceded from a light-like interval define a second (graphically inverted, which is to say "*pastward*") light cone.

Space-like interval

$$c^2 \Delta t^2 < \Delta r^2$$
$$s^2 > 0$$

When a space-like interval separates two events, not enough time passes between their occurrences for there to exist a causal relationship crossing the spatial distance between the two events at the speed of light or slower. Generally, the events are considered not to occur in each other's future or past. There exists a reference frame such that the two events are observed to occur at the same time, but there is no reference frame in which the two events can occur in the same spatial location.

For these space-like event pairs with a positive spacetime interval ($s^2 > 0$), the measurement of space-like separation is the proper distance, $\Delta \sigma$:

$$\Delta \sigma = \sqrt{s^2} = \sqrt{\Delta r^2 - c^2 \Delta t^2} \text{ (proper distance)}.$$

Like the proper time of time-like intervals, the proper distance of space-like spacetime intervals is a real number value.

9.3.2 Interval as area

The interval has been presented as the area of an oriented rectangle formed by two events and isotropic lines through them. Time-like or space-like separations correspond to oppositely oriented rectangles, one type considered to have rectangles of negative area. The case of two events separated by light corresponds to the rectangle degenerating to the segment between the events and zero area.[19] The transformations leaving interval-length invariant are the area-preserving squeeze mappings.

The parameters traditionally used rely on quadrature of the hyperbola, which is the natural logarithm. This transcendental function is essential in mathematical analysis as its inverse unites circular functions and hyperbolic functions: The exponential function, e^t, t a real number, used in the hyperbola (e^t, e^{-t}), generates hyperbolic sectors and the hyperbolic angle parameter. The functions cosh and sinh, used with rapidity as hyperbolic angle, provide the common representation of squeeze in the form $\begin{pmatrix} \cosh \phi & \sinh \phi \\ \sinh \phi & \cosh \phi \end{pmatrix}$, or as the split-complex unit $e^{j\phi} = \cosh \phi + j \sinh \phi$.

9.4 Mathematics of spacetimes

For physical reasons, a spacetime continuum is mathematically defined as a four-dimensional, smooth, connected Lorentzian manifold (M, g) . This means the smooth Lorentz metric g has signature $(3, 1)$. The metric determines the geometry of spacetime, as well as determining the geodesics of particles and light beams. About each point (event) on this manifold, coordinate charts are used to represent observers in reference frames. Usually, Cartesian coordinates (x, y, z, t) are used. Moreover, for simplicity's sake, units of measurement are usually chosen such that the speed of light c is equal to 1.

A reference frame (observer) can be identified with one of these coordinate charts; any such observer can describe any event p . Another reference frame may be identified by a second coordinate chart about p . Two observers (one in each reference frame) may describe the same event p but obtain different descriptions.

Usually, many overlapping coordinate charts are needed to cover a manifold. Given two coordinate charts, one containing p (representing an observer) and another containing q (representing another observer), the intersection of the charts represents the region of spacetime in which both observers can measure physical quantities and hence compare results. The relation between the two sets of measurements is given by a non-singular coordinate transformation on this intersection. The idea of coordinate charts as local observers who can perform measurements in their vicinity also makes good physical sense, as this is how one actually collects physical data—locally.

For example, two observers, one of whom is on Earth, but the other one who is on a fast rocket to Jupiter, may observe a comet crashing into Jupiter (this is the event p). In general, they will disagree about the exact location and timing of this impact, i.e., they will have different 4-tuples

(x, y, z, t) (as they are using different coordinate systems). Although their kinematic descriptions will differ, dynamical (physical) laws, such as momentum conservation and the first law of thermodynamics, will still hold. In fact, relativity theory requires more than this in the sense that it stipulates these (and all other physical) laws must take the same form in all coordinate systems. This introduces tensors into relativity, by which all physical quantities are represented.

Geodesics are said to be time-like, null, or space-like if the tangent vector to one point of the geodesic is of this nature. Paths of particles and light beams in spacetime are represented by time-like and null (light-like) geodesics, respectively.

9.4.1   Topology

Main article: Spacetime topology

The assumptions contained in the definition of a spacetime are usually justified by the following considerations.

The connectedness assumption serves two main purposes. First, different observers making measurements (represented by coordinate charts) should be able to compare their observations on the non-empty intersection of the charts. If the connectedness assumption were dropped, this would not be possible. Second, for a manifold, the properties of connectedness and path-connectedness are equivalent, and one requires the existence of paths (in particular, geodesics) in the spacetime to represent the motion of particles and radiation.

Every spacetime is paracompact. This property, allied with the smoothness of the spacetime, gives rise to a smooth linear connection, an important structure in general relativity. Some important theorems on constructing spacetimes from compact and non-compact manifolds include the following:

- A compact manifold can be turned into a spacetime if, and only if, its Euler characteristic is 0. (Proof idea: the existence of a Lorentzian metric is shown to be equivalent to the existence of a nonvanishing vector field.)

- Any non-compact 4-manifold can be turned into a spacetime.[20]

9.4.2   Spacetime symmetries

Main article: Spacetime symmetries

Often in relativity, spacetimes that have some form of symmetry are studied. As well as helping to classify spacetimes, these symmetries usually serve as a simplifying assumption in specialized work. Some of the most popular ones include:

- Axisymmetric spacetimes

- Spherically symmetric spacetimes

- Static spacetimes

- Stationary spacetimes

9.4.3   Causal structure

Main article: Causal structure
See also: Causality (physics) and Causality

The causal structure of a spacetime describes causal relationships between pairs of points in the spacetime based on the existence of certain types of curves joining the points.

9.5   Spacetime in special relativity

Main article: Minkowski space

The geometry of spacetime in special relativity is described by the Minkowski metric on R^4. This spacetime is called Minkowski space. The Minkowski metric is usually denoted by η and can be written as a four-by-four matrix:

$$\eta_{ab} = \mathrm{diag}(1, -1, -1, -1)$$

where the Landau–Lifshitz space-like convention is being used. A basic assumption of relativity is that coordinate transformations must leave spacetime intervals invariant. Intervals are invariant under Lorentz transformations. This invariance property leads to the use of four-vectors (and other tensors) in describing physics.

Strictly speaking, one can also consider events in Newtonian physics as a single spacetime. This is Galilean–Newtonian relativity, and the coordinate systems are related by Galilean transformations. However, since these preserve spatial and temporal distances independently, such a spacetime can be decomposed into spatial coordinates plus temporal coordinates, which is not possible in the general case.

9.6 Spacetime in general relativity

In general relativity, it is assumed that spacetime is curved by the presence of matter (energy), this curvature being represented by the Riemann tensor. In special relativity, the Riemann tensor is identically zero, and so this concept of "non-curvedness" is sometimes expressed by the statement *Minkowski spacetime is flat*.

The earlier discussed notions of time-like, light-like and space-like intervals in special relativity can similarly be used to classify one-dimensional curves through curved spacetime. A time-like curve can be understood as one where the interval between any two infinitesimally close events on the curve is time-like, and likewise for light-like and space-like curves. Technically the three types of curves are usually defined in terms of whether the tangent vector at each point on the curve is time-like, light-like or space-like. The world line of a slower-than-light object will always be a time-like curve, the world line of a massless particle such as a photon will be a light-like curve, and a space-like curve could be the world line of a hypothetical tachyon. In the local neighborhood of any event, time-like curves that pass through the event will remain inside that event's past and future light cones, light-like curves that pass through the event will be on the surface of the light cones, and space-like curves that pass through the event will be outside the light cones. One can also define the notion of a three-dimensional "space-like hypersurface", a continuous three-dimensional "slice" through the four-dimensional property with the property that every curve that is contained entirely within this hypersurface is a space-like curve.[21]

Many spacetime continua have physical interpretations which most physicists would consider bizarre or unsettling. For example, a compact spacetime has closed timelike curves, which violate our usual ideas of causality (that is, future events could affect past ones). For this reason, mathematical physicists usually consider only restricted subsets of all the possible spacetimes. One way to do this is to study "realistic" solutions of the equations of general relativity. Another way is to add some additional "physically reasonable" but still fairly general geometric restrictions and try to prove interesting things about the resulting spacetimes. The latter approach has led to some important results, most notably the Penrose–Hawking singularity theorems.

9.7 Quantized spacetime

Main article: Quantum spacetime

In general relativity, spacetime is assumed to be smooth and continuous—and not just in the mathematical sense. In the theory of quantum mechanics, there is an inherent discreteness present in physics. In attempting to reconcile these two theories, it is sometimes postulated that spacetime should be quantized at the very smallest scales. Current theory is focused on the nature of spacetime at the Planck scale. Causal sets, loop quantum gravity, string theory, causal dynamical triangulation, and black hole thermodynamics all predict a quantized spacetime with agreement on the order of magnitude. Loop quantum gravity makes precise predictions about the geometry of spacetime at the Planck scale.

9.8 See also

- Anthropic_principle § Applications of the principle §§ Spacetime
- Basic introduction to the mathematics of curved spacetime
- Four-vector
- Frame-dragging
- Global spacetime structure
- Hole argument
- List of mathematical topics in relativity
- Local spacetime structure
- Lorentz invariance
- Manifold
- Mathematics of general relativity
- Metric space
- Philosophy of space and time
- Relativity of simultaneity
- Strip photography
- World manifold

9.9 References

[1] Kopeikin, Sergei; Efroimsky, Michael; Kaplan, George (2011). *Relativistic Celestial Mechanics of the Solar System*. John Wiley & Sons. p. 157. ISBN 3527634576., Extract of page 157

[2] Atuq Eusebio Manga Qespi, Instituto de lingüística y Cultura Amerindia de la Universidad de Valencia. *Pacha: un concepto andino de espacio y tiempo*. Revista española de Antropología Americana, 24, p. 155–189. Edit. Complutense, Madrid. 1994

[3] Paul Richard Steele, Catherine J. Allen, *Handbook of Inca mythology*, p. 86, (ISBN 1-57607-354-8)

[4] Shirley Ardener, University of Oxford, *Women and space: ground rules and social maps*, p. 36 (ISBN 0-85496-728-1)

[5] Jean d'Alembert (1754) Dimension from ARTFL Encyclopedie project

[6] R.C. Archibald (1914) *Time as a fourth dimension Bulletin of the American Mathematical Society* 20:409.

[7] Daniel M. Siegel (2014) "Maxwell's contributions to electricity and magnetism", chapter 10 in *James Clerk Maxwell: Perspectives on his Life and Work*, Raymond Flood, Mark McCartney, Andrew Whitaker, editors, Oxford University Press ISBN 978-0-19-966437-5

[8] Pierre Duhem (1954) *The Aim and Structure of Physical Theory*, page 98, Princeton University Press

[9] Siegel 2014 p 191

[10] Minkowski, Hermann (1909), "Raum und Zeit", *Physikalische Zeitschrift* **10**: 75–88

    • Various English translations on Wikisource: Space and Time.

[11] James Jeans (1947) The Growth of Physical Science, "Space-time", pp. 205–301, link from Internet Archive

[12] Matolcsi, Tamás (1994). *Spacetime Without Reference Frames*. Budapest: Akadémiai Kiadó.

[13] Ellis, G. F. R.; Williams, Ruth M. (2000). *Flat and curved space–times* (2nd ed.). Oxford University Press. p. 9. ISBN 0-19-850657-0.

[14] Petkov, Vesselin (2010). *Minkowski Spacetime: A Hundred Years Later*. Springer. p. 70. ISBN 90-481-3474-9., Section 3.4, p. 70

[15] Note that the term *spacetime interval* is applied by several authors to the quantity s^2 and not to s. The reason that the quantity s^2 is used and not s is that s^2 can be positive, zero or negative, and is a more generally convenient and useful quantity than the Minkowski norm with a timelike/null/spacelike distinguisher: the pair $(\sqrt{|s^2|}, \operatorname{sgn}(s^2))$. Despite the notation, it should not be regarded as the square of a number, but as a symbol. The cost for this convenience is that this "interval" is quadratic in linear separation along a straight line.

[16] More generally the spacetime interval in flat space can be written as $s^2 = g_{\alpha\beta}\Delta x^\alpha \Delta x^\beta$ with metric tensor g independent of spacetime position.

[17] This characterization is not universal: both the arcs between two points of a great circle on a sphere are geodesics.

[18] Berry, Michael V. (1989). *Principles of Cosmology and Gravitation*. CRC Press. p. 58. ISBN 0-85274-037-9., Extract of page 58, caption of Fig. 25

[19] I. M. Yaglom (1979) *A Simple Non-Euclidean Geometry and its Physical Basis*, page 178, Springer, ISBN 0387-90332-1, MR 520230

[20] Geroch, Robert; Horowitz, Gary T. (1979). "Chapter 5. Global structure of spacetimes". In Hawking, S.W.; Israel, W. *General Relativity An Einstein Centenary Survey*. Cambridge University Press. p. 219. ISBN 0521299284.

[21] See "Quantum Spacetime and the Problem of Time in Quantum Gravity" by Leszek M. Sokolowski, where on this page he writes "Each of these hypersurfaces is spacelike, in the sense that every curve, which entirely lies on one of such hypersurfaces, is a spacelike curve." More commonly a spacelike hypersurface is defined technically as a surface such that the normal vector at every point is time-like, but the definition above may be somewhat more intuitive.

9.10 Further Reading

- Albert Einstein on Space-Time 13th edition Encyclopedia Britannica Historical: Albert Einstein's 1926 article

- Ehrenfest, Paul (1920) "How do the fundamental laws of physics make manifest that Space has 3 dimensions?" *Annalen der Physik 366*: 440.

- George F. Ellis and Ruth M. Williams (1992) *Flat and curved space–times*. Oxford Univ. Press. ISBN 0-19-851164-7

- Encyclopedia of Space-time and gravitation Scholarpedia Expert articles

9.11 External links

- http://universaltheory.org

- Barrow, John D.; Tipler, Frank J. (1988). *The Anthropic Cosmological Principle*. Oxford University Press. ISBN 978-0-19-282147-8. LCCN 87028148.

- Isenberg, J. A. (1981). "Wheeler–Einstein–Mach spacetimes". *Phys. Rev. D* **24** (2): 251–256. Bibcode:1981PhRvD..24..251I. doi:10.1103/PhysRevD.24.251.

- Kant, Immanuel (1929) "Thoughts on the true estimation of living forces" in J. Handyside, trans., *Kant's Inaugural Dissertation and Early Writings on Space*. Univ. of Chicago Press.

- Lorentz, H. A., Einstein, Albert, Minkowski, Hermann, and Weyl, Hermann (1952) *The Principle of Relativity: A Collection of Original Memoirs*. Dover.

- Lucas, John Randolph (1973) *A Treatise on Time and Space*. London: Methuen.

- Penrose, Roger (2004). *The Road to Reality*. Oxford: Oxford University Press. ISBN 0-679-45443-8. Chpts. 17–18.

- Poe, Edgar A. (1848). *Eureka; An Essay on the Material and Spiritual Universe*. Hesperus Press Limited. ISBN 1-84391-009-8.

- Robb, A. A. (1936). *Geometry of Time and Space*. University Press.

- Erwin Schrödinger (1950) *Space–time structure*. Cambridge Univ. Press.

- Schutz, J. W. (1997). *Independent axioms for Minkowski Space–time*. Addison-Wesley Longman. ISBN 0-582-31760-6.

- Tangherlini, F. R. (1963). "Schwarzschild Field in n Dimensions and the Dimensionality of Space Problem". *Nuovo Cimento* **14** (27): 636.

- Taylor, E. F.; Wheeler, John A. (1963). *Spacetime Physics*. W. H. Freeman. ISBN 0-7167-2327-1.

- Wells, H.G. (2004). *The Time Machine*. New York: Pocket Books. ISBN 0-671-57554-6. (pp. 5–6)

- Stanford Encyclopedia of Philosophy: "Space and Time: Inertial Frames" by Robert DiSalle.

Chapter 10

Effective field theory

In physics, an **effective field theory** is a type of approximation to (or effective theory for) an underlying physical theory, such as a quantum field theory or a statistical mechanics model. An effective field theory includes the appropriate degrees of freedom to describe physical phenomena occurring at a chosen length scale or energy scale, while ignoring substructure and degrees of freedom at shorter distances (or, equivalently, at higher energies). Intuitively, one averages over the behavior of the underlying theory at shorter length scales to derive a hopefully simplified model at longer length scales. Effective field theories typically work best when there is a large separation between length scale of interest and the length scale of the underlying dynamics. Effective field theories have found use in particle physics, statistical mechanics, condensed matter physics, general relativity, and hydrodynamics. They simplify calculations, and allow treatment of Dissipation and Radiation effects .[1][2]

10.1 The renormalization group

Presently, effective field theories are discussed in the context of the renormalization group (RG) where the process of *integrating out* short distance degrees of freedom is made systematic. Although this method is not sufficiently concrete to allow the actual construction of effective field theories, the gross understanding of their usefulness becomes clear through a RG analysis. This method also lends credence to the main technique of constructing effective field theories, through the analysis of symmetries. If there is a single mass scale M in the *microscopic* theory, then the effective field theory can be seen as an expansion in $1/M$. The construction of an effective field theory accurate to some power of $1/M$ requires a new set of free parameters at each order of the expansion in $1/M$. This technique is useful for scattering or other processes where the maximum momentum scale k satisfies the condition $k/M \ll 1$. Since effective field theories are not valid at small length scales, they need not be renormalizable. Indeed, the ever expanding number of parameters at each order in $1/M$ required for an effective field theory means that they are generally not renormalizable in the same sense as quantum electrodynamics which requires only the renormalization of two parameters.

10.2 Examples of effective field theories

10.2.1 Fermi theory of beta decay

The best-known example of an effective field theory is the Fermi theory of beta decay. This theory was developed during the early study of weak decays of nuclei when only the hadrons and leptons undergoing weak decay were known. The typical reactions studied were:

$$n \to p + e^- + \overline{\nu}_e$$
$$\mu^- \to e^- + \overline{\nu}_e + \nu_\mu.$$

This theory posited a pointlike interaction between the four fermions involved in these reactions. The theory had great phenomenological success and was eventually understood to arise from the gauge theory of electroweak interactions, which forms a part of the standard model of particle physics. In this more fundamental theory, the interactions are mediated by a flavour-changing gauge boson, the $W^\pm$. The immense success of the Fermi theory was because the W particle has mass of about 80 GeV, whereas the early experiments were all done at an energy scale of less than 10 MeV. Such a separation of scales, by over 3 orders of magnitude, has not been met in any other situation as yet.

10.2.2 BCS theory of superconductivity

Another famous example is the BCS theory of superconductivity. Here the underlying theory is of electrons in a metal interacting with lattice vibrations called phonons. The phonons cause attractive interactions between some electrons, causing them to form Cooper pairs. The length scale of these pairs is much larger than the wavelength of phonons, making it possible to neglect the dynamics of phonons and construct a theory in which two electrons effectively interact at a point. This theory has had remarkable success in describing and predicting the results of experiments on superconductivity.

10.2.3 Effective Field Theories in Gravity

General relativity itself is expected to be the low energy effective field theory of a full theory of quantum gravity, such as string theory. The expansion scale is the Planck mass. Effective field theories have also been used to simplify problems in General Relativity, in particular in calculating the gravitational wave signature of inspiralling finite-sized objects.[3] The most common EFT in GR is "Non-Relativistic General Relativity" (NRGR),[4][5][6] which is similar to the post-Newtonian expansion.[7] Another common GR EFT is the Extreme Mass Ratio (EMR), which in the context of the inspiralling problem is called EMRI.

10.2.4 Other examples

Presently, effective field theories are written for many situations.

- One major branch of nuclear physics is quantum hadrodynamics, where the interactions of hadrons are treated as a field theory, which should be derivable from the underlying theory of quantum chromodynamics. Due to the smaller separation of length scales here, this effective theory has some classificatory power, but not the spectacular success of the Fermi theory.

- In particle physics the effective field theory of QCD called chiral perturbation theory has had better success.[8] This theory deals with the interactions of hadrons with pions or kaons, which are the Goldstone bosons of spontaneous chiral symmetry breaking. The expansion parameter is the pion energy/momentum.

- For hadrons containing one heavy quark (such as the bottom or charm), an effective field theory which expands in powers of the quark mass, called the heavy-quark effective theory (HQET), has been found useful.

- For hadrons containing two heavy quarks, an effective field theory which expands in powers of the relative velocity of the heavy quarks, called non-relativistic QCD (NRQCD), has been found useful, especially when used in conjunctions with lattice QCD.

- For hadron reactions with light energetic (collinear) particles, the interactions with low-energetic (soft) degrees of freedom are described by the soft-collinear effective theory (SCET).

- Much of condensed matter physics consists of writing effective field theories for the particular property of matter being studied.

- Hydrodynamics can also be treated using Effective Field Theories[9]

10.3 See also

- Form factor (quantum field theory)

- Renormalization group

- Quantum field theory

- Quantum triviality

- Ginzburg–Landau theory

10.4 References

[1] "Classical Mechanics of Nonconservative Systems" by Chad Galley

[2] "Radiation reaction at the level of the action" by Ofek Birnholtz, Shahar Hadar, and Barak Kol

[3] "An Effective Field Theory of Gravity for Extended Objects" by Walter D. Goldberger, Ira Z. Rothstein

[4]

[5] "Non-Relativistic Gravitation: From Newton to Einstein and Back" by Barak Kol & Michael Smolkin

[6]

[7] "Theory of post-Newtonian radiation and reaction" by Ofek Birnholtz, Shahar Hadar, and Barak Kol

[8] On the foundations of chiral perturbation theory, H. Leutwyler (Annals of Physics, v 235, 1994, p 165-203)

[9] "Dissipation in the effective field theory for hydrodynamics: First order effects" by Solomon Endlich, Alberto Nicolis, Rafael A. Porto, Junpu Wang

10.5   External links

- Effective Field Theory, A. Pich, Lectures at the 1997 Les Houches Summer School "Probing the Standard Model of Particle Interactions."

- Effective field theories, reduction and scientific explanation, by S. Hartmann, *Studies in History and Philosophy of Modern Physics* **32B**, 267-304 (2001).

- Aspects of heavy quark theory, by I. Bigi, M. Shifman and N. Uraltsev (Annual Reviews of Nuclear and Particle Science, v 47, 1997, p 591-661)

- Effective field theory (Interactions, Symmetry Breaking and Effective Fields - from Quarks to Nuclei. an Internet Lecture by Jacek Dobaczewski)

Chapter 11

Graviton

This article is about the hypothetical particle. For other uses, see Graviton (disambiguation).

In physics, the **graviton** is a hypothetical elementary particle that mediates the force of gravitation in the framework of quantum field theory. If it exists, the graviton is expected to be massless (because the gravitational force appears to have unlimited range) and must be a spin−2 boson. The spin follows from the fact that the source of gravitation is the stress–energy tensor, a second-rank tensor (compared to electromagnetism's spin-1 photon, the source of which is the four-current, a first-rank tensor). Additionally, it can be shown that any massless spin-2 field would give rise to a force indistinguishable from gravitation, because a massless spin-2 field must couple to (interact with) the stress–energy tensor in the same way that the gravitational field does. Seeing as the graviton is hypothetical, its discovery would unite quantum theory with gravity.[4] This result suggests that, if a massless spin-2 particle is discovered, it must be the graviton, so that the only experimental verification needed for the graviton may simply be the discovery of a massless spin-2 particle.[5]

11.1 Theory

The four other known forces of nature are mediated by elementary particles: electromagnetism by the photon, the strong interaction by the gluons, the Higgs field by the Higgs Boson, and the weak interaction by the W and Z bosons. The hypothesis is that the gravitational interaction is likewise mediated by an – as yet undiscovered – elementary particle, dubbed as *the graviton*. In the classical limit, the theory would reduce to general relativity and conform to Newton's law of gravitation in the weak-field limit.[6][7][8]

11.1.1 Gravitons and renormalization

When describing graviton interactions, the classical theory (i.e., the tree diagrams) and semiclassical corrections (one-loop diagrams) behave normally, but Feynman diagrams with two (or more) loops lead to ultraviolet divergences; that is, infinite results that cannot be removed because the quantized general relativity is not renormalizable, unlike quantum electrodynamics. That is, the usual ways physicists calculate the probability that a particle will emit or absorb a graviton give nonsensical answers and the theory loses its predictive power. These problems, together with some conceptual puzzles, led many physicists to believe that a theory more complete than quantized general relativity must describe the behavior near the Planck scale.

11.1.2 Comparison with other forces

Unlike the force carriers of the other forces, gravitation plays a special role in general relativity in defining the spacetime in which events take place. In some descriptions, matter modifies the 'shape' of spacetime itself, and gravity is a result of this shape, an idea which at first glance may appear hard to match with the idea of a force acting between particles.[9] Because the diffeomorphism invariance of the theory does not allow any particular space-time background to be singled out as the "true" space-time background, general relativity is said to be background independent. In contrast, the Standard Model is *not* background independent, with Minkowski space enjoying a special status as the fixed background space-time.[10] A theory of quantum gravity is needed in order to reconcile these differences.[11] Whether this theory should be background independent is an open question. The answer to this question will determine our understanding of what specific role gravitation plays in the fate of the universe.[12]

11.1.3   Gravitons in speculative theories

String theory predicts the existence of gravitons and their well-defined interactions. A graviton in perturbative string theory is a closed string in a very particular low-energy vibrational state. The scattering of gravitons in string theory can also be computed from the correlation functions in conformal field theory, as dictated by the AdS/CFT correspondence, or from matrix theory.

A feature of gravitons in string theory is that, as closed strings without endpoints, they would not be bound to branes and could move freely between them. If we live on a brane (as hypothesized by brane theories) this "leakage" of gravitons from the brane into higher-dimensional space could explain why gravitation is such a weak force, and gravitons from other branes adjacent to our own could provide a potential explanation for dark matter. However, if gravitons were to move completely freely between branes this would dilute gravity too much, causing a violation of Newton's inverse square law. To combat this, Lisa Randall found that a three-brane (such as ours) would have a gravitational pull of its own, preventing gravitons from drifting freely, possibly resulting in the diluted gravity we observe while roughly maintaining Newton's inverse square law.[13] See brane cosmology.

A theory by Ahmed Farag Ali and Saurya Das adds quantum mechanical corrections (using Bohm trajectories) to general relativistic geodesics. If gravitons are given a small but non-zero mass, it could explain the cosmological constant without need for dark energy and solve the smallness problem.[14]

11.2   Experimental observation

Unambiguous detection of individual gravitons, though not prohibited by any fundamental law, is impossible with any physically reasonable detector.[15] The reason is the extremely low cross section for the interaction of gravitons with matter. For example, a detector with the mass of Jupiter and 100% efficiency, placed in close orbit around a neutron star, would only be expected to observe one graviton every 10 years, even under the most favorable conditions. It would be impossible to discriminate these events from the background of neutrinos, since the dimensions of the required neutrino shield would ensure collapse into a black hole.[15]

However, experiments to detect gravitational waves, which may be viewed as coherent states of many gravitons, are underway (such as LIGO and VIRGO). Although these experiments cannot detect individual gravitons, they might provide information about certain properties of the graviton.[16]

For example, if gravitational waves were observed to propagate slower than c (the speed of light in a vacuum), that would imply that the graviton has mass (however, gravitational waves must propagate slower than "c" in a region with non-zero mass density if they are to be detectable).[17] Astronomical observations of the kinematics of galaxies, especially the galaxy rotation problem and modified Newtonian dynamics, might point toward gravitons having non-zero mass.[18]

11.3   Difficulties and outstanding issues

Most theories containing gravitons suffer from severe problems. Attempts to extend the Standard Model or other quantum field theories by adding gravitons run into serious theoretical difficulties at high energies (processes involving energies close to or above the Planck scale) because of infinities arising due to quantum effects (in technical terms, gravitation is nonrenormalizable). Since classical general relativity and quantum mechanics seem to be incompatible at such energies, from a theoretical point of view, this situation is not tenable. One possible solution is to replace particles with strings. String theories are quantum theories of gravity in the sense that they reduce to classical general relativity plus field theory at low energies, but are fully quantum mechanical, contain a graviton, and are believed to be mathematically consistent.[19]

11.4   See also

- Gravitomagnetism
- Gravitational wave
- Planck mass
- Gravitation
- Static forces and virtual-particle exchange
- Multiverse
- Gravitino

11.5   References

[1]   G is used to avoid confusion with gluons (symbol g)

[2]   Rovelli, C. (2001). "Notes for a brief history of quantum gravity". arXiv:gr-qc/0006061 [gr-qc].

[3] Blokhintsev. D. I.; Gal'perin, F. M. (1934). "Gipoteza neitrino i zakon sokhraneniya energii" [Neutrino hypothesis and conservation of energy]. *Pod Znamenem Marxisma* (in Russian) **6**: 147–157.

[4] Lightman, A. P.; Press, W. H.; Price, R. H.; Teukolsky, S. A. (1975). "Problem 12.16". *Problem book in Relativity and Gravitation*. Princeton University Press. ISBN 0-691-08162-X.

[5] For a comparison of the geometric derivation and the (non-geometric) spin-2 field derivation of general relativity, refer to box 18.1 (and also 17.2.5) of Misner, C. W.; Thorne, K. S.; Wheeler, J. A. (1973). *Gravitation*. W. H. Freeman. ISBN 0-7167-0344-0.

[6] Feynman, R. P.; Morinigo, F. B.; Wagner, W. G.; Hatfield, B. (1995). *Feynman Lectures on Gravitation*. Addison-Wesley. ISBN 0-201-62734-5.

[7] Zee, A. (2003). *Quantum Field Theory in a Nutshell*. Princeton University Press. ISBN 0-691-01019-6.

[8] Randall, L. (2005). *Warped Passages: Unraveling the Universe's Hidden Dimensions*. Ecco Press. ISBN 0-06-053108-8.

[9] See the other articles on General relativity, Gravitational field, Gravitational wave, etc

[10] Colosi, D.; et al. (2005). "Background independence in a nutshell: The dynamics of a tetrahedron". *Classical and Quantum Gravity* **22** (14): 2971. arXiv:gr-qc/0408079. Bibcode:2005CQGra..22.2971C. doi:10.1088/0264-9381/22/14/008.

[11] Witten, E. (1993). "Quantum Background Independence In String Theory". arXiv:hep-th/9306122 [hep-th].

[12] Smolin, L. (2005). "The case for background independence". arXiv:hep-th/0507235 [hep-th].

[13] Kaku, Michio (2006). *Parallel Worlds - The science of alternative universes and our future in the Cosmos*. pp. 218–221.

[14] Ali, Ahmed Farang (2014). "Cosmology from quantum potential". *Physical Letters B* **741**: 276–279. arXiv:1404.3093v3. doi:10.1016/j.physletb.2014.12.057.

[15] Rothman, T.; Boughn, S. (2006). "Can Gravitons be Detected?". *Foundations of Physics* **36** (12): 1801–1825. arXiv:gr-qc/0601043. Bibcode:2006FoPh...36.1801R. doi:10.1007/s10701-006-9081-9.

[16] Dyson, Freeman (8 October 2013). "Is a graviton detectable?". *International Journal of Modern Physics A* **28** (25): 1330041–1–1330035–14. Bibcode:2013IJMPA..2830041D. doi:10.1142/S0217751X1330041X.

[17] Will, C. M. (1998). "Bounding the mass of the graviton using gravitational-wave observations of inspiralling compact binaries". *Physical Review D* **57** (4): 2061–2068. arXiv:gr-qc/9709011. Bibcode:1998PhRvD..57.2061W. doi:10.1103/PhysRevD.57.2061.

[18] Trippe, S. (2013). "A Simplified Treatment of Gravitational Interaction on Galactic Scales", J. Kor. Astron. Soc. **46**, 41. arXiv:1211.4692

[19] Sokal, A. (July 22, 1996). "Don't Pull the String Yet on Superstring Theory". *The New York Times*. Retrieved March 26, 2010.

11.6 External links

-

- Graviton on *In Our Time* at the BBC. (listen now)

Chapter 12

Force carrier

In particle physics, **force carriers** are particles that give rise to forces between other particles. These particles are bundles of energy (quanta) of a particular kind of field. There is one kind of field for every species of elementary particle. For instance, there is an electron field whose quanta are electrons, and an electromagnetic field whose quanta are photons.[1] The force carrier particles that mediate the electromagnetic, weak, and strong interactions are called gauge bosons.

12.1 Particle and field viewpoints

Main article: Wave–particle duality

In particle physics, quantum field theories such as the Standard Model describe nature in terms of fields. Each field has a complementary description as the set of particles of a particular type. A force between two particles can be described either as the action of a force field generated by one particle on the other, or in terms of the exchange of virtual force carrier particles between them.

The energy of a wave in a field (for example, electromagnetic waves in the electromagnetic field) is quantized, and the quantum excitations of the field can be interpreted as particles. The Standard Model contains the following particles, each of which is an excitation of a particular field:

- Gluons, excitations of the strong gauge field.

- Photons, W bosons, and Z bosons, excitations of the electroweak gauge fields.

- Higgs bosons, excitations of one component of the Higgs field, which gives mass to fundamental particles.

- Several types of fermions, described as excitations of fermionic fields.

In addition, composite particles such as mesons can be described as excitations of an effective field.

Gravity is not a part of the Standard Model, but it is thought that there may be particles called gravitons which are the excitations of gravitational waves. The status of this particle is still tentative, because the theory is incomplete and because the interactions of *single* gravitons may be too weak to be detected.[2]

12.2 Forces from the particle viewpoint

Main article: Static forces and virtual-particle exchange
When one particle scatters off another, altering its trajec-

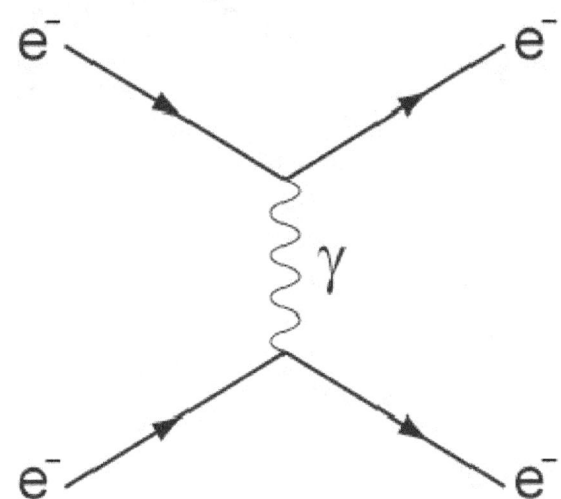

A Feynman diagram of scattering between two electrons by emission of a virtual photon.

tory, there are two ways to think about the process. In the field picture, we imagine that the field generated by one particle caused a force on the other. Alternatively, we can imagine one particle emitting a virtual particle which is absorbed by the other. The virtual particle transfers momentum from one particle to the other. This particle viewpoint is especially helpful when there are a large number of complicated quantum corrections to the calculation since these corrections can be visualized as Feynman diagrams containing additional virtual particles.

The description of forces in terms of virtual particles is limited by the applicability of the perturbation theory from which it is derived. In certain situations, such as low-energy QCD and the description of bound states, perturbation theory breaks down.

12.3 Examples

- The electromagnetic force can be described by the exchange of virtual photons.

- The nuclear force binding protons and neutrons can be described by an effective field of which mesons are the excitations.

- At sufficiently large energies, the strong interaction between quarks can be described by the exchange of virtual gluons.

- Beta decay is an example of an interaction due to the exchange of a W boson, but not an example of a force.

- Gravitation may be due to the exchange of virtual gravitons.

12.4 History

The concept of messenger particles dates back to the 18th century when the French physicist Charles Coulomb showed that the electrostatic force between electrically charged objects follows a law similar to Newton's Law of Gravitation. In time, this relationship became known as Coulomb's law. By 1862, Hermann von Helmholtz had described a ray of light as the "quickest of all the messengers". In 1905, Albert Einstein proposed the existence of a light-particle in answer to the question: "what are light quanta?"

In 1923, at the Washington University in St. Louis, Arthur Holly Compton demonstrated an effect now known as Compton scattering. This effect is only explainable if light can behave as a stream of particles and it convinced the physics community of the existence of Einstein's light-particle. Lastly, in 1926, one year before the theory of quantum mechanics was published, Gilbert N. Lewis introduced the term "photon", which soon became the name for Einstein's light particle. From there, the concept of messenger particles developed further.

12.5 See also

- Virtual particle
- Fundamental interaction
- Particle physics

12.6 References

[1] Steven Weinberg, Dreams of a Final Theory, Hutchinson, 1993

[2] Rothman, Tony; Stephen Boughn (November 2006). "Can Gravitons be Detected?". *Foundations of Physics* **36** (12): 1801–1825. arXiv:gr-qc/0601043. Bibcode:2006FoPh...36.1801R. doi:10.1007/s10701-006-9081-9.

12.7 External links

- Messenger Particles - Cern Interactive Slide Show

Chapter 13

Dilaton

In particle physics, the **dilaton** is a hypothetical particle that appears in theories with extra dimensions when the volume of the compactified dimensions is allowed to vary. It appears for instance in Kaluza–Klein theory's compactifications of extra dimensions. It is a particle of a scalar field Φ, a scalar field that always comes with gravity. For comparison, in standard general relativity, Newton's constant, or equivalently the Planck mass is a constant. If this constant is promoted to a dynamical field, the result is the dilaton.

In Kaluza–Klein theories, after dimensional reduction, the effective Planck mass varies as some power of the volume of compactified space. This is why volume can turn out as a dilaton in the lower-dimensional effective theory.

Although string theory naturally incorporates Kaluza–Klein theory (which first introduced the dilaton), perturbative string theories, such as type I string theory, type II string theory and heterotic string theory, already contain the dilaton in the maximal number of 10 dimensions. However, on the other hand, M-theory in 11 dimensions does not include the dilaton in its spectrum unless it is compactified. In fact, the dilaton in type IIA string theory is actually the radion of M-theory compactified over a circle, while the dilaton in $E_8 \times E_8$ string theory is the radion for the Hořava–Witten model. (For more on the M-theory origin of the dilaton, see .)

In string theory, there is also a dilaton in the worldsheet CFT(Conformal field theory). The exponential of its vacuum expectation value determines the coupling constant g, as $\int R = 2\pi\chi$ for compact worldsheets by the Gauss–Bonnet theorem and the Euler characteristic $\chi = 2 - 2g$, where g is the genus that counts the number of handles and thus the number of loops or string interactions described by a specific worldsheet.

$$g = \exp(\langle \phi \rangle)$$

Therefore the coupling constant is a dynamical variable in string theory, unlike the case of quantum field theory where it is constant. As long as supersymmetry is unbroken, such scalar fields can take arbitrary values (they are moduli). However, supersymmetry breaking usually creates a potential energy for the scalar fields and the scalar fields localize near a minimum whose position should in principle be calculable in string theory.

The dilaton acts like a Brans–Dicke scalar, with the effective Planck scale depending upon *both* the string scale and the dilaton field.

In supersymmetry, the superpartner of the dilaton is called the **dilatino**, and the dilaton combines with the axion to form a complex scalar field.

13.1 Dilaton action

The dilaton-gravity action is

$$\int d^D x \sqrt{-g} \left[\frac{1}{2\kappa} \left(\Phi R - \omega[\Phi] \frac{g^{\mu\nu} \partial_\mu \Phi \partial_\nu \Phi}{\Phi} \right) - V[\Phi] \right]$$

This is more general than Brans–Dicke in vacuum in that we have a dilaton potential.

13.2 See also

- CGHS model
- R=T model
- Quantum gravity

13.3 References

- Fujii, Y. (2003). "Mass of the dilaton and the cosmological constant". *Prog.*

Theor. Phys. **110** (3): 433–439. arXiv:gr-qc/0212030. Bibcode:2003PThPh.110..433F. doi:10.1143/PTP.110.433.

- Hayashi, M.; Watanabe, T.; Aizawa, I. & Aketo, K. (2003). "Dilatonic Inflation and SUSY Breaking in String-inspired Supergravity". *Modern Physics Letters A* **18** (39): 2785–2793. arXiv:hep-ph/0303029. Bibcode:2003MPLA...18.2785H. doi:10.1142/S0217732303012465.

- Alvarenge, F.; Batista, A. & Fabris, J. (2005). "Does Quantum Cosmology Predict a Constant Dilatonic Field". *International Journal of Modern Physics D* **14** (2): 291–307. arXiv:gr-qc/0404034. Bibcode:2005IJMPD..14..291A. doi:10.1142/S0218271805005955.

- Lu, H.; Huang, Z.; Fang, W. & Zhang, K. (2004). "Dark Energy and Dilaton Cosmology". arXiv:hep-th/0409309 [hep-th].

- Wesson, Paul S. (1999). *Space-Time-Matter, Modern Kaluza-Klein Theory*. Singapore: World Scientific. p. 31. ISBN 981-02-3588-7.

Chapter 14

Quantum electrodynamics

In particle physics, **quantum electrodynamics (QED)** is the relativistic quantum field theory of electrodynamics. In essence, it describes how light and matter interact and is the first theory where full agreement between quantum mechanics and special relativity is achieved. QED mathematically describes all phenomena involving electrically charged particles interacting by means of exchange of photons and represents the quantum counterpart of classical electromagnetism giving a complete account of matter and light interaction.

In technical terms, QED can be described as a perturbation theory of the electromagnetic quantum vacuum. Richard Feynman called it "the jewel of physics" for its extremely accurate predictions of quantities like the anomalous magnetic moment of the electron and the Lamb shift of the energy levels of hydrogen.[1]:Ch1

14.1 History

Main article: History of quantum mechanics

The first formulation of a quantum theory describing radiation and matter interaction is attributed to British scientist Paul Dirac, who (during the 1920s) was able to compute the coefficient of spontaneous emission of an atom.[2]

Dirac described the quantization of the electromagnetic field as an ensemble of harmonic oscillators with the introduction of the concept of creation and annihilation operators of particles. In the following years, with contributions from Wolfgang Pauli, Eugene Wigner, Pascual Jordan, Werner Heisenberg and an elegant formulation of quantum electrodynamics due to Enrico Fermi,[3] physicists came to believe that, in principle, it would be possible to perform any computation for any physical process involving photons and charged particles. However, further studies by Felix Bloch with Arnold Nordsieck,[4] and Victor Weisskopf,[5] in 1937 and 1939, revealed that such computations were reliable only at a first order of perturbation theory, a problem already pointed out by Robert Oppenheimer.[6] At higher

Paul Dirac

orders in the series infinities emerged, making such computations meaningless and casting serious doubts on the internal consistency of the theory itself. With no solution for this problem known at the time, it appeared that a fundamental incompatibility existed between special relativity and quantum mechanics.

Difficulties with the theory increased through the end of 1940. Improvements in microwave technology made it possible to take more precise measurements of the shift of the levels of a hydrogen atom,[7] now known as the Lamb shift and magnetic moment of the electron.[8] These experiments unequivocally exposed discrepancies which the theory was unable to explain.

A first indication of a possible way out was given by Hans

Hans Bethe

Feynman (center) and Oppenheimer (right) at Los Alamos.

Bethe. In 1947, while he was traveling by train to reach Schenectady from New York,[9] after giving a talk at the conference at Shelter Island on the subject, Bethe completed the first non-relativistic computation of the shift of the lines of the hydrogen atom as measured by Lamb and Retherford.[10] Despite the limitations of the computation, agreement was excellent. The idea was simply to attach infinities to corrections of mass and charge that were actually fixed to a finite value by experiments. In this way, the infinities get absorbed in those constants and yield a finite result in good agreement with experiments. This procedure was named renormalization.

Based on Bethe's intuition and fundamental papers on the subject by Sin-Itiro Tomonaga,[11] Julian Schwinger,[12][13] Richard Feynman[14][15][16] and Freeman Dyson,[17][18] it was finally possible to get fully covariant formulations that were finite at any order in a perturbation series of quantum electrodynamics. Sin-Itiro Tomonaga, Julian Schwinger and Richard Feynman were jointly awarded with a Nobel prize in physics in 1965 for their work in this area.[19] Their contributions, and those of Freeman Dyson, were about covariant and gauge invariant formulations of quantum electrodynamics that allow computations of observables at any order of perturbation theory. Feynman's math-

ematical technique, based on his diagrams, initially seemed very different from the field-theoretic, operator-based approach of Schwinger and Tomonaga, but Freeman Dyson later showed that the two approaches were equivalent.[17] Renormalization, the need to attach a physical meaning at certain divergences appearing in the theory through integrals, has subsequently become one of the fundamental aspects of quantum field theory and has come to be seen as a criterion for a theory's general acceptability. Even though renormalization works very well in practice, Feynman was never entirely comfortable with its mathematical validity, even referring to renormalization as a "shell game" and "hocus pocus".[1]:128

QED has served as the model and template for all subsequent quantum field theories. One such subsequent theory is quantum chromodynamics, which began in the early 1960s and attained its present form in the 1975 work by H. David Politzer, Sidney Coleman, David Gross and Frank Wilczek. Building on the pioneering work of Schwinger, Gerald Guralnik, Dick Hagen, and Tom Kibble,[20][21] Peter Higgs, Jeffrey Goldstone, and others, Sheldon Glashow, Steven Weinberg and Abdus Salam independently showed how the weak nuclear force and quantum electrodynamics could be merged into a single electroweak force.

14.2 Feynman's view of quantum electrodynamics

14.2.1 Introduction

Near the end of his life, Richard P. Feynman gave a series of lectures on QED intended for the lay public. These lectures were transcribed and published as Feynman (1985), *QED: The strange theory of light and matter*,[1] a classic nonmathematical exposition of QED from the point of view articulated below.

The key components of Feynman's presentation of QED are three basic actions.[1]:85

- A photon goes from one place and time to another place and time.

- An electron goes from one place and time to another place and time.

- An electron emits or absorbs a photon at a certain place and time.

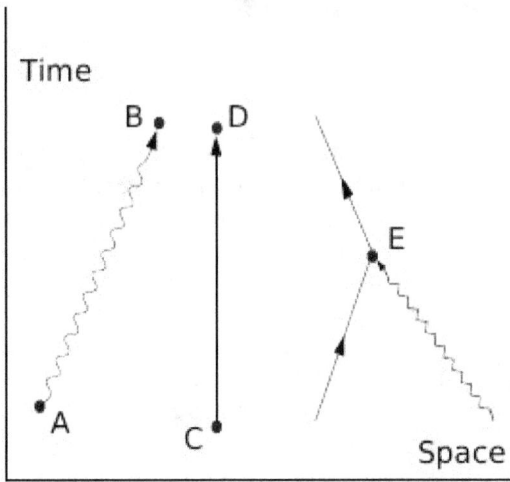

Feynman diagram elements

These actions are represented in a form of visual shorthand by the three basic elements of Feynman diagrams: a wavy line for the photon, a straight line for the electron and a junction of two straight lines and a wavy one for a vertex representing emission or absorption of a photon by an electron. These can all be seen in the adjacent diagram.

It is important not to over-interpret these diagrams. Nothing is implied about *how* a particle gets from one point to another. The diagrams do *not* imply that the particles are moving in straight or curved lines. They do *not* imply that the particles are moving with fixed speeds. The fact that the photon is often represented, by convention, by a wavy line and not a straight one does *not* imply that it is thought that it is more wavelike than is an electron. The images are just symbols to represent the actions above: photons and electrons do, somehow, move from point to point and electrons, somehow, emit and absorb photons. We do not know how these things happen, but the theory tells us about the probabilities of these things happening.

As well as the visual shorthand for the actions Feynman introduces another kind of shorthand for the numerical quantities called probability amplitudes. The probability is the

square of the total probability amplitude. If a photon moves from one place and time—in shorthand, A—to another place and time—in shorthand, B—the associated quantity is written in Feynman's shorthand as P(A to B). The similar quantity for an electron moving from C to D is written E(C to D). The quantity which tells us about the probability amplitude for the emission or absorption of a photon he calls 'j'. This is related to, but not the same as, the measured electron charge 'e'.[1]:91

QED is based on the assumption that complex interactions of many electrons and photons can be represented by fitting together a suitable collection of the above three building blocks, and then using the probability amplitudes to calculate the probability of any such complex interaction. It turns out that the basic idea of QED can be communicated while making the assumption that the square of the total of the probability amplitudes mentioned above (P(A to B), E(A to B) and 'j') acts just like our everyday probability. (A simplification made in Feynman's book.) Later on, this will be corrected to include specifically quantum-style mathematics, following Feynman.

The basic rules of probability amplitudes that will be used are that a) if an event can happen in a variety of different ways then its probability amplitude is the **sum** of the probability amplitudes of the possible ways and b) if a process involves a number of independent sub-processes then its probability amplitude is the **product** of the component probability amplitudes.[1]:93

14.2.2 Basic constructions

Suppose we start with one electron at a certain place and time (this place and time being given the arbitrary label A) and a photon at another place and time (given the label B). A typical question from a physical standpoint is: 'What is the probability of finding an electron at C (another place and a later time) and a photon at D (yet another place and time)?'. The simplest process to achieve this end is for the electron to move from A to C (an elementary action) and for the photon to move from B to D (another elementary action). From a knowledge of the probability amplitudes of each of these sub-processes – E(A to C) and P(B to D) – then we would expect to calculate the probability amplitude of both happening together by multiplying them, using rule b) above. This gives a simple estimated overall probability amplitude, which is squared to give an estimated probability.

But there are other ways in which the end result could come about. The electron might move to a place and time E where it absorbs the photon; then move on before emitting another photon at F; then move on to C where it is detected, while the new photon moves on to D. The probability of this com-

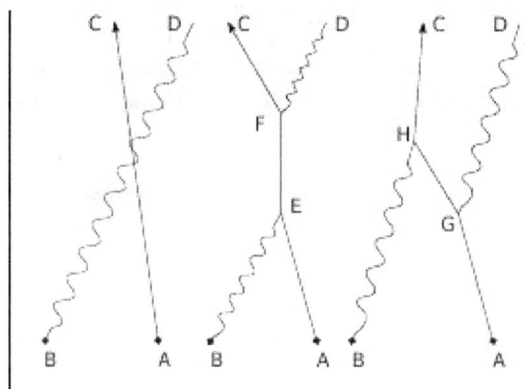

Compton scattering

plex process can again be calculated by knowing the probability amplitudes of each of the individual actions: three electron actions, two photon actions and two vertexes – one emission and one absorption. We would expect to find the total probability amplitude by multiplying the probability amplitudes of each of the actions, for any chosen positions of E and F. We then, using rule a) above, have to add up all these probability amplitudes for all the alternatives for E and F. (This is not elementary in practice, and involves integration.) But there is another possibility, which is that the electron first moves to G where it emits a photon which goes on to D, while the electron moves on to H, where it absorbs the first photon, before moving on to C. Again we can calculate the probability amplitude of these possibilities (for all points G and H). We then have a better estimation for the total probability amplitude by adding the probability amplitudes of these two possibilities to our original simple estimate. Incidentally the name given to this process of a photon interacting with an electron in this way is Compton scattering.

There are an *infinite number* of other intermediate processes in which more and more photons are absorbed and/or emitted. For each of these possibilities there is a Feynman diagram describing it. This implies a complex computation for the resulting probability amplitudes, but provided it is the case that the more complicated the diagram the less it contributes to the result, it is only a matter of time and effort to find as accurate an answer as one wants to the original question. This is the basic approach of QED. To calculate the probability of *any* interactive process between electrons and photons it is a matter of first noting, with Feynman diagrams, all the possible ways in which the process can be constructed from the three basic elements. Each diagram involves some calculation involving definite rules to find the associated probability amplitude.

That basic scaffolding remains when one moves to a quan-

tum description but some conceptual changes are needed. One is that whereas we might expect in our everyday life that there would be some constraints on the points to which a particle can move, that is *not* true in full quantum electrodynamics. There is a possibility of an electron at A, or a photon at B, moving as a basic action to *any other place and time in the universe*. That includes places that could only be reached at speeds greater than that of light and also *earlier times*. (An electron moving backwards in time can be viewed as a positron moving forward in time.)[1:89, 98–99]

14.2.3   Probability amplitudes

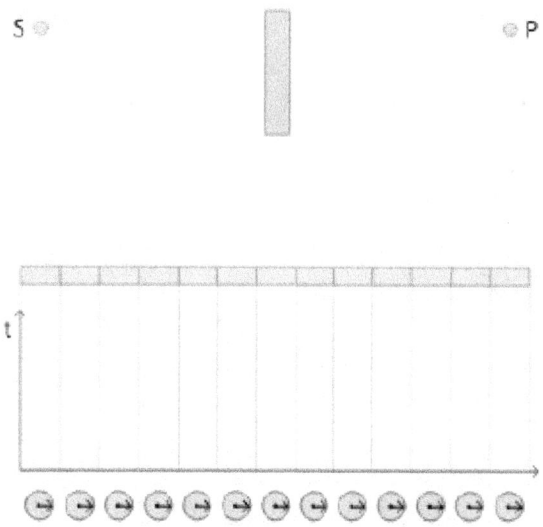

*Feynman replaces complex numbers with spinning arrows, which start at emission and end at detection of a particle. The sum of all resulting arrows represents the total probability of the event. In this diagram, light emitted by the source **S** bounces off a few segments of the mirror (in blue) before reaching the detector at **P**. The sum of all paths must be taken into account. The graph below depicts the total time spent to traverse each of the paths above.*

Quantum mechanics introduces an important change in the way probabilities are computed. Probabilities are still represented by the usual real numbers we use for probabilities in our everyday world, but probabilities are computed as the square of probability amplitudes. Probability amplitudes are complex numbers.

Feynman avoids exposing the reader to the mathematics of complex numbers by using a simple but accurate representation of them as arrows on a piece of paper or screen. (These must not be confused with the arrows of Feynman diagrams which are actually simplified representations in two dimensions of a relationship between points in three dimensions of space and one of time.) The amplitude arrows are fundamental to the description of the world given by quantum theory. No satisfactory reason has been given for *why* they are needed. But pragmatically we have to accept that they are an essential part of our description of all quantum phenomena. They are related to our everyday ideas of probability by the simple rule that the probability of an event is the **square** of the length of the corresponding amplitude arrow. So, for a given process, if two probability amplitudes, **v** and **w**, are involved, the probability of the process will be given either by

$$P = |\mathbf{v} + \mathbf{w}|^2$$

or

$$P = |\mathbf{v}\,\mathbf{w}|^2.$$

The rules as regards adding or multiplying, however, are the same as above. But where you would expect to add or multiply probabilities, instead you add or multiply probability amplitudes that now are complex numbers.

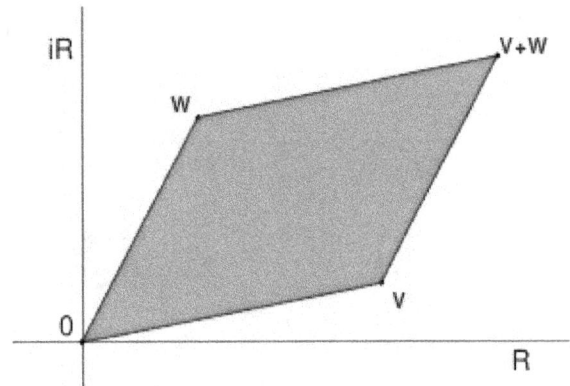

Addition of probability amplitudes as complex numbers

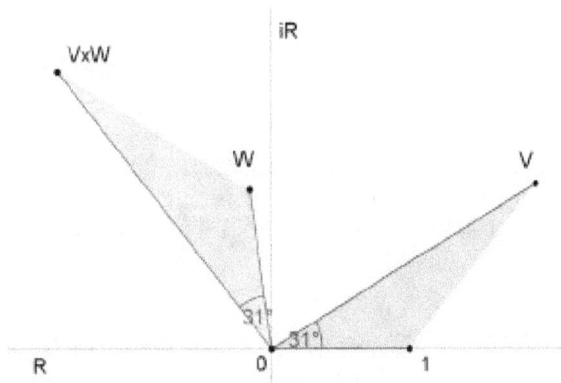

Multiplication of probability amplitudes as complex numbers

Addition and multiplication are familiar operations in the theory of complex numbers and are given in the figures. The

sum is found as follows. Let the start of the second arrow be at the end of the first. The sum is then a third arrow that goes directly from the start of the first to the end of the second. The product of two arrows is an arrow whose length is the product of the two lengths. The direction of the product is found by adding the angles that each of the two have been turned through relative to a reference direction: that gives the angle that the product is turned relative to the reference direction.

That change, from probabilities to probability amplitudes, complicates the mathematics without changing the basic approach. But that change is still not quite enough because it fails to take into account the fact that both photons and electrons can be polarized, which is to say that their orientations in space and time have to be taken into account. Therefore, P(A to B) actually consists of 16 complex numbers, or probability amplitude arrows.[1]:120–121 There are also some minor changes to do with the quantity "j", which may have to be rotated by a multiple of 90° for some polarizations, which is only of interest for the detailed bookkeeping.

Associated with the fact that the electron can be polarized is another small necessary detail which is connected with the fact that an electron is a fermion and obeys Fermi–Dirac statistics. The basic rule is that if we have the probability amplitude for a given complex process involving more than one electron, then when we include (as we always must) the complementary Feynman diagram in which we just exchange two electron events, the resulting amplitude is the reverse – the negative – of the first. The simplest case would be two electrons starting at A and B ending at C and D. The amplitude would be calculated as the "difference", E(A to D) × E(B to C) − E(A to C) × E(B to D), where we would expect, from our everyday idea of probabilities, that it would be a sum.[1]:112–113

14.2.4 Propagators

Finally, one has to compute P (A to B) and E (C to D) corresponding to the probability amplitudes for the photon and the electron respectively. These are essentially the solutions of the Dirac Equation which describes the behavior of the electron's probability amplitude and the Klein–Gordon equation which describes the behavior of the photon's probability amplitude. These are called Feynman propagators. The translation to a notation commonly used in the standard literature is as follows:

$$P(\text{A to B}) \to D_F(x_B - x_A), \quad E(\text{C to D}) \to S_F(x_D - x_C)$$

where a shorthand symbol such as x_A stands for the four real numbers which give the time and position in three dimensions of the point labeled A.

14.2.5 Mass renormalization

Main article: Self-energy

A problem arose historically which held up progress for

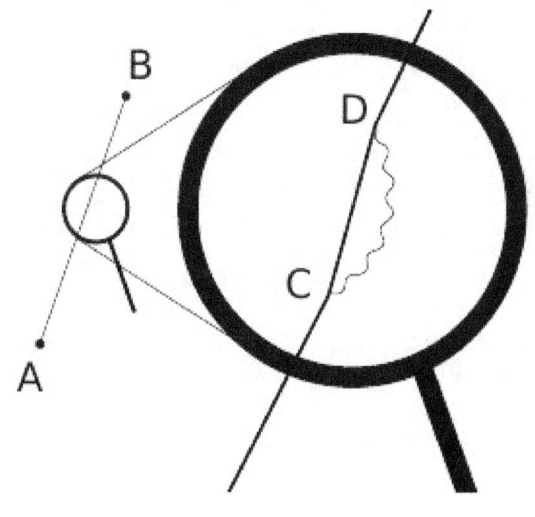

Electron self-energy loop

twenty years: although we start with the assumption of three basic "simple" actions, the rules of the game say that if we want to calculate the probability amplitude for an electron to get from A to B we must take into account **all** the possible ways: all possible Feynman diagrams with those end points. Thus there will be a way in which the electron travels to C, emits a photon there and then absorbs it again at D before moving on to B. Or it could do this kind of thing twice, or more. In short we have a fractal-like situation in which if we look closely at a line it breaks up into a collection of "simple" lines, each of which, if looked at closely, are in turn composed of "simple" lines, and so on *ad infinitum*. This is a very difficult situation to handle. If adding that detail only altered things slightly then it would not have been too bad, but disaster struck when it was found that the simple correction mentioned above led to *infinite* probability amplitudes. In time this problem was "fixed" by the technique of renormalization. However, Feynman himself remained unhappy about it, calling it a "dippy process".[1]:128

14.2.6 Conclusions

Within the above framework physicists were then able to calculate to a high degree of accuracy some of the properties of electrons, such as the anomalous magnetic dipole moment. However, as Feynman points out, it fails totally to explain why particles such as the electron have the masses they do. "There is no theory that adequately explains these

numbers. We use the numbers in all our theories, but we don't understand them – what they are, or where they come from. I believe that from a fundamental point of view, this is a very interesting and serious problem."[1]:152

14.3 Mathematics

Mathematically, QED is an abelian gauge theory with the symmetry group U(1). The gauge field, which mediates the interaction between the charged spin-1/2 fields, is the electromagnetic field. The QED Lagrangian for a spin-1/2 field interacting with the electromagnetic field is given by the real part of[22]:78

where

γ^μ are Dirac matrices;

ψ a bispinor field of spin-1/2 particles (e.g. electron–positron field);

$\bar{\psi} \equiv \psi^\dagger \gamma^0$, called "psi-bar", is sometimes referred to as the Dirac adjoint;

$D_\mu \equiv \partial_\mu + ieA_\mu + ieB_\mu$ is the gauge covariant derivative;

e is the coupling constant, equal to the electric charge of the bispinor field;

m is the mass of the electron or positron;

A_μ is the covariant four-potential of the electromagnetic field generated by the electron itself;

B_μ is the external field imposed by external source;

$F_{\mu\nu} = \partial_\mu A_\nu - \partial_\nu A_\mu$ is the electromagnetic field tensor.

14.3.1 Equations of motion

To begin, substituting the definition of D into the Lagrangian gives us

$$\mathcal{L} = i\bar{\psi}\gamma^\mu \partial_\mu \psi - e\bar{\psi}\gamma_\mu (A^\mu + B^\mu)\psi - m\bar{\psi}\psi - \frac{1}{4}F_{\mu\nu}F^{\mu\nu}.$$

Next, we can substitute this Lagrangian into the Euler–Lagrange equation of motion for a field:

to find the field equations for QED.

The two terms from this Lagrangian are then

$$\partial_\mu \left(\frac{\partial \mathcal{L}}{\partial(\partial_\mu \psi)} \right) = \partial_\mu \left(i\bar{\psi}\gamma^\mu \right),$$

$$\frac{\partial \mathcal{L}}{\partial \psi} = -e\bar{\psi}\gamma_\mu (A^\mu + B^\mu) - m\bar{\psi}.$$

Substituting these two back into the Euler–Lagrange equation (2) results in

$$i\partial_\mu \bar{\psi}\gamma^\mu + e\bar{\psi}\gamma_\mu(A^\mu + B^\mu) + m\bar{\psi} = 0$$

with complex conjugate

$$i\gamma^\mu \partial_\mu \psi - e\gamma_\mu(A^\mu + B^\mu)\psi - m\psi = 0.$$

Bringing the middle term to the right-hand side transforms this second equation into

The left-hand side is like the original Dirac equation and the right-hand side is the interaction with the electromagnetic field.

One further important equation can be found by substituting the above Lagrangian into another Euler–Lagrange equation, this time for the field A^μ :

The two terms this time are

$$\partial_\nu \left(\frac{\partial \mathcal{L}}{\partial(\partial_\nu A_\mu)} \right) = \partial_\nu (\partial^\mu A^\nu - \partial^\nu A^\mu),$$

$$\frac{\partial \mathcal{L}}{\partial A_\mu} = -e\bar{\psi}\gamma^\mu \psi$$

and these two terms, when substituted back into (3) give us

Now, if we impose the Lorenz gauge condition, that the divergence of the four potential vanishes

$$\partial_\mu A^\mu = 0$$

then we get

$$\Box A^\mu = e\bar{\psi}\gamma^\mu \psi,$$

which is a wave equation for the four potential, the QED version of the classical Maxwell equations in the Lorenz gauge. (In the above equation, the square represents the D'Alembert operator.)

14.3.2   Interaction picture

This theory can be straightforwardly quantized by treating bosonic and fermionic sectors as free. This permits us to build a set of asymptotic states which can be used to start a computation of the probability amplitudes for different processes. In order to do so, we have to compute an evolution operator that, for a given initial state $|i\rangle$, will give a final state $\langle f|$ in such a way to have[22]:5

$$M_{fi} = \langle f|U|i\rangle.$$

This technique is also known as the S-matrix. The evolution operator is obtained in the interaction picture where time evolution is given by the interaction Hamiltonian, which is the integral over space of the second term in the Lagrangian density given above:[22]:123

$$V = e \int d^3 x\, \bar{\psi}\gamma^\mu \psi A_\mu$$

and so, one has[22]:86

$$U = T \exp\left[-\frac{i}{\hbar} \int_{t_0}^{t} dt' V(t') \right]$$

where T is the time ordering operator. This evolution operator only has meaning as a series, and what we get here is a perturbation series with the fine structure constant as the development parameter. This series is called the Dyson series.

14.3.3   Feynman diagrams

Despite the conceptual clarity of this Feynman approach to QED, almost no early textbooks follow him in their presentation. When performing calculations it is much easier to work with the Fourier transforms of the propagators. Quantum physics considers particles' momenta rather than their positions, and it is convenient to think of particles as being created or annihilated when they interact. Feynman diagrams then *look* the same, but the lines have different interpretations. The electron line represents an electron with a given energy and momentum, with a similar interpretation of the photon line. A vertex diagram represents the annihilation of one electron and the creation of another together with the absorption or creation of a photon, each having specified energies and momenta.

Using Wick theorem on the terms of the Dyson series, all the terms of the S-matrix for quantum electrodynamics can be computed through the technique of Feynman diagrams. In this case rules for drawing are the following[22]:801–802

Incoming fermion: $\alpha \longrightarrow$ $\rightarrow$ $u_\alpha(\vec{p}, s)$

Incoming antifermion: $\alpha \longleftarrow$ $\rightarrow$ $\bar{v}_\alpha(\vec{p}, s)$

Outgoing fermion: $\longrightarrow \alpha$ $\rightarrow$ $\bar{u}_\alpha(\vec{p}, s)$

Outgoing antifermion: $\longleftarrow \alpha$ $\rightarrow$ $v_\alpha(p, s)$

Incoming photon: $\mu \sim\!\!\sim\!\!\sim$ $\rightarrow$ $\epsilon_\mu(\vec{k}, \lambda)$

Outgoing photon: $\sim\!\!\sim\!\!\sim \mu$ $\rightarrow$ $\epsilon_\mu(\vec{k}, \lambda)^*$

$\alpha \longrightarrow \beta$ $\rightarrow$ $\left(\dfrac{i}{\slashed{p} - m + i\varepsilon}\right)_{\beta\alpha}$

$\mu \sim\!\!\sim\!\!\sim \nu$ $\rightarrow$ $\dfrac{-i\eta_{\mu\nu}}{p^2 + i\varepsilon}$

$$\beta \diagdown \!\!\diagup \mu \quad \rightarrow \quad -ie\gamma^\mu_{\beta\alpha}(2\pi)^4\delta^{(4)}(p_1 + p_2 + p_3).$$

To these rules we must add a further one for closed loops that implies an integration on momenta $\int d^4p/(2\pi)^4$, since these internal ("virtual") particles are not constrained to any specific energy–momentum – even that usually required by special relativity (see this article for details). From them, computations of probability amplitudes are straightforwardly given. An example is Compton scattering, with an electron and a photon undergoing elastic scattering. Feynman diagrams are in this case[22]:158–159

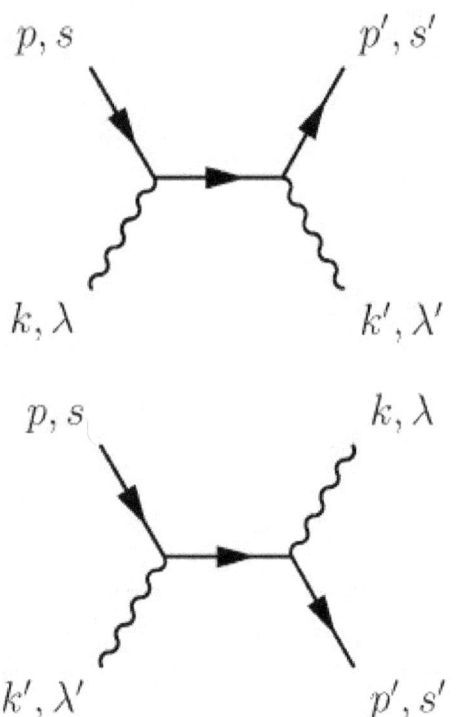

and so we are able to get the corresponding amplitude at the first order of a perturbation series for the S-matrix:

$$M_{fi} = (ie)^2 \overline{u}(\vec{p}',s') \not{\epsilon}'(\vec{k}',\lambda')^* \frac{\not{p}+\not{k}+m_e}{(p+k)^2-m_e^2} \not{\epsilon}(\vec{k},\lambda)$$

$$\text{X} \quad u(\vec{p},s) + (ie)^2 \overline{u}(\vec{p}',s') \not{\epsilon}(\vec{k},\lambda) \frac{\not{p}-\not{k}'+m_e}{(p-k')^2-m_e^2}$$

$$\text{X} \quad \not{\epsilon}'(\vec{k}',\lambda')^* u(\vec{p},s)$$

from which we are able to compute the cross section for this scattering.

14.4   Renormalizability

Higher order terms can be straightforwardly computed for the evolution operator but these terms display diagrams containing the following simpler ones[22]:ch 10

- One-loop contribution to the vacuum polarization function

- One-loop contribution to the electron self-energy function

- One-loop contribution to the vertex function

that, being closed loops, imply the presence of diverging integrals having no mathematical meaning. To overcome

this difficulty, a technique called renormalization has been devised, producing finite results in very close agreement with experiments. It is important to note that a criterion for theory being meaningful after renormalization is that the number of diverging diagrams is finite. In this case the theory is said to be **renormalizable**. The reason for this is that to get observables renormalized one needs a finite number of constants to maintain the predictive value of the theory untouched. This is exactly the case of quantum electrodynamics displaying just three diverging diagrams. This procedure gives observables in very close agreement with experiment as seen e.g. for electron gyromagnetic ratio.

Renormalizability has become an essential criterion for a quantum field theory to be considered as a viable one. All the theories describing fundamental interactions, except gravitation whose quantum counterpart is presently under very active research, are renormalizable theories.

14.5   Nonconvergence of series

An argument by Freeman Dyson shows that the radius of convergence of the perturbation series in QED is zero.[23] The basic argument goes as follows: if the coupling constant were negative, this would be equivalent to the Coulomb force constant being negative. This would "reverse" the electromagnetic interaction so that *like* charges would *attract* and *unlike* charges would *repel*. This would render the vacuum unstable against decay into a cluster of electrons on one side of the universe and a cluster of positrons on the other side of the universe. Because the theory is 'sick' for any negative value of the coupling constant, the series do not converge, but are an asymptotic series.

From a modern perspective, we say that QED is not well defined as a quantum field theory to arbitrarily high energy.[24] The coupling constant runs to infinity at finite energy, signalling a Landau pole. The problem is essentially that QED appears to suffer from quantum triviality issues. This is one of the motivations for embedding QED within a Grand Unified Theory.

14.6   See also

14.7   References

[1] Feynman, Richard (1985). *QED: The Strange Theory of Light and Matter*. Princeton University Press. ISBN 978-0-691-12575-6.

[2] P.A.M. Dirac (1927). "The Quantum Theory of the Emission and Absorption of Radiation".

Proceedings of the Royal Society of London A **114** (767): 243–265. Bibcode:1927RSPSA.114..243D. doi:10.1098/rspa.1927.0039.

[3] E. Fermi (1932). "Quantum Theory of Radiation". *Reviews of Modern Physics* **4**: 87–132. Bibcode:1932RvMP....4...87F. doi:10.1103/RevModPhys.4.87.

[4] Bloch, F.; Nordsieck, A. (1937). "Note on the Radiation Field of the Electron". *Physical Review* **52** (2): 54–59. Bibcode:1937PhRv...52...54B. doi:10.1103/PhysRev.52.54.

[5] V. F. Weisskopf (1939). "On the Self-Energy and the Electromagnetic Field of the Electron". *Physical Review* **56**: 72–85. Bibcode:1939PhRv...56...72W. doi:10.1103/PhysRev.56.72.

[6] R. Oppenheimer (1930). "Note on the Theory of the Interaction of Field and Matter". *Physical Review* **35** (5): 461–477. Bibcode:1930PhRv...35..461O. doi:10.1103/PhysRev.35.461.

[7] Lamb, Willis; Retherford, Robert (1947). "Fine Structure of the Hydrogen Atom by a Microwave Method,". *Physical Review* **72** (3): 241–243. Bibcode:1947PhRv...72..241L. doi:10.1103/PhysRev.72.241.

[8] Foley, H.; Kusch, P. (1948). "On the Intrinsic Moment of the Electron". *Physical Review* **73** (3): 412. Bibcode:1948PhRv...73..412F. doi:10.1103/PhysRev.73.412.

[9] Schweber, Silvan (1994). "Chapter 5". *QED and the Men Who Did it: Dyson, Feynman, Schwinger, and Tomonaga*. Princeton University Press. p. 230. ISBN 978-0-691-03327-3.

[10] H. Bethe (1947). "The Electromagnetic Shift of Energy Levels". *Physical Review* **72** (4): 339–341. Bibcode:1947PhRv...72..339B. doi:10.1103/PhysRev.72.339.

[11] S. Tomonaga (1946). "On a Relativistically Invariant Formulation of the Quantum Theory of Wave Fields". *Progress of Theoretical Physics* **1** (2): 27–42. doi:10.1143/PTP.1.27.

[12] J. Schwinger (1948). "On Quantum-Electrodynamics and the Magnetic Moment of the Electron". *Physical Review* **73** (4): 416–417. Bibcode:1948PhRv...73..416S. doi:10.1103/PhysRev.73.416.

[13] J. Schwinger (1948). "Quantum Electrodynamics. I. A Covariant Formulation". *Physical Review* **74** (10): 1439–1461. Bibcode:1948PhRv...74.1439S. doi:10.1103/PhysRev.74.1439.

[14] R. P. Feynman (1949). "Space–Time Approach to Quantum Electrodynamics". *Physical Review* **76** (6): 769–789. Bibcode:1949PhRv...76..769F. doi:10.1103/PhysRev.76.769.

[15] R. P. Feynman (1949). "The Theory of Positrons". *Physical Review* **76** (6): 749–759. Bibcode:1949PhRv...76..749F. doi:10.1103/PhysRev.76.749.

[16] R. P. Feynman (1950). "Mathematical Formulation of the Quantum Theory of Electromagnetic Interaction". *Physical Review* **80** (3): 440–457. Bibcode:1950PhRv...80..440F. doi:10.1103/PhysRev.80.440.

[17] F. Dyson (1949). "The Radiation Theories of Tomonaga, Schwinger, and Feynman". *Physical Review* **75** (3): 486–502. Bibcode:1949PhRv...75..486D. doi:10.1103/PhysRev.75.486.

[18] F. Dyson (1949). "The S Matrix in Quantum Electrodynamics". *Physical Review* **75** (11): 1736–1755. Bibcode:1949PhRv...75.1736D. doi:10.1103/PhysRev.75.1736.

[19] "The Nobel Prize in Physics 1965". Nobel Foundation. Retrieved 2008-10-09.

[20] Guralnik, G. S.; Hagen, C. R.; Kibble, T. W. B. (1964). "Global Conservation Laws and Massless Particles". *Physical Review Letters* **13** (20): 585–587. Bibcode:1964PhRvL..13..585G. doi:10.1103/PhysRevLett.13.585.

[21] Guralnik, G. S. (2009). "The History of the Guralnik, Hagen and Kibble development of the Theory of Spontaneous Symmetry Breaking and Gauge Particles". *International Journal of Modern Physics A* **24** (14): 2601–2627. arXiv:0907.3466. Bibcode:2009IJMPA..24.2601G. doi:10.1142/S0217751X09045431.

[22] Peskin, Michael; Schroeder, Daniel (1995). *An introduction to quantum field theory* (Reprint ed.). Westview Press. ISBN 978-0201503975.

[23] Kinoshita, Toichiro. "Quantum Electrodynamics has Zero Radius of Convergence Summarized from Toichiro Kinoshita". Retrieved 06-10-2010. Check date values in: |access-date= (help)

[24] Espriu and Tarrach. "Ambiguities in QED: Renormalons versus Triviality". arXiv:hep-ph/9604431.

14.8 Further reading

14.8.1 Books

- De Broglie, Louis (1925). *Recherches sur la theorie des quanta [Research on quantum theory]*. France: Wiley-Interscience.

- Feynman, Richard Phillips (1998). *Quantum Electrodynamics* (New ed.). Westview Press. ISBN 978-0-201-36075-2.

- Jauch, J.M.; Rohrlich, F. (1980). *The Theory of Photons and Electrons*. Springer-Verlag. ISBN 978-0-387-07295-1.

- Greiner, Walter; Bromley, D.A.; Müller, Berndt (2000). *Gauge Theory of Weak Interactions*. Springer. ISBN 978-3-540-67672-0.

- Kane, Gordon, L. (1993). *Modern Elementary Particle Physics*. Westview Press. ISBN 978-0-201-62460-1.

- Miller, Arthur I. (1995). *Early Quantum Electrodynamics: A Sourcebook*. Cambridge University Press. ISBN 978-0-521-56891-3.

- Milonni, Peter W., (1994) *The quantum vacuum - an introduction to quantum electrodynamics*. Academic Press. ISBN 0-12-498080-5

- Schweber, Silvan S. (1994). *QED and the Men Who Made It*. Princeton University Press. ISBN 978-0-691-03327-3.

- Schwinger, Julian (1958). *Selected Papers on Quantum Electrodynamics*. Dover Publications. ISBN 978-0-486-60444-2.

- Tannoudji-Cohen, Claude; Dupont-Roc, Jacques; Grynberg, Gilbert (1997). *Photons and Atoms: Introduction to Quantum Electrodynamics*. Wiley-Interscience. ISBN 978-0-471-18433-1.

14.8.2 Journals

- Dudley, J.M.; Kwan, A.M. (1996). "Richard Feynman's popular lectures on quantum electrodynamics: The 1979 Robb Lectures at Auckland University". *American Journal of Physics* **64** (6): 694–698. Bibcode:1996AmJPh..64..694D. doi:10.1119/1.18234.

14.9 External links

- Feynman's Nobel Prize lecture describing the evolution of QED and his role in it

- Feynman's New Zealand lectures on QED for non-physicists

- http://qed.wikina.org/ - Animations demonstrating QED

Chapter 15

Background independence

Background independence is a condition in theoretical physics, that requires the defining equations of a theory to be independent of the actual shape of the spacetime and the value of various fields within the spacetime. In particular this means that it must be possible to not refer to a specific coordinate system—the theory must be coordinate-free. In addition, the different spacetime configurations (or backgrounds) should be obtained as different solutions of the underlying equations.

15.1 What is background-independence?

Background-independence is a loosely defined property of a theory of physics. Roughly speaking, it limits the number of mathematical structures used to describe space and time that are put in place "by hand". Instead, these structures are the result of dynamical equations, such as Einstein field equations, so that one can determine from first principles what form they should take. Since the form of the metric determines the result of calculations, a theory with background independence is more predictive than a theory without it, since the theory requires fewer inputs to make its predictions. This is analogous to desiring fewer free parameters in a fundamental theory. So background-independence can be seen as extending the mathematical objects that should be predicted from theory to include not just the parameters, but also geometrical structures. Summarizing this, Rickles writes: "Background structures are contrasted with dynamical ones, and a background independent theory only possesses the latter type—obviously, background dependent theories are those possessing the former type in addition to the latter type."[1]

In general relativity, background-independence is identified with the property that the metric of space-time is the solution of a dynamical equation.[2] In classical mechanics, this is not the case, the metric is fixed by the physicist to match experimental observations. This is undesirable, since the form of the metric impacts the physical predictions, but is not itself predicted by the theory.

15.2 Manifest background-independence

Manifest background-independence is primarily an aesthetic rather than a physical requirement. It is analogous, and closely related, to requiring in differential geometry that equations be written in a form that is independent of the choice of charts and coordinate embeddings. If a background-independent formalism is present, it can lead to simpler and more elegant equations. However there is no physical content in requiring that a theory be **manifestly background-independent** – for example, the equations of general relativity can be rewritten in local coordinates without affecting the physical implications.

Although making a property manifest is only aesthetic, it is a useful tool for making sure the theory actually has that property. For example, if a theory is written in a manifestly Lorentz invariant way, one can check at every step to be sure that Lorentz invariance is preserved. Making a property manifest also makes it clear whether or not the theory actually has that property. The inability to make classical mechanics manifestly Lorentz invariant does not reflect a lack of imagination on the part of the theorist, but rather a physical feature of the theory. The same goes for making classical mechanics, or electromagnetism background independent.

15.3 Theories of quantum gravity

Because of the speculative nature of quantum gravity research, there is much debate as to the correct implementation of background-independence. Ultimately, the answer is to be decided by experiment, but until experiments can probe quantum gravity phenomena, physicists have to settle for debate. Below is a brief summary of the two largest quantum gravity approaches.

Physicists have studied models of 3D quantum gravity, which is a much simpler problem than 4D quantum gravity (this is because in 3D, quantum gravity has no local degrees of freedom). In these models, there are non-zero transition amplitudes between two different topologies,[3] or in other words, the topology changes. This and other similar results lead physicists to believe that any consistent quantum theory of gravity should include topology change as a dynamical process.

15.3.1 String theory

String theory is usually formulated with perturbation theory around a fixed background. While it is possible that the theory defined this way is background-invariant, if so it is not manifest. One attempt to formulate string theory in a manifestly background-independent fashion is string field theory, but little progress has been made in understanding it.

Another approach is the AdS/CFT duality, which is believed to provide a full, non-perturbative definition of string theory in spacetimes with anti-de Sitter asymptotics. If so, this could describe a kind of superselection sector of the putative full, background-independent theory. A full non-perturbative definition of the theory in arbitrary space-time backgrounds is still lacking.

Topology change is an established process in string theory.

15.3.2 Loop quantum gravity

A very different approach to quantum gravity called loop quantum gravity has been claimed to be background-independent, at least in the sense that geometric quantities, such as area, are predicted without reference to a background metric. However, one could say that the physics of loop quantum gravity is only background-independent in a weak sense. This is because it requires a fixed choice of topology for the space-time, which could be seen as a background structure.

15.4   See also

- General relativity
- String theory
- Causal dynamical triangulation
- Loop quantum gravity

- Quantum field theory
- Coordinate-free

15.5   References

[1]   D. Rickles, Who's Afraid of Background Independence?, p. 4

[2]   John Baez, Higher-Dimensional Algebra and Planck-Scale Physics

[3]   Hiroshi Ooguri, Partition Functions and Topology-Changing Amplitudes in the 3D Lattice Gravity of Ponzano and Regge

15.6   Further reading

- Rozali, M. (2009). "Comments on Background Independence and Gauge Redundancies". *Advanced Science Letters* **2** (2): 244. arXiv:0809.3962. doi:10.1166/asl.2009.1031.

- Smolin, L. (2005). "The case for background independence". arXiv:hep-th/0507235 [hep-th].

- Colosi, D.; et al. (2005). "Background independence in a nutshell". *Classical and Quantum Gravity* **22** (14): 2971–2989. arXiv:gr-qc/0408079. Bibcode:2005CQGra..22.2971C. doi:10.1088/0264-9381/22/14/008.

- Witten, E. (1993). "Quantum Background Independence in String Theory". arXiv:hep-th/9306122 [hep-th].

- Stachel, J. (1993). "The Meaning of General Covariance: The Hole Story". In J. Earman, A. Janis, G. Massey and N. Rescher. *Philosophical Problems of the Internal and External Worlds: Essays on the Philosophy of Adolf Grünbaum*. University of Pittsburgh Press. pp. 129–160. ISBN 0-8229-3738-7.

- Stachel, J. (1994). "Changes in the Concepts of Space and Time Brought About by Relativity". In C. C. Gould and R. S. Cohen. *Artifacts, Representations and Social Practice*. Kluwer Academic. pp. 141–162. ISBN 0-7923-2481-1.

- Zahar, E. (1989). *Einstein's Revolution: A Study in Heuristic*. Open Court Publishing Company. ISBN 0-8126-9066-4.

Chapter 16

Minkowski space

For spacetime graphics, see Minkowski diagram. For Minkowski space associated to a number field, see Minkowski space (number field). For geometry of the Minkowski plane, see Minkowski plane .

In mathematical physics, **Minkowski space** or **Minkowski spacetime** is a combination of Euclidean space and time into a four-dimensional manifold where the spacetime interval between any two events is independent of the inertial frame of reference in which they are recorded. Although initially developed by mathematician Hermann Minkowski for Maxwell's equations of electromagnetism, the mathematical structure of Minkowski spacetime was shown to be an immediate consequence of the postulates of special relativity.[1]

Minkowski space is closely associated with Einstein's theory of special relativity, and is the most common mathematical structure on which special relativity is formulated. While the individual components in Euclidean space and time will often differ due to length contraction and time dilation, in Minkowski spacetime, all frames of reference will agree on the total distance in spacetime between events.[nb 1] Because it treats time differently than the three spatial dimensions, Minkowski space differs from four-dimensional Euclidean space.[nb 2]

In Euclidean space, the isometry group (the maps preserving the regular inner product) is the Euclidean group. The analogous isometry group for Minkowski space, preserving intervals of spacetime equipped with the associated non-positive definite bilinear form (here called the **Minkowski inner product**,[nb 3]) is the Poincaré group. The Minkowski inner product is defined as to yield the spacetime interval between two events when given their coordinate difference vector as argument.

Hermann Minkowski (1864 – 1909) was a German mathematician. He found that the theory of special relativity, introduced by his former student Albert Einstein, could best be understood in a four-dimensional space, since known as the Minkowski spacetime.

16.1 History

16.1.1 Four-dimensional Euclidean space-time

See also: Four-dimensional space

In 1905, and later published in 1906, Henri Poincaré showed that by taking time to be an imaginary fourth spacetime coordinate ($\sqrt{-1}\, c\, t$), a Lorentz transformation can be regarded as a rotation of coordinates in a four-dimensional Euclidean space with three real coordinates representing space, and one imaginary coordinate, representing time, as the fourth dimension. Since the space is then a pseudo-Euclidean space, the rotation is a representation of a hyperbolic rotation, although Poincaré did not give this interpretation, his purpose being only to explain the Lorentz transformation in terms of the familiar Euclidean rotation.[2]

This idea was elaborated by Hermann Minkowski,[3] who used it to restate the Maxwell equations in four dimensions, showing directly their invariance under the Lorentz transformation. He further reformulated in four dimensions the then-recent theory of special relativity of Einstein. From this he concluded that time and space should be treated equally, and so arose his concept of events taking place in a unified four-dimensional spacetime continuum.

16.1.2   Minkowski space

In a further development,[4] he gave an alternative formulation of this idea that used a real time coordinate instead of an imaginary one, representing the four variables (x, y, z, t) of space and time in coordinate form in a four dimensional affine space. Points in this space correspond to events in spacetime. In this space, there is a defined light-cone associated with each point (see diagram above), and events not on the light-cone are classified by their relation to the apex as *spacelike* or *timelike*. It is principally this view of spacetime that is current nowadays, although the older view involving imaginary time has also influenced special relativity. Minkowski, aware of the fundamental restatement of the theory which he had made, said

> The views of space and time which I wish to lay before you have sprung from the soil of experimental physics, and therein lies their strength. They are radical. Henceforth space by itself, and time by itself, are doomed to fade away into mere shadows, and only a kind of union of the two will preserve an independent reality.
> — Hermann Minkowski, 1907[4]

For further historical information see references Galison (1979), Corry (1997) and Walter (1999).

16.2   Mathematical structure

For an overview, Minkowski space is a 4-dimensional real vector space equipped with a nondegenerate, symmetric bi-

linear form on the tangent space at each point in spacetime, here simply called the Minkowski inner product, with signature either $(-,+,+,+)$ or $(+,-,-,-)$. In practice, one need not be concerned with the tangent spaces. The vector space nature of Minkowski space allows for the canonical identification of vectors in tangent spaces at points (events) with vectors (points, events) in Minkowski space itself.[5] For some purposes it is desirable to identify tangent vectors at a point p with *displacement vectors* at p, which is, of course, admissible by essentially the same canonical identification.[6]

The signature refers to which sign the Minkowski inner product yields when given space and time basis vectors as arguments. In general, mathematicians and general relativists prefer the former while particle physicists tend to use the latter. Arguments for the former (pure space vectors yield positive "norm-squared") include "continuity" from the Euclidean case corresponding to the non-relativistic limit $c \to \infty$. Arguments for the latter (pure space vectors yield negative "norm-squared") include that otherwise ubiquitous minus signs in particle physics go away.

Mathematically associated to this bilinear form is a tensor of type $(0,2)$ at each point in spacetime, called the Minkowski metric. The Minkowski metric, the bilinear form, and the Minkowski inner product are actually all the very same object. In coordinates, this is the 4×4 matrix representing the bilinear form. Keeping this in mind may facilitate reading what follows.

For comparison, in general relativity, a Lorentzian manifold L is likewise equipped with a metric tensor g, which is a nondegenerate symmetric bilinear form on the tangent space T_pL at each point p of L. In coordinates, it may be represented by a 4×4 matrix *depending on spacetime position*. Minkowski space is thus a comparatively simple special case of a Lorentzian manifold. Its metric tensor, called the Minkowski metric, is in coordinates the same symmetric matrix at every point of M, and its arguments can, per above, be taken as vectors in spacetime itself.

Introducing more terminology (but not more structure), Minkowski space is thus a pseudo-Euclidean space with total dimension $n = 4$ and signature (3, 1) or (1, 3). Elements of Minkowski space are called events. Minkowski space is often denoted $\mathbf{R}^{3,1}$ or $\mathbf{R}^{1,3}$ to emphasize the chosen signature, or just M. It is perhaps the simplest example of a pseudo-Riemannian manifold.

16.2.1   Pseudo-Euclidean metric generalities

Main article: Pseudo-Euclidean space

The Minkowski metric[nb 4] η is the metric tensor of

Minkowski space. It is a Pseudo-Euclidean metric. As such it is a nondegenerate symmetric bilinear form, a type $(0,2)$ tensor. It accepts two arguments up, vp, vectors in $TpM, p \in M$, the tangent space at p in M. Due to the above-mentioned canonical identification of TpM with M itself, it accepts arguments u, v with both u and v in M.

As a notational convention, vectors v in M, called 4-vectors, are denoted in sans-serif italics, and not, as is common in the Eucliedean setting, with boldface **v**. The latter is generally reserved for the 3-vector part (to be introduced below) of a 4-vector.

The definition

$$u \cdot v = \eta(u, v)$$

yields an inner product-like structure on M, previously and also henceforth, called the Minkowski inner product, similar to the Euclidean inner product, but it describes a different geometry. It has the following properties.

- $\eta(au + v, w) = a\eta(u, w) + \eta(v, w), \quad \forall u, v \in M, \forall a \in \mathbb{R}$ slot) first in (linearity

- $\eta(u, v) = \eta(v, u) \quad$ (symmetry)

- $\eta(u, v) = 0 \quad \forall v \in M \Rightarrow u = 0 \quad$ (non-degeneracy)

The first two conditions imply bilinearity. The defining *difference* between a pseudo-inner product and an inner product proper is that the former is *not* required to be positive definite, that is, $\eta(u, u) < 0$ is allowed.

Two vectors v and w are said to be orthogonal if $\eta(v, w) = 0$.

A vector e is called a unit vector if $\eta(e, e) = \pm 1$. A basis for M consisting of mutually orthogonal unit vectors is called an orthonormal basis.

For a given inertial frame, an orthonormal basis in space, combined by the unit time vector, forms an orthonormal basis in Minkowski space. The number of positive and negative unit vectors in any such basis is a fixed pair of numbers, equal to the signature of the bilinear form associated with the inner product. This is Sylvester's law of inertia.

More terminology (but not more structure): The Minkowski metric is a pseudo-Riemannian metric, more specifically, a Lorentzian metric, even more specifically, *the* Lorentz metric, reserved for 4-dimensional flat spacetime with the remaining ambiguity only being the signature convention.

16.2.2 Minkowski metric

From the two postulates of special relativity follows that the spacetime interval between two events 1, 2,

$$\pm \left[c^2(t_1 - t_2)^2 - (x_1 - x_2)^2 - (y_1 - y_2)^2 - (z_1 - z_2)^2 \right],$$

is independent of the inertial frame chosen. The factor $\pm$ simply means that the choice of signature is left open. The numerical values of η, viewed as a matrix representing the Minkowski inner product, follow from the theory of bilinear forms.

Just as the signature of the metric is differently defined in the literature, this quantity is not consistently named. The interval (as defined here) is sometimes referred to as the interval squared.[7] Even the square root of the present interval occurs.[8] When signature and interval are fixed, ambiguity still remains as which coordinate is the time coordinate. It may be the fourth, or it may be the zeroth. This is not an exhaustive list of notational inconsistencies. It is a fact of life that one has to check out the definitions first thing when one consults the relativity literature.

The invariance of the interval under coordinate transformations between inertial frames follows from the invariance of

$$\pm \left[c^2 t^2 - x^2 - y^2 - z^2 \right]$$

(with either sign $\pm$ preserved), provided the transformations are linear. This quadratic form can be used to define a bilinear form

$$u \cdot v = \pm \left[c^2 t_1 t_2 - x_1 x_2 - y_1 y_2 - z_1 z_2 \right].$$

via the polarization identity. This bilinear form can in turn be written as

$$u \cdot v = u^{\mathrm{T}}[\eta]v,$$

where $[\eta]$ is a 4×4 matrix associated with η. Possibly confusingly, denote $[\eta]$ with just η as is common practice. The matrix is read off from the explicit bilinear form as

$$\eta = \pm \begin{pmatrix} -1 & 0 & 0 & 0 \\ 0 & 1 & 0 & 0 \\ 0 & 0 & 1 & 0 \\ 0 & 0 & 0 & 1 \end{pmatrix},$$

and the bilinear form

$$u \cdot v = \eta(u, v),$$

with which this section started by assuming its existence, is now identified.

For definiteness and shorter presentation, the signature $(-,+,+,+)$ is adopted below. The choice has no (known) physical implications. The symmetry group preserving the bilinear form with one choice of signature is isomorphic (under the map given here) with the symmetry group preserving the other choice of signature. This means that both choices are in accord with the two postulates of relativity.

16.2.3 Standard basis

A standard basis for Minkowski space is a set of four mutually orthogonal vectors $\{ e_0, e_1, e_2, e_3 \}$ such that

$$-\eta(e_0, e_0) = \eta(e_1, e_1) = \eta(e_2, e_2) = \eta(e_3, e_3) = 1.$$

These conditions can be written compactly in the form

$$\eta(e_\mu, e_\nu) = \eta_{\mu\nu}.$$

Relative to a standard basis, the components of a vector v are written (v^0, v^1, v^2, v^3) where the Einstein notation is used to write $v = v^\mu e\mu$. The component v^0 is called the **timelike component** of v while the other three components are called the **spatial components**. The spatial components of a 4-vector v may be identified with a 3-vector $\mathbf{v} = (v_1, v_2, v_3)$.

In terms of components, the Minkowski inner product between two vectors v and w is given by

$$\eta(v, w) = \eta_{\mu\nu}v^\mu w^\nu = v^0 w_0 + v^1 w_1 + v^2 w_2 + v^3 w_3 = v^\mu w_\mu$$

and

$$\eta(v, v) = \eta_{\mu\nu}v^\mu v^\nu = v^0 v_0 + v^1 v_1 + v^2 v_2 + v^3 v_3 = v^\mu v_\mu.$$

Here **lowering of an index** with the metric was used. Technically, a non-degenerate bilinear form provides a map between a vector space and its dual, in this context, the map is between the tangent spaces of M and the cotangent spaces of M. At a point in M, the tangent and cotangent spaces are dual. Just as an authentic inner product on a vector space with one argument fixed, by Riesz representation theorem, may be expressed as the action of a linear functional on the

vector space, the same holds for the Minkowski inner product of Minkowski space.

Thus if v^μ are the components of a vector in a tangent space, then $\eta_{\mu\nu}v^\mu = v\nu$ are the components of a vector in the cotangent space (a linear functional). Due to the identification of vectors in tangent spaces with vectors in M itself, this is mostly ignored, and vectors with lower indices are referred to as **covariant vectors**. In this latter interpretation, the covariant vectors are (almost always implicitly) identified with vectors (linear functionals) in the dual of Minkowski space. The ones with upper indices are **contravariant vectors**. In the same fashion, the inverse of the map from tangent to cotangent spaces, explicitly given by the inverse of η in matrix representation, can be used to define **raising of an index**. The components of this inverse are denoted $\eta^{\mu\nu}$. It happens that $\eta^{\mu\nu} = \eta_{\mu\nu}$. These maps between a vector space and its dual can be denoted $\eta^\flat$ (eta-flat) and $\eta^\sharp$ (eta-sharp) by the musical analogy.[9]

The time-proven robustness of the formalism itself, sometimes referred to as index gymnastics, ensures that moving vectors around and changing from contravariant to covariant vectors and vice versa is mathematically sound. Incorrect expressions tend to reveal themselves quickly.

16.2.4 Geometry

16.3 Lorentz transformations and symmetry

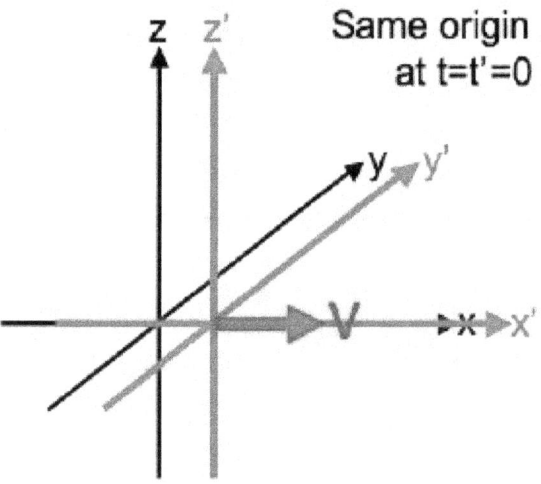

Standard configuration of coordinate systems for Lorentz transformations.

The Poincaré group is the group of all transformations preserving the interval. The interval is quite easily seen to be preserved by the translation group in 4 dimensions. The other transformations are those that preserve the interval and leave the origin fixed. Given the bilinear form associated with the Minkowski metric, the appropriate group follows directly from the theory (in particular the definition) of classical groups. In the linked article, one should identify η (in its a matrix representation) with the matrix Φ.

The appropriate group is O(3,1), in this context called the Lorentz group. Its elements are called (homogeneous) Lorentz transformations. For other methods of derivation, with a more physical twist, see derivations of the Lorentz transformations.

Among the simplest Lorentz transformations is a Lorentz boost. For reference, a boost in the x-direction is given by

$$\begin{bmatrix} U_0' \\ U_1' \\ U_2' \\ U_3' \end{bmatrix} = \begin{bmatrix} \gamma & -\beta\gamma & 0 & 0 \\ -\beta\gamma & \gamma & 0 & 0 \\ 0 & 0 & 1 & 0 \\ 0 & 0 & 0 & 1 \end{bmatrix} \begin{bmatrix} U_0 \\ U_1 \\ U_2 \\ U_3 \end{bmatrix},$$

where

$$\gamma = \frac{1}{\sqrt{1 - \frac{v^2}{c^2}}}$$

is the Lorentz factor, and

$$\beta = \frac{v}{c}.$$

Other Lorentz transformations are pure rotations, and hence elements of the SO(3) subgroup of O(3,1). A general homogeneous Lorentz transformation is a product of a pure boost and a pure rotation. An *inhomogeneous* Lorentz transformation is a homogeneous transformation followed by a translation in space and time. Special transformations are those that invert the space coordinates (P) and time coordinate (T) respectively, or both (PT).

All four-vectors in Minkowski space transform, by definition, according to the same formula under Lorentz transformations. Minkowski diagrams illustrate Lorentz transformations.

16.4 Causal structure

Main article: Causal structure

Vectors $v = (ct, x, y, z) = (ct, \mathbf{r})$ are classified according to the sign of $c^2 t^2 - r^2$. A vector is **timelike** if $c^2 t^2 > r^2$, **spacelike**

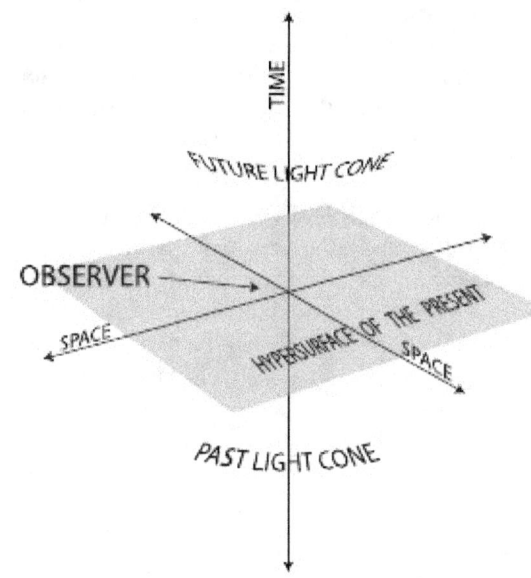

*Subdivision of Minkowski spacetime with respect to an event in four disjoint sets. The light cone, the **absolute future**, the **absolute past**, and **elsewhere**. The terminology is from Sard (1970).*

if $c^2 t^2 < r^2$, and **null** or **lightlike** if $c^2 t^2 = r^2$. This can be expressed in terms of the sign of $\eta(v,v)$ as well, but depends on the signature. The classification of any vector will be the same in all frames of reference, because of the invariance of the interval.

The set of all null vectors at an event[nb 5] of Minkowski space constitutes the light cone of that event. Given a timelike vector v, there is a worldline of constant velocity associated with it, represented by a straight line in a Minkowski diagram.

Once a direction of time is chosen,[nb 6] timelike and null vectors can be further decomposed into various classes. For timelike vectors one has

1. future-directed timelike vectors whose first component is positive, (tip of vector located in absolute future in figure) and

2. past-directed timelike vectors whose first component is negative (absolute past).

Null vectors fall into three classes:

1. the zero vector, whose components in any basis are (0,0,0,0) (origin),

2. future-directed null vectors whose first component is positive (upper light cone), and

3. past-directed null vectors whose first component is negative (lower light cone).

Spacelike vectors are in elsewhere. The terminology stems from the fact that spacelike separated events are connected by vectors requiring faster-than-light travel, and so cannot possibly influence each other. Together with spacelike and lightlike vectors there are 7 classes in all.

An orthonormal basis for Minkowski space necessarily consists of one timelike and three spacelike unit vectors. If one wishes to work with non-orthonormal bases it is possible to have other combinations of vectors. For example, one can easily construct a (non-orthonormal) basis consisting entirely of null vectors, called a **null basis**. Over the reals, if two null vectors are orthogonal (zero Minkowski tensor value), then they must be proportional. However, allowing complex numbers, one can obtain a null tetrad, which is a basis consisting of null vectors, some of which are orthogonal to each other.

Vector fields are called timelike, spacelike or null if the associated vectors are timelike, spacelike or null at each point where the field is defined.

16.4.1 Chronological and causality relations

Let $x, y \in M$. We say that

1. x *chronologically precedes* y if $y - x$ is future-directed timelike. This relation has the transitive property and so can be written x < y.

2. x *causally precedes* y if $y - x$ is future-directed null or future-directed timelike. It gives a partial ordering of space-time and so can be written x ≤ y.

16.4.2 Reversed triangle inequality

If v and w are both future-directed timelike four-vectors, then in the (+ - - -) sign convention for norm,

$$\|v + w\| \geq \|v\| + \|w\| \,.$$

16.5 Relationships to other formulations

16.5.1 Different number of dimensions

Strictly speaking, Minkowski space refers to a mathematical formulation in four dimensions. However, the mathematics can easily be extended or simplified to create an analogous "Minkowski space" in any number of dimensions. If $n \geq 2$, n-dimensional Minkowski space is a vector space of real

dimension n on which there is a constant Lorentz metric of signature $(n - 1, 1)$ or $(1, n - 1)$. These generalizations are used in theories where spacetime is assumed to have more or less than 4 dimensions. String theory and M-theory are two examples where $n > 4$. In string theory, there appears conformal field theories with $1 + 1$ spacetime dimensions.

16.5.2 Flat versus curved space

As a *flat spacetime*, the three spatial components of Minkowski spacetime always obey the Pythagorean Theorem. Minkowski space is a suitable basis for special relativity, a good description of physical systems over finite distances in systems without significant gravitation. However, in order to take gravity into account, physicists use the theory of general relativity, which is formulated in the mathematics of a non-Euclidean geometry. When this geometry is used as a model of physical space, it is known as curved space.

Even in curved space, Minkowski space is still a good description in an infinitesimal region surrounding any point (barring gravitational singularities).[nb 7] More abstractly, we say that in the presence of gravity spacetime is described by a curved 4-dimensional manifold for which the tangent space to any point is a 4-dimensional Minkowski space. Thus, the structure of Minkowski space is still essential in the description of general relativity.

16.6 See also

- Causal structure
- Euclidean space
- Four vector
- Hyperboloid model
- Introduction to mathematics of general relativity
- Lorentzian manifold
- Metric tensor
- Minkowski diagram
- Minkowski plane
- Speed of light
- Super Minkowski space
- World line

16.7 Remarks

[1] This makes spacetime distance an invariant.

[2] Minkowski space can be formulated as an equivalent 4-D Euclidean space if you assume time is always an imaginary number. This is how the spacetime was first formulated, but since Minkowski reworked the structure, time is almost always required to be a real number.

[3] Consistent use of the term "Minkowski inner product" is intended for the bilinear form here, since it is in widespread use. It is by no means "standard" in the literature, but no such standard seems to exist.

[4] The Minkowski inner product is not an inner product, since it is not positive-definite, i.e. the quadratic form $\eta(v, v)$ need not be positive for nonzero v. The positive-definite condition has been replaced by the weaker condition of non-degeneracy. The bilinear form is said to be *indefinite*.

[5] Translate the coordinate system so that the event is the new origin.

[6] This corresponds to the time coordinate either increasing or decreasing when proper time for any particle increases. An application of T flips this direction.

[7] This similarity between flat and curved space at infinitesimally small distance scales is foundational to the definition of a manifold in general.

16.8 Notes

[1] Landau & Lifshitz 2002, p. 5

[2] Poincaré 1905–1906, pp. 129–176 Wikisource translation: On the Dynamics of the Electron

[3] Minkowski 1907–1908, pp. 53–111 *Wikisource translation: The Fundamental Equations for Electromagnetic Processes in Moving Bodies.

[4] Minkowski 1907–1909, pp. 75–88 Various English translations on Wikisource: Space and Time.

[5] Lee 2003, Proposition 3.8. The identification is routinely done in mathematics.

[6] Lee 2003, See Lee's discussion on geometric tangent vectors early in chapter 3.

[7] Sard 1970, p. 71

[8] Landau & Lifshitz 2002, p. 4

[9] Lee 2003. The tangent-cotangent isomorphism p. 282.

16.9 References

- Corry, L. (1997). "Hermann Minkowski and the postulate of relativity". *Arch. Hist. Exact Sci.* (Springer-Verlag) **51** (4): 273–314. doi:10.1007/BF00518231. ISSN 0003-9519. (subscription required (help)).

- Catoni, F.; et al. (2008). *Mathematics of Minkowski Space.* Frontiers in Mathematics. Basel: Birkhäuser Verlag. doi:10.1007/978-3-7643-8614-6. ISBN 978-3-7643-8613-9. ISSN 1660-8046.

- Galison, P. L. (1979). R McCormach; et al., eds. *Minkowski's Space-Time: from visual thinking to the absolute world.* Historical Studies in the Physical Sciences **10**. Johns Hopkins University Press. pp. 85–121. doi:10.2307/27757388. (subscription required (help)).

- Landau, L.D.; Lifshitz, E.M. (2002) [1939]. *The Classical Theory of Fields.* Course of Theoretical Physics **2** (4th ed.). Butterworth–Heinemann. ISBN 0 7506 2768 9.

- Lee, J. M. (2003). *Introduction to Smooth manifolds.* Springer Graduate Texts in Mathematics **218**. ISBN 0-387-95448-1.

- Minkowski, Hermann (1907–1908), "Die Grundgleichungen für die elektromagnetischen Vorgänge in bewegten Körpern" [The Fundamental Equations for Electromagnetic Processes in Moving Bodies], *Nachrichten von der Gesellschaft der Wissenschaften zu Göttingen, Mathematisch-Physikalische Klasse*: 53–111 *Wikisource translation: The Fundamental Equations for Electromagnetic Processes in Moving Bodies

- Minkowski, Hermann (1907–1909), "Raum und Zeit" [Space and Time], *Physikalische Zeitschrift* **10**: 75–88 Various English translations on Wikisource: Space and Time

- Naber, G. L. (1992). *The Geometry of Minkowski Spacetime.* New York: Springer-Verlag. ISBN 0-387-97848-8.

- Penrose, Roger (2005). "18 Minkowskian geometry". *Road to Reality : A Complete Guide to the Laws of the Universe.* Alfred A. Knopf. ISBN 9780679454434.

- Poincaré, Henri (1905–1906), "Sur la dynamique de l'électron" [On the Dynamics of the Electron], *Rendiconti del Circolo matematico di Palermo* **21**: 129–176, doi:10.1007/BF03013466 Wikisource translation: On the Dynamics of the Electron

- Sard, R. D. (1970). *Relativistic Mechanics - Special Relativity and Classical Particle Dynamics.* New York: W. A. Benjamin. ISBN 978-0805384918.

- Shaw, R. (1982). "§ 6.6 Minkowski space, § 6.7,8 Canonical forms pp 221–242". *Linear Algebra and Group Representations.* Academic Press. ISBN 0-12-639201-3.

- Walter, Scott (1999). "Minkowski, Mathematicians, and the Mathematical Theory of Relativity". In Goenner, Hubert *et al.* (ed.). *The Expanding Worlds of General Relativity.* Boston: Birkhäuser. pp. 45–86. ISBN 0-8176-4060-6.

16.10   External links

Media related to Minkowski diagrams at Wikimedia Commons

- Animation clip on YouTube visualizing Minkowski space in the context of special relativity.

- The Geometry of Special Relativity: The Minkowski Space - Time Light Cone

Chapter 17

Color confinement

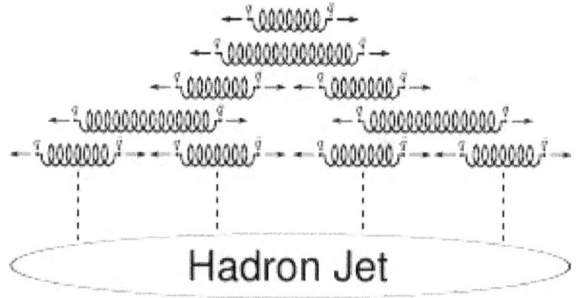

Hadron Jet

The color force favors confinement because at a certain range it is more energetically favorable to create a quark-antiquark pair than to continue to elongate the color flux tube. This is analoguous to the behavior of an elongated rubber-band.

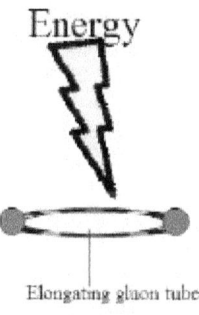

Energy

Elongating gluon tube

An animation of color confinement. Energy is supplied to the quarks, and the gluon tube elongates until it reaches a point where it "snaps" and forms a quark-antiquark pair.

Color confinement, often simply called **confinement**, is the phenomenon that color charged particles (such as quarks) cannot be isolated singularly, and therefore cannot be directly observed.[1] Quarks, by default, clump together to form groups, or hadrons. The two types of hadrons are the mesons (one quark, one antiquark) and the baryons (three quarks).

The constituent quarks in a group cannot be separated from their parent hadron, and this is why quarks currently cannot be studied or observed in any more direct way than at a

hadron level.[2]

17.1 Origin

The reasons for quark confinement are somewhat complicated; no analytic proof exists that quantum chromodynamics should be confining. The current theory is that confinement is due to the force-carrying gluons having color charge. As any two electrically charged particles separate, the electric fields between them diminish quickly, allowing (for example) electrons to become unbound from atomic nuclei. However, as a quark-antiquark pair separates, the gluon field forms a narrow tube (or string) of color field between them. This is quite different from the behavior of the electric field of a pair of positive and negative electric charges, which extends into the whole surrounding space and diminishes at large distances. Because of this behavior of the gluon field, a strong force between the quark pair acts constantly—regardless of their distance[3][4]—with a force of around 10,000 newtons. [5]

When two quarks become separated, as happens in particle accelerator collisions, at some point it is more energetically favorable for a new quark–antiquark pair to spontaneously appear, than to allow the tube to extend further. As a result of this, when quarks are produced in particle accelerators, instead of seeing the individual quarks in detectors, scientists see "jets" of many color-neutral particles (mesons and baryons), clustered together. This process is called *hadronization*, *fragmentation*, or *string breaking*, and is one of the least understood processes in particle physics.

The confining phase is usually defined by the behavior of the action of the Wilson loop, which is simply the path in spacetime traced out by a quark–antiquark pair created at one point and annihilated at another point. In a non-confining theory, the action of such a loop is proportional to its perimeter. However, in a confining theory, the action of the loop is instead proportional to its area. Since the area will be proportional to the separation of the quark–antiquark pair, free quarks are suppressed. Mesons are allowed in such a picture, since a loop containing another loop in the opposite direction will have only a small area between the two loops.

17.2 Models exhibiting confinement

Besides QCD in four spacetime dimensions, another model which exhibits confinement is the Schwinger model.[6] Compact Abelian gauge theories also exhibit confinement in 2 and 3 spacetime dimensions.[7] Confinement has recently been found in elementary excitations of magnetic systems called spinons.[8]

17.3 Models of fully screened quarks

Besides the quark confinement idea, there is a potential possibility, that color charge of quarks gets fully screened by the gluonic color, surrounding the quark. Exact solutions of SU(3) classical Yang–Mills theory, which provide full screening (by gluon fields) of the color charge of a quark have been found.[9] However, such classical solutions do not take into account non-trivial properties of QCD vacuum. Therefore, a significance of such full gluonic screening solutions for a separated quark is not clear.

17.4 See also

- Gluon field strength tensor
- Asymptotic freedom
- Center vortices
- Deconfining phase
- Quantum mechanics
- Particle physics
- Fundamental force
- Dual superconducting model
- Beta-function
- Infrared safety

17.5 References

[1] V. Barger, R. Phillips (1997). *Collider Physics*. Addison–Wesley. ISBN 0-201-14945-1.

[2] T.-Y. Wu, W.-Y. Pauchy Hwang (1991). *Relativistic quantum mechanics and quantum fields*. World Scientific. p. 321. ISBN 981-02-0608-9.

[3] T. Muta (2009). *Foundations of quantum chromodynamics: an introduction to perturbative methods in gauge theories* (3rd ed.). World Scientific. ISBN 978-981-279-353-9.

[4] A. Smilga (2001). *Lectures on quantum chromodynamics*. World Scientific. ISBN 978-981-02-4331-9.

[5] Fritzsch, op. cite. p. 164. The author states that the force between differently coloured quarks remains constant at any distance after they travel only a tiny distance from each other, and is equal to that need to raise one ton, which is 1000 kg x 9.8 m/s^2 = ~10,000 N.

[6] Wilson, Kenneth G. (1974-10-15). "Confinement of Quarks". *Physical Review D* (College Park, MD, USA: American Physical Society) **10**: 2445–2459. Bibcode:1974PhRvD..10.2445W. doi:10.1103/PhysRevD.10.2445. ISSN 1550-2368. OCLC 55589778. Retrieved 2014-04-12.

[7] Schön, Verena; Michael, Thies (2000-08-22). "2d Model Field Theories at Finite Temperature and Density (Section 2.5)". arXiv:hep-th/0008175v1 [hep-th].

[8] Lake, Bella; Tsvelik, Alexei M.; Notbohm, Susanne; Tennant, D. Alan; Perring, Toby G.; Reehuis, Manfred; Sekar, Chinnathambi; Krabbes, Gernot; Büchner, Bernd (2009-11-29). "Confinement of fractional quantum number particles in a condensed-matter system". *Nature Physics* (London, UK: Nature Publishing Group) **6** (1): 50–55. arXiv:0908.1038. Bibcode:2010NatPh...6...50L. doi:10.1038/nphys1462. ISSN 1745-2481. OCLC 150143123. Retrieved 2014-04-12. (subscription required (help)).

[9] Cahill, Kevin (1978-08-28). "Example of Color Screening". *Physical Review Letters* (American Physical Society) **41** (9): 599–601. Bibcode:1978PhRvL..41..599C. doi:10.1103/PhysRevLett.41.599. ISSN 1079-7114. OCLC 31492939. Retrieved 2014-04-12. (subscription required (help)).

17.6 External links

- Quarks

Chapter 18

Perturbation theory

This article is about perturbation theory as a general mathematical method. For perturbation theory as applied to quantum mechanics, see Perturbation theory (quantum mechanics).

Perturbation theory comprises mathematical methods for finding an approximate solution to a problem, by starting from the exact solution of a related, simpler problem. A critical feature of the technique is a middle step that breaks the problem into "solvable" and "perturbation" parts.[1] Perturbation theory is applicable if the problem at hand cannot be solved exactly, but can be formulated by adding a "small" term to the mathematical description of the exactly solvable problem.

Perturbation theory leads to an expression for the desired solution in terms of a formal power series in some "small" parameter – known as a **perturbation series** – that quantifies the deviation from the exactly solvable problem. The leading term in this power series is the solution of the exactly solvable problem, while further terms describe the deviation in the solution, due to the deviation from the initial problem. Formally, we have for the approximation to the full solution A, a series in the small parameter (here called ε), like the following:

$$A = A_0 + \varepsilon^1 A_1 + \varepsilon^2 A_2 + \cdots$$

In this example, A_0 would be the known solution to the exactly solvable initial problem and A_1, A_2, ... represent the **higher-order terms** which may be found iteratively by some systematic procedure. For small ε these higher-order terms in the series become successively smaller.

An approximate "perturbation solution" is obtained by truncating the series, usually by keeping only the first two terms, the initial solution and the "first-order" perturbation correction

$$A \approx A_0 + \varepsilon A_1 .$$

18.1 General description

Perturbation theory is closely related to methods used in numerical analysis. The earliest use of what would now be called *perturbation theory* was to deal with the otherwise unsolvable mathematical problems of celestial mechanics: for example the orbit of the Moon, which moves noticeably differently from a simple Keplerian ellipse because of the competing gravitation of the Earth and the Sun.

Perturbation methods start with a simplified form of the original problem, which is *simple enough* to be solved exactly. In celestial mechanics, this is usually a Keplerian ellipse. Under non-relativistic gravity, an ellipse is exactly correct when there are only two gravitating bodies (say, the Earth and the Moon) but not quite correct when there are three or more objects (say, the Earth, Moon, Sun, and the rest of the solar system) and not quite correct when the gravitational interaction is stated using formulas from General relativity.

The solved, but simplified problem is then *"perturbed"* to make the conditions that the perturbed solution actually satisfies closer to the real problem, such as including the gravitational attraction of a third body (the Sun). The "conditions" are a formula (or several) that represent reality, often something arising from a physical law like Newton's second law, the force-acceleration equation,

$$\mathbf{F} = m\mathbf{a} .$$

In the case of the example, the force **F** is calculated based on the number of gravitationally relevant bodies; the acceleration **a** is obtained, using calculus, from the path of the Moon in its orbit. Both of these come in two forms: approximate values for force and acceleration, which result from simplifications, and hypothetical exact values for force and acceleration, which would require the complete answer to calculate.

The slight changes that result from accommodating the perturbation, which themselves may have been simplified yet again, are used as corrections to the approximate solution. Because of simplifications introduced along every step of the way, the corrections are never perfect, and the conditions met by the corrected solution do not perfectly match the equation demanded by reality. However, even only one cycle of corrections often provides an excellent approximate answer to what the real solution should be.

There is no requirement to stop at only one cycle of corrections. A partially corrected solution can be re-used as the new starting point for yet another cycle of perturbations and corrections. In principle, cycles of finding increasingly better corrections could go on indefinitely. In practice, one typically stops at one or two cycles of corrections. The usual difficulty with the method is that the corrections progressively make the new solutions very much more complicated, so each cycle is much more difficult to manage than the previous cycle of corrections. Isaac Newton is reported to have said, regarding the problem of the Moon's orbit, that *"It causeth my head to ache."*[2]

This general procedure is a widely used mathematical tool in advanced sciences and engineering: start with a simplified problem and gradually add corrections that make the formula that the corrected problem matches closer and closer to the formula that represents reality. It is the natural extension to mathematical functions of the "guess, check, and fix" method used by older civilisations to compute certain numbers, such as square roots.

18.2   Examples

Examples for the "mathematical description" are: an algebraic equation, a differential equation (e.g., the equations of motion in celestial mechanics or a wave equation), a free energy (in statistical mechanics), a Hamiltonian operator (in quantum mechanics).

Examples for the kind of solution to be found perturbatively: the solution of the equation (e.g., the trajectory of a particle), the statistical average of some physical quantity (e.g., average magnetization), the ground state energy of a quantum mechanical problem.

Examples for the exactly solvable problems to start with: linear equations, including linear equations of motion (harmonic oscillator, linear wave equation), statistical or quantum-mechanical systems of non-interacting particles (or in general, Hamiltonians or free energies containing only terms quadratic in all degrees of freedom).

Examples of "perturbations" to deal with: Nonlinear contributions to the equations of motion, interactions between particles, terms of higher powers in the Hamiltonian/Free Energy.

For physical problems involving interactions between particles, the terms of the perturbation series may be displayed (and manipulated) using Feynman diagrams.

18.3   History

Perturbation theory was first devised to solve otherwise intractable problems in the calculation of the motions of planets in the solar system. The gradually increasing accuracy of astronomical observations led to incremental demands in the accuracy of solutions to Newton's gravitational equations, which led several notable 18th and 19th century mathematicians to extend and generalize the methods of perturbation theory. These well-developed perturbation methods were adopted and adapted to solve new problems arising during the development of Quantum Mechanics in 20th century atomic and subatomic physics.

18.3.1   Beginnings in the study of planetary motion

Since the planets are very remote from each other, and since their mass is small as compared to the mass of the Sun, the gravitational forces between the planets can be neglected, and the planetary motion is considered, to a first approximation, as taking place along Kepler's orbits, which are defined by the equations of the two-body problem, the two bodies being the planet and the Sun.[3]

Since astronomic data came to be known with much greater accuracy, it became necessary to consider how the motion of a planet around the Sun is affected by other planets. This was the origin of the three-body problem; thus, in studying the system Moon–Earth–Sun the mass ratio between the Moon and the Earth was chosen as the small parameter. Lagrange and Laplace were the first to advance the view that the constants which describe the motion of a planet around the Sun are "perturbed", as it were, by the motion of other planets and vary as a function of time; hence the name "perturbation theory".[3]

Perturbation theory was investigated by the classical scholars — Laplace, Poisson, Gauss — as a result of which the computations could be performed with a very high accuracy. The discovery of the planet Neptune in 1848 by Urbain Le Verrier, based on the deviations in motion of the planet Uranus (he sent the coordinates to Johann Gottfried Galle who successfully observed Uranus through his telescope), represented a triumph of perturbation theory.[3]

18.3.2 Rise of understanding of chaotic systems

The development of basic perturbation theory for differential equations was fairly complete by the middle of the 19th century. It was at that time that Charles-Eugène Delaunay was studying the perturbative expansion for the Earth-Moon-Sun system, and discovered the so-called "problem of small denominators". Here, the denominator appearing in the n-th term of the perturbative expansion could become arbitrarily small, causing the n-th correction to be as large or larger than the first-order correction.

At the turn of the 20th century, this problem led Henri Poincaré to make one of the first deductions of the existence of chaos, and what is poetically called the "butterfly effect": that even a very small perturbation can ultimately have a very large effect on non-dissipative or "friction-free" dynamic systems.

A partial resolution of the small-divisor problem was given by the statement of the KAM theorem in 1954. Developed by Andrey Kolmogorov, Vladimir Arnold and Jürgen Moser, this theorem stated the conditions under which a system of partial differential equations will have only mildly chaotic behaviour under small perturbations.

18.3.3 Application to new problems in 20th century physics

Perturbation theory saw a particularly dramatic expansion and evolution with the arrival of quantum mechanics. Although perturbation theory was used in the semi-classical theory of the Bohr atom, the calculations were monstrously complicated, and subject to somewhat ambiguous interpretation. The discovery of Heisenberg's matrix mechanics allowed a vast simplification of the application of perturbation theory. Notable examples are the Stark effect and the Zeeman effect, which have a simple enough theory to be included in standard undergraduate textbooks in quantum mechanics. Other early applications include the fine structure and the hyperfine structure in the hydrogen atom.

In modern times, perturbation theory underlies much of quantum chemistry and quantum field theory. In chemistry, perturbation theory was used to obtain the first solutions for the helium atom.

In the middle of the 20th century, Richard Feynman realized that the perturbative expansion could be given a dramatic and beautiful graphical representation in terms of what are now called Feynman diagrams. Although originally applied only in quantum field theory, such diagrams now find increasing use in any area where perturbative expansions are studied.

18.3.4 Search for better methods for quantum mechanics

In the late 20th century, broad dissatisfaction with perturbation theory in the quantum physics community, including not only the difficulty of going beyond second order in the expansion, but also questions about whether the perturbative expansion is even convergent, has led to a strong interest in the area of non-perturbative analysis, that is, the study of exactly solvable models.

Much of the theoretical work in non-perturbative analysis goes under the name of quantum groups and non-commutative geometry. The prototypical model is the Korteweg–de Vries equation, a highly non-linear equation for which the interesting solutions, the solitons, cannot be reached by perturbation theory, even if the perturbations were carried out to infinite order.

18.4 Perturbation orders

The standard exposition of perturbation theory is given in terms of the *order* to which the perturbation is carried out: **first-order perturbation theory** or **second-order perturbation theory**, and whether the perturbed states are degenerate, which requires **singular perturbation**. In the singular case extra care must be taken, and the theory is slightly more elaborate.

18.4.1 First-order, non-singular perturbation theory

This section develops, in simple terms,[4] the general theory for the perturbative solution to a differential equation to the first order. To keep the exposition simple, a crucial assumption is made: that the solutions to the unperturbed system are not *degenerate*, so that the perturbation series can be inverted. There are ways of dealing with the degenerate (or *singular*) case; these require extra care.

Suppose one wants to solve a differential equation of the form

$$Dg(x) = \lambda g(x) \,.$$

where D is some specific differential operator, and λ is an eigenvalue. Many problems involving ordinary or partial differential equations can be cast in this form.

It is presumed that the differential operator can be written in the form

$$D = D^{(0)} + \varepsilon D^{(1)}$$

where ε is presumed to be small, and that, furthermore, the complete set of solutions for $D^{(0)}$ are known.

That is, one has a set of solutions $f_n^{(0)}(x)$, labelled by some arbitrary index n, such that

$$D^{(0)} f_n^{(0)}(x) = \lambda_n^{(0)} f_n^{(0)}(x).$$

Furthermore, one assumes that the set of solutions $\{f_n^{(0)}(x)\}$ form an orthonormal set,

$$\int f_m^{(0)}(x) f_n^{(0)}(x)\, dx = \delta_{mn}$$

with δmn the Kronecker delta function.

To zeroth order, one expects that the solutions $g(x)$ are then somehow "close" to one of the unperturbed solutions $f_n^{(0)}(x)$. That is,

$$g(x) = f_n^{(0)}(x) + \mathcal{O}(\varepsilon)$$

and

$$\lambda = \lambda_n^{(0)} + \mathcal{O}(\varepsilon).$$

where $\mathcal{O}$ denotes the relative size, in big-O notation, of the perturbation.

To solve this problem, one assumes that the solution $g(x)$ can be written as a linear combination of the $f_n^{(0)}(x)$,

$$g(x) = \sum_m c_m f_m^{(0)}(x)$$

with all of the constants $c_m = \mathcal{O}(\varepsilon)$ except for n, where $c_n = \mathcal{O}(1)$.

Substituting this last expansion into the differential equation, taking the inner product of the result with $f_n^{(0)}(x)$, and making use of orthogonality, one obtains

$$c_n \lambda_n^{(0)} + \varepsilon \sum_m c_m \int f_n^{(0)}(x) D^{(1)} f_m^{(0)}(x)\, dx = \lambda c_n.$$

This can be trivially rewritten as a simple linear algebra problem of finding the eigenvalue of a matrix, where

$$\sum_m A_{nm} c_m = \lambda c_n$$

where the matrix elements Anm are given by

$$A_{nm} = \delta_{nm} \lambda_n^{(0)} + \varepsilon \int f_n^{(0)}(x) D^{(1)} f_m^{(0)}(x)\, dx \, .$$

Rather than solving this full matrix equation, one notes that, of all the cm in the linear equation, only one, namely cn, is not small. Thus, to the *first order* in ε, the linear equation may be solved trivially as

$$\lambda = \lambda_n^{(0)} + \varepsilon \int f_n^{(0)}(x) D^{(1)} f_n^{(0)}(x)\, dx$$

since all of the other terms in the linear equation are of order $\mathcal{O}(\varepsilon^2)$. The above gives the solution of the eigenvalue to first order in perturbation theory.

The function $g(x)$ to first order is obtained through similar reasoning. Substituting

$$g(x) = f_n^{(0)}(x) + \varepsilon f_n^{(1)}(x)$$

so that

$$\left(D^{(0)} + \varepsilon D^{(1)}\right) \left(f_n^{(0)}(x) + \varepsilon f_n^{(1)}(x)\right) = \left(\lambda_n^{(0)} + \varepsilon \lambda_n^{(1)}\right)$$
$$\times \left(f_n^{(0)}(x) + \varepsilon f_n^{(1)}(x)\right)$$

gives an equation for $f_n^{(1)}(x)$.

It may be solved integrating with the partition of unity

$$\delta(x - y) = \sum_n f_n^{(0)}(x) f_n^{(0)}(y)$$

to give

$$f_n^{(1)}(x) = \sum_{m\,(\neq n)} \frac{f_m^{(0)}(x)}{\lambda_n^{(0)} - \lambda_m^{(0)}} \int f_m^{(0)}(y) D^{(1)} f_n^{(0)}(y)\, dy$$

which finally gives the exact solution to the perturbed differential equation to first order in the perturbation ε.

Several observations may be made about the form of this solution. First, the sum over functions with differences of eigenvalues in the denominator evokes the resolvent in Fredholm theory. This is no accident; the resolvent acts essentially as a kind of Green's function or propagator, passing the perturbation along. Higher-order perturbations resemble this form, with an additional sum over a resolvent appearing at each order.

The form of this solution also illustrates the idea behind the small-divisor problem. If, for whatever reason, two eigenvalues are close, so that the difference $\lambda_n^{(0)} - \lambda_m^{(0)}$ becomes small, the corresponding term in the above sum will become disproportionately large. In particular, if this happens in higher-order terms, the higher-order perturbation may become as large or larger in magnitude than the first-order perturbation. Such a situation calls into question the validity of utilizing a perturbative analysis to begin with, which can be understood to be a fairly catastrophic situation; it is frequently encountered in chaotic dynamical systems, and requires the development of techniques other than perturbation theory to solve the problem.

Curiously, the situation is not at all bad if two or more eigenvalues are *exactly equal*. This case is referred to as singular or degenerate perturbation theory, addressed below. The degeneracy of eigenvalues indicates that the unperturbed system has some sort of symmetry, and that the generators of that symmetry commute with the unperturbed differential operator. Typically, the perturbing term does not possess the symmetry, and so the full solutions do not, either; one says that the perturbation *lifts* or *breaks* the degeneracy. In this case, the perturbation can still be performed, as in following sections; however, care must be taken to work in a basis for the unperturbed states, so that these map one-to-one to the perturbed states, rather than being a mixture.

18.4.2 Perturbation theory of degenerate states

One may note that a problem occurs in the above first order perturbation theory when two or more eigenfunctions of the unperturbed system correspond to the same eigenvalue, i.e. when the eigenvalue equation becomes

$$D^{(0)} f_{n,i}^{(0)}(x) = \lambda_n^{(0)} f_{n,i}^{(0)}(x) \,.$$

and the index i labels *several states with the same eigenvalue* $\lambda_n^{(0)}$. The expression for the eigenfunctions which has energy differences in the denominators becomes infinite. In that case, degenerate perturbation theory must be applied.

The degeneracy must first be removed for higher order perturbation theory. First, consider the eigenfunction which is a linear combination of eigenfunctions with the same eigenvalue only,

$$g(x) = \sum_k c_{n,k} f_{n,k}^{(0)}(x) \,.$$

which, again from the orthogonality of $f_{n,k}^{(0)}$, leads to the following equation,

$$c_{n,i} \lambda_n^{(0)} + \varepsilon \sum_k c_{n,k} \int f_{n,i}^{(0)}(x) D^{(1)} f_{n,k}^{(0)}(x) \, dx = \lambda c_{n,i}$$

for each n.

As for the majority of low quantum numbers n, i changes over a *small range of integers*, so often the later equation can be solved analytically as an at most 4×4 matrix equation. Once the degeneracy is removed, the first and any order of the above perturbation theory may be further applied relying on the new eigenfunctions.

18.4.3 An example of second-order singular perturbation theory

Main article: Singular perturbation

Consider the following equation for the unknown variable x:

$$x = 1 + \varepsilon x^5.$$

For the initial problem with $\varepsilon = 0$, the solution is $x_0 = 1$. For small ε the lowest-order approximation may be found by inserting the ansatz

$$x = x_0 + \varepsilon x_1 + \cdots$$

into the equation and demanding the equation to be fulfilled up to terms that involve powers of ε higher than the first. This yields $x_1 = 1$. In the same way, the higher orders may be found. However, even in this simple example it may be observed that for (arbitrarily) small positive ε there are four other solutions to the equation (with very large magnitude). The reason we don't find these solutions in the above perturbation method is because these solutions diverge when $\varepsilon \rightarrow 0$ while the ansatz assumes regular behavior in this limit.

The four additional solutions can be found using the methods of **singular perturbation theory**. In this case this works as follows. Since the four solutions diverge at $\varepsilon = 0$, it makes sense to rescale x. We put

$$x = y \varepsilon^{-\nu}$$

such that in terms of y the solutions stay finite. This means that we need to choose the exponent ν to match the rate at which the solutions diverge. In terms of y the equation reads:

$$\varepsilon^{-\nu} y = 1 + \varepsilon^{1-5\nu} y^5$$

The 'right' value for ν is obtained when the exponent of ε in the prefactor of the term proportional to y is equal to the exponent of ε in the prefactor of the term proportional to y^5, i.e. when $\nu = 1/4$. This is called 'significant degeneration'. If we choose ν larger, then the four solutions will collapse to zero in terms of y and they will become degenerate with the solution we found above. If we choose ν smaller, then the four solutions will still diverge to infinity.

Putting $\nu = 1/4$ in the above equation yields:

$$y = \varepsilon^{\frac{1}{4}} + y^5$$

This equation can be solved using ordinary perturbation theory in the same way as regular expansion for x was obtained. Since the expansion parameter is now $\varepsilon^{1/4}$ we put:

$$y = y_0 + \varepsilon^{\frac{1}{4}} y_1 + \varepsilon^{\frac{1}{2}} y_2 + \cdots$$

There are five solutions for y_0: $\{0, \pm 1, \pm i\}$. We must disregard the solution $y = 0$ since it corresponds to the original regular solution which appears to be at zero for $\varepsilon = 0$, because in the limit $\varepsilon \to 0$ we are rescaling by an infinite amount. The next term is $y_1 = -1/4$. In terms of x the four solutions are thus given as:

$$x = \varepsilon^{-\frac{1}{4}} \left[y_0 - \frac{1}{4}\varepsilon^{\frac{1}{4}} + \cdots \right]$$

18.4.4 Example of degenerate perturbation theory – Stark effect in resonant rotating wave

Let us consider a hydrogen atom rotating with a constant angular frequency ω in an electric field. The Hamiltonian is given by:

$$H = H_0 + \varepsilon x$$

where the unperturbed Hamiltonian is

$$H_0 = \frac{\mathbf{p}^2}{2} - \frac{1}{r} - \omega L_z,$$

and L_z is the operator for the z-component of angular momentum: $L_z = i\partial/\partial\varphi$. The perturbation εx can be seen as the strength of the applied electric field multiplied by one of the space coordinates (This calculation is in atomic units, so that every quantity is dimensionless).

The eigenvalues of H_0 are

$$E_{n,m} = -\frac{1}{2}n^2 - m\omega$$

For the lowest energy eigenstates of Hydrogen $|n, l, m\rangle$, $|1, 0, 0\rangle$ and $|2, 1, 1\rangle$ in the resonance $E_{2,1} - E_{1,0} = 0$ their energies are therefore equal $E_{1,0} = E_{2,1} = -1/2$, while the eigenstates are different.

The eigenvalue equation for the Hamiltonian takes the form

$$\begin{bmatrix} E_{1,0} & \varepsilon d \\ \varepsilon d & E_{1,0} \end{bmatrix} \begin{bmatrix} a \\ b \end{bmatrix} = E \begin{bmatrix} a \\ b \end{bmatrix}$$

where

$$d = \frac{128}{243} a_0$$

which leads to the quadratic equation which can be readily solved

$$(E_{1,0} - E)^2 - d^2 \varepsilon^2 = 0$$

with the solution

$$|\chi 1\rangle = \frac{1}{\sqrt{2}}(|1, 0, 0\rangle + |2, 1, 1\rangle)$$
$$E(1) = E_{1,0} + d\varepsilon$$

$$|\chi 2\rangle = \frac{1}{\sqrt{2}}(|1, 0, 0\rangle - |2, 1, 1\rangle)$$
$$E(2) = E_{1,0} - d\varepsilon$$

These states are the Stark states in the rotating frame, they are Trojan (higher eigenvalue) and anti-Trojan wavepackets.

18.5 Some modern applications and limitations

Both regular and singular perturbation theory are frequently used in physics and engineering. Regular perturbation theory may only be used to find those solutions of a problem that evolve smoothly out of the initial solution when changing the parameter (that are "adiabatically connected" to the initial solution).

A well-known example from physics where regular perturbation theory fails is in fluid dynamics when one treats the viscosity as a small parameter. Close to a boundary, the fluid velocity goes to zero, even for very small viscosity (the no-slip condition). For zero viscosity, it is not possible to impose this boundary condition and a regular perturbative expansion amounts to an expansion about an unrealistic physical solution. Singular perturbation theory can, however, be applied here and this amounts to 'zooming in' at the boundaries (using the method of matched asymptotic expansions).

Perturbation theory can fail when the system can transition to a different "phase" of matter, with a qualitatively different behaviour, that cannot be modelled by the physical formulas put into the perturbation theory (e.g., a solid crystal melting into a liquid). In some cases, this failure manifests itself by divergent behavior of the perturbation series. Such divergent series can sometimes be resummed using techniques such as Borel resummation.

Perturbation techniques can be also used to find approximate solutions to non-linear differential equations. Examples of techniques used to find approximate solutions to these types of problems are the Lindstedt–Poincaré technique and the method of multiple time scales.

There is absolutely no guarantee that perturbative methods result in a convergent solution. In fact, asymptotic series are the norm.

18.6 Perturbation theory in chemistry

Many of the ab initio quantum chemistry methods use perturbation theory directly or are closely related methods. Implicit perturbation theory[5] works with the complete Hamiltonian from the very beginning and never specifies a perturbation operator as such. Møller–Plesset perturbation theory uses the difference between the Hartree–Fock Hamiltonian and the exact non-relativistic Hamiltonian as the perturbation. The zero-order energy is the sum of orbital energies. The first-order energy is the Hartree–Fock energy and electron correlation is included at second-order or higher. Calculations to second, third or fourth order are very common and the code is included in most ab initio quantum chemistry programs. A related but more accurate method is the coupled cluster method.

18.7 See also

- Cosmological perturbation theory

- Dynamic nuclear polarisation

- Alternative approach to perturbation theory[6]

- Eigenvalue perturbation

- Interval FEM

- Orders of approximation

- Structural stability

- Lyapunov stability

18.8 References

[1] William E. Wiesel (2010). *Modern Astrodynamics*. Ohio: Aphelion Press. p. 107. ISBN 978-145378-1470.

[2] Cropper, William H. (2004), *Great Physicists: The Life and Times of Leading Physicists from Galileo to Hawking*, Oxford University Press, p. 34, ISBN 978-0-19-517324-6.

[3] Perturbation theory. N. N. Bogolyubov, jr. (originator), Encyclopedia of Mathematics. URL: http://www.encyclopediaofmath.org/index.php?title=Perturbation_theory&oldid=11676

[4] • Sakurai, J.J., and Napolitano, J. (1964,2011). *Modern quantum mechanics* (2nd ed.), Addison Wesley ISBN 978-0-8053-8291-4 . Chapter 5

[5] King, Matcha (1976). "Theory of the Chemical Bond". *JACS* **98** (12): 3415–3420. doi:10.1021/ja00428a004.

[6] Martinez-Carranza, J.; Soto-Eguibar, F.; Moya-Cessa, H. (2012). "Alternative analysis to perturbation theory in quantum mechanics". *The European Physical Journal D* **66**. doi:10.1140/epjd/e2011-20654-5.

18.9 External links

- Introduction to regular perturbation theory by Eric Vanden-Eijnden (PDF)

- Perturbation Method of Multiple Scales

Chapter 19

Topological quantum field theory

A **topological quantum field theory** (or **topological field theory** or **TQFT**) is a quantum field theory which computes topological invariants.

Although TQFTs were invented by physicists, they are also of mathematical interest, being related to, among other things, knot theory and the theory of four-manifolds in algebraic topology, and to the theory of moduli spaces in algebraic geometry. Donaldson, Jones, Witten, and Kontsevich have all won Fields Medals for work related to topological field theory.

In condensed matter physics, topological quantum field theories are the low energy effective theories of topologically ordered states, such as fractional quantum Hall states, string-net condensed states, and other strongly correlated quantum liquid states.

19.1 Overview

In a topological field theory, the correlation functions do not depend on the metric of spacetime. This means that the theory is not sensitive to changes in the shape of spacetime; if the spacetime warps or contracts, the correlation functions do not change. Consequently, they are topological invariants.

Topological field theories are not very interesting on the flat Minkowski spacetime used in particle physics. Minkowski space can be contracted to a point, so a TQFT on Minkowski space computes only trivial topological invariants. Consequently, TQFTs are usually studied on curved spacetimes, such as, for example, Riemann surfaces. Most of the known topological field theories are defined on spacetimes of dimension less than five. It seems that a few higher-dimensional theories exist, but they are not very well understood.

Quantum gravity is believed to be background-independent (in some suitable sense), and TQFTs provide examples of background independent quantum field theories. This has

prompted ongoing theoretical investigation of this class of models.

(Caveat: It is often said that TQFTs have only finitely many degrees of freedom. This is not a fundamental property. It happens to be true in most of the examples that physicists and mathematicians study, but it is not necessary. A topological sigma model with target infinite-dimensional projective space, if such a thing could be defined, would have countably infinitely many degrees of freedom.)

19.2 Specific models

The known topological field theories fall into two general classes: Schwarz-type TQFTs and Witten-type TQFTs. Witten TQFTs are also sometimes referred to as cohomological field theories. See (Schwarz 2000).

19.2.1 Schwarz-type TQFTs

In Schwarz-type TQFTs, the correlation functions or partition functions of the system are computed by the path integral of metric independent action functionals. For instance, in the BF model, the spacetime is a two-dimensional manifold M, the observables are constructed from a two-form F, an auxiliary scalar B, and their derivatives. The action (which determines the path integral) is

$$S = \int_M BF$$

The spacetime metric does not appear anywhere in the theory, so the theory is explicitly topologically invariant. The first example appeared in 1977 and is due to A. Schwarz, its action functional is:

$$\int_M A \wedge dA.$$

Another more famous example is Chern–Simons theory, which can be used to compute knot invariants. In general partition functions depend on a metric but the above examples are shown to be metric-independent.

19.2.2 Witten-type TQFTs

The first example of the topological field theories of Witten-type appeared in Witten's paper in 1988 (Witten 1988a), i.e. topological Yang–Mills theory in four dimensions. Though its action functional contains the spacetime metric $g\alpha_\beta$, after a topological twist it turns out to be metric independent. The independence of the stress-energy tensor $T^{\alpha\beta}$ of the system from the metric depends on whether BRST-operator is closed. Following Witten's example a lot of examples are found in string theory.

Witten-type TQFTs arise if the following conditions are satisfied:

1. The action S of the TQFT has a symmetry, i.e. if δ denotes a symmetry transformation (e.g. a Lie derivative) then it holds $\delta S = 0$

2. The symmetry transformation is exact, i.e. $\delta^2 = 0$

3. There are existing observables $O_1, \ldots, O_n$ which satisfy $\delta O_i = 0$ for all $i \in \{1, \ldots, n\}$.

4. The stress-energy-tensor (or similar physical quantities) is of the form $T^{\alpha\beta} = \delta G^{\alpha\beta}$ for an arbitrary tensor $G^{\alpha\beta}$.

As an example (Linker 2015) given a 2-form field B with the differential operator δ which satisfies $\delta^2 = 0$. Then the action $S = \int_M B \wedge \delta B$ has a symmetry if $\delta B \wedge \delta B = 0$ since

$$\delta S = \int_M \delta(B \wedge \delta B) = \int_M \delta B \wedge \delta B + \int_M B \wedge \delta^2 B = 0$$

Further it holds (under the condition that δ is independent on B and acts similarly to a functional derivative):

$$\frac{\delta}{\delta B^{\alpha\beta}} S = \int_M \frac{\delta}{\delta B^{\alpha\beta}} B \wedge \delta B + \int_M B \wedge \delta \frac{\delta}{\delta B^{\alpha\beta}} \quad B =$$

$$\int_M \frac{\delta}{\delta B^{\alpha\beta}} B \wedge \delta B - \int_M \delta B \wedge \frac{\delta}{\delta B^{\alpha\beta}} B = -2 \int_M \delta B \wedge \frac{\delta}{\delta B^{\alpha\beta}} B$$

The expression $\frac{\delta}{\delta B^{\alpha\beta}} S$ is proportional to δG with another 2-form G .

Now any averages of observables $< O_i >:= \int d\mu O_i e^{iS}$ for the corresponding Haar measure μ are independent on the "geometric" field B and therefore topological:

$$\frac{\delta}{\delta B} < O_i >= \int d\mu O_i i \frac{\delta}{\delta B} S e^{iS} \propto \int d\mu O_i \delta G e^{iS}$$

$$= \delta(\int d\mu O_i G e^{iS}) = 0$$

In the third equality it was used the fact that $\delta O_i = \delta S = 0$ and the invariance of the Haar measure under symmetry transformations. Since $\int d\mu O_i G e^{iS}$ is only a number, the Lie derivative applied on it vanishes.

19.3 Mathematical formulations

19.3.1 The original Atiyah–Segal axioms

Atiyah suggested a set of axioms for topological quantum field theory (Atiyah 1989) which was inspired by Segal's proposed axioms for conformal field theory(afterday, Segal's idea was sumarized in (Segal 2001)), and Witten's idea of the geometric meaning of supersymmetry, (Witten 1982). Atiyah's axioms are constructed on gluing the boundary with differentiable (topological or continuous) transformation, while Segal's are with conformal transformation. These axioms have been relatively useful for mathematical treatments of Schwarz-type QFTs, although it isn't clear that they capture the whole structure of Witten-type QFTs. The basic idea is that a TQFT is a functor from a certain category of cobordisms to the category of vector spaces.

There are in fact two different sets of axioms which could reasonably be called the Atiyah axioms. These axioms differ basically in whether or not they study a TQFT defined on a single fixed n-dimensional Riemannian / Lorentzian spacetime M or a TQFT defined on all n-dimensional spacetimes at once.

Let Λ be a commutative ring with 1 (for almost all real-world purposes we will have $\Lambda = \mathbf{Z}$, $\mathbf{R}$ or $\mathbf{C}$). Atiyah originally proposed the axioms of a topological quantum field theory (TQFT) in dimension d defined over a ground ring Λ as following:

- A finitely generated Λ-module $Z(\Sigma)$ associated to each oriented closed smooth d-dimensional manifold Σ (corresponding to the *homotopy* axiom),

- An element $Z(M) \in Z(\partial M)$ associated to each oriented smooth $(d+1)$-dimensional manifold (with boundary) M (corresponding to an *additive* axiom).

These data are subject to the following axioms (4 and 5 were added by Atiyah):

1. Z is *functorial* with respect to orientation preserving diffeomorphisms of Σ and M,

2. Z is *involutory*, i.e. $Z(\Sigma^*) = Z(\Sigma)^*$ where Σ^* is Σ with opposite orientation and $Z(\Sigma)^*$ denotes the dual module.

3. Z is *multiplicative*.

4. $Z(\varphi) = \Lambda$ for the d-dimensional empty manifold and $Z(\varphi) = 1$ for the $(d+1)$-dimensional empty manifold.

5. $Z(M^*) = \overline{Z(M)}$ (the *hermitian* axiom). Equivalently, $Z(M^*)$ is the disjoint of $Z(M)$

Remark. If for a closed manifold M we view $Z(M)$ as a numerical invariant, then for a manifold with boundary we should think of $Z(M) \in Z(\partial M)$ as a "relative" invariant. Let $f : \Sigma \times I \to \Sigma \times I$ be an orientation preserving diffeomorphism, and identify opposite ends of $\Sigma \times I$ by f. This gives a manifold Σf and our axioms imply

$$Z(\Sigma_f) = \text{Trace } \Sigma(f)$$

where $\Sigma(f)$ is the induced automorphism of $Z(\Sigma)$.

Remark. For a manifold M with boundary Σ we can always form the double $M \cup_\Sigma M^*$ which is a closed manifold. The fifth shows that

$$Z(M \cup_\Sigma M^*) = |Z(M)|^2$$

where on the right we compute the norm in the hermitian (possibly indefinite) metric.

19.3.2 The relation to physics

Physically (2)+(4) is related to relativistic invariance while (3)+(5) is indicative of the quantum nature of the theory.

Σ is meant to indicate the physical space (usually, $d = 3$ for standard physics) and the extra dimension in $\Sigma \times I$ is "imaginary" time. The space $Z(M)$ is the Hilbert space of the quantum theory and a physical theory, with a Hamiltonian H, will have an time evolution operator e^{itH} or an "imaginary time" operator e^{-tH}. The main feature of *topological* QFTs is that $H = 0$, which implies that there is no real dynamics or propagation, along the cylinder $\Sigma \times I$. However, there can be non-trivial "propagation" (or tunneling amplitudes) from Σ_0 to Σ_1 through an intervening manifold M with $\partial M = \Sigma_0^* \cup \Sigma_1$; this reflects the topology of M.

If $\partial M = \Sigma$, then the distinguished vector $Z(M)$ in the Hilbert space $Z(\Sigma)$ is thought of as the *vacuum state* defined by M. For a closed manifold M the number $Z(M)$ is the vacuum expectation value. In analogy with statistical mechanics it is also called the partition function.

The reason why a theory with zero Hamiltonian can be sensibly formulated in the Feynman path integral approach to QFT. This incorporates relativistic invariance (which caters for general $(d+1)$-dimensional "spacetimes") and the theory is formally defined by writing down a suitable Lagrangian - a functional of the classical fields of the theory. A Lagrangian which involves only first derivatives in time formally leads to a zero Hamiltonian, but the Lagrangian itself may have non-trivial features which relate it to the topology of M.

19.3.3 Atiyah's examples

In 1988, M. Atiyah published a paper in which he described many new examples of topological quantum field theory that were considered at that time. (Atiyah 1988) It contains some new topological invariants and the new ideas, which are Casson invariant, Donaldson invariant, Gromov's theory, Floer homology and Jones-Witten's theory.

$d = 0$

In this case Σ consists of finitely many points. To single point we associate a vector space $V = Z(\text{point})$ and to n-points the n-fold tensor product: $V^{\otimes n} = V \otimes ... \otimes V$. The symmetric group Sn acts on $V^{\otimes n}$. A standard way to get the quantum Hilbert space is to give a classical symplectic manifold (or phase space) and then quantize it. Let us extend Sn to compact Lie group G and consider "integrable" orbits for which the symplectic structure comes from a line bundle then quantization leads to the irreducible representations V of G. This is the physical interpretation of the Borel-Weil theorem or the Borel-Weil-Bott theorem. The Lagrangian of these theories is the classical action (holonomy of the line bundle). Thus topological QFT's with $d = 0$ relate naturally to the classical representation theory of Lie groups and symmetric groups.

$d = 1$

We should consider periodic boundary conditions given by closed loops in a compact symplectic manifold X. Along to (Witten 1982) holonomy round such loops used in the case of $d = 0$ as a Lagrangian is used to modify the Hamiltonian. For a closed surface M the invariant $Z(M)$ of the theory is the number of pseudo holomorphic maps $f : M \to X$ in the sense of Gromov (they are ordinary holomorphic maps if X is a Kähler manifold). If this number becomes to infinite i.e. if there are "moduli", then we must fix further data on M. This can be done by picking some points Pi and then looking at holomorphic maps $f : M \to X$ with $f(Pi)$ constrained to lie on a fixed hyperplane. (Witten 1988b) has written down the relevant Lagrangian for this theory.

Floer has given a rigorous treatment, i.e. Floer homology, based on (Witten 1982)'s Morse theory ideas, for the case when the boundary conditions are the interval instead of periodic, the initial and end-points of paths lie on two fixed Lagrangian submanifolds. This theory has been developed as Gromov–Witten invariant theory.

Another example is Holomorphic Conformal Field Theory. This might not be strictly topological quantum field theory at that time because Hilbert spaces are infinite dimensional. The conformal field theories are also related to compact Lie group G in which the classical phase consists of a central extension of the loop group LG. Quantizing these produces the Hilbert spaces of the theory of irreducible (projective) representations of LG. The group $\text{Diff}_+(\mathbf{S}^1)$ now substitutes for the symmetric group and play an important role. The partition function in such theories depends on complex structure: it is not purely topological.

$d = 2$

Jones-Witten theory is the most important theory in this case. Here the classical phase space, associated to a closed surface Σ is the moduli space of flat G-bundle over Σ. The Lagrangian is an integer multiple of the Chern–Simons function of a G-connection on a 3-manifold (which has to be "framed"). The integer multiple k, called the level, is a parameter of the theory and $k \to \infty$ gives the classical limit. This theory can be naturally coupled with the $d = 0$ theory to produce a "relative" theory. The details have been described by Witten who shows that partition function for a (framed) link in the 3-sphere is just the value of the Jones polynomial for a suitable root of unity. The theory can be defined over the relevant cyclotomic field. By considering Riemann surface with boundary, we can couple it to the $d = 1$ conformal theory instead of coupling $d = 2$ theory to $d = 0$. This theory has been developed as the Jones-Witten theory and turned out to be the trigger binding the knot theory and the quantum theory.

$d = 3$

Donaldson has defined integer invariant of smooth 4-manifolds by using moduli spaces of SU(2)-instantons. These invariants are polynomials on the second homology. Thus 4-manifolds should have extra data consisting of the symmetric algebra of H_2. (Witten 1988a) has produced a super-symmetric Lagrangian which formally reproduces the Donaldson theory. Witten's formula might be understood as an infinite-dimensional analogue of the Gauss–Bonnet theorem. At a later date, this theory was further developed and became the Seiberg–Witten gauge theory which reduces SU(2) to U(1) in $N = 2$, $d = 4$ gauge the-

ory. The Hamiltonian version of the theory has been developed by Floer in terms of the space of connections on a 3-manifold. Floer uses the Chern–Simons function, which is the Lagrangian of the Jones-Witten theory to modify the Hamiltonian. For details, see (Atiyah 1988). (Witten 1988a) has also shown how one can couple the $d = 3$ and $d = 1$ theories together: this is quite analogous to the coupling between $d = 2$ and $d = 0$ in the Jones-Witten theory.

Now, it isn't considered on a fixed dimension but on all the dimensions at the same time, namely, topological field theory is viewed as a functor.

19.3.4   The case of a fixed spacetime

Let *BordM* be the category whose morphisms are n-dimensional submanifolds of M and whose objects are connected components of the boundaries of such submanifolds. Regard two morphisms as equivalent if they are homotopic via submanifolds of M, and so form the quotient category *hBordM*: The objects in *hBordM* are the objects of *BordM*, and the morphisms of *hBordM* are homotopy equivalence classes of morphisms in *BordM*. A TQFT on M is a symmetric monoidal functor from *hBordM* to the category of vector spaces.

Note that cobordisms can, if their boundaries match up, be sewn together to form a new bordism. This is the composition law for morphisms in the cobordism category. Since functors are required to preserve composition, this says that the linear map corresponding to a sewn together morphism is just the composition of the linear map for each piece.

There is an equivalence of categories between the category of 2-dimensional topological quantum field theories and the category of commutative Frobenius algebras.

19.3.5   All n-dimensional spacetimes at once

To consider all spacetimes at once, it is necessary to replace *hBordM* by a larger category. So let *Bordn* be the category of bordisms, i.e. the category whose morphisms are n-dimensional manifolds with boundary, and whose objects are the connected components of the boundaries of n-dimensional manifolds. (Note that any $(n-1)$-dimensional manifold may appear as an object in *Bordn*.) As above, regard two morphisms in *Bordn* as equivalent if they are homotopic, and form the quotient category *hBordn*. *Bordn* is a monoidal category under the operation which takes two bordisms to the bordism made from their disjoint union. A TQFT on n-dimensional manifolds is then a functor from *hBordn* to the category of vector spaces, which takes disjoint unions of bordisms to the tensor product of them.

For example, for (1+1)-dimensional bordisms (2-

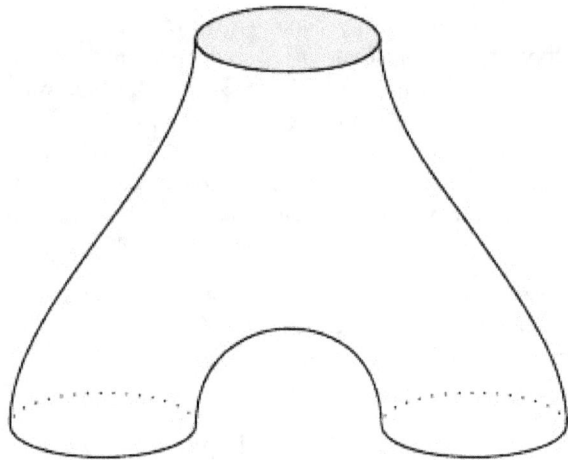

The pair of pants is a (1+1)-dimensional bordism, which corresponds to a product or coproduct in a 2-dimensional TQFT.

dimensional bordisms between 1-dimensional manifolds), the map associated with a pair of pants gives a product or coproduct, depending on how the boundary components are grouped – which is commutative or cocommutative, while the map associated with a disk gives a counit (trace) or unit (scalars), depending on grouping of boundary, and thus (1+1)-dimension TQFTs correspond to Frobenius algebras.

Furthermore, we consider simultaneously 4-dimensional, 3-dimensional and 2-dimensional manifolds that are related by the above bordisms, then obtain ample and important examples.

19.3.6   Development at a later time

Looking at the development of topological quantum field theory we should consider that it has many applications to Seiberg–Witten gauge theory, topological string theory, the relationship between knot theory and quantum theory, and quantum knot invariants. Furthermore it has provided objects of great interest to both mathematics and physics. Also of important recent interest are non-local operators in TQFT. (Gukov & Kapustin (2013)). If string theory is viewed as the fundamental, then non-local TQFTs can be viewed as non-physical models that provide a computationally efficient approximation to local string theory.

19.4   See also

- Quantum topology
- Topological defect
- Topological entropy in physics
- Topological order
- Topological quantum number
- Topological string theory
- Arithmetic topology
- Cobordism hypothesis

19.5   References

- Atiyah, Michael (1989), "Topological quantum field theories", *Publications Mathématiques de l'IHÉS* **68** (68): 175–186, doi:10.1007/BF02698547, MR 1001453

- Linker, Patrick (2015). "Topological Dipole Field Theory". *The Winnower* **2**: e144311.19292. doi:10.15200/winn.144311.19292.

- Witten, Edward (1982), "Super-symmetry and Morse Theory", *J. Diff. Geom.* **17**: 661–692

- Schwarz, Albert (2000), *TOPOLOGICAL QUANTUM FIELD THEORIES* (PDF)

- Segal, Graeme (2001), "Topological structures in string theory", *The Royal Society* **359**: 1389-1398, doi:10.1098/rsta.2001.0841

- Lurie, Jacob, *On the Classification of Topological Field Theories* (PDF)

- Witten, Edward (1988a), "Topological quantum field theory", *Communications in Mathematical Physics* **117** (3): 353–386, Bibcode:1988CMaPh.117..353W, doi:10.1007/BF01223371, MR 953828

- Witten, Edward (1988b), "Topological sigma models", *Communications in Mathematical Physics* **118** (3): 411–449, Bibcode:1988CMaPh.118..411W, doi:10.1007/bf01466725

- Atiyah, Michael (1988). "New invariants of three and four dimensional manifolds". *Proc. Symp. Pure Math.*, 48, American Math. Soc. **48**: 285–299.

- Gukov, Sergei; Kapustin, Anton (2013). "Topological Quantum Field Theory, Nonlocal Operators, and Gapped Phases of Gauge Theories". *JHEP*. On arxiv url=http://arxiv.org/abs/1307.4793

Chapter 20

Unruh effect

The hypothetical **Unruh effect** (or sometimes **Fulling–Davies–Unruh effect**) is the prediction that an accelerating observer will observe black-body radiation where an inertial observer would observe none. In other words, the background appears to be warm from an accelerating reference frame; in layman's terms, a thermometer waved around in empty space, subtracting any other contribution to its temperature, will record a non-zero temperature. The ground state for an inertial observer is seen as in thermodynamic equilibrium with a non-zero temperature by the uniformly accelerating observer.

The Unruh effect was first described by Stephen Fulling in 1973, Paul Davies in 1975 and W. G. Unruh in 1976.[1][2][3] It is currently not clear whether the Unruh effect has actually been observed, since the claimed observations are under dispute. There is also some doubt about whether the Unruh effect implies the existence of **Unruh radiation**.

20.1   The equation

The **Unruh temperature**, derived by William Unruh in 1976, is the effective temperature experienced by a uniformly accelerating detector in a vacuum field. It is given by[4]

$$T = \frac{\hbar a}{2\pi c k_B},$$

where a is the local acceleration, k_B is the Boltzmann constant, $\hbar$ is the reduced Planck constant, and c is the speed of light. Thus, for example, a proper acceleration of 2.5×10^{20} m·s^{-2} corresponds approximately to a temperature of 1 K.

The Unruh temperature has the same form as the Hawking temperature $T_H = \hbar g/(2\pi c k_B)$ of a black hole, which was derived (by Stephen Hawking) independently around the same time. It is, therefore, sometimes called the Hawking–Unruh temperature.[5]

20.2   Explanation

Unruh demonstrated theoretically that the notion of vacuum depends on the path of the observer through spacetime. From the viewpoint of the accelerating observer, the vacuum of the inertial observer will look like a state containing many particles in thermal equilibrium—a warm gas.[6]

Although the Unruh effect would initially be perceived as counter-intuitive, it makes sense if the word *vacuum* is interpreted in a specific way.

In modern terms, the concept of "vacuum" is not the same as "empty space": space is filled with the quantized fields that make up the universe. Vacuum is simply the lowest *possible* energy state of these fields.

The energy states of any quantized field are defined by the Hamiltonian, based on local conditions, including the time coordinate. According to special relativity, two observers moving relative to each other must use different time coordinates. If those observers are accelerating, there may be no shared coordinate system. Hence, the observers will see different quantum states and thus different vacua.

In some cases, the vacuum of one observer is not even in the space of quantum states of the other. In technical terms, this comes about because the two vacua lead to unitarily inequivalent representations of the quantum field canonical commutation relations. This is because two mutually accelerating observers may not be able to find a globally defined coordinate transformation relating their coordinate choices.

An accelerating observer will perceive an apparent event horizon forming (see Rindler spacetime). The existence of **Unruh radiation** could be linked to this apparent event horizon, putting it in the same conceptual framework as Hawking radiation. On the other hand, the theory of the Unruh effect explains that the definition of what constitutes a "particle" depends on the state of motion of the observer.

The free field needs to be decomposed into positive and negative frequency components before defining the creation and annihilation operators. This can only be done in spacetimes with a timelike Killing vector field. This decomposition happens to be different in Cartesian and Rindler coordinates (although the two are related by a Bogoliubov transformation). This explains why the "particle numbers", which are defined in terms of the creation and annihilation operators, are different in both coordinates.

The Rindler spacetime has a horizon, and locally any non-extremal black hole horizon is Rindler. So the Rindler spacetime gives the local properties of black holes and cosmological horizons. The Unruh effect would then be the near-horizon form of the Hawking radiation.

20.3   Calculations

In special relativity, an observer moving with uniform proper acceleration a through Minkowski spacetime is conveniently described with Rindler coordinates. The line element in Rindler coordinates is

$$ds^2 = -\rho^2 d\sigma^2 + d\rho^2,$$

where $\rho = 1/a$, and where σ is related to the observer's proper time τ by $\sigma = a\tau$ (here $c = 1$). Rindler coordinates are related to the standard (Cartesian) Minkowski coordinates by

$$x = \rho \cosh \sigma$$

$$t = \rho \sinh \sigma.$$

An observer moving with fixed ρ traces out a hyperbola in Minkowski space.

An observer moving along a path of constant ρ is uniformly accelerating, and is coupled to field modes which have a definite steady frequency as a function of σ. These modes are constantly Doppler shifted relative to ordinary Minkowski time as the detector accelerates, and they change in frequency by enormous factors, even after only a short proper time.

Translation in σ is a symmetry of Minkowski space: It is a boost around the origin. For a detector coupled to modes with a definite frequency in σ, the boost operator is then the Hamiltonian. In the Euclidean field theory, these boosts analytically continue to rotations, and the rotations close after 2π. So

$$e^{2\pi iH} = 1.$$

The path integral for this Hamiltonian is closed with period 2π which guarantees that the H modes are thermally occupied with temperature $(2\pi)^{-1}$. This is not an actual temperature, because H is dimensionless. It is conjugate to the timelike polar angle σ which is also dimensionless. To restore the length dimension, note that a mode of fixed frequency f in σ at position ρ has a frequency which is determined by the square root of the (absolute value of the) metric at ρ, the redshift factor. From the equation for the line element given above, it is easily seen that this is just ρ. The actual inverse temperature at this point is therefore

$$\beta = 2\pi\rho.$$

Since the acceleration of a trajectory at constant ρ is equal to $1/a$, the actual inverse temperature observed is

$$\beta = \frac{2\pi}{a}.$$

Restoring units yields

$$k_{\mathrm{B}}T = \frac{\hbar a}{2\pi c}.$$

The temperature of the vacuum, seen by an isolated observer accelerating at the Earth's gravitational acceleration of $g = 9.81$ m·s^{-2}, is only 4×10^{-20} K. For an experimental test of the Unruh effect it is planned to use accelerations up to 10^{26} m·s^{-2}, which would give a temperature of about 400,000 K.[7][8]

To put this in perspective, at a vacuum Unruh temperature of 3.978×10^{-20} K, an electron would have a de Broglie wavelength of $h/\sqrt{(3m_ekT)} = 540.85$ m, and a proton at that temperature would have a wavelength of 12.62 m. If electrons and protons were in intimate contact in a very cold vacuum, they would have rather long wavelengths and interaction distances.

At one astronomical unit from the sun, the acceleration is $\frac{GM_S}{(1\,\mathrm{AU})^2} = 0.005932$ m · s^{-2}. This gives an Unruh temperature of 2.41×10^{-23} K. At that temperature, the electron and proton wavelengths are 21.994 km and 513 m, respectively. Even a uranium atom will have a wavelength of 2.2 m at such a low temperature.

20.4   Other implications

The Unruh effect would also cause the decay rate of accelerating particles to differ from inertial particles. Stable particles like the electron could have nonzero transition rates to higher mass states when accelerating at a high enough rate.[9][10][11]

20.5 Unruh radiation

Although Unruh's prediction that an accelerating detector would see a thermal bath is not controversial, the interpretation of the transitions in the detector in the non-accelerating frame are. It is widely, although not universally, believed that each transition in the detector is accompanied by the emission of a particle, and that this particle will propagate to infinity and be seen as *Unruh radiation*.

The existence of Unruh radiation is not universally accepted. Some claim that it has already been observed,[12] while others claim that it is not emitted at all.[13] While the skeptics accept that an accelerating object thermalises at the Unruh temperature, they do not believe that this leads to the emission of photons, arguing that the emission and absorption rates of the accelerating particle are balanced.

20.6 Experimental observation of the Unruh effect

Researchers claim experiments that successfully detected the Sokolov–Ternov effect[14] may also detect the Unruh effect under certain conditions.[15]

Theoretical work in 2011 suggests that accelerating detectors might be used for the direct detection of the Unruh effect with current technology.[16]

20.7 See also

- Dynamical Casimir effect
- Hawking radiation
- Pair production
- Quantum information
- Stochastic electrodynamics
- Superradiance
- Virtual particle
- Woodward effect

20.8 References

[1] S.A. Fulling (1973). "Nonuniqueness of Canonical Field Quantization in Riemannian Space-Time". *Physical Review D* **7** (10): 2850. Bibcode:1973PhRvD...7.2850F. doi:10.1103/PhysRevD.7.2850.

[2] P.C.W. Davies (1975). "Scalar production in Schwarzschild and Rindler metrics". *Journal of Physics A* **8** (4): 609. Bibcode:1975JPhA....8..609D. doi:10.1088/0305-4470/8/4/022.

[3] W.G. Unruh (1976). "Notes on black-hole evaporation". *Physical Review D* **14** (4): 870. Bibcode:1976PhRvD..14..870U. doi:10.1103/PhysRevD.14.870.

[4] See equation 7.6 in W.G. Unruh (2001). "Black Holes, Dumb Holes, and Entropy". *Physics meets Philosophy at the Planck Scale.* Cambridge University Press. pp. 152–173.

[5] P.M. Alsing, P.W. Milonni (2004). "Simplified derivation of the Hawking-Unruh temperature for an accelerated observer in vacuum". *American Journal of Physics* **72** (12): 1524. arXiv:quant-ph/0401170v2. Bibcode:2004AmJPh..72.1524A. doi:10.1119/1.1761064.

[6] Reinhold A. Bertlmann & Anton Zeilinger (2002). *Quantum (un)speakables: From Bell to Quantum Information.* Springer. p. 401 *ff*. ISBN 3-540-42756-2.

[7] M. Visser (2001). "Experimental Unruh radiation?". *Newsletter of the APS Topical Group on Gravitation* **17**: 2044. arXiv:gr-qc/0102044. Bibcode:2001gr.qc.....2044P.

[8] H.C. Rosu (2001). "Hawking-like effects and Unruh-like effects: Toward experiments?". *Gravitation and Cosmology* 7: 1. arXiv:gr-qc/9406012. Bibcode:1994gr.qc.....6012R.

[9] R. Mueller (1997). "Decay of accelerated particles". *Physical Review D* **56** (2): 953–960. arXiv:hep-th/9706016. Bibcode:1997PhRvD..56..953M. doi:10.1103/PhysRevD.56.953.

[10] D.A.T. Vanzella, G.E.A. Matsas (2001). "Decay of accelerated protons and the existence of the Fulling-Davies-Unruh effect". *Physical Review Letters* **87** (15): 151301. arXiv:gr-qc/0104030. Bibcode:2001PhRvL..87o1301V. doi:10.1103/PhysRevLett.87.151301.

[11] H. Suzuki, K. Yamada (2003). "Analytic Evaluation of the Decay Rate for Accelerated Proton". *Physical Review D* **67** (6): 065002. arXiv:gr-qc/0211056. Bibcode:2003PhRvD..67f5002S. doi:10.1103/PhysRevD.67.065002.

[12] I.I. Smolyaninov (2005). "Photoluminescence from a gold nanotip as an example of tabletop Unruh-Hawking radiation". *Physics Letters A* **372** (47): 7043–7045. arXiv:cond-mat/0510743. Bibcode:2008PhLA..372.7043S. doi:10.1016/j.physleta.2008.10.061.

[13] G.W. Ford, R.F. O'Connell (2005). "Is there Unruh radiation?". *Physics Letters A* **350**: 17–26. arXiv:quant-ph/0509151. Bibcode:2006PhLA..350...17F. doi:10.1016/j.physleta.2005.09.068.

[14] Bell, J. S.; Leinaas, J. M. (7 February 1983). "Electrons as accelerated thermometers". *Nuclear Physics B* **212** (1): 131–150. Bibcode:1983NuPhB.212..131B. doi:10.1016/0550-3213(83)90601-6.

[15] E.T. Akhmedov, D. Singleton (2007). "On the physical meaning of the Unruh effect". *JETP Letters* **86** (9): 615–619. arXiv:0705.2525. Bibcode:2007JETPL..86..615A. doi:10.1134/S0021364007210138.

[16] E. Martin-Martinez, I. Fuentes, R. B. Mann (2011). "Using Berry's Phase to Detect the Unruh Effect at Lower Accelerations". *Physical Review Letters* **107** (13): 131301. arXiv:1012.2208. Bibcode:2011PhRvL.107m1301M. doi:10.1103/PhysRevLett.107.131301.

20.9 Further reading

- K.P. Thorne (1995). "Black holes evaporate". *Black Holes and Time Warps* (Reprint ed.). W. W. Norton & Company. ISBN 0-393-31276-3. See especially box 12.5 on p. 444.

- R.M. Wald (1994). "Chapter 5". *Quantum Field Theory in Curved Spacetime and Black Hole Thermodynamics*. University of Chicago Press. ISBN 0-226-87027-8.

- L.C.B. Crispino, A. Higuchi, G.E.A. Matsas (2008). "The Unruh effect and its applications". *Reviews of Modern Physics* **80** (3): 787. arXiv:0710.5373. Bibcode:2008RvMP...80..787C. doi:10.1103/RevModPhys.80.787.

20.10 External links

- Unruh effect - A Scholarpedia article about the same topic written by Stephen Fulling and George Matsas.

Chapter 21

Quantum field theory in curved spacetime

In particle physics, **quantum field theory in curved spacetime** is an extension of standard, Minkowski space quantum field theory to curved spacetime. A general prediction of this theory is that particles can be created by time-dependent gravitational fields (multigraviton pair production), or by time-independent gravitational fields that contain horizons.

21.1 Description

Interesting new phenomena occur; owing to the equivalence principle the quantization procedure locally resembles that of normal coordinates where the affine connection at the origin is set to zero and a nonzero Riemann tensor in general once the proper (covariant) formalism is chosen; however, even in flat spacetime quantum field theory, the number of particles is not well-defined locally. For nonzero cosmological constants, on curved spacetimes quantum fields lose their interpretation as asymptotic particles. Only in certain situations, such as in asymptotically flat spacetimes (zero cosmological curvature), can the notion of incoming and outgoing particle be recovered, thus enabling one to define an S-matrix. Even then, as in flat spacetime, the asymptotic particle interpretation depends on the observer (i.e., different observers may measure different numbers of asymptotic particles on a given spacetime).

Another observation is that unless the background metric tensor has a global timelike Killing vector, there is no way to define a vacuum or ground state canonically. The concept of a vacuum is not invariant under diffeomorphisms. This is because a mode decomposition of a field into positive and negative frequency modes is not invariant under diffeomorphisms. If $t'(t)$ is a diffeomorphism, in general, the Fourier transform of $\exp[ikt'(t)]$ will contain negative frequencies even if $k > 0$. Creation operators correspond to positive frequencies, while annihilation operators correspond to negative frequencies. This is why a state which looks like a vacuum to one observer cannot look like a vacuum state to another observer; it could even appear as a heat bath under suitable hypotheses.

Since the end of the eighties, the local quantum field theory approach due to Rudolf Haag and Daniel Kastler has been implemented in order to include an algebraic version of quantum field theory in curved spacetime. Indeed, the viewpoint of local quantum physics is suitable to generalize the renormalization procedure to the theory of quantum fields developed on curved backgrounds. Several rigorous results concerning QFT in the presence of a black hole have been obtained. In particular the algebraic approach allows one to deal with the problems, above mentioned, arising from the absence of a preferred reference vacuum state, the absence of a natural notion of particle and the appearance of unitarily inequivalent representations of the algebra of observables. (See these lecture notes [1] for an elementary introduction to these approaches and the more advanced review [2])

21.2 Applications

The most striking application of the theory is Hawking's prediction that Schwarzschild black holes radiate with a thermal spectrum. A related prediction is the Unruh effect: accelerated observers in the vacuum measure a thermal bath of particles.

This formalism is also used to predict the primordial density perturbation spectrum arising from cosmic inflation, i.e. the Bunch–Davies vacuum. Since this spectrum is measured by a variety of cosmological measurements—such as the CMB - if inflation is correct this particular prediction of the theory has already been verified.

The Dirac equation can be formulated in curved spacetime, see Dirac equation in curved spacetime for details.

21.3 Approximation to quantum gravity

The theory of quantum field theory in curved spacetime can be considered as a first approximation to quantum gravity. A second step towards that theory would be semiclassical gravity, which would include the influence of particles created by a strong gravitational field on the spacetime (which is still considered classical and the equivalence principle still holds). However gravity is not renormalizable in QFT,[3] so merely formulating QFT in curved spacetime is not a theory of quantum gravity.

21.4 See also

- Field (physics)
- Statistical field theory
- Topological quantum field theory
- Local quantum field theory
- General relativity
- Quantum geometry
- Quantum spacetime
- Quantum field theory (history)

21.5 References

21.5.1 Notes

[1] C. J. Fewster (2008). "Lectures on quantum field theory in curved spacetime (Lecture Note 39/2008 Max Planck Institute for Mathematics in the Natural Sciences (2008))" (PDF). York, UK.

[2] I. Khavkine and V. Moretti (2015). "Algebraic QFT in Curved Spacetime and quasifree Hadamard states: an introduction" (PDF). Trento, Italy. arXiv:1412.5945.

[3] A. Shomer (2007). "A pedagogical explanation for the non-renormalizability of gravity". arXiv:0709.3555.

21.5.2 Further reading: books and relevant papers

- N.D. Birrell & P.C.W. Davies. *Quantum fields in curved space.* CUP (1982).

- S.A. Fulling. *Aspects of quantum field theory in curved space-time.* CUP (1989).

- B.S. Kay & R.M. Wald. *Theorems on the Uniqueness and Thermal Properties of Stationary, Nonsingular, Quasifree States on Space-Times with a Bifurcate Killing Horizon.* Physics Reports 207 (1991) 49-136

- R.M. Wald. *Quantum field theory in curved space-time and black hole thermodynamics.* Chicago U. (1995).

- L. H. Ford. Quantum Field Theory in Curved Space-time (1997).

- S. Hollands, R.M. Wald. *Local Wick polynomials and time ordered products of quantum fields in curved space-time.* Commun. Math. Phys. 223 (2001) 289-326

- R. Verch. *A spin statistics theorem for quantum fields on curved space-time manifolds in a generally covariant framework.* Commun.Math.Phys. 223 (2001) 261-288

- S. Hollands, R.M. Wald. *On the renormalization group in curved space-time.* Commun.Math.Phys. 237 (2003) 123-160

- A. Bytsenko, G. Cognola, E. Elizalde, V. Moretti and S. Zerbini. *Analytic Aspects of Quantum Fields.* World Scientific (2003)

- V. Moretti. *Comments on the stress-energy tensor operator in curved spacetime* Commun. Math. Phys. 232, (2003) 189-222.

- R. Brunetti, K. Fredenhagen, R.Verch. *The Generally covariant locality principle: A New paradigm for local quantum field theory.* Commun. Math. Phys. 237 (2003) 31-68.

- T. Jacobson Introduction to Quantum Fields in Curved Spacetime and the Hawking Effect (2004).

- V. Mukhanov and S. Winitzki. *Introduction to Quantum Effects in Gravity.* CUP (2007).

- L. Parker & D. Toms. *Quantum Field Theory in Curved Spacetime.* (2009).

- T.-P. Hack. On the Backreaction of Scalar and Spinor Quantum Fields in Curved Spacetimes (2010) Ph.D.Thesis Hamburg U. (Advisors: K. Fredenhagen, V. Moretti, R. M. Wald)

- C. Dappiaggi, V. Moretti, N. Pinamonti. *Rigorous construction and Hadamard property of the Unruh state in Schwarzschild spacetime.* Adv. Theor. Math. Phys. 15, vol 2, (2011) 355-448

Chapter 22

Scale relativity

Scale relativity is a geometrical and fractal space-time theory. The idea of a fractal space-time theory was first introduced by Garnet Ord,[1] and by Laurent Nottale in a paper with Jean Schneider.[2] The proposal to combine fractal space-time theory with relativity principles was made by Laurent Nottale.[3][4] The resulting scale relativity theory is an extension of the concept of relativity found in special relativity and general relativity to physical scales (time, length, energy, or momentum scales). In physics, relativity theories have shown that position, orientation, movement and acceleration cannot be defined in an absolute way, but only relative to a system of reference.

Noticing the relativity of scales, as noticing the other forms of relativity is just a first step. Scale relativity theory proposes to make the next step by translating this simple insight formally in physical theory, by introducing explicitly in coordinate systems the "state of scale".

To describe scale transformations requires the use of fractal geometries, which are typically concerned with scale changes. Scale relativity is thus an extension of relativity theory to the concept of scale, using fractal geometries to study scale transformations.

The construction of the theory is similar to previous relativity theories, with three different levels: Galilean, special and general. The development of a full general scale relativity is not finished yet. However, the existing progress and results already have consequences for the foundations of quantum mechanics, particle physics, and high energy physics. Furthermore, empirical predictions in physics, astrophysics, and cosmology have already been validated, most often with a high precision, or highly statistically significant results.

22.1 History

22.1.1 Feynman's paths in quantum mechanics

Richard Feynman developed a path integral formulation of quantum mechanics.[5] Searching for the most important paths relevant for quantum particles, Feynman noticed that such paths were very irregular on small scales, i.e. infinite and non-differentiable. This means that in between two points, a particle can have not one path, but an infinity of potential paths.

This can be illustrated with a concrete example. Imagine that you are hiking in the mountains, and that you are free to walk wherever you like. To go from point A to point B, there is not one shortest path, but an infinity of possible paths, each going through different valleys and hills.[6]

Scale relativity hypothesizes that quantum behavior comes from the fractal nature of spacetime. Indeed, fractal geometries allow to study such non-differentiable paths. This fractal interpretation of quantum mechanics has been further specified by Abbot and Wise,[7] showing that the paths have a fractal dimension 2. Scale relativity goes one step further by asserting that the fractality of these paths is a consequence of the fractality of space-time.

There are other pioneers who saw the fractal nature of quantum mechanical paths.[8][9] Also, as much as the development of general relativity required the mathematical tools of non-Euclidean (Riemannian) geometries, the development of a fractal space-time theory would not have been possible without the concept of fractal geometries developed and popularized by Benoît Mandelbrot. Fractals are usually associated with the self-similar case of a fractal curve, but other more complicated fractals are possible, e.g. considering not only curves, but also fractal surfaces or fractal volumes, as well as investigating fractal dimensions which have other values than 2, and which also vary with scale.

22.1.2 Independent discovery

Garnet Ord[10] and Laurent Nottale[11] both connected fractal space-time with quantum mechanics. Nottale coined the term "scale relativity" in 1992.[12] He developed the theory and its applications with more than one hundred scientific papers (see here), two technical books in English,[13][14] and three popular books in French.[15][16][17]

22.2 Basic concepts

22.2.1 Principle of scale relativity

The principle of relativity says that physical laws should be valid in all coordinate systems. This principle has been applied to states of position (the origin and orientation of axes), as well as to the states of movement of coordinate systems (speed, acceleration). Such states are never defined in an absolute manner, but relatively to one another. For example, there is no absolute movement, in the sense that it can only be defined in a relative way between one body and another. Scale relativity proposes in a similar manner to define a scale relative to another one, and not in an absolute way. Only scale ratios have a physical meaning, never an absolute scale, in the same way as there exists no absolute position or velocity, but only position or velocity differences.

The concept of *resolution* is re-interpreted as the "state of scale" of the system, in the same way as velocity characterizes the state of movement. The principle of scale relativity can thus be formulated as:

> "the laws of physics must be such that they apply to coordinate systems whatever their state of scale."[18]

The main goal of scale relativity is to find laws which mathematically respect this new principle of relativity. Mathematically, this can be expressed through the principle of covariance applied to scales, that is, the invariance of the form of physics equations under transformations of resolutions (dilations and contractions).

22.2.2 Including resolutions in coordinate systems

Galileo introduced explicitly velocity parameters in the observational referential. Then, Einstein introduced explicitly acceleration parameters. In a similar way, Nottale introduces scale parameters explicitly in the observational referential. The core idea of scale-relativity is thus to include res-

olutions explicitly in coordinate systems, thereby integrating measure theory explicitly in the formulation of physical laws.

An important consequence is that coordinates are not numbers anymore, but *functions*, which depend on the resolution.[19] For example, the length of the Brittany coast is explicitly dependent on the resolution at which one measures it.[20]

If we measure a pen with a ruler graduated at a millimetric scale, we should write that it is 15 ± 0.1 cm. The error bar indicates the resolution of our measure. If we had measured the pen at another resolution, for example with a ruler graduated at the centimeter scale, we would have found another result, 15 ± 1 cm. In scale relativity, this resolution defines the "state of scale". In the relativity of movement, this is similar to the concept of speed, which defines the "state of movement".

The relative state of scale is fundamental to know about for any physical description. For example, if we want to describe the movement and properties of a sphere, we may as well use classical mechanics or quantum mechanics depending on the size of the sphere in question.[21]

In particular, information on resolution is essential to understand quantum mechanical systems, and in scale relativity, resolutions are included in coordinate systems, so it seems a logical and promising approach to account for quantum phenomena.

22.2.3 Dropping the hypothesis of differentiability

Scientific theories usually do not improve by adding complexity, but rather by starting from a more and more simple basis. This fact can be observed throughout the history of science. The reason is that starting from a less constrained basis provides more freedom and therefore allows richer phenomena to be included in the scope of the theory. Therefore, new theories usually do not contradict the old ones, but widen their domain of validity and include previous knowledge as special cases. For example, releasing the constraint of rigidity of space led Einstein to derive his theory of general relativity and to understand gravitation. As expected, this theory naturally includes Newton's theory, which is recovered as a linear approximation under weak fields.

The same type of approach has been followed by Nottale to build the theory of scale relativity. The basis of current theories is a continuous and two-times differentiable space. Space is by definition a continuum, but the assumption of differentiability is not supported by any fundamental reason. It is usually assumed only because it is observed that the first two derivatives of position with respect to time are needed to describe motion. Scale relativity theory is rooted in the idea that the constraint of differentiability can be relaxed and that this allows quantum laws to be derived.

In terms of geometry, differentiability means that a curve is sufficiently smooth and can be approximated by a tangent. Mathematically, two points are placed on this curve and one observes the slope of the straight line joining them as they become closer and closer. If the curve is smooth enough, this process converges (almost) everywhere and the curve is said to be differentiable. It is often believed that this property is common in nature. However, most natural objects have instead a very rough surface, or contour. For example the bark of trees and snowflakes have a detailed structure that does not become smoother when the scale is refined. For such curves, the slope of the tangent fluctuates endlessly or diverges. The derivative is then undefined (almost) everywhere and the curve is said to be non-differentiable.Therefore, when the *assumption* of space differentiability is abandoned, there is an additional degree of freedom that allows the geometry of space to be extremely rough. The difficulty in this approach is that new mathematical tools are needed to model this geometry because the classical derivative cannot be used. Nottale found a solution to this problem by using the fact that nondifferentiability implies scale dependence and therefore the use of fractal geometry. Scale dependence means that the distances on a nondifferentiable curve depend on the scale of observation. It is therefore possible to maintain differential calculus provided that the scale at which derivatives are calculated is given, and that their definition includes no limit. It amounts to saying that nondifferentiable curves have a whole set of tangents in one point instead of one, and that there is a specific tangent at each scale.

To abandon the hypothesis of differentiability does not mean abandoning differentiability. Instead, this leads to a more general framework, where *both* differentiable and non-differentiable cases are included. Combined with motion relativity, scale relativity by definition thus extends and contains general relativity. As much as general relativity is possible when we drop the hypothesis of euclidian space-time, allowing the possibility of curved space-time, scale relativity is possible when we abandon the hypothesis of differentiability, allowing the possibility of a fractal space-time. The objective is then to describe a continuous space-time which is not everywhere differentiable, as it was in general relativity. Abandoning differentiability doesn't mean abandoning differential equations. The concept of fractal allows work with the nondifferentiable case with differential equations. In differential calculus, we can see the concept of limit as a zoom, but in this generalization of differential calculus, one doesn't look only at the limit zooms (zero and infinity) but also everything in between, that is,

all possible zooms. In sum, we can drop the hypothesis of the differentiability of space-time, keeping differential equations, provided that fractal geometries are used. With them, we can still deal with the nondifferentiable case with the tools of differential equations. This leads to a double differential equation treatment: in space-time and in scale space.

22.2.4   Fractal space-time

If Einstein showed that space-time was curved, Nottale shows that it is not only curved, but also fractal. Nottale has proven a key theorem which shows that a space which is continuous and non-differentiable is necessarily fractal.[22] It means that such a space depends on scale.

Importantly, the theory does not merely describe fractal objects in a given space. Instead, it is *space itself which is fractal*. To understand what a fractal space means requires to study not just fractal curves, but also fractal surfaces, fractal volumes, etc.

Mathematically, a fractal space-time is defined as a nondifferentiable generalization of Riemannian geometry[23] ,[24] Such a fractal space-time geometry is the natural choice to develop this new principle of relativity, in the same way that curved geometries were needed to develop Einstein's theory of general relativity.[25]

In the same way that general relativistic effects are not felt in a typical human life, the most radical effects of the fractality of spacetime appear only at the extreme limits of scales: micro scales or at cosmological scales. This approach therefore proposes to bridge not only the quantum and the classical, but also the classical and the cosmological, with fractal to non-fractal transitions (see Fig. 1). More plots of this transition can be seen in the literature.[26][27]

22.2.5   Minimum and maximum invariant scales

A fundamental and elegant result of scale relativity is to propose a minimum and maximum scale in physics, invariant under dilations, in a very similar way as the speed of light is an upper limit for speed.

Minimum invariant scale

In special relativity, there is an unreachable speed, the speed of light. We can add speeds without end, but they will always be less than the speed of light. The sums of all speeds are limited by the speed of light. Additionally, the composition of two velocities is inferior to the sum of those two speeds.

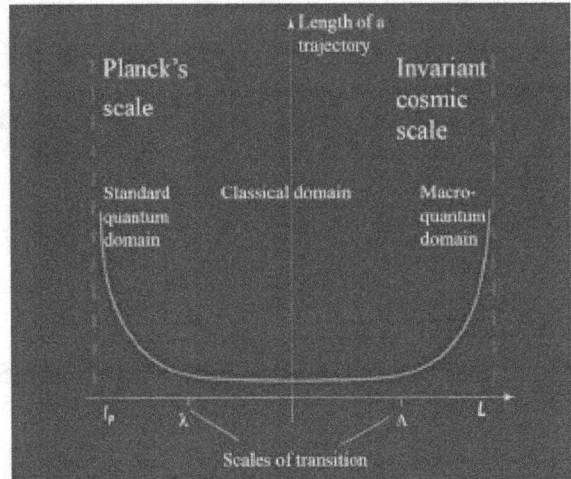

Fig. 1. Variation of the fractal dimension of space-time geodesics (trajectories), according to the resolution, in the framework of special scale relativity. The scale symmetry is broken in two transitioning scales λ and Λ (non-absolute), which divide the scale space in three domains: (1) a classical domain, intermediary, where space-time doesn't depend on resolutions because the laws of movement dominate over scale laws; and two asymptotical domains towards (2) very small and (3) very large scales where scale laws dominate over the laws of movement, which makes explicit the fractal structure of space-time.

In special scale relativity, similar unreachable observational scales are proposed, the Planck length scale (lP) and the Planck time scale (tP). Dilations are borned by lP and tP, which means that we can divide spatial or temporal intervals without end, but they will always be superior to Planck's length and time scales. This is a result of special scale relativity (see section 2.7 below). Similarly, the composition of two scale changes is inferior to the product of these two scales.[28]

Maximum invariant scale

The choice of the maximum scale (noted L) is less easy to explain, but it mostly consists to identify it with the cosmological constant: $L = 1/(\Lambda^2)$[29][30] [31][32] . This is motivated in parts because a dimensional analysis shows that the cosmological constant is the inverse of the square of a length, i.e. a curvature.

22.2.6   Galilean scale relativity

The theory of scale relativity follows a similar construction as the one of the relativity of movement, which took place in three steps: galilean, special and general relativity.

This is not surprising, as in both cases the goal is to find laws satisfying transformation laws including one parameter that is relative: the speed in the case of the relativity of movement; the resolution in the case of the relativity of scales.

Galilean scale relativity involves linear transformations a constant fractal dimension, self-similarity and scale invariance. This situation is best illustrated with self-similar fractals. Here, the length of geodesics varies constantly with resolution. The fractal dimensions of free particles doesn't change with zooms. These are self-similar curves.

In galilean relativity, recall that the laws of motion are the same in all inertial frames. Galileo famously concluded that "the movement is like nothing".[33] In the case of self-similar fractals, paraphrasing Galileo, one could say that "scaling is like nothing". Indeed, the same patterns occur at different scales, so scaling is not noticeable, it is like nothing.

In the relativity of movement, Galileo's theory is an additive galilean group:

$$X' = X - VT$$

$$T' = T$$

However, if we consider scale transformations (dilations and contractions), the laws are products, and not sums. This can be seen by the necessity to use units of measurements. Indeed, when we say that an object measures 10 meters, we actually mean the object measures 10 **times** the definite predetermined length called "meter". The number 10 is actually a scale ratio of two lengths 10/1m, where 10 is the measured quantity, and 1m is the arbitrary defining unit. This is the reason why the group is multiplicative.

Moreover, an arbitrary scale e doesn't have any physical meaning in itself (like the number 10), only scale ratios r=e'/e have a meaning, in our example, r=10/1. Using the Gell-Mann-Lévy method,[34] we can use a more relevant scale variable, V=ln (e'/e), and then find back an additive group for scale transformations by taking the logarithm – which converts products into sums.

Interestingly, when, in addition to the principle of scale relativity, one adds the principle of relativity of movement, there is a transition of the structure of geodesics at large scales, where trajectories do not depend on the resolution anymore, where trajectories become classical. This explains the shift of behavior from quantum to classical[35] [36] See also Fig. 1.

22.2.7   Special scale relativity

Special scale relativity can be seen as a correction of galilean scale relativity, where Galilean transformations are replaced by Lorentz transformations.[37][38] Interestingly, the "corrections remain small at "large" scale (i.e. around the Compton scale of particles) and increase when going to smaller length scales (i.e. large energies) in the same way as motion-relativistic corrections increase when going to large speeds".[39]

In Galilean relativity, it was considered "obvious" that we could add speeds without limit ($w = u + v$). This composition laws for speed was not challenged. However, Poincaré and Einstein did challenge it with special relativity, setting a maximum speed on movement, the speed of light. Formally, if v is a velocity, $v + c = c$. The status of the speed of light in special relativity is a horizon, unreachable, impassable, invariant under changes of movement.

Regarding scale, we are still within a Galilean kind of thinking. Indeed, we assume without justification that the composition of two dilations is $\varrho * \varrho = \varrho^2$. Written with logarithms, this equality becomes $ln\varrho + ln\varrho = 2ln\varrho$. However, nothing guarantees that this law should hold at quantum or cosmic scales.[40] As a matter of fact, this dilation law is corrected in special scale relativity, and becomes: $ln\varrho + ln\varrho = 2 \ln \varrho / (1 + \ln \varrho^2)$.

More generally, in special relativity the composition law for velocities differs from the Galilean approximation and becomes (with the speed of light $c=1$):

$$u \oplus v = (u + v) / (1 + u*v)$$

Similarly, in special scale relativity, the composition law for dilations differs from our Galilean intuitions and becomes (in a logarithm of base K which includes a possible constant $C = \ln K$, which plays the same role as c):

$$\log_\varrho 1 \oplus \log_\varrho 2 = (\log_\varrho 1 + \log_\varrho 2) / (1 + \log_\varrho 1 * \log_\varrho 2)$$

The status of the Planck scale in special scale relativity plays a similar role as the speed of light in special relativity. It is a horizon for small scales, unreachable, impassable, invariant under scale changes, i.e. dilations and contractions. The consequence for special scale relativity is that applying two times the same contraction ϱ to an object, the result is a contraction less strong than contraction ϱ x ϱ. Formally, if ϱ is a contraction, $\varrho * lP = lP$.

As noted above, there is also an unreachable, impassable maximum scale, invariant under scale changes, which is the cosmic length L.[41] In particular, it is invariant under the expansion of the universe.

22.2.8   General scale relativity

In Galilean scale relativity, spacetime was fractal with *constant* fractal dimensions. In special scale relativity, frac-

tal dimensions can vary. This varying fractal dimension remains however constrained by a log-Lorentz law. This means that the laws satisfy a logarithmic version of the Lorentz transformation. The varying fractal dimension is covariant, in a similar way as proper time is covariant in special relativity.

In general scale relativity, the fractal dimension is not constrained anymore, and can take any value. In other words, it is the situation where there is *curvature in scale space*. Einstein's curved space-time becomes a particular case of the more general fractal spacetime.

General scale relativity is much more complicated, technical, and less developed than its Galilean and special versions. It involves non-linear laws, scale dynamics and gauge fields. In the case of non self-similarity, changing scales generates a new scale-force or scale-field which needs to be taken into account in a scale dynamics approach. Quantum mechanics then needs to be analyzed in scale space.[42][43]

Finally, in general scale relativity, we need to take into account both movement *and* scale transformations, where scale variables depend on space-time coordinates. More details about the implications for abelian gauge fields[44][45] and non-abelian gauge fields[46] can be found in the literature. Nottale's 2011 book[47] provides the state of the art.

To sum up, one can see some structural similarities between the relativity of movement and the relativity of scales in Table 1.

Table 1. Comparison between relativity of movement and relativity of scales. In both cases, there are two kinds of variables linked to the coordinate systems: variables which define the coordinate system, and variables that characterize the state of the coordinate system. In this analogy, the resolution can be assimilated to a speed; acceleration to a scale acceleration; space to the length of a fractal; and time, to the variable fractal dimension.[48] Table adapted from this paper.[49]

22.3   Consequences for quantum mechanics

22.3.1   Introduction

The fractality of space-time implies an infinity of virtual geodesics. This remark already means that a fluid mechanics is needed. Note that this view is not new, as many authors have noticed fractal properties at quantum scales, thereby suggesting that typical quantum mechanical paths are fractal. See this article for a review.[50] However, the idea to consider a fluid of geodesics in a fractal spacetime is an original proposal from Nottale.

In scale relativity, quantum mechanical effects appear as effects of fractal structures on the movement. The fundamental indeterminism and nonlocality of quantum mechanics are deduced from the fractal geometry itself.

There is an analogy between the interpretation of gravitation in general relativity and quantum effects in scale relativity. Indeed, if gravitation is a manifestation of space-time curvature in general relativity, quantum effects are manifestations of a fractal space-time in scale relativity.

To sum up, there are two aspects which allows scale relativity to better understand quantum mechanics. On the one side, fractal fluctuations themselves are hypothesized to lead to quantum effects. On the other side, non-differentiability leads to a local irreversibility of the dynamics and therefore to the use of complex numbers.

Quantum mechanics thus receives not only a new interpretation, but a firm foundation in relativity principles.

22.3.2   Quantum-classical transition

As Philip Turner summarized:

> "the structure of space has both a smooth (differentiable) component at the macro-scale and a chaotic, fractal (non-differentiable) component at the micro-scale, the transition taking place at the de Broglie length scale."[51]

This transition is explained with Galilean scale relativity[52] (see also above).

22.3.3   Derivation of quantum mechanics' postulates

Starting from scale relativity, it is possible to derive the fundamental "postulates" of quantum mechanics.[53] More specifically, building on the result of the key theorem showing that a space which is continuous and non-differentiable is necessarily fractal (see section 2.4), Schrödinger's equation, Born's and von Neumann's postulate are derived.

To derive Schrödinger's equation, Nottale[54] started with Newton's second law of motion, and used the result of the key theorem. Many subsequent works then confirmed the derivation.[55][56][57][58][59][60][61]

Actually, the Schrödinger equation derived becomes generalized in scale relativity, and opens the way to a macroscopic

quantum mechanics (see below for validated empirical predictions in astrophysics). This may also help to better understand macroscopic quantum phenomena in the future.

Reasoning about fractal geodesics and non-differentiability, it is also possible to derive von Neumann's postulate[62] and Born's postulate.[63]

With the hypothesis of a fractal space-time, the Klein-Gordon, and the Dirac equation can then be derived.[64][65][66]

The significance of these fundamental results is immense, as the foundations of quantum mechanics which were up to now axiomatic, are now logically derived from more primary relativity theory principles and methods.

22.3.4   Gauge transformations

Gauge fields appear when scale and movements are combined. Scale relativity proposes a geometric theory of gauge fields. As Turner explains:

> The theory offers a new interpretation of gauge transformations and gauge fields (both Abelian and non-Abelian), which are manifestations of the fractality of space-time, in the same way that gravitation is derived from its curvature.[67]

The relationships between fractal space-time, gauge fields and quantum mechanics are technical and advanced subject-matters elaborated in details in Nottale's latest book.[68]

22.4   Consequences for elementary particles physics

22.4.1   Introduction

Scale relativity gives a geometric interpretation to charges, which are now "defined as the conservative quantities that are built from the new scale symmetries".[69] Relations between mass scales and coupling constants can be theoretically established, and some of them empirically validated. This is possible because in scale relativity, the problem of divergences in quantum field theory is resolved. Indeed, in the new framework, masses and charges become finite, even at infinite energy. In special scale relativity, the possible scale ratios become limited, constraining in a geometric way the quantization of charges. Let us compare a few theoretical predictions with their experimental measures.

22.4.2 Fine-structure constant

Main article: Fine-structure constant

Nottale's latest theoretical prediction of the fine-structure constant at the Z_0 scale is:[70]

$$\alpha^{-1}(mZ) = 128.92$$

By comparison, a recent experimental measure gives:[71]

$$\alpha^{-1}(mZ) = 128.91 \pm 0.02$$

At low energy, the theoretical fine-structure constant prediction is:[72]

$$\alpha^{-1} = 137.01 \pm 0.035;$$

which is within the range of the experimental precision:

$$\alpha^{-1} = 137.036$$

22.4.3 SU (2) coupling at Z scale

Here the SU(2) coupling corresponds to rotations in a three-dimensional scale-space. The theoretical estimate of the SU (2) coupling at Z scale is:[73]

$$\alpha^{-1}_2 Z = 29.8169 \pm 0.0002$$

While the experimental value gives:[74]

$$\alpha^{-1}_2 Z = 29.802 \pm 0.027.$$

22.4.4 Strong nuclear force at Z scale

Special scale relativity predicts the value of the strong nuclear force with great precision, as later experimental measurements confirmed. The first prediction of the strong nuclear force at the Z energy level was made in 1992:[75]

$$\alpha S\ (mZ) = 0.1165 \pm 0.0005$$

A recent and refined theoretical estimate gives:[76]

$$\alpha S\ (mZ) = 0.1173 \pm 0.0004,$$

which fits very well with the experimental measure:[77]

$$\alpha S\ (mZ) = 0.1176 \pm 0.0009$$

22.4.5 Mass of the electron

As an application from this new approach to gauge fields, a theoretical estimate of the mass of the electron (m_e) is possible, from the experimental value of the fine-structure constant. This leads a very good agreement:[78]

$$m_e(\text{theoretical}) = 1.007\ m_e\ (\text{experimental})$$

22.5 Astrophysical applications

22.5.1 Macroquantum mechanics

Some chaotic systems can be analyzed thanks to a macroquantum mechanics. The main tool here is the generalized Schrödinger equation, which brings statistical predictability characteristic of quantum mechanics into other scales in nature. The equation predicts probability density peaks. For example, the position of exoplanets can be predicted in a statistical manner. The theory predicts that planets have more chances to be found at such or such distance from their star. As Baryshev and Teerikorpi[79] write:

> "With his equation for the probability density of planetary orbits around a star, Nottale has seemingly come close to the old analogy which saw a similarity between our solar system and an atom in which electrons orbit the nucleus. But now the analogy is deeper and mathematically and physically supported: it comes from the suggestion that chaotic planetary orbits on very long time scales have preferred sizes, the roots of which go to fractal space-time and generalized Newtonian equation of motion which assumes the form of the quantum Schrödinger equation."

However, as Nottale[80] acknowledges, this general approach is not totally new:

> "The suggestion to use the formalism of quantum mechanics for the treatment of macroscopic problems, in particular for understanding structures in the solar system, dates back to the beginnings of the quantum theory"

22.5.2 Gravitational systems

Space debris

Main article: Space debris

At the scale of Earth's orbit, space debris probability peaks at 718 km and 1475 km have been predicted with scale relativity,[81] which is in agreement with observations at 850 km and 1475 km.[82] Da Rocha and Nottale suggest that the dynamical braking of the Earth's atmosphere may be responsible for the difference between the theoretical prediction and the observational data of the first peak.

Solar system

Scale relativity predicts a new law for interplanetary distances, proposing an alternative to the nowadays falsified Titius-bode "law".[83][84] However, the predictions here are statistical and not deterministic as in Newtonian dynamics. In addition to being statistical, the scale relativistic law has a different theoretical form, and is more reliable than the original Titius-Bode version:[85]

> "The Titius-Bode "law" of planetary distance is of the form $a + b \times c^{\,n}$, with $a = 0.4$ AU, $b = 0.3$ AU and $c = 2$ in its original version. It is partly inconsistent — Mercury corresponds to $n = -\infty$, Venus to $n = 0$, the Earth to $n = 1$, etc. It therefore "predicts" an infinity of orbits between Mercury and Venus and fails for the main asteroid belt and beyond Saturn. It has been shown by Herrmann (1997) that its agreement with the observed distances is not statistically significant. [...] [I]n the scale relativity framework, the predicted law of distance is not a Titius-Bode-like power law but a more constrained and statistically significant quadratic law of the form $an = a_0 n^2$."

Extrasolar systems

Main article: Exoplanet

The method also applies to other extrasolar systems. Let us illustrate this with the first exoplanets found around the pulsar PSR B1257+12.[86] Three planets, A, B and C have been found. Their orbital period ratios (noted PA/PC for the period ratio of planet A to C) can be estimated and compared to observations.[87] Using the macroscopic Schrödinger equation, the recent theoretical estimates are:[88]

$$(PA/PC)^{1/3} = 0.63593 \text{ (predicted)}$$

$$(PB/PC)^{1/3} = 0.8787 \text{ (predicted)},$$

which fit the observed values[89] with great precision:

$$(PA/PC)^{1/3} = 0.63597 \text{ (observed)}$$

$$(PB/PC)^{1/3} = 0.8783 \text{ (observed)}.$$

Interestingly, the puzzling fact that many exoplanets (e.g. hot jupiters) are so close to their parent stars receives a natural explanation in this framework. Indeed, it corresponds to the fundamental orbital of the model, where (exo)planets are at 0.04 UA / solar mass of their parent star.[90]

More validated predictions can be found regarding orbital periods and the distances of planets from their parent star.[91][92][93]

Galaxy pairs

Daniel da Rocha studied the velocity of about 2000 galaxy pairs,[94] which gave statistically significant results when compared to the theoretical structuration in phase space from scale relativity. The method and tools here are similar to the one used for explaining the structure in solar systems.

Similar successful results apply at other extragalactic scales: the local group of galaxies, clusters of galaxies, the local supercluster and other very large scale structures.[95]

22.5.3 Dark matter

Main article: Dark matter

Scale relativity suggests that the fractality of matter contributes to the phenomenon of dark matter. Indeed, some of the dynamical and gravitational effects which seem to require unseen matter are suggested to be consequences of the fractality of space on very large scales.[96]

In the same way as quantum physics differs from the classical at very small scales because of fractal effects, symmetrically, at very large scales, scale relativity also predicts that corrections from the fractality of space-time must be taken into account (see also Fig. 1).

Such an interpretation is somehow similar in spirit to Modified Newtonian dynamics (MOND), although here the approach is founded on relativity principles. Indeed, in MOND, Newtonian dynamics is modified in an *ad hoc* manner to account for the new effects, while in scale relativity, it is the new fractal geometric field taken into consideration which leads to the emergence of a dark potential.

On the largest scale, scale relativity offers a new perspective on the issue of redshift quantization. With a reasoning similar to the one which allows to predict probability peaks for the velocity of planets, this can be generalized to larger intergalactic scales. Nottale writes:[97]

"In the same way as there are well-established structures in the position space (stars, clusters of stars, galaxies, groups of galaxies, clusters of galaxies, large scale structures), the velocity probability peaks are simply the manifestation of structuration in the velocity space. In other words, as it is already well-known in classical mechanics, a full view of the structuring can be obtained in phase space."

22.6 Cosmological applications

22.6.1 Large numbers hypothesis

Main article: Dirac large numbers hypothesis

Nottale noticed that reasoning about scales was a promising road to explain the large numbers hypothesis.[98] This was elaborated in more details in a working paper.[99] The scale-relativistic way to explain the large numbers hypothesis was later discussed by Nottale[100][101] and by Sidharth.[102]

22.6.2 Prediction of the cosmological constant

Main article: Cosmological constant

In scale relativity, the cosmological constant is interpreted as a curvature. If one does a dimensional analysis, it is indeed the inverse of the square of a length. The predicted value of the cosmological constant, back in 1993 was:[103]

$$\Omega \Lambda \, h^2 = 0.36$$

Depending on model choices, the most recent predictions give the following range:[104]

$$0.311 < \Omega \Lambda \, h^2 \text{ (predicted)} < 0.356,$$

while the measured cosmological constant from the Planck satellite[105] is:

$$\Omega \Lambda \, h^2 \text{ (measured)} = 0.318 \pm 0.012.$$

Given the improvements of the empirical measures from 1993 until 2011, Nottale commented:[106]

"The convergence of the observational values towards the theoretical estimate, despite an improvement of the precision by a factor of more than 20, is striking."

Dark energy can be considered as a measurement of the cosmological constant. In scale relativity, dark energy would come from a potential energy manifested by the fractal geometry of the universe at large scales,[107] in the same way as the Newtonian potential is a manifestation of its curved geometry in general relativity.

22.6.3 Horizon problem

Main article: horizon problem

Scale relativity offers a new perspective on the old horizon problem in cosmology. The problem states that different regions of the universe have not had contact with each others because of the great distances between them, but nevertheless they have the same temperature and other physical properties. This should not be possible, given that the transfer of information (or energy, heat, etc.) can occur, at most, at the speed of light.

Nottale[108] writes that special scale relativity "naturally solves the problem because of the new behaviour it implies for light cones. Though there is no inflation in the usual sense, since the scale factor time dependence is unchanged with respect to standard cosmology, there is an inflation of the light cone as t -> Λ/c", where Λ is the Planck length scale $(\hbar G/c^3)^{1/2}$. This inflation of the light cones makes them flare and cross themselves, thereby allowing a causal connection between any two points, and solving the horizon problem (see also[109]).

22.7 Applications to other fields

Although scale relativity started as a spacetime theory, its methods and concepts can and have been used in other fields. For example, quantum-classical kinds of transitions can be at play at intermediate scales, provided that there exists a fractal medium which is locally nondifferentiable. Such a fractal medium then plays a role similar to that played by fractal spacetime for particles. Objects and particles embedded in such a medium will acquire macroquantum properties. As examples, we can mention gravitational structuring in astrophysics (see section 5), turbulence,[110] superconductivity at laboratory scales (see section 7.1, and also modeling in geography (section 7.4).

What follows are *not* strict applications of scale relativity, but rather models constructed with the general idea of relativity of scales.[111][112] Fractal models, and in particular self-similar fractal laws have been applied to describe numerous biological systems such as trees, blood networks, or plants. It is thus to be expected that the mathematical tools developed through a fractal space-time theory can have a wider variety of applications to describe fractal systems.

22.7.1 Superconductivity and macroquantum phenomena

Main article: Macroscopic quantum phenomena

The generalized Schrödinger equation, under certain conditions, can apply to macroscopic scales. This leads to the proposal that quantum-like phenomena need not to be only at quantum scales. In a recent paper, Turner and Nottale[113] proposed new ways to explore the origins of macroscopic quantum coherence in high temperature super conductivity.

22.7.2 Morphogenesis

If we assume that morphologies come from a growth process, we can model this growth as an infinite family of virtual, fractal, and locally irreversible trajectories. This allows to write a growth equation in a form which can be integrated into a Schrödinger-like equation.

The structuring implied by such a generalized Schrödinger equation provides a new basis to study, with a purely energetic approach, the issues of formation, duplication, bifurcation and hierarchical organization of structures.[114]

An inspiring example is the solution describing growth from a center, which bears similarities with the problem of particle scattering in quantum mechanics. Searching for some of the simplest solutions (with a central potential and a spherical symmetry), a solution leads to a flower shape, the common Platycodon flower (see Fig. 2). In honor to Erwin Schrödinger, Nottale, Chaline and Grou named their book "Flowers for Schrödinger" (Des fleurs pour Schrödinger[115]).

22.7.3 Biology

In a short paper,[116] researchers inspired by scale relativity proposed a log-periodic law for the development of the human embryo, which fits pretty well with the steps of the human embryo development.

With scale-relativistic models, Nottale and Auffray did tackle the issue of multiple-scale integration in systems biology.[117][118]

Other studies suggest that many living systems processes, because embedded in a fractal medium, are expected to display wave-like and quantized structuration.[119]

Fig. 2. Schrödinger's flower. Morphogenesis of a flower-like structure, solution of a growth process equation that takes the form of a Schrödinger equation under fractal conditions. For more details, see: Nottale 2007.

22.7.4 Geography

The mathematical tools of scale relativity have also been applied to geographical problems.[120][121]

22.7.5 Singularity and evolutionary trees

In their review of approaches to technological singularities, Magee and Devezas[122] included the work of Nottale, Chaline and Grou:[123]

"Utilizing the fractal mathematics due to Mandlebrot (1983) these authors develop a model based upon a fractal tree of the time sequences of major evolutionary leaps at various scales (log-periodic law of acceleration – deceleration). The application of the model to the evolution of western civilization shows evidence of an acceleration in the succession (pattern) of economic crisis/non-crisis, which point to a next crisis in the period 2015–2020, with a critical point $T_c = 2080$. The meaning of T_c in this approach is the limit of the evolutionary capacity of the analyzed group and is biologically analogous with the end of a species and emergence of a new species."

The interpretation of this emergence of a new species remains open to debate, whether it will take the form of the emergence of transhumans, cyborgs, superintelligent AI, or a global brain.

22.8 Reception and critique

22.8.1 Scale relativity and other approaches

It may help to understand scale relativity by comparing it to various other approaches to unifying quantum and classical theories.

String theory

Main article: String theory

Although string theory and scale relativity start from different assumptions to tackle the issue of reconciling quantum mechanics and relativity theory, the two approaches need not to be opposed. Indeed, Castro[124] suggested to combine string theory with the principle of scale relativity:

> "It was emphasized by Nottale in his book that a full motion plus scale relativity including all spacetime components, angles and rotations remains to be constructed. In particular the general theory of scale relativity. Our aim is to show that string theory provides an important step in that direction and vice versa: the scale relativity principle must be operating in string theory."

Quantum gravity

Main article: quantum gravity

Scale relativity is based on a geometrical approach, and thereby recovers the quantum laws, instead of assuming them. This distinguishes it from other quantum gravity approaches. Nottale comments:[125]

> "The main difference is that these quantum gravity studies assume the quantum laws to be set as fundamental laws. In such a framework, the fractal geometry of space-time at the Planck scale is a consequence of the quantum nature of physical laws, so that the fractality and the quantum nature co-exist as two different things. In the scale relativity theory, there are not two things (in analogy with Einstein's general relativity theory in which gravitation is a manifestation of the curvature of space-time): the quantum laws are considered as manifestations of the fractality and nondifferentiability of space-time, so that they do not have to be added to the geometric description."

Loop quantum gravity

Main article: Loop quantum gravity

They have in common to start from relativity theory and principles, and to fulfill the condition of background independence.

El Naschie's E-Infinity theory

El Naschie has developed a similar, yet different fractal space-time theory, because he gives up differentiability *and* continuity. El Naschie thus uses a "Cantorian" space-time, and uses mostly number theory (see[126]). This is to be contrasted with scale relativity, which keeps the hypothesis of continuity, and thus works preferentially with mathematical analysis and fractals.

Causal dynamical triangulation

Main article: Causal dynamical triangulation

Through computer simulations of causal dynamical triangulation theory, a fractal to nonfractal transition was found from quantum scales to larger scales.[127] This result seems to be compatible with quantum-classical transition deduced in an other way, from the theoretical framework of scale relativity.

Noncommutative geometry

Main article: Noncommutative geometry

For both scale relativity and non-commutative geometries, particles are geometric properties of space-time. The intersection of both theories seems fruitful and still to be explored. In particular, Nottale[128] further generalized this non-communicativity, saying that it "is now at the level of the fractal space-time itself, which therefore fundamentally comes under Connes's noncommutative geometry.[129][130] Moreover, this noncommutativity might be considered as a key for a future better understanding of the parity and CP violations, which will not be developed here."

Doubly special relativity

Main article: Doubly special relativity

Both theories have identified the Planck length as a fundamental minimum scale. However, as Nottale comments:[131]

> "the main difference between the "Doubly-Special-Relativity" approach and the scale relativity one is that we have identified the question of defining an invariant length-scale as coming under a relativity of scales. Therefore the new group to be constructed is a multiplicative group that becomes additive only when working with the logarithms of scale ratios, which are definitely the physically relevant scale variables, as we have shown by applying the Gell-Mann–Levy method to the construction of the dilation operator (see Sec. 4.2.1)."

22.8.2   Cognitive aspects

Special and general relativity theory are notoriously hard to understand for non-specialists. This is partly because our psychological and sociological use of the concepts of space and time are not the same as the one in physics. Yet, the relativity of scales is still harder to apprehend than other relativity theories. Indeed humans can change their positions and velocities but have virtually no experience of shrinking or dilating themselves.

Such transformations appear in fiction however, such as in *Alice's Adventures in Wonderland* or in the movie *Honey, I Shrunk the Kids*.

22.8.3   Sociological analysis

Sociologists Bontems and Gingras did a detailed bibliometrical analysis of scale relativity and showed the difficulty for such a theory with a different theoretical starting point to compete with well-established paradigms such as String theory.[132]

Back in 2007, they considered the theory to be neither mainstream, that is, there are not many people working on it compared to other paradigms; but also neither controversial, as there is very little informed and academic discussion around the theory. The two sociologists thus qualified the theory as "marginal", in the sense that the theory is developed inside academics, but is not controversial.

They also show that Nottale has a double career. First, a classical one, working on gravitational lensing, and a second one, about scale relativity. Nottale first secured his scientific reputation with important publications about gravitational lensing,[133][134] then obtained a stable academic position, giving him more freedom to explore the foundations of spacetime and quantum mechanics.

A possible obstacle to the growth in popularity of scale relativity is that fractal geometries necessary to deal with special and general scale relativity are less well known and developed mathematically than the simple and well-known self-similar fractals. This technical difficulty may make the advanced concepts of the theory harder to learn. Physicists interested in scale relativity need to invest some time into understanding fractal geometries. The situation is similar to the need to learn non-euclidian geometries in order to work with Einstein's general relativity.[135] Similarly, the generality and transdiciplinary nature of the theory also made Auffray and Noble comment:[136] "The scale relativity theory and tools extend the scope of current domain-specific theories, which are naturally recovered, not replaced, in the new framework. This may explain why the community of physicists has been slow to recognize its potential and even to challenge it."

Nottale's popular book,[137] written in French, has been compared with Einstein's popular book[138] "Relativity: The Special and the General Theory". A future translation of this book from French into English might help the popularization of the theory.

22.8.4   Reactions

The reactions from scientists to scale relativity are generally positive. For example, Baryshev and Teerikorpi write:[139]

> "Though Nottale's theory is still developing and not yet a generally accepted part of physics, there are already many exciting views and predictions surfacing from the new formalism. It is concerned in particular with the frontier domains of modern physics, i.e. small length- and time-scales (microworld, elementary particles), large length-scales (cosmology), and long time-scales."

Regarding the predictions of planetary spacings, Potter and Jargodzki commented:[140]

> In the 1990s, applying chaos theory to gravitationally bound systems, L. Nottale found that statistical fits indicate that the planet orbital distances, including that of Pluto, and the major satellites of the Jovian planets, follow a numerical scheme with their orbital radii proportional to the squares of integers n^2 extremely well!

Auffray and Noble gave an overview:[141]

> Scale relativity has implications for every aspect of physics, from elementary particle physics

to astrophysics and cosmology. It provides numerous examples of theoretical predictions of standard model parameters, a theoretical expectation for the Higgs boson mass which will be potentially assessed in the coming years by the Large Hadron Collider, and a prediction of the cosmological constant which remains within the range of increasingly refined observational data. Strikingly, many predictions in astrophysics have already been validated through observations such as the distribution of exoplanets or the formation of extragalactic structures.

Although many applications have led to validated predictions (see above), Peter[142] criticized a provisionally estimated value of the Higgs boson made by Nottale in his 2011 book:[143]

"a prediction for the Higgs boson that should have been observed at mH ≃113.7GeV...it would appear, according to the book itself, that the theory it describes would be already ruled out by LHC data!"

However, this prediction was initially made at a time when the Higgs boson mass was totally unknown.[144] Additionally, the prediction does not rely on scale relativity itself, but on a new suggested form of the electroweak theory. The final LHC result is mH=125.6 ± 0.3 GeV,[145] and lies therefore at about 110% of this early estimate.

Particle physicist and skeptic Victor Stenger also noticed that the theory "predicts a nonzero value of the cosmological constant in the right ballpark".[146] He also acknowledged that the theory "makes a number of other remarkable predictions".

22.9 See also

- Fractal cosmology
- General relativity
- Special relativity
- Galilean invariance
- Multifractal system

22.10 References

Abbott, Laurence F., and Mark B. Wise. 1981. "Dimension of a Quantum-Mechanical Path." *American Journal of Physics* 49 (1): 37–39. http://www.columbia.edu/cu/neurotheory/Larry/AbbottAmJPhys81.pdf

Allen, A. D. 1983. "Fractals and Quantum Mechanics." *Speculations in Science and Technology* 6 (2): 165–70.

Anz, Meador P. 2000. "A Decade of Growth." *The Orbital Debris Quarterly News*, NASA JSC, 5 (4): 1–2. http://orbitaldebris.jsc.nasa.gov/newsletter/pdfs/ODQNv5i4.pdf.

Auffray, C., and D. Noble. 2010. "Scale Relativity: An Extended Paradigm for Physics and Biology?" *Foundations of Science* 16 (4): 303–5. doi:10.1007/s10699-010-9203-x. http://www.arxiv.org/abs/0912.5508.

Auffray, C., and L. Nottale. 2008. "Scale Relativity Theory and Integrative Systems Biology: 1 Founding Principles and Scale Laws." *Progress in Biophysics and Molecular Biology* 97 (1): 79–114. http://luth2.obspm.fr/~{}luthier/nottale/arPBMB08AN.pdf.

Baryshev, Yurij, and Pekka Teerikorpi. 2002. *Discovery of Cosmic Fractals*. River Edge, N.J: World Scientific.

Ben Adda, Fayçal, and Jacky Cresson. 2004. "Quantum Derivatives and the Schrödinger Equation." *Chaos, Solitons & Fractals* 19 (5): 1323–34. doi:10.1016/S0960-0779(03)00339-4.

———. 2005. "Fractional Differential Equations and the Schrödinger Equation." *Applied Mathematics and Computation* 161 (1): 323–45. doi:10.1016/j.amc.2003.12.031.

Beringer, J., et. al., and (Particle Data Group). 2013. "Status of Higgs Boson Physics." *PR D86*, no. 010001. http://pdg.lbl.gov/2013/reviews/rpp2013-rev-higgs-boson.pdf.

Bontems, Vincent, and Yves Gingras. 2007. "De La Science Normale a La Science Marginale. Analyse D'une Bifurcation de Trajectoire Scientifique: Le Cas de La Theorie de La Relativité d'Echelle." *Social Science Information* 46 (4): 607–53. doi:10.1177/0539018407082595. http://ssi.sagepub.com/cgi/content/abstract/46/4/607.

Campesino-Romeo, E., J. C. D'Olivo, and M. Socolovsky. 1982. "Hausdorff Dimension for the Quantum Harmonic Oscillator." *Physics Letters A* 89 (7): 321–24. doi:10.1016/0375-9601(82)90182-7.

Cash, R., J. Chaline, L. Nottale, and P. Grou. 2002. "Développement Humain et Loi Log-Périodique." *Comptes Rendus Biologies* 325 (5): 585–90. doi:10.1016/S1631-0691(02)01468-3. http://luth2.obspm.fr/~{}luthier/nottale/arcash.pdf.

Castro, Carlos. 1997. "String Theory, Scale Relativity and the Generalized Uncertainty Principle." *Foundations of Physics Letters* 10 (3): 273–93. doi:10.1007/BF02764209. http://link.springer.com/article/10.1007%2FBF02764209.

Célérier, Marie-Noëlle, and Laurent Nottale. 2003. "A Scale-Relativistic Derivation of the Dirac Equation." *Electromagnetic Phenomena* 3: 70–80. http://arxiv.org/abs/hep-th/0210027.

———. 2004. "Quantum–classical Transition in Scale Relativity." *Journal of Physics A: Mathematical and General* 37 (3): 931. doi:10.1088/0305-4470/37/3/026. http://arxiv.org/abs/quant-ph/0609161.

———. 2010. "Electromagnetic Klein–Gordon and Dirac Equations in Scale Relativity." *International Journal of Modern Physics A* 25 (22): 4239–53. doi:10.1142/S0217751X10050615. http://arxiv.org/abs/1009.2934.

Connes, Alain. 1994. *Noncommutative Geometry.* San Diego: Academic Press.

Cresson, Jacky. 2003. "Scale Calculus and the Schrödinger Equation." *Journal of Mathematical Physics* 44 (11): 4907–38. doi:10.1063/1.1618923. http://scitation.aip.org/content/aip/journal/jmp/44/11/10.1063/1.1618923.

Da Rocha, D. 2004. "Structuration gravitationnelle en Relativité d'échelle." Phdthesis, Observatoire de Paris. https://tel.archives-ouvertes.fr/tel-00010204/document.

Da Rocha, D., and L Nottale. 2003. "Gravitational Structure Formation in Scale Relativity." *Chaos, Solitons & Fractals* 16 (4): 565–95. doi:10.1016/S0960-0779(02)00223-0. http://arxiv.org/abs/astro-ph/0310036.

Dubois, Daniel M. 2000. "Computational Derivation of Quantum Relativist Electromagnetic Systems with Forward-Backward Space-Time Shifts." In *AIP Conference Proceedings*, 517:417–29. AIP Publishing. doi:10.1063/1.1291279. http://scitation.aip.org/content/aip/proceeding/aipcp/10.1063/1.1291279.

Dubrulle, B. 2000. "Finite Size Scale Invariance." *The European Physical Journal B - Condensed Matter and Complex Systems* 14 (4): 757–71. doi:10.1007/s100510051087. http://link.springer.com/article/10.1007/s100510051087.

Feynman, Richard P., and Albert R. Hibbs. 1965. *Quantum Mechanics and Path Integrals.* International Series in Pure and Applied Physics. New York: McGraw-Hill.

Forriez, M., P. Martin, and L. Nottale. 2010. "Lois d'échelle et transitions fractal-non fractal en géographie." *L'Espace géographique* Vol. 39 (2): 97–112. http://www.cairn.info/resume.php?ID_ARTICLE=EG_392_0097.

Galileo, Galilei. 1991. *Dialogues Concerning Two New Sciences.* Great Minds Series. Buffalo, N.Y: Prometheus Books.

Gell-Mann, M., and M. Lévy. 1960. "The Axial Vector Current in Beta Decay." *Il Nuovo Cimento* 16 (4): 705–26. doi:10.1007/BF02859738.

Groom, D. E., M. Aguillar-Benitez, C. Amsler, and et. al. 2000. "Review of Particle Physics. Particle Data Group." http://inis.iaea.org/Search/search.aspx?orig_q=RN:31041263.

Herrmann, Felix Johan. 1997. "A Scaling Medium Representation: A Discussion on Well-Logs, Fractals and Waves." PhD thesis, TU Delft, Delft University of Technology. http://repository.tudelft.nl/assets/uuid:5423a23f-814e-4260-ba9b-f60e16cd2f14/as_herrmann_1997.pdf.

Jumarie, G. 2001. "Schrödinger Equation for Quantum Fractal Space–Time of Order N Via the Complex-Valued Fractional Brownian Motion." *International Journal of Modern Physics A* 16 (31): 5061–84. doi:10.1142/S0217751X01005468. http://www.worldscientific.com/doi/abs/10.1142/S0217751X01005468.

———. 2006. "Modified Riemann-Liouville Derivative and Fractional Taylor Series of Nondifferentiable Functions Further Results." *Computers & Mathematics with Applications* 51 (9–10): 1367–76. doi:10.1016/j.camwa.2006.02.001. http://www.sciencedirect.com/science/article/pii/S0898122106000861.

———. 2007. "The Minkowski's Space–time Is Consistent with Differential Geometry of Fractional Order." *Physics Letters A* 363 (1–2): 5–11. doi:10.1016/j.physleta.2006.10.085. http://www.sciencedirect.com/science/article/pii/S0375960106017038.

Karoji, H., and L. Nottale. 1976. "Possible Implications of the Rubin-Ford Effect." *Nature* 259 (5538): 31–33. doi:10.1038/259031a0. http://dx.doi.org/10.1038/259031a0.

Konacki, Maciej, and Alex Wolszczan. 2003. "Masses and Orbital Inclinations of Planets in the PSR B1257+12 System." *The Astrophysical Journal Letters* 591 (2): L147. doi:10.1086/377093. http://iopscience.iop.org/1538-4357/591/2/L147.

Kröger, H. 1997. "Proposal for an Experiment to Measure the Hausdorff Dimension of Quantum-Mechanical Trajectories." *Physical Review A* 55 (2): 951–66. doi:10.1103/PhysRevA.55.951.

Lapidus, Michel L. 2008. *In Search of the Riemann Zeros: Strings, Fractal Membranes and Noncommutative Spacetimes.* Providence, R.I: American Mathematical Society.

Loll, R. 2008. "The Emergence of Spacetime or Quantum Gravity on Your Desktop." *Classical and Quantum Gravity* 25 (11): 114006. doi:10.1088/0264-9381/25/11/114006. http://arxiv.org/abs/0711.0273.

Magee, Christopher L., and Tessaleno C. Devezas. 2011. "How Many Singularities Are Near and How Will They Disrupt Human History?" *Technological Forecasting and Social Change* 78 (8): 1365–78. doi:10.1016/

j.techfore.2011.07.013. http://web.mit.edu/cmagee/www/documents/29-singularitysdarticle.pdf.

Mandelbrot, Benoit B. 1983. *The Fractal Geometry of Nature*. Macmillan.

Merker, Joël. 1999. "Deux infinis cousus main." *Revue de synthèse* 120 (1): 165–74. doi:10.1007/BF03182083. http://link.springer.com/article/10.1007/BF03182083.

Müller, Xavier. 2005. "La Relativité d'Echelle - Et Si Le Monde N'était Qu'un Océan Fractal ?" *Science et Vie*.

Nottale, L. 1989. "Fractals and the Quantum Theory of Spacetime." *International Journal of Modern Physics A* 04 (19): 5047–5117. doi:10.1142/S0217751X89002156. http://www.worldscientific.com/doi/abs/10.1142/S0217751X89002156.

———. 1992. "The Theory of Scale Relativity." *International Journal of Modern Physics A* 7 (20): 4899–4936. doi:10.1142/S0217751X92002222. http://luth.obspm.fr/~{}luthier/nottale/arIJMP2.pdf.

———. 1993a. *Fractal Space-Time and Microphysics: Towards a Theory of Scale Relativity*. World Scientific. http://luth2.obspm.fr/~{}luthier/nottale/LIWOS7-1cor.pdf.

———. 1993b. "Mach's Principle, Dirac's Large Numbers and the Cosmological Constant Problem." Working Paper. http://luth2.obspm.fr/~{}luthier/nottale/arlambda.pdf.

———. 1994. "Scale Relativity: First Steps Toward a Field Theory." In *Relativity in General*, 1:121. http://luth2.obspm.fr/~{}luthier/nottale/arSalas.pdf.

———. 1996a. "Scale Relativity and Fractal Space-Time: Applications to Quantum Physics, Cosmology and Chaotic Systems." *Chaos, Solitons & Fractals* 7 (6): 877–938. doi:10.1016/0960-0779(96)00002-1. http://luth2.obspm.fr/~{}luthier/nottale/arRevFST.pdf.

———. 1996b. "Scale-Relativity and Quantization of Extra-Solar Planetary Systems." *Astronomy and Astrophysics* 315: L9–12. http://luth2.obspm.fr/~{}luthier/nottale/arA&A315.pdf.

———. 1997. "Scale Relativity." In *Scale Invariance and Beyond*, edited by B. Dubrulle, F. Graner, and D. Sornette. Berlin, Heidelberg: Springer Berlin Heidelberg. http://luth2.obspm.fr/~{}luthier/nottale/arhouche.pdf.

———. 1998a. *La Relativité Dans Tous Ses États: Au Delà de l'Espace-Temps*. Hachette.

———. 1998b. "Scale Relativity and Quantization of the Planetary System around the Pulsar PSR B1257 + 12." *Chaos, Solitons & Fractals* 9 (7): 1043–50. doi:10.1016/S0960-0779(97)00079-9. http://luth2.obspm.fr/~{}luthier/nottale/arPSR.pdf.

———. 1998c. "Scale Relativity and Quantization of Planet Obliquities." *Chaos, Solitons & Fractals* 9 (7): 1035–41. doi:10.1016/S0960-0779(97)00078-7. http://luth2.obspm.fr/~{}luthier/nottale/arobliq.pdf.

———. 2001. "Scale Relativity and Non-Differentiable Fractal Space-Time." In *Frontiers of Fundamental Physics 4*, edited by B. G. Sidharth and M. V. Altaisky, 65–79. Springer US. http://luth.obspm.fr/%7Eluthier/nottale/arBirla00.pdf.

———. 2003a. "La relativité d'échelle à l'épreuve des faits." *Pour la Science*. http://www.pourlascience.fr/ewb_pages/a/article-la-relativite-d-echelle-a-l-epreuve-des-faits-24365.php.

———. 2003b. "Scale-Relativistic Cosmology." *Chaos, Solitons and Fractals* 16 (4): 539–64. http://luth2.obspm.fr/~{}luthier/nottale/arScRCosmo.pdf.

———. 2004a. *Relativité D'échelle et Structuration de L'univers*. https://www.youtube.com/watch?v=JwaKLTitwtc.

———. 2004b. "The Theory of Scale Relativity: Non-Differentiable Geometry and Fractal Space-Time." In *AIP Conference Proceedings*, 718:68–95. AIP Publishing. doi:10.1063/1.1787313. http://luth2.obspm.fr/~{}luthier/nottale/arcasys03.pdf.

———. 2007. "Scale Relativity: A Fractal Matrix for Organization in Nature." *Electronic Journal of Theoretical Physics* 4 (16): 187–274. http://www.luth.obspm.fr/%7Eluthier/nottale/arEJTP.pdf.

———. 2010. "Scale Relativity and Fractal Space-Time: Theory and Applications." *Foundations of Science* 15 (2): 101–52. doi:10.1007/s10699-010-9170-2. http://www.arxiv.org/abs/0912.5508.

———. 2011. *Scale Relativity and Fractal Space-Time: A New Approach to Unifying Relativity and Quantum Mechanics*. World Scientific Publishing Company.

———. 2012. "Nature et Valeur de La Constante Cosmologique." In *Premières Rencontres d'Avignon (2007-2009) Autour de La Relativité d'Echelle*, edited by L. Nottale and P. Martin, 31–39. Actes d'Avignon. Avignon. http://luth.obspm.fr/%7Eluthier/nottale/arAvLambda.pdf.

———. 2013a. *Laurent Nottale : La Tête dans les Étoiles* Interview by Sophie Sendra. http://www.bscnews.fr/201310283220/Exclusivites/laurent-nottale-la-tete-dans-les-etoiles.html.

———. 2013b. "Macroscopic Quantum-Type Potentials in Theoretical Systems Biology." *Cells* 3 (1): 1–35. doi:10.3390/cells3010001. http://www.ncbi.nlm.nih.gov/pmc/articles/PMC3980741/.

Nottale, L., and C. Auffray. 2008. "Scale Relativity Theory

and Integrative Systems Biology: 2 Macroscopic Quantum-Type Mechanics." *Progress in Biophysics and Molecular Biology* 97 (1): 115–57. http://luth2.obspm.fr/~{}luthier/nottale/arPBMB08NA.pdf.

Nottale, L., and M. N. Célérier. 2007. "Derivation of the Postulates of Quantum Mechanics from the First Principles of Scale Relativity." *Journal of Physics A-Mathematical and Theoretical* 40 (48): 14471–98. doi:10.1088/1751-8113/40/48/012. http://arxiv.org/abs/0711.2418.

Nottale, L., M.N. Célérier, and T. Lehner. 2006. "Non-Abelian Gauge Field Theory in Scale Relativity." *Journal of Mathematical Physics* 47 (3): 032303. http://arxiv.org/abs/hep-th/0605280.

Nottale, L., J. Chaline, and P. Grou. 2000. *Les Arbres de L'évolution: Univers, Vie, Sociétés.* Hachette littératures.

———. 2002. "On the Fractal Structure of Evolutionary Trees." In *Fractals in Biology and Medicine*, edited by Gabriele A. Losa, Theo F. Nonnenmacher, Danilo Merlini, and Ewald R. Weibel, 247–58. Birkhäuser. http://luth2.obspm.fr/%7Eluthier/nottale/arbiomed.pdf.

———. 2009. *Des Fleurs Pour Schrödinger: La Relativité d'Echelle Et Ses Applications.* Ellipses.

Nottale, L., P. Martin, and M. Forriez. 2012. "Analyse En Relativité D'échelle Du Bassin Versant Du Gardon (Gard, France)." In *Premières Rencontres d'Avignon (2007-2009) Autour de La Relativité d'Echelle*, edited by L. Nottale and P. Martin, 267–95. Actes d'Avignon. Avignon. http://luth.obspm.fr/%7Eluthier/nottale/arAvGardon.pdf.

Nottale, L., and J. Schneider. 1984. "Fractals and Nonstandard Analysis." *Journal of Mathematical Physics* 25 (5): 1296–1300. http://luth2.obspm.fr/~{}luthier/nottale/arFNSAtex.pdf.

Nottale, L., G. Schumacher, and J. Gay. 1997. "Scale Relativity and Quantization of the Solar System." *Astronomy and Astrophysics* 322: 1018–25. http://luth2.obspm.fr/~{}luthier/nottale/arA&A322.pdf.

Nottale, L., G. Schumacher, and E. T. Lefevre. 2000. "Scale-Relativity and Quantization of Exoplanet Orbital Semi-Major Axes." *Astronomy and Astrophysics* 361: 379–87. http://luth2.obspm.fr/~{}luthier/nottale/arA&A361.pdf.

Nottale, L., and J. P. Vigier. 1977. "Continuous Increase of Hubble Modulus behind Clusters of Galaxies." *Nature* 268 (5621): 608–10. doi:10.1038/268608a0. http://dx.doi.org/10.1038/268608a0.

Ord, G. N. 1983. "Fractal Space-Time: A Geometric Analogue of Relativistic Quantum Mechanics." *Journal of Physics A: Mathematical and General* 16 (9): 1869. doi:10.1088/0305-4470/16/9/012. http://iopscience.iop.org/0305-4470/16/9/012.

Peter, Patrick. 2013. "Laurent Nottale: Scale Relativity and Fractal Space-Time." *General Relativity and Gravitation* 45 (7): 1459–61. doi:10.1007/s10714-013-1535-8.

Planck Collaboration, P. A. R. Ade, N. Aghanim, C. Armitage-Caplan, M. Arnaud, M. Ashdown, F. Atrio-Barandela, et al. 2013. "Planck 2013 Results. XVI. Cosmological Parameters." *arXiv:1303.5076 [astro-Ph]*, March. http://arxiv.org/abs/1303.5076.

Potter, Frank, and Christopher Jargodzki. 2005. *Mad About Modern Physics: Braintwisters, Paradoxes, and Curiosities.* Hoboken, N.J: J. Wiley.

Sidharth, B. G. 2001. *Chaotic Universe: From Planck to the Hubble Scale.* Contemporary Fundamental Physics. Huntington, N.Y: Nova Science Publishers.

Stenger, V. J. 2011. *The Fallacy of Fine-Tuning: Why the Universe Is Not Designed for Us.* Prometheus Books.

Turner, Philip. 2013. "A Review of Scale Relativity and Fractal Space-Time." http://www.researchgate.net/publication/259373681_A_Review_of_Scale_Relativity_and_fractal_space-time.

Turner, Philip, and L. Nottale. 2015. "The Origins of Macroscopic Quantum Coherence in High Temperature Super Conductivity." *Physica C: Superconductivity and Its Applications.* doi:10.1016/j.physc.2015.04.006. http://arxiv.org/abs/1410.3659.

Vidal, C. 2010. "Introduction to the Special Issue on the Evolution and Development of the Universe." *Foundations of Science* 15 (2): 95–99. doi:10.1007/s10699-010-9176-9. http://www.arxiv.org/abs/0912.5508.

Wolszczan, A., and D. A. Frail. 1992. "A Planetary System Around the Millisecond Pulsar PSR1257 + 12." *Nature* 355 (6356): 145–47. doi:10.1038/355145a0.

Yao, W.-M., and et. al. 2006. "Review of Particle Physics." *Journal of Physics G: Nuclear and Particle Physics* 33 (1): 1. doi:10.1088/0954-3899/33/1/001. http://iopscience.iop.org/0954-3899/33/1/001.

22.11 Notes

[1] Ord, G. N. (1983). "Fractal space-time: a geometric analogue of relativistic quantum mechanics". *Journal of Physics A: Mathematical and General* **16** (9): 1869. doi:10.1088/0305-4470/16/9/012. ISSN 0305-4470. Retrieved 2015-03-22.

[2] Nottale, L.; Schneider, J. (1984). "Fractals and Nonstandard Analysis" (PDF). *Journal of mathematical physics* **25** (5): 1296–1300. Retrieved 2015-03-10.

[3] Nottale, L. (1989). "Fractals and the quantum theory of spacetime". *International Journal of Modern Physics A* **4** (19): 5047–5117. doi:10.1142/S0217751X89002156. ISSN 0217-751X. Retrieved 2015-03-10.

[4] Nottale, L. (1992). "The Theory of Scale Relativity" (PDF). *International Journal of Modern Physics A* **7** (20): 4899–4936. doi:10.1142/S0217751X92002222. Retrieved 2015-03-10.

[5] Feynman, Richard P.; Hibbs, Albert R. (1965). *Quantum Mechanics and Path Integrals*. International series in pure and applied physics. New York: McGraw-Hill.

[6] Müller, Xavier (2005). "La Relativité d'Echelle - Et si le monde n'était qu'un océan fractal ?". *Science et Vie* **1051**. pp. 70–73.; p. 71

[7] Abbott, Laurence F.; Wise, Mark B. (1981). "Dimension of a Quantum-Mechanical Path" (PDF). *American Journal of Physics* **49** (1): 37–39. Retrieved 2015-04-12.

[8] Campesino-Romeo, E.; D'Olivo, J. C.; Socolovsky, M. (1982). "Hausdorff dimension for the quantum harmonic oscillator". *Physics Letters A* **89** (7): 321–324. doi:10.1016/0375-9601(82)90182-7. ISSN 0375-9601.

[9] Allen, A. D. (1983). "Fractals and Quantum Mechanics". *Speculations in Science and Technology* **6** (2): 165–170. ISSN 0155-7785.

[10] Ord, G. N. (1983). "Fractal space-time: a geometric analogue of relativistic quantum mechanics". *Journal of Physics A: Mathematical and General* **16** (9): 1869. doi:10.1088/0305-4470/16/9/012. ISSN 0305-4470. Retrieved 2015-03-22.

[11] Nottale, L.; Schneider, J. (1984). "Fractals and Nonstandard Analysis" (PDF). *Journal of mathematical physics* **25** (5): 1296–1300. Retrieved 2015-03-10.

[12] Nottale, L. (1992). "The Theory of Scale Relativity" (PDF). *International Journal of Modern Physics A* **7** (20): 4899–4936. doi:10.1142/S0217751X92002222. Retrieved 2015-03-10.

[13] Nottale, L. (1993). *Fractal Space-Time and Microphysics: Towards a Theory of Scale Relativity*. World Scientific.

[14] Nottale, L. (2011). *Scale Relativity And Fractal Space-Time: A New Approach to Unifying Relativity and Quantum Mechanics*. World Scientific Publishing Company. ISBN 1848166508.

[15] Nottale, Laurent (1998-12-24). *L'Univers et la Lumière. Cosmologie Classique Et Mirages Gravitationnels*. Flammarion. ISBN 2080813838.

[16] Nottale, L.; Chaline, J.; Grou, P. (2000). *Les Arbres de L'évolution: Univers, Vie, Sociétés*. Hachette littératures.

[17] Nottale, L.; Chaline, J.; Grou, P. (2009). *Des Fleurs pour Schrödinger: La Relativité d'Echelle Et Ses Applications*. Ellipses. ISBN 978-2-7298-5182-8.

[18] Nottale, L. (2011). *Scale Relativity and Fractal Space-Time: A New Approach to Unifying Relativity and Quantum Mechanics*. World Scientific Publishing Company. ISBN 1848166508.;p. 8

[19] Nottale, L. (1998). *La Relativité dans tous ses États: au delà de l'Espace-Temps*. Hachette.;p. 218

[20] Mandelbrot, Benoit B. (1983). *The Fractal Geometry of Nature*. Macmillan.

[21] Nottale, L. (Director) (2004). *Relativité d'échelle et structuration de l'univers*. Event occurs at 5667 seconds. Retrieved 2015-04-12.

[22] Nottale, L. (1993). *Fractal Space-Time and Microphysics: Towards a Theory of Scale Relativity*. World Scientific.; p.82

[23] Nottale, L. (1993). *Fractal Space-Time and Microphysics: Towards a Theory of Scale Relativity*. World Scientific.; p.84

[24] Nottale, L. (1998). *La Relativité dans tous ses États: au delà de l'Espace-Temps*. Hachette.; p.188

[25] Nottale, L. (2013). *Laurent Nottale : La Tête dans les Étoiles*. Interview with Sophie Sendra. Retrieved 2013-11-13.

[26] Nottale, L. (1993). *Fractal Space-Time and Microphysics: Towards a Theory of Scale Relativity*. World Scientific.; p.304

[27] Nottale, L. (1996). "Scale Relativity and Fractal Space-Time: Applications to Quantum Physics, Cosmology and Chaotic Systems" (PDF). *Chaos, Solitons & Fractals* **7** (6): 877–938. doi:10.1016/0960-0779(96)00002-1. Retrieved 2015-03-10.;p. 915

[28] Nottale, L. (1998). *La Relativité dans tous ses États: au delà de l'Espace-Temps*. Hachette.; p.161

[29] Nottale, L. (1993). *Fractal Space-Time and Microphysics: Towards a Theory of Scale Relativity*. World Scientific.; p.299

[30] Nottale, L. (1996). "Scale Relativity and Fractal Space-Time: Applications to Quantum Physics, Cosmology and Chaotic Systems" (PDF). *Chaos, Solitons & Fractals* **7** (6): 877–938. doi:10.1016/0960-0779(96)00002-1. Retrieved 2015-03-10.

[31] Nottale, L. (2003). "Scale-Relativistic Cosmology". *Chaos, Solitons and Fractals* **16** (4): 539–564.

[32] Nottale, L. (2011). *Scale Relativity and Fractal Space-Time: A New Approach to Unifying Relativity and Quantum Mechanics*. World Scientific Publishing Company. ISBN 1848166508.; Chap. 12.6

[33] Galileo, Galilei (1991). *Dialogues Concerning Two New Sciences*. Great minds series. Buffalo, N.Y: Prometheus Books. ISBN 0879757078.

[34] Gell-Mann, M.; Lévy, M. (1960-05-01). "The Axial Vector Current in Beta Decay". *Il Nuovo Cimento* **16** (4): 705–726. doi:10.1007/BF02859738. ISSN 0029-6341.

[35] Nottale, L. (1993). *Fractal Space-Time and Microphysics: Towards a Theory of Scale Relativity*. World Scientific. ISBN 9810208782.

[36] Nottale, L. (1996). "Scale Relativity and Fractal Space-Time: Applications to Quantum Physics, Cosmology and Chaotic Systems" (PDF). *Chaos, Solitons & Fractals* **7** (6): 877–938. doi:10.1016/0960-0779(96)00002-1. Retrieved 2015-03-10.

[37] Nottale, L. (1992). "The Theory of Scale Relativity" (PDF). *International Journal of Modern Physics A* **7** (20): 4899–4936. doi:10.1142/S0217751X92002222. Retrieved 2015-03-10.

[38] Nottale, L. (1993). *Fractal Space-Time and Microphysics: Towards a Theory of Scale Relativity*. World Scientific. ISBN 9810208782.

[39] Nottale, L. (2011). *Scale Relativity and Fractal Space-Time: A New Approach to Unifying Relativity and Quantum Mechanics*. World Scientific Publishing Company. ISBN 1848166508.; p. 460

[40] Nottale, L. (1998). *La Relativité dans tous ses États: au delà de l'Espace-Temps*. Hachette.; p. 212

[41] Nottale, L. (1993). *Fractal Space-Time and Microphysics: Towards a Theory of Scale Relativity*. World Scientific. ISBN 9810208782.

[42] Nottale, L. (1997). "Scale Relativity" (PDF). In B. Dubrulle, F. Graner, D. Sornette (eds.). *Scale Invariance and Beyond*. Berlin, Heidelberg: Springer Berlin Heidelberg. ISBN 978-3-540-64000-4. Retrieved 2015-03-10.

[43] Nottale, L. (2004). "The Theory of Scale Relativity: Non-Differentiable Geometry and Fractal Space-Time" (PDF). *AIP Conference Proceedings*. COMPUTING ANTICIPATORY SYSTEMS: CASYS'03 - Sixth International Conference **718**. AIP Publishing. pp. 68–95. doi:10.1063/1.1787313. Retrieved 2015-04-13.

[44] Nottale, L. (1994). "Scale Relativity: First Steps Toward a Field Theory" (PDF). *Relativity in General* **1**. p. 121. Retrieved 2015-03-10.

[45] Nottale, L. (1996). "Scale Relativity and Fractal Space-Time: Applications to Quantum Physics, Cosmology and Chaotic Systems" (PDF). *Chaos, Solitons & Fractals* **7** (6): 877–938. doi:10.1016/0960-0779(96)00002-1. Retrieved 2015-03-10.

[46] Nottale, L.; Célérier, M.N.; Lehner, T. (2006). "Non-Abelian Gauge Field Theory in Scale Relativity". *Journal of mathematical physics* **47** (3): 032303. Retrieved 2015-03-10.

[47] Nottale, L. (2011). *Scale Relativity and Fractal Space-Time: A New Approach to Unifying Relativity and Quantum Mechanics*. World Scientific Publishing Company. ISBN 1848166508.

[48] Nottale, L. (1992). "The Theory of Scale Relativity" (PDF). *International Journal of Modern Physics A* **7** (20): 4899–4936. doi:10.1142/S0217751X92002222. Retrieved 2015-03-10.

[49] Forriez, M.; Martin, P.; Nottale, L. (2010-07-07). "Lois d'échelle et transitions fractal-non fractal en géographie". *L'Espace géographique* **39** (2): 97–112. ISSN 0046-2497. Retrieved 2015-04-10.

[50] Kröger, H. (1997). "Proposal for an experiment to measure the Hausdorff dimension of quantum-mechanical trajectories". *Physical Review A* **55** (2): 951–966. doi:10.1103/PhysRevA.55.951.

[51] Turner, Philip (2013). "A Review of Scale Relativity and fractal space-time".

[52] Célérier, Marie-Noëlle; Nottale, Laurent (2004). "Quantum–classical transition in scale relativity". *Journal of Physics A: Mathematical and General* **37** (3): 931. doi:10.1088/0305-4470/37/3/026. Retrieved 2015-03-10.

[53] Nottale, L.; Célérier, M. N. (2007). "Derivation of the Postulates of Quantum Mechanics from the First Principles of Scale Relativity". *Journal of Physics A-Mathematical and Theoretical* **40** (48): 14471–14498. doi:10.1088/1751-8113/40/48/012.

[54] Nottale, L. (1993). *Fractal Space-Time and Microphysics: Towards a Theory of Scale Relativity*. World Scientific. ISBN 9810208782.; sec. 5.6

[55] Dubois, Daniel M. (2000-05-26). "Computational Derivation of Quantum Relativist Electromagnetic Systems with Forward-Backward Space-Time Shifts". *AIP Conference Proceedings*. Computing Anticipatory Systems: CASYS'99 - Third International Conference **517**. AIP Publishing. pp. 417–429. doi:10.1063/1.1291279. Retrieved 2015-03-21.

[56] Jumarie, G. (2001-12-20). "Schrödinger Equation for Quantum Fractal Space–Time of Order n Via the Complex-Valued Fractional Brownian Motion". *International Journal of Modern Physics A* **16** (31): 5061–5084. doi:10.1142/S0217751X01005468. ISSN 0217-751X. Retrieved 2015-03-21.

[57] Cresson, Jacky (2003-11-01). "Scale Calculus and the Schrödinger Equation". *Journal of Mathematical Physics* **44** (11): 4907–4938. doi:10.1063/1.1618923. ISSN 0022-2488. Retrieved 2015-03-21.

[58] Ben Adda, Fayçal; Cresson, Jacky (2004). "Quantum Derivatives and the Schrödinger Equation". *Chaos, Solitons & Fractals* **19** (5): 1323–1334. doi:10.1016/S0960-0779(03)00339-4. ISSN 0960-0779.

[59] Ben Adda, Fayçal; Cresson, Jacky (2005). "Fractional differential equations and the Schrödinger equation". *Applied Mathematics and Computation* **161** (1): 323–345. doi:10.1016/j.amc.2003.12.031. ISSN 0096-3003.

[60] Jumarie, G. (2006-05). "Modified Riemann-Liouville derivative and fractional Taylor series of nondifferentiable functions further results". *Computers & Mathematics with Applications* **51** (9–10): 1367–1376. doi:10.1016/j.camwa.2006.02.001. ISSN 0898-1221. Retrieved 2015-03-21. Check date values in: |date= (help)

[61] Jumarie, G. (2007-03-19). "The Minkowski's space-time is consistent with differential geometry of fractional order". *Physics Letters A* **363** (1–2): 5–11. doi:10.1016/j.physleta.2006.10.085. ISSN 0375-9601. Retrieved 2015-03-21.

[62] Nottale, L. (2011). *Scale Relativity and Fractal Space-Time: A New Approach to Unifying Relativity and Quantum Mechanics.* World Scientific Publishing Company. ISBN 1848166508.; sec. 5.7.2

[63] Nottale, L. (2011). *Scale Relativity and Fractal Space-Time: A New Approach to Unifying Relativity and Quantum Mechanics.* World Scientific Publishing Company. ISBN 1848166508.; sec. 5.7.3

[64] Célérier, Marie-Noëlle; Nottale, Laurent (2003). "A Scale-Relativistic Derivation of the Dirac Equation". *Electromagnetic Phenomena* **3**: 70–80. Retrieved 2015-03-10.

[65] Célérier, Marie-Noëlle; Nottale, Laurent (2010). "Electromagnetic Klein–Gordon and Dirac Equations in Scale Relativity". *International Journal of Modern Physics A* **25** (22): 4239–4253. doi:10.1142/S0217751X10050615.

[66] Nottale, L.; Célérier, M. N. (2007). "Derivation of the Postulates of Quantum Mechanics from the First Principles of Scale Relativity". *Journal of Physics A-Mathematical and Theoretical* **40** (48): 14471–14498. doi:10.1088/1751-8113/40/48/012.

[67] Turner, Philip (2013). "A Review of Scale Relativity and fractal space-time".

[68] Nottale, L. (2011). *Scale Relativity and Fractal Space-Time: A New Approach to Unifying Relativity and Quantum Mechanics.* World Scientific Publishing Company. ISBN 1848166508.

[69] Nottale, L. (2011). *Scale Relativity and Fractal Space-Time: A New Approach to Unifying Relativity and Quantum Mechanics.* World Scientific Publishing Company. ISBN 1848166508.; p.297

[70] Nottale, L. (2011). *Scale Relativity and Fractal Space-Time: A New Approach to Unifying Relativity and Quantum Mechanics.* World Scientific Publishing Company. ISBN 1848166508.; p.490

[71] Yao, W.-M.; et. al. (2006-07-01). "Review of Particle Physics". *Journal of Physics G: Nuclear and Particle Physics* **33** (1): 1. doi:10.1088/0954-3899/33/1/001. ISSN 0954-3899. Retrieved 2015-04-16.

[72] Nottale, L. (2011). *Scale Relativity and Fractal Space-Time: A New Approach to Unifying Relativity and Quantum Mechanics.* World Scientific Publishing Company. ISBN 1848166508.; p.490

[73] Nottale, L. (2011). *Scale Relativity and Fractal Space-Time: A New Approach to Unifying Relativity and Quantum Mechanics.* World Scientific Publishing Company. ISBN 1848166508.; p.499

[74] Groom, D. E.; Aguillar-Benitez, M.; Amsler, C.; et. al. (2000). "Review of particle physics. Particle data group". Retrieved 2015-04-16.

[75] Nottale, L. (1992). "The Theory of Scale Relativity" (PDF). *International Journal of Modern Physics A* **7** (20): 4899–4936. doi:10.1142/S0217751X92002222. Retrieved 2015-03-10.

[76] Nottale, L. (2010). "Scale Relativity and Fractal Space-Time: Theory and Applications". *Foundations of Science* **15** (2): 101–152. doi:10.1007/s10699-010-9170-2. ISSN 1233-1821. Retrieved 2010-08-25.; p.123-124

[77] Yao, W.-M.; et. al. (2006-07-01). "Review of Particle Physics". *Journal of Physics G: Nuclear and Particle Physics* **33** (1): 1. doi:10.1088/0954-3899/33/1/001. ISSN 0954-3899. Retrieved 2015-04-16.

[78] Nottale, L. (2011). *Scale Relativity and Fractal Space-Time: A New Approach to Unifying Relativity and Quantum Mechanics.* World Scientific Publishing Company. ISBN 1848166508.; p.483

[79] Baryshev, Yurij; Teerikorpi, Pekka (2002). *Discovery of Cosmic Fractals.* River Edge, NJ: World Scientific. ISBN 9810248717.; p.256

[80] Nottale, L. (2011). *Scale Relativity and Fractal Space-Time: A New Approach to Unifying Relativity and Quantum Mechanics.* World Scientific Publishing Company. ISBN 1848166508.;p. 589

[81] da Rocha, D.; Nottale, L (2003). "Gravitational Structure Formation in Scale Relativity". *Chaos, Solitons & Fractals* **16** (4): 565–595. doi:10.1016/S0960-0779(02)00223-0. ISSN 0960-0779.; p. 577

[82] Anz, Meador P. (2000). "A Decade of Growth" (PDF). *The Orbital Debris Quarterly News.* NASA JSC **5** (4): 1–2.

[83] Nottale, L. (1993). *Fractal Space-Time and Microphysics: Towards a Theory of Scale Relativity.* World Scientific. ISBN 9810208782.;p.311-321

[84] Nottale, L.; Schumacher, G.; Gay, J. (1997). "Scale Relativity and Quantization of the Solar System" (PDF). *Astronomy and Astrophysics* **322**: 1018–1025. ISSN 0004-6361. Retrieved 2013-11-04.

[85] Nottale, L. (2011).   *Scale Relativity and Fractal Space-Time: A New Approach to Unifying Relativity and Quantum Mechanics.* World Scientific Publishing Company. ISBN 1848166508.:p. 559

[86] Wolszczan, A.; Frail, D. A. (1992). "A Planetary System Around the Millisecond Pulsar PSR1257 + 12". *Nature* **355** (6356): 145–147. doi:10.1038/355145a0.

[87] Nottale, L. (1998). "Scale relativity and quantization of the planetary system around the pulsar PSR B1257 + 12" (PDF). *Chaos, Solitons & Fractals* **9** (7): 1043–1050. doi:10.1016/S0960-0779(97)00079-9. ISSN 0960-0779. Retrieved 2015-04-06.

[88] Nottale, L. (2011).   *Scale Relativity and Fractal Space-Time: A New Approach to Unifying Relativity and Quantum Mechanics.* World Scientific Publishing Company. ISBN 1848166508.:p. 622

[89] Konacki, Maciej; Wolszczan, Alex (2003). "Masses and Orbital Inclinations of Planets in the PSR B1257+12 System". *The Astrophysical Journal Letters* **591** (2): –147. doi:10.1086/377093. ISSN 1538-4357. Retrieved 2015-04-06.

[90] Nottale, L. (2011).   *Scale Relativity and Fractal Space-Time: A New Approach to Unifying Relativity and Quantum Mechanics.* World Scientific Publishing Company. ISBN 1848166508.; sec. 13.5

[91] Nottale, L. (1996). "Scale-relativity and quantization of extra-solar planetary systems" (PDF). *Astronomy and Astrophysics* **315**: –9–L12. ISSN 0004-6361. Retrieved 2015-04-17.

[92] Nottale, L. (1998). "Scale-relativity and Quantization of Planet Obliquities" (PDF). *Chaos, Solitons & Fractals* **9** (7): 1035–1041. doi:10.1016/S0960-0779(97)00078-7. ISSN 0960-0779. Retrieved 2015-04-17.

[93] Nottale, L.; Schumacher, G.; Lefevre, E. T. (2000). "Scale-Relativity and Quantization of Exoplanet Orbital Semi-Major Axes" (PDF). *Astronomy and Astrophysics* **361**: 379–387.

[94] da Rocha, D. (2004-12-02). "Structuration gravitationnelle en Relativité d'échelle". Observatoire de Paris. Retrieved 2015-04-10.

[95] Nottale, L. (2011).   *Scale Relativity and Fractal Space-Time: A New Approach to Unifying Relativity and Quantum Mechanics.* World Scientific Publishing Company. ISBN 1848166508.;sec.13.8

[96] Nottale, L. (2011).   *Scale Relativity and Fractal Space-Time: A New Approach to Unifying Relativity and Quantum Mechanics.* World Scientific Publishing Company. ISBN 1848166508.:p. 520

[97] Nottale, L. (2011).   *Scale Relativity and Fractal Space-Time: A New Approach to Unifying Relativity and Quantum Mechanics.* World Scientific Publishing Company. ISBN 1848166508.:p. 656

[98] Nottale, L. (1993). *Fractal Space-Time and Microphysics: Towards a Theory of Scale Relativity.* World Scientific. ISBN 9810208782.:p. 303

[99] Nottale, L. (1993). *Mach's principle, Dirac's large numbers and the cosmological constant problem.*

[100] Nottale, L. (1996). "Scale Relativity and Fractal Space-Time: Applications to Quantum Physics, Cosmology and Chaotic Systems" (PDF). *Chaos, Solitons & Fractals* **7** (6): 877–938. doi:10.1016/0960-0779(96)00002-1. Retrieved 2015-03-10.

[101] Nottale, L. (2011).   *Scale Relativity and Fractal Space-Time: A New Approach to Unifying Relativity and Quantum Mechanics.* World Scientific Publishing Company. ISBN 1848166508.:p. 543-545

[102] Sidharth, B. G. (2001). *Chaotic universe: from Planck to the Hubble scale.* Contemporary fundamental physics. Huntington, N.Y: Nova Science Publishers. ISBN 1560729775.

[103] Nottale, L. (1993). *Fractal Space-Time and Microphysics: Towards a Theory of Scale Relativity* (PDF). World Scientific. ISBN 9810208782.:p. 305

[104] Nottale, L. (2012). "Nature et Valeur de la Constante Cosmologique" (PDF). In L. Nottale, P. Martin (eds.). *Premières Rencontres d'Avignon (2007-2009) autour de la Relativité d'Echelle.* Actes d'Avignon. Avignon. pp. 31–39. Retrieved 2015-05-30.:p. 39

[105] Planck Collaboration; Ade, P. A. R.; Aghanim, N.; Armitage-Caplan, C.; Arnaud, M.; Ashdown, M.; Atrio-Barandela, F.; Aumont, J.; Baccigalupi, C.; Banday, A. J.; Barreiro, R. B.; Bartlett, J. G.; Battaner, E.; Benabed, K.; Benoît, A.; Benoit-Lévy, A.; Bernard, J.-P.; Bersanelli, M.; Bielewicz, P.; Bobin, J.; Bock, J. J.; Bonaldi, A.; Bond, J. R.; Borrill, J.; Bouchet, F. R.; Bridges, M.; Bucher, M.; Burigana, C.; Butler, R. C.; Calabrese, E.; Cappellini, B.; Cardoso, J.-F.; Catalano, A.; Challinor, A.; Chamballu, A.; Chary, R.-R.; Chen, X.; Chiang, H. C.; Chiang, L.-Y.; Christensen, P. R.; Church, S.; Clements, D. L.; Colombi, S.; Colombo, L. P. L.; Couchot, F.; Coulais, A.; Crill, B. P.; Curto, A.; Cuttaia, F.; Danese, L.; Davies, R. D.; Davis, R. J.; de Bernardis, P.; de Rosa, A.; de Zotti, G.; Delabrouille, J.; Delouis, J.-M.; Désert, F.-X.; Dickinson, C.; Diego, J. M.; Dolag, K.; Dole, H.; Donzelli, S.; Doré, O.; Douspis, M.; Dunkley, J.; Dupac, X.; Efstathiou, G.; Elsner, F.; Enßlin, T. A.; Eriksen, H. K.; Finelli, F.; Forni, O.; Frailis, M.; Fraisse, A. A.; Franceschi, E.; Gaier, T. C.; Galeotta, S.; Galli, S.; Ganga, K.; Giard, M.; Giardino, G.; Giraud-Héraud, Y.; Gjerløw, E.; González-Nuevo, J.; Górski, K. M.; Gratton, S.; Gregorio, A.; Gruppuso, A.; Gudmundsson, J. E.; Haissinski, J.; Hamann, J.; Hansen, F. K.; Hanson, D.; Harrison, D.; Henrot-Versillé, S.; Hernández-Monteagudo, C.; Herranz, D.; Hildebrandt, S. R.; Hivon, E.; Hobson, M.; Holmes, W. A.; Hornstrup, A.; Hou, Z.; Hovest, W.; Huffenberger, K. M.; Jaffe, A. H.; Jaffe, T. R.; Jewell, J.; Jones, W. C.; Juvela, M.; Keihänen, E.; Keskitalo, R.; Kisner, T.

S.; Kneissl, R.; Knoche, J.; Knox, L.; Kunz, M.; Kurki-Suonio, H.; Lagache, G.; Lähteenmäki, A.; Lamarre, J.-M.; Lasenby, A.; Lattanzi, M.; Laureijs, R. J.; Lawrence, C. R.; Leach, S.; Leahy, J. P.; Leonardi, R.; León-Tavares, J.; Lesgourgues, J.; Lewis, A.; Liguori, M.; Lilje, P. B.; Linden-Vørnle, M.; López-Caniego, M.; Lubin, P. M.; Macías-Pérez, J. F.; Maffei, B.; Maino, D.; Mandolesi, N.; Maris, M.; Marshall, D. J.; Martin, P. G.; Martinez-González, E.; Masi, S.; Massardi, M.; Matarrese, S.; Matthai, F.; Mazzotta, P.; Meinhold, P. R.; Melchiorri, A.; Melin, J.-B.; Mendes, L.; Menegoni, E.; Mennella, A.; Migliaccio, M.; Millea, M.; Mitra, S.; Miville-Deschênes, M.-A.; Moneti, A.; Montier, L.; Morgante, G.; Mortlock, D.; Moss, A.; Munshi, D.; Murphy, J. A.; Naselsky, P.; Nati, F.; Natoli, P.; Netterfield, C. B.; Nørgaard-Nielsen, H. U.; Noviello, F.; Novikov, D.; Novikov, I.; O'Dwyer, I. J.; Osborne, S.; Oxborrow, C. A.; Paci, F.; Pagano, L.; Pajot, F.; Paoletti, D.; Partridge, B.; Pasian, F.; Patanchon, G.; Pearson, D.; Pearson, T. J.; Peiris, H. V.; Perdereau, O.; Perotto, L.; Perrotta, F.; Pettorino, V.; Piacentini, F.; Piat, M.; Pierpaoli, E.; Pietrobon, D.; Plaszczynski, S.; Platania, P.; Pointecouteau, E.; Polenta, G.; Ponthieu, N.; Popa, L.; Poutanen, T.; Pratt, G. W.; Prézeau, G.; Prunet, S.; Puget, J.-L.; Rachen, J. P.; Reach, W. T.; Rebolo, R.; Reinecke, M.; Remazeilles, M.; Renault, C.; Ricciardi, S.; Riller, T.; Ristorcelli, I.; Rocha, G.; Rosset, C.; Roudier, G.; Rowan-Robinson, M.; Rubiño-Martín, J. A.; Rusholme, B.; Sandri, M.; Santos, D.; Savelainen, M.; Savini, G.; Scott, D.; Seiffert, M. D.; Shellard, E. P. S.; Spencer, L. D.; Starck, J.-L.; Stolyarov, V.; Stompor, R.; Sudiwala, R.; Sunyaev, R.; Sureau, F.; Sutton, D.; Suur-Uski, A.-S.; Sygnet, J.-F.; Tauber, J. A.; Tavagnacco, D.; Terenzi, L.; Toffolatti, L.; Tomasi, M.; Tristram, M.; Tucci, M.; Tuovinen, J.; Türler, M.; Umana, G.; Valenziano, L.; Valiviita, J.; Van Tent, B.; Vielva, P.; Villa, F.; Vittorio, N.; Wade, L. A.; Wandelt, B. D.; Wehus, I. K.; White, M.; White, S. D. M.; Wilkinson, A.; Yvon, D.; Zacchei, A.; Zonca, A. (2013-03-20). "Planck 2013 results. XVI. Cosmological parameters". *arXiv:1303.5076 [astro-ph]*. Retrieved 2015-05-30.

[106] Nottale, L. (2011). *Scale Relativity and Fractal Space-Time: A New Approach to Unifying Relativity and Quantum Mechanics*. World Scientific Publishing Company. ISBN 1848166508.:p. 554

[107] Nottale, L. (2011). *Scale Relativity and Fractal Space-Time: A New Approach to Unifying Relativity and Quantum Mechanics*. World Scientific Publishing Company. ISBN 1848166508.:p. 543

[108] Nottale, L. (1993). *Fractal Space-Time and Microphysics: Towards a Theory of Scale Relativity*. World Scientific. ISBN 9810208782.:p. 292

[109] Nottale, L. (2003). "Scale-Relativistic Cosmology" (PDF). *Chaos, Solitons and Fractals* 16 (4): 539–564.

[110] Dubrulle, B. (2000-04-01). "Finite Size Scale Invariance". *The European Physical Journal B - Condensed Matter and Complex Systems* 14 (4): 757–771.

doi:10.1007/s100510051087. ISSN 1434-6028. Retrieved 2015-05-30.

[111] Nottale, L.; Chaline, J.; Grou, P. (2000). *Les Arbres de L'évolution: Univers, Vie, Sociétés*. Hachette littératures.

[112] Nottale, L.; Chaline, J.; Grou, P. (2002). "On the Fractal Structure of Evolutionary Trees". In Gabriele A. Losa, Theo F. Nonnenmacher, Danilo Merlini, Ewald R. Weibel (eds.). *Fractals in Biology and Medicine* (PDF). Birkhäuser. pp. 247–258. ISBN 9783764364748.

[113] Turner, Philip; Nottale, L. (2015). "The Origins of Macroscopic Quantum Coherence in High Temperature Super Conductivity". *Physica C: Superconductivity and its Applications*. doi:10.1016/j.physc.2015.04.006. ISSN 0921-4534. Retrieved 2015-04-16.

[114] Nottale, L. (2007). "Scale Relativity: A Fractal Matrix for Organization in Nature" (PDF). *Electronic Journal of Theoretical Physics* 4 (16): 187–274.

[115] Nottale, L.; Chaline, J.; Grou, P. (2009). *Des Fleurs pour Schrödinger: La Relativité d'Echelle Et Ses Applications*. Ellipses. ISBN 978-2-7298-5182-8.

[116] Cash, R.; Chaline, J.; Nottale, L.; Grou, P. (2002-05). "Développement Humain et Loi Log-Périodique" (PDF). *Comptes Rendus Biologies* 325 (5): 585–590. doi:10.1016/S1631-0691(02)01468-3. ISSN 1631-0691. Retrieved 2015-04-10. Check date values in: |date= (help)

[117] Auffray, C.; Nottale, L. (2008). "Scale Relativity Theory and Integrative Systems Biology: 1 Founding Principles and Scale Laws" (PDF). *Progress in biophysics and molecular biology* 97 (1): 79–114. ISSN 0079-6107.

[118] Nottale, L.; Auffray, C. (2008). "Scale Relativity Theory and Integrative Systems Biology: 2 Macroscopic Quantum-Type Mechanics" (PDF). *Progress in biophysics and molecular biology* 97 (1): 115–157. ISSN 0079-6107.

[119] Nottale, L. (2013-12-30). "Macroscopic Quantum-Type Potentials in Theoretical Systems Biology". *Cells* 3 (1): 1–35. doi:10.3390/cells3010001. ISSN 2073-4409. PMC 3980741. PMID 24709901. Retrieved 2015-05-31.

[120] Forriez, M.; Martin, P.; Nottale, L. (2010-07-07). "Lois d'échelle et transitions fractal-non fractal en géographie". *L'Espace géographique* 39 (2): 97–112. ISSN 0046-2497. Retrieved 2015-04-10.

[121] Nottale, L.; Martin, P.; Forriez, M. (2012). "Analyse en relativité d'échelle du bassin versant du Gardon (Gard, France)" (PDF). In L. Nottale, P. Martin (eds.). *Premières Rencontres d'Avignon (2007-2009) autour de la Relativité d'Echelle*. Actes d'Avignon. Avignon. pp. 267–295. ISBN 2-910545-07-5. Retrieved 2015-05-30.

[122] Magee, Christopher L.; Devezas, Tessaleno C. (2011). "How Many Singularities Are Near and How Will They Disrupt Human History?" (PDF). *Technological*

Forecasting and Social Change **78** (8): 1365–1378. doi:10.1016/j.techfore.2011.07.013. ISSN 0040-1625. Retrieved 2013-10-02.: p. 1370

[123] Nottale, L.; Chaline, J.; Grou, P. (2000). *Les Arbres de L'évolution: Univers, Vie, Sociétés*. Hachette littératures.

[124] Castro, Carlos (1997-06-01). "String Theory, Scale Relativity and the Generalized Uncertainty Principle". *Foundations of Physics Letters* **10** (3): 273–293. doi:10.1007/BF02764209. ISSN 0894-9875. Retrieved 2015-03-23.:p. 275

[125] Nottale, L. (2011). *Scale Relativity and Fractal Space-Time: A New Approach to Unifying Relativity and Quantum Mechanics*. World Scientific Publishing Company. ISBN 1848166508.:p. 458

[126] Nottale, L. (2011). *Scale Relativity and Fractal Space-Time: A New Approach to Unifying Relativity and Quantum Mechanics*. World Scientific Publishing Company. ISBN 1848166508.:p. 7

[127] Loll, R. (2008-06-07). "The Emergence of Spacetime or Quantum Gravity on Your Desktop". *Classical and Quantum Gravity* **25** (11): 114006. doi:10.1088/0264-9381/25/11/114006. ISSN 0264-9381. Retrieved 2015-04-19.

[128] Nottale, L. (2011). *Scale Relativity and Fractal Space-Time: A New Approach to Unifying Relativity and Quantum Mechanics*. World Scientific Publishing Company. ISBN 1848166508.:p. 277

[129] Connes, Alain (1994). *Noncommutative Geometry*. San Diego: Academic Press. ISBN 012185860X.

[130] Lapidus, Michel L. (2008). *In Search of the Riemann Zeros: Strings, Fractal Membranes and Noncommutative Spacetimes*. Providence, R.I: American Mathematical Society. ISBN 9780821842225.

[131] Nottale, L. (2011). *Scale Relativity and Fractal Space-Time: A New Approach to Unifying Relativity and Quantum Mechanics*. World Scientific Publishing Company. ISBN 1848166508.:p. 459

[132] Bontems, Vincent; Gingras, Yves (2007-12-01). "De la science normale a la science marginale. Analyse d'une bifurcation de trajectoire scientifique: le cas de la Theorie de la Relativité d'Echelle". *Social Science Information* **46** (4): 607–653. doi:10.1177/0539018407082595. Retrieved 2008-12-11.

[133] Karoji, H.; Nottale, L. (1976-01-01). "Possible implications of the Rubin-Ford effect". *Nature* **259** (5538): 31–33. doi:10.1038/259031a0. Retrieved 2009-07-29.

[134] Nottale, L.; Vigier, J. P. (1977). "Continuous increase of Hubble modulus behind clusters of galaxies". *Nature* **268** (5621): 608–610. doi:10.1038/268608a0. Retrieved 2009-07-29.

[135] Vidal, C. (2010). "Introduction to the Special Issue on the Evolution and Development of the Universe". *Foundations of Science* **15** (2): 95–99. doi:10.1007/s10699-010-9176-9. Retrieved 2010-05-20.

[136] Auffray, C.; Noble, D. (2010-12-30). "Scale Relativity: an Extended Paradigm for Physics and Biology?". *Foundations of Science* **16** (4): 303–305. doi:10.1007/s10699-010-9203-x. ISSN 1233-1821. Retrieved 2015-04-19.

[137] Nottale, L. (1998). *La Relativité dans tous ses États: au delà de l'Espace-Temps*. Hachette.

[138] Merker, Joël (1999-01-01). "Deux infinis cousus main". *Revue de synthèse* **120** (1): 165–174. doi:10.1007/BF03182083. ISSN 0035-1776. Retrieved 2013-11-13.:p. 166

[139] Baryshev, Yurij; Teerikorpi, Pekka (2002). *Discovery of Cosmic Fractals*. River Edge, N.J: World Scientific. ISBN 9810248717.:p. 255

[140] Potter, Frank; Jargodzki, Christopher (2005). *Mad About Modern Physics: Braintwisters, Paradoxes, and Curiosities*. Hoboken, N.J: J. Wiley. ISBN 0471448559.:p. 113

[141] Auffray, C.; Noble, D. (2010-12-30). "Scale Relativity: an Extended Paradigm for Physics and Biology?". *Foundations of Science* **16** (4): 303–305. doi:10.1007/s10699-010-9203-x. ISSN 1233-1821. Retrieved 2015-04-19.:p. 303

[142] Peter, Patrick (2013). "Laurent Nottale: Scale relativity and fractal space-time". *General Relativity and Gravitation* **45** (7): 1459–1461. doi:10.1007/s10714-013-1535-8. ISSN 0001-7701.

[143] Nottale, L. (2011). *Scale Relativity and Fractal Space-Time: A New Approach to Unifying Relativity and Quantum Mechanics*. World Scientific Publishing Company. ISBN 1848166508.

[144] Nottale, L. (2001). "Scale Relativity and Non-Differentiable Fractal Space-Time". *Frontiers of Fundamental Physics 4*. Springer. pp. 65–79. Retrieved 2015-03-10.

[145] Beringer, J.; (Particle Data Group) (2013). "Status of Higgs Boson Physics". *PR D86* (010001). Missing |last2= in Authors list (help):p. 33

[146] Stenger, V. J. (2011). *The Fallacy of Fine-Tuning: Why the Universe Is Not Designed for Us*. Prometheus Books. ISBN 1616144432.:p. 100

22.12   External links

- Laurent Nottale's site

- List of downloadable papers from Nottale

Chapter 23

Weinberg–Witten theorem

In theoretical physics, the **Weinberg–Witten (WW) theorem**, proved by Steven Weinberg and Edward Witten, states that massless particles (either composite or elementary) with spin $j > 1/2$ cannot carry a Lorentz-covariant current, while massless particles with spin $j > 1$ cannot carry a Lorentz-covariant stress-energy. The theorem is usually interpreted to mean that the graviton ($j = 2$) cannot be a composite particle in a relativistic quantum field theory.

23.1 Background

During the 1980s, preon theories, technicolor and the like were very popular and some people speculated that gravity might be an emergent phenomenon or that gluons might be composite. Weinberg and Witten, on the other hand, developed a no-go theorem that excludes, under very general assumptions, the hypothetical composite and emergent theories. Decades later new theories of emergent gravity are proposed and mainstream high-energy physicists are still using this theorem to "debunk" such theories. Because most of these emergent theories aren't Lorentz covariant, the WW theorem doesn't apply. The violation of Lorentz covariance, however, usually leads to other problems.

23.2 Theorem

Weinberg and Witten proved two separate results. According to them, the first is due to Sidney Coleman, who did not publish it:

- A 3 + 1D QFT (quantum field theory) with a conserved 4-vector current J^μ (see four-current) which is Poincaré covariant (and gauge invariant if there happens to be any gauge symmetry which hasn't been gauge-fixed) does not admit massless particles with helicity $|h| > 1/2$ that also have nonzero charges associated with the conserved current in question.

- A 3 + 1D QFT with a conserved stress–energy tensor $T^{\mu\nu}$ which is Poincaré covariant (and gauge invariant if there happens to be any gauge symmetry which hasn't been gauge-fixed) does not admit massless particles with helicity $|h| > 1$.

23.3 A sketch of the proof

The conserved charge Q is given by $\int d^3x\, J^0$. We shall consider the matrix elements of the charge and of the current J^μ for one-particle asymptotic states, of equal helicity, $|p\rangle$ and $|p'\rangle$, labeled by their lightlike 4-momenta. We shall consider the case in which $(p - p')$ isn't null, which means that the momentum transfer is spacelike. Let q be the eigenvalue of those states for the charge operator Q, so that:

$$q\delta^3(\vec{p'} - \vec{p}) = \langle p'|Q|p\rangle = \int d^3x\, \langle p'|J^0(\vec{x}, 0)|p\rangle$$
$$= \int d^3x\, \langle p'|e^{-i\vec{P}\cdot\vec{x}}J^0(0,0)e^{i\vec{P}\cdot\vec{x}}|p\rangle$$
$$= \int d^3x\, e^{i(\vec{p}-\vec{p'})\cdot\vec{x}}\langle p'|J^0(0,0)|p\rangle$$
$$= (2\pi)^3\delta^3(\vec{p'} - \vec{p})\langle p'|J^0(0,0)|p\rangle$$

where we have now made used of translational covariance, which is part of the Poincaré covariance. Thus:

$$\langle p'|J^0(0)|p\rangle = \frac{q}{(2\pi)^3}$$

with $q \neq 0$.

Let's transform to a reference frame where p moves along the positive z-axis and p' moves along the negative z-axis. This is always possible for any spacelike momentum transfer.

In this reference frame, $\langle p'|J^0(0)|p\rangle$ and $\langle p'|J^3(0)|p\rangle$ change by the phase factor $e^{i(h-(-h))\theta} = e^{2ih\theta}$ under rotations by θ counterclockwise about the z-axis whereas

$\langle p'|J^1(0) + iJ^2(0)|p\rangle$ and $\langle p'|J^1(0) - iJ^2(0)|p\rangle$ change by the phase factors $e^{i(2h+1)\theta}$ and $e^{i(2h-1)\theta}$ respectively.

If h is nonzero, we need to specify the phases of states. In general, this can't be done in a Lorentz-invariant way (see Thomas precession), but the one particle Hilbert space *is* Lorentz-covariant. So, if we make any arbitrary but fixed choice for the phases, then each of the matrix components in the previous paragraph has to be invariant under the rotations about the z-axis. So, unless $|h| = 0$ or $1/2$, all of the components have to be zero.

Weinberg and Witten *did not* assume the continuity

$$\langle p|J^0(0)|p\rangle = \lim_{p'\to p} \langle p'|J^0(0)|p\rangle$$

Rather, the authors argue that the *physical* (i.e., the measurable) quantum numbers of a massless particle are always defined by the matrix elements in the limit of zero momentum, defined for a sequence of spacelike momentum transfers. Also, $\delta^3(\vec{p'} - \vec{p})$ in the first equation can be replaced by "smeared out" Dirac delta function, which corresponds to performing the d^3x volume integral over a finite box.

The proof of the second part of theorem is completely analogous, replacing the matrix elements of the current with the matrix elements of the stress–energy tensor $T^{\mu\nu}$:

$$p^\mu = \int d^3x\, T^{\mu 0}(\vec{x}, 0)$$

$$\langle p|T^{00}(0)|p\rangle = \frac{E}{(2\pi)^3}$$

with $E \neq 0$.

For spacelike momentum transfers, we can go to the reference frame where $p' + p$ is along the t-axis and $p' - p$ is along the z-axis. In this reference frame, the components of $\langle p'|\mathbf{T}(0)|p\rangle$ transforms as $e^{i(2h-2)\theta}$, $e^{i(2h-1)\theta}$, $e^{i(2h)\theta}$, $e^{i(2h+1)\theta}$ or $e^{i(2h+2)\theta}$ under a rotation by θ about the z-axis. Similarly, we can conclude that $|h| = 0, \frac{1}{2}, 1$

Note that this theorem also applies to free field theories. If they contain massless particles with the "wrong" helicity/charge, they have to be gauge theories.

23.4   Ruling out emergent theories

What has this theorem got to do with emergence/composite theories?

If let's say gravity is an emergent theory of a fundamentally flat theory over a flat Minkowski spacetime, then by Noether's theorem, we have a conserved stress–energy tensor which is Poincaré covariant. If the theory has an internal gauge symmetry (of the Yang–Mills kind), we may pick the Belinfante–Rosenfeld stress–energy tensor which is gauge-invariant. As there is no fundamental diffeomorphism symmetry, we don't have to worry about that this tensor isn't BRST-closed under diffeomorphisms. So, the Weinberg–Witten theorem applies and we can't get a massless spin-2 (i.e. helicity ±2) composite/emergent graviton.

If we have a theory with a fundamental conserved 4-current associated with a global symmetry, then we can't have emergent/composite massless spin-1 particles which are charged under that global symmetry.

23.5   Theories where the theorem is inapplicable

23.5.1   Nonabelian gauge theories

There are a number of ways to see why nonabelian Yang–Mills theories in the Coulomb phase don't violate this theorem. Yang–Mills theories don't have any conserved 4-current associated with the Yang–Mills charges that are both Poincaré covariant and gauge invariant. Noether's theorem gives a current which is conserved and Poincaré covariant, but not gauge invariant. As $|p\rangle$ is really an element of the BRST cohomology, i.e. a quotient space, it is really an equivalence class of states. As such, $\langle p'|J|p\rangle$ is only well defined if J is BRST-closed. But if J isn't gauge-invariant, then J isn't BRST-closed in general. The current defined as $J^\mu(x) \equiv \frac{\delta}{\delta A_\mu(x)} S_{\text{matter}}$ is not conserved because it satisfies $D_\mu J^\mu = 0$ instead of $\partial_\mu J^\mu = 0$ where D is the covariant derivative. The current defined after a gauge-fixing like the Coulomb gauge is conserved but isn't Lorentz covariant.

23.5.2   Spontaneously broken gauge theories

The gauge bosons associated with spontaneously broken symmetries are massive. For example, in QCD, we have electrically charged rho mesons which can be described by an emergent hidden gauge symmetry which is spontaneously broken. Therefore, there is nothing in principle stopping us from having composite preon models of W and Z bosons.

On a similar note, even though the photon is charged under the SU(2) weak symmetry (because it is the gauge boson associated with a linear combination of weak isospin and hypercharge), it is also moving through a condensate of such charges, and so, isn't an exact eigenstate of the weak charges and this theorem doesn't apply either.

23.5.3 Massive gravity

On a similar note, it is possible to have a composite/emergent theory of massive gravity.

23.5.4 General relativity

In GR, we have diffeomorphisms and A|ψ> (over an element |ψ> of the BRST cohomology) only makes sense if A is BRST-closed. There are no local BRST-closed operators and this includes any stress–energy tensor that we can think of.

23.5.5 Induced gravity

In induced gravity, the fundamental theory is also diffeomorphism invariant and the same comment applies.

23.5.6 Seiberg duality

If we take N=1 chiral super QCD with N_c colors and N_f flavors with $N_f - 2 \geq N_c > \frac{2}{3}N_f$, then by the Seiberg duality, this theory is dual to a nonabelian $SU(N_f - N_c)$ gauge theory which is trivial (i.e. free) in the infrared limit. As such, the dual theory doesn't suffer from any infraparticle problem or a continuous mass spectrum. Despite this, the dual theory is still a nonabelian Yang–Mills theory. Because of this, the dual magnetic current still suffers from all the same problems even though it is an "emergent current". Free theories aren't exempt from the Weinberg–Witten theorem.

23.5.7 Conformal field theory

In a conformal field theory, the only truly massless particles are noninteracting singletons (see singleton field). The other "particles"/bound states have a continuous mass spectrum which can take on any arbitrarily small nonzero mass. So, we can have spin-3/2 and spin-2 bound states with arbitrarily small masses but still not violate the theorem. In other words, they are infraparticles.

23.5.8 Infraparticles

Two otherwise identical charged infraparticles moving with different velocities belong to different superselection sectors. Let's say they have momenta p' and p respectively. Then as $J^\mu(0)$ is a local neutral operator, it does not map between different superselection sectors. So, $< p'|J^\mu(0)|p >$ is zero. The only way |p''> and |p> can belong in the same

sector is if they have the same velocity, which means that they are proportional to each other, i.e. a null or zero momentum transfer, which isn't covered in the proof. So, infraparticles violate the continuity assumption

$$\langle p|J^0(0)|p \rangle = \lim_{p' \to p} \langle p'|J^0(0)|p \rangle$$

This doesn't mean of course that the momentum of a charge particle can't change by some spacelike momentum. It only means that if the incoming state is a one infraparticle state, then the outgoing state contains an infraparticle together with a number of soft quanta. This is nothing other than the inevitable bremsstrahlung. But this also means that the outgoing state isn't a one particle state.

23.5.9 Theories with nonlocal charges

Obviously, a nonlocal charge does not have a local 4-current and a theory with a nonlocal 4-momentum does not have a local stress–energy tensor.

23.5.10 Acoustic metric theories and analog model of gravity

These theories are not Lorentz covariant. However, some of these theories can give rise to an approximate emergent Lorentz symmetry at low energies so that we can both have the cake and eat it too.

23.5.11 Superstring theory

Superstring theory defined over a background metric (possibly with some fluxes) over a 10D space which is the product of a flat 4D Minkowski space and a compact 6D space has a massless graviton in its spectrum. This is an emergent particle coming from the vibrations of a superstring. Let's look at how we would go about defining the stress–energy tensor. The background is given by g (the metric) and a couple of other fields. The effective action is a functional of the background. The VEV of the stress–energy tensor is then defined as the functional derivative

$$T^{MN}(x) \equiv \frac{1}{\sqrt{-g}} \frac{\delta}{\delta g_{MN}(x)} \Gamma[\text{background}].$$

The stress-energy operator is defined as a vertex operator corresponding to this infinitesimal change in the background metric.

Not all backgrounds are permissible. Superstrings have to
have superconformal symmetry, which is a super generalization of Weyl symmetry, in order to be consistent but they
are only superconformal when propagating over some special backgrounds (which satisfy the Einstein field equations
plus some higher order corrections). Because of this, the effective action is only defined over these special backgrounds
and the functional derivative is not well-defined. The vertex
operator for the stress–energy tensor at a point also doesn't
exist.

23.6   References

- Weinberg, Steven; Witten, Edward (1980). "Limits on massless particles". *Physics Letters B* **96**
  (1–2):   59–62.   Bibcode:1980PhLB...96...59W.
  doi:10.1016/0370-2693(80)90212-9.

- Jenkins, Alejandro (2006). *Topics in particle physics
  and cosmology beyond the standard model* (Thesis).
  arXiv:hep-th/0607239. Bibcode:2006PhDT........96J.
  (see Ch. 2 for a detailed review)

Chapter 24

Composite gravity

In theoretical physics, **composite gravity** refers to models that attempted to derive general relativity in a framework where the graviton is constructed as a composite bound state of more elementary particles, usually fermions. A theorem by Steven Weinberg and Edward Witten shows that this is not possible in Lorentz covariant theories: massless particles with spin greater than one are forbidden. The AdS/CFT correspondence may be viewed as a loophole in their argument. However, in this case not only the graviton is emergent; a whole spacetime dimension is emergent, too.[1]

24.1 See also

- Weinberg–Witten theorem

24.2 References

[1] "Probing composite gravity in colliders". scitation.aip.org. Retrieved 2008-07-08.

Chapter 25

Induced gravity

Induced gravity (or **emergent gravity**) is an idea in quantum gravity that space-time curvature and its dynamics emerge as a mean field approximation of underlying microscopic degrees of freedom, similar to the fluid mechanics approximation of Bose–Einstein condensates. The concept was originally proposed by Andrei Sakharov in 1967.

Sakharov observed that many condensed matter systems give rise to emergent phenomena that are analogous to general relativity. For example, crystal defects can look like curvature and torsion in an Einstein–Cartan spacetime. This allows one to create a theory of gravity with torsion from a World Crystal model of spacetime[1] in which the lattice spacing is of the order of a Planck length. Sakharov's idea was to start with an arbitrary background pseudo-Riemannian manifold (in modern treatments, possibly with torsion) and introduce quantum fields (matter) on it but not introduce any gravitational dynamics explicitly. This gives rise to an effective action which to one-loop order contains the Einstein–Hilbert action with a cosmological constant. In other words, general relativity arises as an emergent property of matter fields and is not put in by hand. On the other hand, such models typically predict huge cosmological constants.

Some argue that the particular models proposed by Sakharov and others have been proven impossible by the Weinberg–Witten theorem. However, models with emergent gravity are possible as long as other things, such as spacetime dimensions, emerge together with gravity. Developments in AdS/CFT correspondence after 1997 suggest that the microphysical degrees of freedom in induced gravity might be radically different. The bulk space-time arises as an emergent phenomenon of the quantum degrees of freedom that live in the boundary of the space-time.

25.1 See also

- Entropic gravity
- Causal dynamical triangulation

25.2 References

[1] H. Kleinert (1987). "Gravity as Theory of Defects in a Crystal with Only Second-Gradient Elasticity". *Annalen der Physik* **44**: 117. Bibcode:1987AnP...499..117K. doi:10.1002/andp.19874990206.

25.3 External links

- Carlos Barcelo, Stefano Liberati, Matt Visser, *Living Rev.Rel.* 8:12, 2005.

- D. Berenstein, *Emergent Gravity from CFT*, online lecture.

- C. J. Hogan *Quantum Indeterminacy of Emergent Spacetime*, preprint

- A.D. Sakharov, *Vacuum Quantum Fluctuations in Curved Space and the Theory of Gravitation*, 1967.

- Matt Visser, *Sakharov's induced gravity: a modern perspective*, 2002.

- H. Kleinert, *Multivalued Fields in Condensed Matter, Electrodynamics, and Gravitation*, 2008.

Chapter 26

Holographic principle

The **holographic principle** is a property of string theories and a supposed property of quantum gravity that states that the description of a volume of space can be thought of as encoded on a boundary to the region—preferably a light-like boundary like a gravitational horizon. First proposed by Gerard 't Hooft, it was given a precise string-theory interpretation by Leonard Susskind[1] who combined his ideas with previous ones of 't Hooft and Charles Thorn.[1][2] As pointed out by Raphael Bousso,[3] Thorn observed in 1978 that string theory admits a lower-dimensional description in which gravity emerges from it in what would now be called a holographic way.

In a larger sense, the theory suggests that the entire universe can be seen as a two-dimensional information on the cosmological horizon, such that the three dimensions we observe are an effective description only at macroscopic scales and at low energies. Cosmological holography has not been made mathematically precise, partly because the particle horizon has a non-zero area and grows with time.[4][5]

The holographic principle was inspired by black hole thermodynamics, which conjectures that the maximal entropy in any region scales with the radius *squared*, and not cubed as might be expected. In the case of a black hole, the insight was that the informational content of all the objects that have fallen into the hole might be entirely contained in surface fluctuations of the event horizon. The holographic principle resolves the black hole information paradox within the framework of string theory.[6] However, there exist classical solutions to the Einstein equations that allow values of the entropy larger than those allowed by an area law, hence in principle larger than those of a black hole. These are the so-called "Wheeler's bags of gold". The existence of such solutions conflicts with the holographic interpretation, and their effects in a quantum theory of gravity including the holographic principle are not yet fully understood.[7]

26.1 Black hole entropy

Main article: Black hole thermodynamics

An object with relatively high entropy is microscopically random, like a hot gas. A known configuration of classical fields has zero entropy: there is nothing random about electric and magnetic fields, or gravitational waves. Since black holes are exact solutions of Einstein's equations, they were thought not to have any entropy either.

But Jacob Bekenstein noted that this leads to a violation of the second law of thermodynamics. If one throws a hot gas with entropy into a black hole, once it crosses the event horizon, the entropy would disappear. The random properties of the gas would no longer be seen once the black hole had absorbed the gas and settled down. One way of salvaging the second law is if black holes are in fact random objects, with an enormous entropy whose increase is greater than the entropy carried by the gas.

Bekenstein assumed that black holes are maximum entropy objects—that they have more entropy than anything else in the same volume. In a sphere of radius R, the entropy in a relativistic gas increases as the energy increases. The only known limit is gravitational; when there is too much energy the gas collapses into a black hole. Bekenstein used this to put an upper bound on the entropy in a region of space, and the bound was proportional to the area of the region. He concluded that the black hole entropy is directly proportional to the area of the event horizon.[8]

Stephen Hawking had shown earlier that the total horizon area of a collection of black holes always increases with time. The horizon is a boundary defined by light-like geodesics; it is those light rays that are just barely unable to escape. If neighboring geodesics start moving toward each other they eventually collide, at which point their extension is inside the black hole. So the geodesics are always moving apart, and the number of geodesics which generate the boundary, the area of the horizon, always increases. Hawking's result was called the second law of black hole thermodynamics, by analogy with the law of entropy increase, but at first, he did not take the analogy too seriously.

Hawking knew that if the horizon area were an actual entropy, black holes would have to radiate. When heat is added to a thermal system, the change in entropy is the increase in mass-energy divided by temperature:

$$dS = \frac{dM}{T}.$$

If black holes have a finite entropy, they should also have a finite temperature. In particular, they would come to equilibrium with a thermal gas of photons. This means that black holes would not only absorb photons, but they would also have to emit them in the right amount to maintain detailed balance.

Time independent solutions to field equations do not emit radiation, because a time independent background conserves energy. Based on this principle, Hawking set out to show that black holes do not radiate. But, to his surprise, a careful analysis convinced him that they do, and in just the right way to come to equilibrium with a gas at a finite temperature. Hawking's calculation fixed the constant of proportionality at 1/4; the entropy of a black hole is one quarter its horizon area in Planck units.[9]

The entropy is proportional to the logarithm of the number of microstates, the ways a system can be configured microscopically while leaving the macroscopic description unchanged. Black hole entropy is deeply puzzling — it says that the logarithm of the number of states of a black hole is proportional to the area of the horizon, not the volume in the interior.[10]

Later, Raphael Bousso came up with a covariant version of the bound based upon null sheets.

26.2 Black hole information paradox

Main article: Black hole information paradox

Hawking's calculation suggested that the radiation which black holes emit is not related in any way to the matter that they absorb. The outgoing light rays start exactly at the edge of the black hole and spend a long time near the horizon, while the infalling matter only reaches the horizon much later. The infalling and outgoing mass/energy only interact when they cross. It is implausible that the outgoing state would be completely determined by some tiny residual scattering.

Hawking interpreted this to mean that when black holes absorb some photons in a pure state described by a wave function, they re-emit new photons in a thermal mixed state described by a density matrix. This would mean that quantum mechanics would have to be modified, because in quantum mechanics, states which are superpositions with probability amplitudes never become states which are probabilistic mixtures of different possibilities.[note 1]

Troubled by this paradox, Gerard 't Hooft analyzed the emission of Hawking radiation in more detail. He noted that when Hawking radiation escapes, there is a way in which incoming particles can modify the outgoing particles. Their gravitational field would deform the horizon of the black hole, and the deformed horizon could produce different outgoing particles than the undeformed horizon. When a particle falls into a black hole, it is boosted relative to an outside observer, and its gravitational field assumes a universal form. 't Hooft showed that this field makes a logarithmic tent-pole shaped bump on the horizon of a black hole, and like a shadow, the bump is an alternate description of the particle's location and mass. For a four-dimensional spherical uncharged black hole, the deformation of the horizon is similar to the type of deformation which describes the emission and absorption of particles on a string-theory world sheet. Since the deformations on the surface are the only imprint of the incoming particle, and since these deformations would have to completely determine the outgoing particles, 't Hooft believed that the correct description of the black hole would be by some form of string theory.

This idea was made more precise by Leonard Susskind, who had also been developing holography, largely independently. Susskind argued that the oscillation of the horizon of a black hole is a complete description[note 2] of both the infalling and outgoing matter, because the world-sheet theory of string theory was just such a holographic description. While short strings have zero entropy, he could identify long highly excited string states with ordinary black holes. This was a deep advance because it revealed that strings have a classical interpretation in terms of black holes.

This work showed that the black hole information paradox is resolved when quantum gravity is described in an unusual string-theoretic way assuming the string-theoretical description is complete, unambiguous and non-redundant.[12] The space-time in quantum gravity would emerge as an effective description of the theory of oscillations of a lower-dimensional black-hole horizon, and suggest that any black hole with appropriate properties, not just strings, would serve as a basis for a description of string theory.

In 1995, Susskind, along with collaborators Tom Banks, Willy Fischler, and Stephen Shenker, presented a formulation of the new M-theory using a holographic description in terms of charged point black holes, the D0 branes of type IIA string theory. The Matrix theory they proposed was first suggested as a description of two branes in 11-dimensional supergravity by Bernard de Wit, Jens Hoppe, and Hermann Nicolai. The later authors reinterpreted the same matrix models as a description of the dynamics of point black holes in particular limits. Holography allowed them to conclude that the dynamics of these black holes give a complete non-perturbative formulation of M-theory. In 1997, Juan Maldacena gave the first holographic descriptions of a higher-dimensional object, the 3+1-dimensional type IIB membrane, which resolved a long-standing problem of finding a string description which describes a gauge theory. These developments simultaneously explained how string theory is related to some forms of supersymmetric quantum field theories.

26.3 Limit on information density

Entropy, if considered as information (see information entropy), is measured in bits. The total quantity of bits is related to the total degrees of freedom of matter/energy.

For a given energy in a given volume, there is an upper limit to the density of information (the Bekenstein bound) about the whereabouts of all the particles which compose matter in that volume, suggesting that matter itself cannot be subdivided infinitely many times and there must be an ultimate level of fundamental particles. As the degrees of freedom of a particle are the product of all the degrees of freedom of its sub-particles, were a particle to have infinite subdivisions into lower-level particles, then the degrees of freedom of the original particle must be infinite, violating the maximal limit of entropy density. The holographic principle thus implies that the subdivisions must stop at some level, and that the fundamental particle is a bit (1 or 0) of information.

The most rigorous realization of the holographic principle is the AdS/CFT correspondence by Juan Maldacena. However, J.D. Brown and Marc Henneaux had rigorously proved already in 1986, that the asymptotic symmetry of 2+1 dimensional gravity gives rise to a Virasoro algebra, whose corresponding quantum theory is a 2-dimensional conformal field theory.[13]

26.4 High-level summary

The physical universe is widely seen to be composed of "matter" and "energy". In his 2003 article published in Scientific American magazine, Jacob Bekenstein summarized a current trend started by John Archibald Wheeler, which suggests scientists may *regard the physical world as made of information, with energy and matter as incidentals.*

Bekenstein asks "Could we, as William Blake memorably penned, 'see a world in a grain of sand,' or is that idea no more than 'poetic license,'"[14] referring to the holographic principle.

26.4.1 Unexpected connection

Bekenstein's topical overview "A Tale of Two Entropies"[15] describes potentially profound implications of Wheeler's trend, in part by noting a previously unexpected connection between the world of information theory and classical physics. This connection was first described shortly after the seminal 1948 papers of American applied mathematician Claude E. Shannon introduced today's most widely used measure of information content, now known as Shannon entropy. As an objective measure of the quantity of information, Shannon entropy has been enormously useful, as the design of all modern communications and data storage devices, from cellular phones to modems to hard disk drives and DVDs, rely on Shannon entropy.

In thermodynamics (the branch of physics dealing with heat), entropy is popularly described as a measure of the "disorder" in a physical system of matter and energy. In 1877 Austrian physicist Ludwig Boltzmann described it more precisely in terms of the *number of distinct microscopic states* that the particles composing a macroscopic "chunk" of matter could be in while still *looking* like the same macroscopic "chunk". As an example, for the air in a room, its thermodynamic entropy would equal the logarithm of the count of all the ways that the individual gas molecules could be distributed in the room, and all the ways they could be moving.

26.4.2 Energy, matter, and information equivalence

Shannon's efforts to find a way to quantify the information contained in, for example, an e-mail message, led him unexpectedly to a formula with the same form as Boltzmann's. In an article in the August 2003 issue of Scientific American titled "Information in the Holographic Universe", Bekenstein summarizes that *"Thermodynamic entropy and Shannon entropy are conceptually equivalent: the number of arrangements that are counted by Boltzmann entropy reflects the amount of Shannon information one would need to implement any particular arrangement..."* of matter and energy. The only salient difference between the thermodynamic entropy of physics and Shannon's entropy of information is in the units of measure; the former is expressed in units of energy divided by temperature, the latter in *essentially dimensionless* "bits" of information.

The holographic principle states that the entropy of *ordinary mass* (not just black holes) is also proportional to surface area and not volume; that volume itself is illusory and the universe is really a hologram which is isomorphic to the information "inscribed" on the surface of its boundary.[10]

26.5   Experimental tests

The Fermilab physicist Craig Hogan claims that the holographic principle would imply quantum fluctuations in spatial position[16] that would lead to apparent background noise or "holographic noise" measurable at gravitational wave detectors, in particular GEO 600.[17] However these claims have not been widely accepted, or cited, among quantum gravity researchers and appear to be in direct conflict with string theory calculations.[18]

Analyses in 2011 of measurements of gamma ray burst GRB 041219A in 2004 by the INTEGRAL space observatory launched in 2002 by the European Space Agency shows that Craig Hogan's noise is absent down to a scale of 10^{-48} meters, as opposed to scale of 10^{-35} meters predicted by Hogan, and the scale of 10^{-16} meters found in measurements of the GEO 600 instrument.[19] Research continues at Fermilab under Hogan as of 2013.[20]

Jacob Bekenstein also claims to have found a way to test the holographic principle with a tabletop photon experiment.[21]

26.6   Tests of Maldacena's conjecture

Main article: Maldacena conjecture

Hyakutake et al. in 2013/4 published two papers[22] that bring computational evidence that Maldacena's conjecture is true. One paper computes the internal energy of a black hole, the position of its event horizon, its entropy and other properties based on the predictions of string theory and the effects of virtual particles. The other paper calculates the internal energy of the corresponding lower-dimensional cosmos with no gravity. The two simulations match. The papers are not an actual proof of Maldacena's conjecture for all cases but a demonstration that the conjecture works for a particular theoretical case and a verification of the AdS/CFT correspondence for a particular situation.[23]

26.7   See also

- Bekenstein bound
- Beyond black holes
- Bousso's holographic bound
- Brane cosmology
- Entropic gravity
- Implicate and explicate order according to David Bohm
- Margolus–Levitin theorem
- Physical cosmology
- Quantum foam
- Simulated reality

26.8   Notes

[1] except in the case of measurements, which the black hole should not be performing

[2] "Complete description" means all the *primary* qualities. For example, John Locke (and before him Robert Boyle) determined these to be *size, shape, motion, number,* and *solidity.* Such *secondary quality* information as *color, aroma, taste* and *sound*,[11] or internal quantum state is not information that is implied to be preserved in the surface fluctuations of the event horizon. (See however "path integral quantization")

26.9   References

General

- Bousso, Raphael (2002). "The holographic principle". *Reviews of Modern Physics* **74** (3): 825–874. arXiv:hep-th/0203101. Bibcode:2002RvMP...74..825B. doi:10.1103/RevModPhys.74.825.

- 't Hooft, Gerard (1993). "Dimensional Reduction in Quantum Gravity". arXiv:gr-qc/9310026.. 't Hooft's original paper.

Citations

[1] Susskind, Leonard (1995). "The World as a Hologram". *Journal of Mathematical Physics* **36** (11): 6377–6396. arXiv:hep-th/9409089. Bibcode:1995JMP....36.6377S. doi:10.1063/1.531249.

[2] Thorn, Charles B. (27–31 May 1991). *Reformulating string theory with the 1/N expansion*. International A.D. Sakharov Conference on Physics. Moscow. pp. 447–54. arXiv:hep-th/9405069. ISBN 978-1-56072-073-7.

[3] Bousso, Raphael (2002). "The Holographic Principle". *Reviews of Modern Physics* **74** (3): 825–874. arXiv:hep-th/0203101. Bibcode:2002RvMP...74..825B. doi:10.1103/RevModPhys.74.825.

[4] Lloyd, Seth (2002-05-24). "Computational Capacity of the Universe". *Physical Review Letters* **88** (23): 237901. arXiv:quant-ph/0110141. Bibcode:2002PhRvL..88w7901L. doi:10.1103/PhysRevLett.88.237901. PMID 12059399.

[5] Davies, Paul. "Multiverse Cosmological Models and the Anthropic Principle". *CTNS*. Retrieved 2008-03-14.

[6] Susskind, L. (2008). *The Black Hole War – My Battle with Stephen Hawking to Make the World Safe for Quantum Mechanics*. Little, Brown and Company.

[7] Marolf, Donald (April 2009). "Black Holes, AdS, and CFTs". *General Relativity and Gravitation* **41** (4): 903–17. arXiv:0810.4886. Bibcode:2009GReGr..41..903M. doi:10.1007/s10714-008-0749-7.

[8] Bekenstein, Jacob D. (January 1981). "Universal upper bound on the entropy-to-energy ratio for bounded systems". *Physical Review D* **23** (215): 287–298. Bibcode:1981PhRvD..23..287B. doi:10.1103/PhysRevD.23.287.

[9] Majumdar, Parthasarathi (1998). "Black Hole Entropy and Quantum Gravity" **73**. p. 147. arXiv:gr-qc/9807045. Bibcode:1999InJPB..73..147M.

[10] Bekenstein, Jacob D. (August 2003). "Information in the Holographic Universe — Theoretical results about black holes suggest that the universe could be like a gigantic hologram". *Scientific American*: p. 59.

[11] Dennett, Daniel (1991). *Consciousness Explained*. New York: Back Bay Books. p. 371. ISBN 0-316-18066-1.

[12] Susskind, L. (February 2003). "The Anthropic landscape of string theory". arXiv:hep-th/0302219.

[13] Brown, J. D. & Henneaux, M. (1986). "Central charges in the canonical realization of asymptotic symmetries: an example from three-dimensional gravity". *Communications in Mathematical Physics* **104** (2): 207–226. Bibcode:1986CMaPh.104..207B. doi:10.1007/BF01211590..

[14] Information in the Holographic Universe

[15] http://webcache.googleusercontent.com/search?q=cache:E360V697cvgJ:ref-sciam.livejournal.com/1190.html&hl=en&gl=us&strip=1

[16] Hogan, Craig J. (2008). "Measurement of quantum fluctuations in geometry". *Physical Review D* **77** (10): 104031. arXiv:0712.3419. Bibcode:2008PhRvD..77j4031H. doi:10.1103/PhysRevD.77.104031..

[17] Chown, Marcus (15 January 2009). "Our world may be a giant hologram". *NewScientist*. Retrieved 2010-04-19.

[18] "Consequently, he ends up with inequalities of the type... Except that one may look at the actual equations of Matrix theory and see that none of these commutators is nonzero... The last displayed inequality above obviously can't be a consequence of quantum gravity because it doesn't depend on G at all! However, in the G→0 limit, one must reproduce nongravitational physics in the flat Euclidean background spacetime. Hogan's rules don't have the right limit so they can't be right." – Lubos Motl, Hogan's holographic noise doesn't exist, Feb 7, 2012

[19] "Integral challenges physics beyond Einstein". European Space Agency. 30 June 2011. Retrieved 3 February 2013.

[20] "Frequently Asked Questions for the Holometer at Fermilab". 6 July 2013. Retrieved 14 February 2014.

[21] Cowen, Ron (22 November 2012). "Single photon could detect quantum-scale black holes". *Nature*. Retrieved 3 February 2013.

[22] Cowen, Ron (10 December 2013). "Simulations back up theory that Universe is a hologram". *Nature News*. doi:10.1038/nature.2013.14328. Hyakutake, Yoshifumi (March 2014). "Quantum near-horizon geometry of a black 0-brane". *Progress of Theoretical and Experimental Physics* **2014** (3): 033B04. arXiv:1311.7526. Bibcode:2014PTEP.2014c3B04H. doi:10.1093/ptep/ptu028. Hanada, Masanori; Hyakutake, Yoshifumi; Ishiki, Goro; Nishimura, Jun (23 May 2014). "Holographic description of a quantum black hole on a computer". *Science* **344** (6186): 882–5. arXiv:1311.5607. Bibcode:2014Sci...344..882H. doi:10.1126/science.1250122.

[23] Yirka, Bob (December 13, 2013). "New work gives credence to theory of universe as a hologram". Phys.org.

26.10 External links

- UC Berkeley's Raphael Bousso gives an introductory lecture on the holographic principle - Video.

- *Scientific American* article on holographic principle by Jacob Bekenstein

Chapter 27

Phenomenological quantum gravity

Phenomenological quantum gravity is a research field in theoretical physics and a subfield of quantum gravity. Its objective is to find observable evidence for the quantization of gravity by the development of phenomenological models. These phenomenological models quantify possible quantum gravitational effects and can ideally be tested experimentally. In many cases predicted effects are too small to be measureable with presently available technology, but examples exist of models that have been ruled out already and others that can be tested in the near future.

The relevance of this research area derives from the fact that presently none of the candidate theories for quantum gravity has made contact to experiment. Phenomenological models are designed to bridge this gap by allowing physicists to test for general properties that the to-be-found theory of quantum gravity has. Even negative results are thus useful guides to the development of the theory by excluding possible properties. Phenomenological models are also necessary to assess the promise of future experiments.

27.1 References

- Amelino-Camelia, Giovanni (2008). "Quantum Gravity Phenomenology". *arXiv*.

- Hossenfelder, Sabine (2010). "Experimental Search for Quantum Gravity". *arXiv*.

Chapter 28

Quantum foam

Quantum foam (also referred to as **space-time foam**) is a concept in quantum mechanics devised by John Wheeler in 1955. The *foam* is supposed to be conceptualized as the foundation of the fabric of the universe.[1] Additionally, quantum foam can be used as a qualitative description of subatomic space-time turbulence at extremely small distances (on the order of the Planck length). At such small scales of time and space, the Heisenberg uncertainty principle allows energy to briefly decay into particles and antiparticles and then annihilate without violating physical conservation laws. As the scale of time and space being discussed shrinks, the energy of the virtual particles increases. According to Einstein's theory of general relativity, energy curves space-time. This suggests that—at sufficiently small scales—the energy of these fluctuations would be large enough to cause significant departures from the smooth space-time seen at larger scales, giving space-time a "foamy" character.

With an incomplete theory of quantum gravity, it is impossible to be certain what space-time would look like at these small scales, because existing theories of gravity do not give accurate predictions in that realm. Therefore, any of the developing theories of quantum gravity may improve our understanding of quantum foam as they are tested. However, observations of radiation from nearby quasars by Floyd Stecker of NASA's Goddard Space Flight Center have placed strong experimental limits on the possible violations of Einstein's special theory of relativity implied by the existence of quantum foam.[2] Thus experimental evidence so far has given a range of values in which scientists can test for quantum foam.

28.1 Experimental evidence (and counter-evidence)

The MAGIC (Major Atmospheric Gamma-ray Imaging Cherenkov) telescopes have detected that among gamma-ray photons arriving from the blazar Markarian 501, some

photons at different energy levels arrived at different times, suggesting that some of the photons had moved more slowly and thus contradicting the theory of general relativity's notion of the speed of light being constant, a discrepancy which could be explained by the irregularity of quantum foam.[3] More recent experiments were however unable to confirm the supposed variation on the speed of light due to graininess of space.[4] Other experiments involving the polarization of light from distant gamma ray bursts have also produced contradictory results.[5] More Earth-based experiments are ongoing[6] or proposed.[7]

28.1.1 Constraints and Limits

X-ray and gamma-ray observations of quasars used data from NASA's Chandra X-ray Observatory, the Fermi Gamma-ray Space Telescope and ground-based gamma-ray observations from the Very Energetic Radiation Imaging Telescope Array (VERITAS) show that space-time is uniform down to distances 1000 times smaller than the nucleus of a hydrogen atom.

Quantum mechanics predicts that space-time is not smooth, instead space-time would have a foamy, jittery nature and would consist of many small, ever-changing, regions in which space and time are not definite, but fluctuate.

The predicted scale of space-time foam is about ten times a billionth of the diameter of a hydrogen atom's nucleus, which cannot be measured directly. A foamy space-time would have limits on the accuracy with which distances can be measured because the size of the many quantum bubbles through which light travels will fluctuate. Depending on the space-time model used, the space-time uncertainties accumulate at different rates as light travels through the vast distances.

Chandra's X-ray detection of quasars at distances of billions of light years rules out the model where photons diffuse randomly through space-time foam, similar to light diffusing passing through fog.

Measurements of quasars at shorter, gamma-ray wavelengths with Fermi, and, shorter wavelengths with VERITAS rule out a second model, called a holographic model with less diffusion.[8][9][10][11]

28.2   Relation to other theories

Quantum foam is theorized to be the 'fabric' of the Universe, but cannot be observed yet because it is too small. Also, quantum foam is theorized to be created by virtual particles of very high energy. Virtual particles appear in quantum field theory, arising briefly and then annihilating during particle interactions in such a way that they affect the measured outputs of the interaction, even though the virtual particles are themselves space. These "vacuum fluctuations" affect the properties of the vacuum, giving it a nonzero energy known as vacuum energy, itself a type of zero-point energy. However, physicists are uncertain about the magnitude of this form of energy.[12]

The Casimir effect can also be understood in terms of the behavior of virtual particles in the empty space between two parallel plates. Ordinarily, quantum field theory does not deal with virtual particles of sufficient energy to curve spacetime significantly, so quantum foam is a speculative extension of these concepts which imagines the consequences of such high-energy virtual particles at very short distances and times. Spin foam theory is a modern attempt to make Wheeler's idea quantitative.

28.3   See also

28.4   Footnotes

[1] "Quantum foam". New Scientist. Retrieved 29 June 2008.

- Prof. Derek Leinweber has created calculations of quantum chromodynamics vacuum structure which Nobel laureate Frank Wilczek displayed during his 2004 Nobel lecture, and reviewed at the 2010 Robert Oppenheimer lecture, UC Berkeley Physics department (see the video starting at minute 32:30, up to minute 34:59)

[2] "Einstein makes extra dimensions toe the line". NASA. Retrieved 9 February 2012.

[3] "Gamma Ray Delay May Be Sign of 'New Physics'".

[4] doi:10.1038/nature.2012.9768

[5] Integral challenges physics beyond Einstein / Space Science / Our Activities / ESA

[6] Moyer, Michael (17 January 2012). "Is Space Digital?:". *Scientific American*. Retrieved 3 February 2013.

[7] Cowen, Ron (22 November 2012). "Single photon could detect quantum-scale black holes". *Nature News*. Retrieved 3 February 2013.

[8] "Chandra Press Room :: NASA Telescopes Set Limits on Space-time Quantum "Foam":: 28 May 15". *chandra.si.edu*. Retrieved 2015-05-29.

[9] "Chandra X-ray Observatory - NASA's flagship X-ray telescope". *chandra.si.edu*. Retrieved 2015-05-29.

[10] "[1411.7262] New Constraints on Quantum Gravity from X-ray and Gamma-Ray Observations". *arxiv.org*. Retrieved 2015-05-29.

[11] "Chandra :: Photo Album :: Space-time Foam :: May 28, 2015". *chandra.si.edu*. Retrieved 2015-05-29.

[12] Baez, John (2006-10-08). "What's the Energy Density of the Vacuum?". Retrieved 2007-12-18.

28.5   References

- John Archibald Wheeler with Kenneth Ford. *Geons, Black Holes, and Quantum Foam*. 1995 ISBN 0-393-04642-7.

- Reginald T. Cahill. *Gravity as Quantum Foam In-Flow*. June 2003.

- Borrowed Time: Interview with Michio Kaku, Scientific American

- Swarup, A 2006, 'Sights set on quantum froth', New Scientist, 189, p. 18, Science Full Text Select (H.W. Wilson), EBSCOhost, viewed 10 February 2012.

Chapter 29

Cosmic microwave background

"CMB" redirects here. For other uses, see CMB (disambiguation).

The **cosmic microwave background** (**CMB**) is the thermal radiation left over from the time of recombination in Big Bang cosmology. In older literature, the CMB is also variously known as cosmic microwave background radiation (CMBR) or "relic radiation." The CMB is a cosmic background radiation that is fundamental to observational cosmology because it is the oldest light in the universe, dating to the epoch of recombination. With a traditional optical telescope, the space between stars and galaxies (the *background*) is completely dark. However, a sufficiently sensitive radio telescope shows a faint background glow, almost exactly the same in all directions, that is not associated with any star, galaxy, or other object. This glow is strongest in the microwave region of the radio spectrum. The accidental discovery of CMB in 1964 by American radio astronomers Arno Penzias and Robert Wilson[1][2] was the culmination of work initiated in the 1940s, and earned the discoverers the 1978 Nobel Prize.

> *The CMB is a snapshot of the oldest light in our Universe, imprinted on the sky when the Universe was just 380,000 years old. It shows tiny temperature fluctuations that correspond to regions of slightly different densities, representing the seeds of all future structure: the stars and galaxies of today.*[3]

The CMB is well explained as radiation left over from an early stage in the development of the universe, and its discovery is considered a landmark test of the Big Bang model of the universe. When the universe was young, before the formation of stars and planets, it was denser, much hotter, and filled with a uniform glow from a white-hot fog of hydrogen plasma. As the universe expanded, both the plasma and the radiation filling it grew cooler. When the universe cooled enough, protons and electrons combined to form neutral atoms. These atoms could no longer absorb the thermal radiation, and so the universe became transparent instead of being an opaque fog.[4] Cosmologists refer to the time period when neutral atoms first formed as the *recombination epoch*, and the event shortly afterwards when photons started to travel freely through space rather than constantly being scattered by electrons and protons in plasma is referred to as photon decoupling. The photons that existed at the time of photon decoupling have been propagating ever since, though growing fainter and less energetic, since the expansion of space causes their wavelength to increase over time (and wavelength is inversely proportional to energy according to Planck's relation). This is the source of the alternative term *relic radiation*. The *surface of last scattering* refers to the set of points in space at the right distance from us so that we are now receiving photons originally emitted from those points at the time of photon decoupling.

Precise measurements of the CMB are critical to cosmology, since any proposed model of the universe must explain this radiation. The CMB has a thermal black body spectrum at a temperature of 2.72548 ± 0.00057 K.[5] The spectral radiance dE_v/dv peaks at 160.2 GHz, in the microwave range of frequencies. (Alternatively if spectral radiance is defined as $dE_\lambda/d\lambda$ then the peak wavelength is 1.871 mm.) The glow is very nearly uniform in all directions, but the tiny residual variations show a very specific pattern, the same as that expected of a fairly uniformly distributed hot gas that has expanded to the current size of the universe. In particular, the spectral radiance at different angles of observation in the sky contains small anisotropies, or irregularities, which vary with the size of the region examined. They have been measured in detail, and match what would be expected if small thermal variations, generated by quantum fluctuations of matter in a very tiny space, had expanded to the size of the observable universe we see today. This is a very active field of study, with scientists seeking both better data (for example, the Planck spacecraft) and better interpretations of the initial conditions of expansion. Although many different processes might produce the general form of a black body spectrum, no model other than the Big

Bang has yet explained the fluctuations. As a result, most cosmologists consider the Big Bang model of the universe to be the best explanation for the CMB.

The high degree of uniformity throughout the observable universe and its faint but measured anisotropy lend strong support for the Big Bang model in general and the ΛCDM ("Lambda Cold Dark Matter") model in particular. Moreover, the fluctuations are coherent on angular scales that are larger than the apparent cosmological horizon at recombination. Either such coherence is acausally fine-tuned, or cosmic inflation occurred.[6][7]

29.1 Features

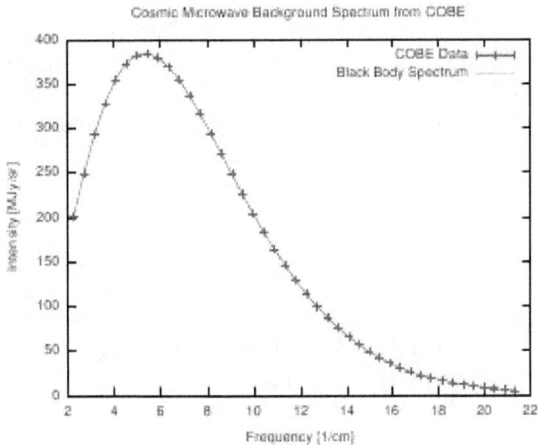

Graph of cosmic microwave background spectrum measured by the FIRAS instrument on the COBE, the most precisely measured black body spectrum in nature.[8] The error bars are too small to be seen even in an enlarged image, and it is impossible to distinguish the observed data from the theoretical curve.

The cosmic microwave background radiation is an emission of uniform, black body thermal energy coming from all parts of the sky. The radiation is isotropic to roughly one part in 100,000: the root mean square variations are only 18 µK,[9] after subtracting out a dipole anisotropy from the Doppler shift of the background radiation. The latter is caused by the peculiar velocity of the Earth relative to the comoving cosmic rest frame as the planet moves at some 371 km/s towards the constellation Leo. The CMB dipole as well as aberration at higher multipoles have been measured, consistent with galactic motion.[10]

In the Big Bang model for the formation of the universe, Inflationary Cosmology predicts that after about 10^{-37} seconds[11] the nascent universe underwent exponential growth that smoothed out nearly all inhomogeneities. The remaining inhomogeneities were caused by quantum fluctu-

ations in the inflaton field that caused the inflation event.[12] After 10^{-6} seconds, the early universe was made up of a hot, interacting plasma of photons, electrons, and baryons. As the universe expanded, adiabatic cooling caused the energy density of the plasma to decrease until it became favorable for electrons to combine with protons, forming hydrogen atoms. This recombination event happened when the temperature was around 3000 K or when the universe was approximately 379,000 years old.[13] At this point, the photons no longer interacted with the now electrically neutral atoms and began to travel freely through space, resulting in the decoupling of matter and radiation.[14]

The color temperature of the ensemble of decoupled photons has continued to diminish ever since; now down to 2.7260±0.0013 K,[5] it will continue to drop as the universe expands. The intensity of the radiation also corresponds to black-body radiation at 2.726 K because red-shifted black-body radiation is just like black-body radiation at a lower temperature. According to the Big Bang model, the radiation from the sky we measure today comes from a spherical surface called *the surface of last scattering*. This represents the set of locations in space at which the decoupling event is estimated to have occurred[15] and at a point in time such that the photons from that distance have just reached observers. Most of the radiation energy in the universe is in the cosmic microwave background,[16] making up a fraction of roughly 6×10^{-5} of the total density of the universe.[17]

Two of the greatest successes of the Big Bang theory are its prediction of the almost perfect black body spectrum and its detailed prediction of the anisotropies in the cosmic microwave background. The CMB spectrum has become the most precisely measured black body spectrum in nature.[8]

Density of energy for CMB is 0.25 eV/cm^{3}[18] (4.005×10^{-14} J/m^3) or (400–500 photons/cm^3[19]).

29.2 History

See also: Discovery of cosmic microwave background radiation

The cosmic microwave background was first predicted in 1948 by Ralph Alpher, and Robert Herman.[20][21][22] Alpher and Herman were able to estimate the temperature of the cosmic microwave background to be 5 K, though two years later they re-estimated it at 28 K. This high estimate was due to a mis-estimate of the Hubble constant by Alfred Behr, which could not be replicated and was later abandoned for the earlier estimate. Although there were several previous estimates of the temperature of space, these suffered from two flaws. First, they were measurements of the *effective* temperature of space and did not suggest that space was filled with a thermal Planck spectrum. Next, they depend on our being at a special spot at the edge of the Milky Way galaxy and they did not suggest the radiation is isotropic. The estimates would yield very different predictions if Earth happened to be located elsewhere in the universe.[23]

The 1948 results of Alpher and Herman were discussed in many physics settings through about 1955, when both left the Applied Physics Laboratory at Johns Hopkins University. The mainstream astronomical community, however, was not intrigued at the time by cosmology. Alpher and Herman's prediction was rediscovered by Yakov Zel'dovich in the early 1960s, and independently predicted by Robert Dicke at the same time. The first published recognition of the CMB radiation as a detectable phenomenon appeared in a brief paper by Soviet astrophysicists A. G. Doroshkevich and Igor Novikov, in the spring of 1964.[24] In 1964, David Todd Wilkinson and Peter Roll, Dicke's colleagues at Princeton University, began constructing a Dicke radiometer to measure the cosmic microwave background.[25] In 1964, Arno Penzias and Robert Woodrow Wilson at the Crawford Hill location of Bell Telephone Laboratories in nearby Holmdel Township, New Jersey had built a Dicke radiometer that they intended to use for radio astronomy and satellite communication experiments. On 20 May 1964 they made their first measurement clearly showing the presence of the microwave background,[26] with their instrument having an excess 4.2K antenna temperature which they could not account for. After receiving a telephone call from Crawford Hill, Dicke famously quipped: "Boys, we've been scooped."[1][27][28] A meeting between the Princeton and Crawford Hill groups determined that the antenna temperature was indeed due to the microwave background. Penzias and Wilson received the 1978 Nobel Prize in Physics for their discovery.[29]

The interpretation of the cosmic microwave background was a controversial issue in the 1960s with some proponents of the steady state theory arguing that the microwave background was the result of scattered starlight from distant galaxies.[30] Using this model, and based on the study of narrow absorption line features in the spectra of stars, the astronomer Andrew McKellar wrote in 1941: "It can be calculated that the 'rotational temperature' of interstellar space is 2 K."[31] However, during the 1970s the consensus was established that the cosmic microwave background is a remnant of the big bang. This was largely because new measurements at a range of frequencies showed that the spectrum was a thermal, black body spectrum, a result that the steady state model was unable to reproduce.[32]

Harrison, Peebles, Yu and Zel'dovich realized that the early universe would have to have inhomogeneities at the level of 10^{-4} or 10^{-5}.[33][34][35] Rashid Sunyaev later calculated the observable imprint that these inhomogeneities would

The Holmdel Horn Antenna on which Penzias and Wilson discovered the cosmic microwave background.

have on the cosmic microwave background.[36] Increasingly stringent limits on the anisotropy of the cosmic microwave background were set by ground based experiments during the 1980s. RELIKT-1, a Soviet cosmic microwave background anisotropy experiment on board the Prognoz 9 satellite (launched 1 July 1983) gave upper limits on the large-scale anisotropy. The NASA COBE mission clearly confirmed the primary anisotropy with the Differential Microwave Radiometer instrument, publishing their findings in 1992.[37][38] The team received the Nobel Prize in physics for 2006 for this discovery.

Inspired by the COBE results, a series of ground and balloon-based experiments measured cosmic microwave background anisotropies on smaller angular scales over the next decade. The primary goal of these experiments was to measure the scale of the first acoustic peak, which COBE did not have sufficient resolution to resolve. This peak corresponds to large scale density variations in the early universe that are created by gravitational instabilities, resulting in acoustical oscillations in the plasma.[39] The first peak in the anisotropy was tentatively detected by the Toco experiment and the result was confirmed by the BOOMERanG and MAXIMA experiments.[40][41][42] These measurements demonstrated that the geometry of the universe is approximately flat, rather than curved.[43] They ruled out cosmic strings as a major component of cosmic structure formation and suggested cosmic inflation was the right theory of structure formation.[44]

The second peak was tentatively detected by several experiments before being definitively detected by WMAP, which has also tentatively detected the third peak.[45] As of 2010, several experiments to improve measurements of the polarization and the microwave background on small angular scales are ongoing. These include DASI, WMAP,

BOOMERanG, QUaD, Planck spacecraft, Atacama Cosmology Telescope, South Pole Telescope and the QUIET telescope.

29.2.1 Timeline

Thermal (non-microwave background) temperature predictions

- 1896 – Charles Édouard Guillaume estimates the "radiation of the stars" to be 5.6K.[46]

- 1926 – Sir Arthur Eddington estimates the non-thermal radiation of starlight in the galaxy "... by the formula $E = \sigma T^4$ the effective temperature corresponding to this density is $3.18°$ absolute ... black body"[47]

- 1930s – Cosmologist Erich Regener calculates that the non-thermal spectrum of cosmic rays in the galaxy has an effective temperature of 2.8 K

- 1931 – Term *microwave* first used in print: "When trials with wavelengths as low as 18 cm. were made known, there was undisguised surprise+that the problem of the micro-wave had been solved so soon." Telegraph & Telephone Journal XVII. 179/1

- 1934 – Richard Tolman shows that black-body radiation in an expanding universe cools but remains thermal

- 1938 – Nobel Prize winner (1920) Walther Nernst reestimates the cosmic ray temperature as 0.75K

- 1941 – Andrew McKellar was attempting to measure the average temperature of the interstellar medium, and used the excitation of CN doublet lines to measure that the "effective temperature of space" (the average bolometric temperature) is about 2.3 K[31][48]

- 1946 – Robert Dicke predicts "... radiation from cosmic matter" at <20 K, but did not refer to background radiation [49]

- 1946 – George Gamow calculates a temperature of 50 K (assuming a 3-billion year old universe),[50] commenting it "... is in reasonable agreement with the actual temperature of interstellar space", but does not mention background radiation.[51]

- 1953 – Erwin Finlay-Freundlich in support of his tired light theory, derives a blackbody temperature for intergalactic space of 2.3K [52] with comment from Max Born suggesting radio astronomy as the arbitrator between expanding and infinite cosmologies.

Microwave background radiation predictions

- 1946 – George Gamow calculates a temperature of 50 K (assuming a 3-billion year old universe),[50] commenting it "... is in reasonable agreement with the actual temperature of interstellar space", but does not mention background radiation.

- 1948 – Ralph Alpher and Robert Herman estimate "the temperature in the universe" at 5 K. Although they do not specifically mention microwave background radiation, it may be inferred.[53]

- 1949 – Ralph Alpher and Robert Herman re-reestimate the temperature at 28 K.

- 1953 – George Gamow estimates 7 K.[49]

- 1956 – George Gamow estimates 6 K.[49]

- 1955 – Émile Le Roux of the Nançay Radio Observatory, in a sky survey at $\lambda = 33$ cm, reported a near-isotropic background radiation of 3 kelvins, plus or minus 2.[49]

- 1957 – Tigran Shmaonov reports that "the absolute effective temperature of the radioemission background ... is 4±3 K".[54] It is noted that the "measurements showed that radiation intensity was independent of either time or direction of observation ... it is now clear that Shmaonov did observe the cosmic microwave background at a wavelength of 3.2 cm"[55][56]

- 1960s – Robert Dicke re-estimates a microwave background radiation temperature of 40 K[49][57]

- 1964 – A. G. Doroshkevich and Igor Dmitrievich Novikov publish a brief paper suggesting microwave searches for the black-body radiation predicted by Gamow, Alpher, and Herman, where they name the CMB radiation phenomenon as detectable.[58]

- 1964–65 – Arno Penzias and Robert Woodrow Wilson measure the temperature to be approximately 3 K. Robert Dicke, James Peebles, P. G. Roll, and D. T. Wilkinson interpret this radiation as a signature of the big bang.

- 1966 – Rainer K. Sachs and Arthur M. Wolfe theoretically predict microwave background fluctuation amplitudes created by gravitational potential variations between observers and the last scattering surface (see Sachs-Wolfe effect)

- 1968 – Martin Rees and Dennis Sciama theoretically predict microwave background fluctuation amplitudes created by photons traversing time-dependent potential wells

- 1969 – R. A. Sunyaev and Yakov Zel'dovich study the inverse Compton scattering of microwave background photons by hot electrons (see Sunyaev-Zel'dovich effect)

- 1983 – Researchers from the Cambridge Radio Astronomy Group and the Owens Valley Radio Observatory first detect the Sunyaev-Zel'dovich effect from clusters of galaxies

- 1983 – RELIKT-1 Soviet CMB anisotropy experiment was launched.

- 1990 – FIRAS on the Cosmic Background Explorer (COBE) satellite measures the black body form of the CMB spectrum with exquisite precision, and shows that the microwave background has a nearly perfect black-body spectrum and thereby strongly constrains the density of the intergalactic medium.

- January 1992 – Scientists that analysed data from the RELIKT-1 report the discovery of anisotropy in the cosmic microwave background at the Moscow astrophysical seminar.[59]

- 1992 – Scientists that analysed data from COBE DMR report the discovery of anisotropy in the cosmic microwave background.[60]

- 1995 – The Cosmic Anisotropy Telescope performs the first high resolution observations of the cosmic microwave background.

- 1999 – First measurements of acoustic oscillations in the CMB anisotropy angular power spectrum from the TOCO, BOOMERANG, and Maxima Experiments. The BOOMERanG experiment makes higher quality maps at intermediate resolution, and confirms that the universe is "flat".

- 2002 – Polarization discovered by DASI.[61]

- 2003 – E-mode polarization spectrum obtained by the CBI.[62] The CBI and the Very Small Array produces yet higher quality maps at high resolution (covering small areas of the sky).

- 2003 – The WMAP spacecraft produces an even higher quality map at low and intermediate resolution of the whole sky (WMAP provides *no* high-resolution data, but improves on the intermediate resolution maps from BOOMERanG).

- 2004 – E-mode polarization spectrum obtained by the CBI.[63]

- 2004 – The Arcminute Cosmology Bolometer Array Receiver produces a higher quality map of the high resolution structure not mapped by WMAP.

- 2005 – The Arcminute Microkelvin Imager and the Sunyaev-Zel'dovich Array begin the first surveys for very high redshift clusters of galaxies using the Sunyaev-Zel'dovich effect.

- 2005 – Ralph A. Alpher is awarded the National Medal of Science for his groundbreaking work in nucleosynthesis and prediction that the universe expansion leaves behind background radiation, thus providing a model for the Big Bang theory.

- 2006 – The long-awaited three-year WMAP results are released, confirming previous analysis, correcting several points, and including polarization data.

- 2006 – Two of COBE's principal investigators, George Smoot and John Mather, received the Nobel Prize in Physics in 2006 for their work on precision measurement of the CMBR.

- 2006-2011 – Improved measurements from WMAP, new supernova surveys ESSENCE and SNLS, and baryon acoustic oscillations from SDSS and WiggleZ, continue to be consistent with the standard Lambda-CDM model.

- 2014 – On March 17, 2014, astrophysicists of the BICEP2 collaboration announced the detection of inflationary gravitational waves in the B-mode power spectrum, which if confirmed, would provide clear experimental evidence for the theory of inflation.[64][65][66][67][68][69] However, on 19 June 2014, lowered confidence in confirming the cosmic inflation findings was reported.[68][70][71]

- 2015 – On January 30, 2015, the same team of astronomers from BICEP2 withdrew the claim made on the previous year. Based on the combined data of BICEP2 and Planck, the European Space Agency announced that the signal can be entirely attributed to dust in the Milky Way.[72]

29.3 Relationship to the Big Bang

The cosmic microwave background radiation and the cosmological redshift-distance relation are together regarded as the best available evidence for the Big Bang theory. Measurements of the CMB have made the inflationary Big Bang theory the Standard Model of Cosmology.[73] The discovery of the CMB in the mid-1960s curtailed interest in alternatives such as the steady state theory.[74]

The CMB essentially confirms the Big Bang theory. In the late 1940s Alpher and Herman reasoned that if there was a big bang, the expansion of the universe would have stretched and cooled the high-energy radiation of the very early universe into the microwave region and down to a temperature of about 5 K. They were slightly off with their estimate, but they had exactly the right idea. They predicted the CMB. It took another 15 years for Penzias and Wilson to stumble into discovering that the microwave background was actually there.[75]

The CMB gives a snapshot of the universe when, according to standard cosmology, the temperature dropped enough to allow electrons and protons to form hydrogen atoms, thus making the universe transparent to radiation. When it originated some 380,000 years after the Big Bang—this time is generally known as the "time of last scattering" or the period of recombination or decoupling—the temperature of the universe was about 3000 K. This corresponds to an energy of about 0.25 eV, which is much less than the 13.6 eV ionization energy of hydrogen.[76]

Since decoupling, the temperature of the background radiation has dropped by a factor of roughly 1,100[77] due to the expansion of the universe. As the universe expands, the CMB photons are redshifted, making the radiation's temperature inversely proportional to a parameter called the universe's scale length. The temperature T_r of the CMB as a function of redshift, z, can be shown to be proportional to the temperature of the CMB as observed in the present day (2.725 K or 0.235 meV):[78]

$$T_r = 2.725(1 + z)$$

For details about the reasoning that the radiation is evidence for the Big Bang, see Cosmic background radiation of the Big Bang.

29.3.1   Primary anisotropy

The anisotropy of the cosmic microwave background is divided into two types: primary anisotropy, due to effects which occur at the last scattering surface and before; and secondary anisotropy, due to effects such as interactions of the background radiation with hot gas or gravitational potentials, which occur between the last scattering surface and the observer.

The structure of the cosmic microwave background anisotropies is principally determined by two effects: acoustic oscillations and diffusion damping (also called collisionless damping or Silk damping). The acoustic oscillations arise because of a conflict in the photon–baryon plasma in the early universe. The pressure of the photons tends to erase anisotropies, whereas the gravitational attraction of the baryons—moving at speeds much slower than light—makes them tend to collapse to form dense

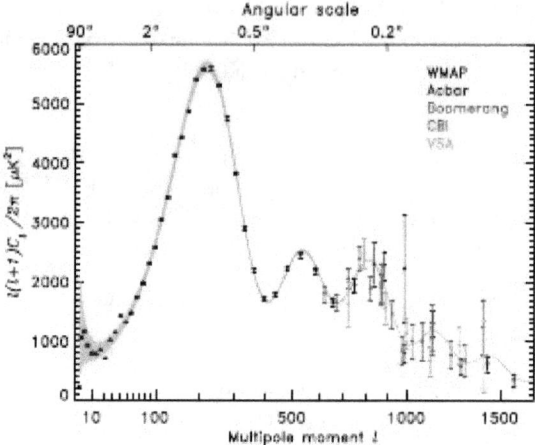

The power spectrum of the cosmic microwave background radiation temperature anisotropy in terms of the angular scale (or multipole moment). The data shown comes from the WMAP (2006), Acbar (2004) Boomerang (2005), CBI (2004), and VSA (2004) instruments. Also shown is a theoretical model (solid line).

haloes. These two effects compete to create acoustic oscillations which give the microwave background its characteristic peak structure. The peaks correspond, roughly, to resonances in which the photons decouple when a particular mode is at its peak amplitude.

The peaks contain interesting physical signatures. The angular scale of the first peak determines the curvature of the universe (but not the topology of the universe). The next peak—ratio of the odd peaks to the even peaks—determines the reduced baryon density.[79] The third peak can be used to get information about the dark matter density.[80]

The locations of the peaks also give important information about the nature of the primordial density perturbations. There are two fundamental types of density perturbations—called *adiabatic* and *isocurvature*. A general density perturbation is a mixture of both, and different theories that purport to explain the primordial density perturbation spectrum predict different mixtures.

- Adiabatic density perturbations

  the fractional additional density of each type of particle (baryons, photons ...) is the same. That is, if at one place there is 1% more energy in baryons than average, then at that place there is also 1% more energy in photons (and 1% more energy in neutrinos) than average. Cosmic inflation predicts that the primordial perturbations are adiabatic.

- Isocurvature density perturbations

  in each place the sum (over different types of particle) of the fractional additional densities is zero. That is, a perturbation where at some spot there is 1% more energy in baryons than average, 1% more energy in photons than average, and 2% *less* energy in neutrinos than average, would be a pure isocurvature perturbation. Cosmic strings would produce mostly isocurvature primordial perturbations.

The CMB spectrum can distinguish between these two because these two types of perturbations produce different peak locations. Isocurvature density perturbations produce a series of peaks whose angular scales (*l*-values of the peaks) are roughly in the ratio 1:3:5:..., while adiabatic density perturbations produce peaks whose locations are in the ratio 1:2:3:...[81] Observations are consistent with the primordial density perturbations being entirely adiabatic, providing key support for inflation, and ruling out many models of structure formation involving, for example, cosmic strings.

Collisionless damping is caused by two effects, when the treatment of the primordial plasma as fluid begins to break down:

- the increasing mean free path of the photons as the primordial plasma becomes increasingly rarefied in an expanding universe

- the finite depth of the last scattering surface (LSS), which causes the mean free path to increase rapidly during decoupling, even while some Compton scattering is still occurring.

These effects contribute about equally to the suppression of anisotropies at small scales, and give rise to the characteristic exponential damping tail seen in the very small angular scale anisotropies.

The depth of the LSS refers to the fact that the decoupling of the photons and baryons does not happen instantaneously, but instead requires an appreciable fraction of the age of the universe up to that era. One method of quantifying how long this process took uses the *photon visibility function* (PVF). This function is defined so that, denoting the PVF by P(t), the probability that a CMB photon last scattered between time t and t+dt is given by P(t)dt.

The maximum of the PVF (the time when it is most likely that a given CMB photon last scattered) is known quite precisely. The first-year WMAP results put the time at which P(t) is maximum as 372,000 years.[82] This is often taken as the "time" at which the CMB formed. However, to figure

out how *long* it took the photons and baryons to decouple, we need a measure of the width of the PVF. The WMAP team finds that the PVF is greater than half of its maximum value (the "full width at half maximum", or FWHM) over an interval of 115,000 years. By this measure, decoupling took place over roughly 115,000 years, and when it was complete, the universe was roughly 487,000 years old.

29.3.2 Late time anisotropy

Since the CMB came into existence, it has apparently been modified by several subsequent physical processes, which are collectively referred to as late-time anisotropy, or secondary anisotropy. When the CMB photons became free to travel unimpeded, ordinary matter in the universe was mostly in the form of neutral hydrogen and helium atoms. However, observations of galaxies today seem to indicate that most of the volume of the intergalactic medium (IGM) consists of ionized material (since there are few absorption lines due to hydrogen atoms). This implies a period of reionization during which some of the material of the universe was broken into hydrogen ions.

The CMB photons are scattered by free charges such as electrons that are not bound in atoms. In an ionized universe, such charged particles have been liberated from neutral atoms by ionizing (ultraviolet) radiation. Today these free charges are at sufficiently low density in most of the volume of the universe that they do not measurably affect the CMB. However, if the IGM was ionized at very early times when the universe was still denser, then there are two main effects on the CMB:

1. Small scale anisotropies are erased. (Just as when looking at an object through fog, details of the object appear fuzzy.)

2. The physics of how photons are scattered by free electrons (Thomson scattering) induces polarization anisotropies on large angular scales. This broad angle polarization is correlated with the broad angle temperature perturbation.

Both of these effects have been observed by the WMAP spacecraft, providing evidence that the universe was ionized at very early times, at a redshift more than 17. The detailed provenance of this early ionizing radiation is still a matter of scientific debate. It may have included starlight from the very first population of stars (population III stars), supernovae when these first stars reached the end of their lives, or the ionizing radiation produced by the accretion disks of massive black holes.

The time following the emission of the cosmic microwave background—and before the observation of the first stars—

is semi-humorously referred to by cosmologists as the dark age, and is a period which is under intense study by astronomers (See 21 centimeter radiation).

Two other effects which occurred between reionization and our observations of the cosmic microwave background, and which appear to cause anisotropies, are the Sunyaev–Zel'dovich effect, where a cloud of high-energy electrons scatters the radiation, transferring some of its energy to the CMB photons, and the Sachs–Wolfe effect, which causes photons from the Cosmic Microwave Background to be gravitationally redshifted or blueshifted due to changing gravitational fields.

29.4   Polarization

The cosmic microwave background is polarized at the level of a few microkelvin. There are two types of polarization, called E-modes and B-modes. This is in analogy to electrostatics, in which the electric field (E-field) has a vanishing curl and the magnetic field (B-field) has a vanishing divergence. The E-modes arise naturally from Thomson scattering in a heterogeneous plasma. The B-modes are not sourced by standard scalar type perturbations. Instead they can be created by two mechanisms: the first one is by gravitational lensing of E-modes, which has been measured by the South Pole Telescope in 2013;[83] the second one is from gravitational waves arising from cosmic inflation. Detecting the B-modes is extremely difficult, particularly as the degree of foreground contamination is unknown, and the weak gravitational lensing signal mixes the relatively strong E-mode signal with the B-mode signal.[84]

29.4.1   E-modes

E-modes were first seen in 2002 by the Degree Angular Scale Interferometer (DASI).

29.4.2   B-modes

Cosmologists predict two types of B-modes, the first generated during cosmic inflation shortly after the big bang,[85][86][87] and the second generated by gravitational lensing at later times.[88]

Primordial gravitational waves

Primordial gravitational waves are gravitational waves that could be observed in the polarisation of the cosmic microwave background and having their origin in the early

universe. Models of cosmic inflation predict that such gravitational waves should appear; thus, their detection supports the theory of inflation, and their strength can confirm and exclude different models of inflation. It is the result of three things: inflationary expansion of space itself, reheating after inflation, and turbulent fluid mixing of matter and radiation. [89]

On 17 March 2014 it was announced that the BICEP2 instrument had detected the first type of B-modes, consistent with inflation and gravitational waves in the early universe at the level of $r = 0.20+0.07$

-0.05, which is the amount of power present in gravitational waves compared to the amount of power present in other scalar density perturbations in the very early universe. Had this been confirmed it would have provided strong evidence of cosmic inflation and the Big Bang,[64][65][66][67][90][91][92] but on 19 June 2014, considerably lowered confidence in confirming the findings was reported[68][68][70][70][71][71] and on 19 September 2014 new results of the Planck experiment reported that the results of BICEP2 can be fully attributed to cosmic dust.[93][94]

Paul Steinhardt is skeptical, suggesting that light scattering from cosmic dust and synchrotron radiation from electrons, both in the Milky Way Galaxy, could have caused the readings.[95]

Gravitational lensing

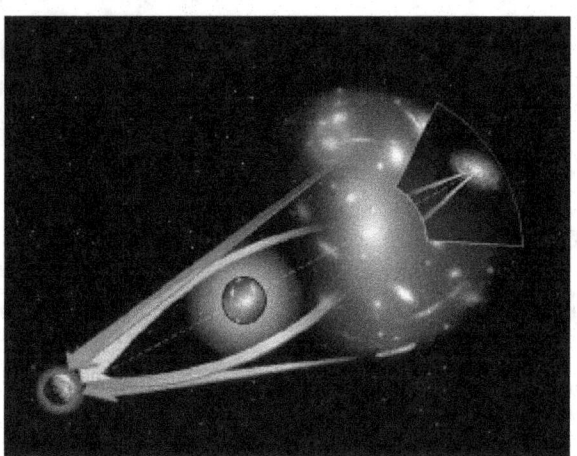

This artist's impression shows how light from the early universe is deflected by the gravitational lensing effect of massive cosmic structures forming B-modes as it travels across the universe. (Credit: ESA)

The second type of B-modes was discovered in 2013 using the South Pole Telescope with help from the Herschel Space Observatory.[96] This discovery may help test theories on the origin of the universe. Scientists are using data from the Planck mission by the European Space Agency, to gain a better understanding of these waves.[97][98][99]

In October 2014, a measurement of the B-mode polarization at 150 GHz was published by the POLARBEAR experiment.[100] Compared to BICEP2, POLARBEAR focuses on a smaller patch of the sky and is less susceptible to dust effects. The team reported that POLARBEAR's measured B-mode polarization was of cosmological origin (and not just due to dust) at a 97.2% confidence level.[101]

29.5 Microwave background observations

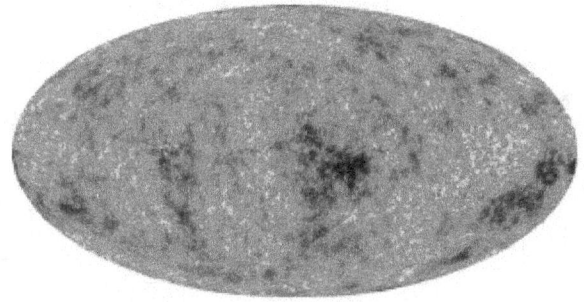

All-sky map of the CMB, created from 9 years of WMAP data.

Main article: List of cosmic microwave background experiments

Subsequent to the discovery of the CMB, hundreds of cosmic microwave background experiments have been conducted to measure and characterize the signatures of the radiation. The most famous experiment is probably the NASA Cosmic Background Explorer (COBE) satellite that orbited in 1989–1996 and which detected and quantified the large scale anisotropies at the limit of its detection capabilities. Inspired by the initial COBE results of an extremely isotropic and homogeneous background, a series of ground- and balloon-based experiments quantified CMB anisotropies on smaller angular scales over the next decade. The primary goal of these experiments was to measure the angular scale of the first acoustic peak, for which COBE did not have sufficient resolution. These measurements were able to rule out cosmic strings as the leading theory of cosmic structure formation, and suggested cosmic inflation was the right theory. During the 1990s, the first peak was measured with increasing sensitivity and by 2000 the BOOMERanG experiment reported that the highest power fluctuations occur at scales of approximately one degree. Together with other cosmological data, these results implied that the geometry of the universe is flat. A number of ground-based interferometers provided measurements of the fluctuations with higher accuracy over the next three

years, including the Very Small Array, Degree Angular Scale Interferometer (DASI), and the Cosmic Background Imager (CBI). DASI made the first detection of the polarization of the CMB and the CBI provided the first E-mode polarization spectrum with compelling evidence that it is out of phase with the T-mode spectrum.

In June 2001, NASA launched a second CMB space mission, WMAP, to make much more precise measurements of the large scale anisotropies over the full sky. WMAP used symmetric, rapid-multi-modulated scanning, rapid switching radiometers to minimize non-sky signal noise.[77] The first results from this mission, disclosed in 2003, were detailed measurements of the angular power spectrum at a scale of less than one degree, tightly constraining various cosmological parameters. The results are broadly consistent with those expected from cosmic inflation as well as various other competing theories, and are available in detail at NASA's data bank for Cosmic Microwave Background (CMB) (see links below). Although WMAP provided very accurate measurements of the large scale angular fluctuations in the CMB (structures about as broad in the sky as the moon), it did not have the angular resolution to measure the smaller scale fluctuations which had been observed by former ground-based interferometers.

A third space mission, the ESA (European Space Agency) Planck Surveyor, was launched in May 2009 and is currently performing an even more detailed investigation. Planck employs both HEMT radiometers and bolometer technology and will measure the CMB at a smaller scale than WMAP. Its detectors were trialled in the Antarctic Viper telescope as ACBAR (Arcminute Cosmology Bolometer Array Receiver) experiment—which has produced the most precise measurements at small angular scales to date—and in the Archeops balloon telescope.

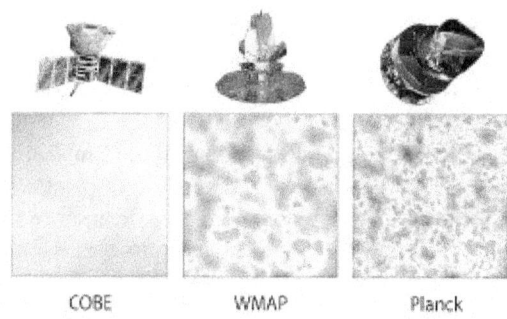

COBE   WMAP   Planck

Comparison of CMB results from COBE, WMAP and Planck – March 21, 2013.

On 21 March 2013, the European-led research team behind the Planck cosmology probe released the mission's all-sky map (565x318 jpeg, 3600x1800 jpeg) of the cosmic microwave background.[102][103] The map suggests the universe is slightly older than researchers thought. According-ing to the map, subtle fluctuations in temperature were imprinted on the deep sky when the cosmos was about 370,000 years old. The imprint reflects ripples that arose as early, as the existence of the universe, as the first nonillionth of a second. Apparently, these ripples gave rise to the present vast cosmic web of galaxy clusters and dark matter. Based on the 2013 data, the universe contains 4.9% ordinary matter, 26.8% dark matter and 68.3% dark energy. On 5 February 2015, new data was released by the Planck mission, according to which the age of the universe is 13.799 ± 0.021 billion years old and the Hubble constant was measured to be 67.74 ± 0.46 (km/s)/Mpc.[104]

Additional ground-based instruments such as the South Pole Telescope in Antarctica and the proposed Clover Project, Atacama Cosmology Telescope and the QUIET telescope in Chile will provide additional data not available from satellite observations, possibly including the B-mode polarization.

29.6   Data reduction and analysis

Raw CMBR data from the space vehicle (i.e. WMAP) contain foreground effects that completely obscure the fine-scale structure of the cosmic microwave background. The fine-scale structure is superimposed on the raw CMBR data but is too small to be seen at the scale of the raw data. The most prominent of the foreground effects is the dipole anisotropy caused by the Sun's motion relative to the CMBR background. The dipole anisotropy and others due to Earth's annual motion relative to the Sun and numerous microwave sources in the galactic plane and elsewhere must be subtracted out to reveal the extremely tiny variations characterizing the fine-scale structure of the CMBR background.

The detailed analysis of CMBR data to produce maps, an angular power spectrum, and ultimately cosmological parameters is a complicated, computationally difficult problem. Although computing a power spectrum from a map is in principle a simple Fourier transform, decomposing the map of the sky into spherical harmonics, in practice it is hard to take the effects of noise and foreground sources into account. In particular, these foregrounds are dominated by galactic emissions such as Bremsstrahlung, synchrotron, and dust that emit in the microwave band; in practice, the galaxy has to be removed, resulting in a CMB map that is not a full-sky map. In addition, point sources like galaxies and clusters represent another source of foreground which must be removed so as not to distort the short scale structure of the CMB power spectrum.

Constraints on many cosmological parameters can be obtained from their effects on the power spectrum, and results are often calculated using Markov Chain Monte Carlo

sampling techniques.

29.6.1   CMBR dipole anisotropy

From the CMB data it is seen that our local group of galaxies (the galactic cluster that includes the Solar System's Milky Way Galaxy) appears to be moving at 627 ± 22 km/s relative to the reference frame of the CMB (also called the CMB rest frame, or the frame of reference in which there is no motion through the CMB) in the direction of galactic longitude $l = 276°\pm3°$, $b = 30°\pm3°$.[105][106] This motion results in an anisotropy of the data (CMB appearing slightly warmer in the direction of movement than in the opposite direction).[107] From a theoretical point of view, the existence of a CMB rest frame breaks Lorentz invariance even in empty space far away from any galaxy.[108] The standard interpretation of this temperature variation is a simple velocity red shift and blue shift due to motion relative to the CMB, but alternative cosmological models can explain some fraction of the observed dipole temperature distribution in the CMB.[109]

29.6.2   Low multipoles and other anomalies

With the increasingly precise data provided by WMAP, there have been a number of claims that the CMB exhibits anomalies, such as very large scale anisotropies, anomalous alignments, and non-Gaussian distributions.[110][111][112][113] The most longstanding of these is the low-l multipole controversy. Even in the COBE map, it was observed that the quadrupole ($l = 2$, spherical harmonic) has a low amplitude compared to the predictions of the Big Bang. In particular, the quadrupole and octupole ($l = 3$) modes appear to have an unexplained alignment with each other and with both the ecliptic plane and equinoxes,[114][115][116] an alignment sometimes referred to as the *axis of evil*.[111] A number of groups have suggested that this could be the signature of new physics at the greatest observable scales; other groups suspect systematic errors in the data.[117][118][119] Ultimately, due to the foregrounds and the cosmic variance problem, the greatest modes will never be as well measured as the small angular scale modes. The analyses were performed on two maps that have had the foregrounds removed as far as possible: the "internal linear combination" map of the WMAP collaboration and a similar map prepared by Max Tegmark and others.[45][77][120] Later analyses have pointed out that these are the modes most susceptible to foreground contamination from synchrotron, dust, and Bremsstrahlung emission, and from experimental uncertainty in the monopole and dipole. A full Bayesian analysis of the WMAP power spectrum demonstrates that the quadrupole prediction of Lambda-CDM cosmology is consistent with the data at the 10% level and that the observed octupole is not remarkable.[121] Carefully accounting for the procedure used to remove the foregrounds from the full sky map further reduces the significance of the alignment by ~5%.[122][123][124][125]

Recent observations with the Planck telescope, which is very much more sensitive than WMAP and has a larger angular resolution, confirm the observation of the axis of evil. Since two different instruments recorded the same anomaly, instrumental error (but not foreground contamination) appears to be ruled out.[126] Coincidence is a possible explanation, chief scientist from WMAP, Charles L. Bennett suggested coincidence and human psychology were involved, *"I do think there is a bit of a psychological effect; people want to find unusual things."*[127]

29.7 Future evolution

Assuming the universe keeps expanding and it does not suffer a Big Crunch, a Big Rip, or another similar fate, the cosmic microwave background will continue redshifting until it will no longer be detectable,[128] and will be overtaken first by the one produced by starlight, and later by the background radiation fields of processes that are assumed will take place in the far future of the universe.[129]. §VD.

29.8 In popular culture

- In the *Stargate Universe* TV series, an Ancient spaceship, *Destiny*, was built to study patterns in the CMBR which indicate that the universe as we know it might have been created by some form of sentient intelligence.[130]

- In *Wheelers*, a novel by Ian Stewart & Jack Cohen, CMBR is explained as the encrypted transmissions of an ancient civilization. This allows the Jovian "blimps" to have a society older than the currently-observed age of the universe.

- In *The Three-Body Problem*, a novel by Liu Cixin, CMBR becomes observable to the naked eye due to interference from an alien civilization.

29.9 See also

- Physical cosmology

- Observational cosmology

- Gravitational wave background

- Cosmic gravitational wave background

- Observation history of galaxies

- Lambda-CDM model

- Heat death of the universe

- Computational packages for Cosmologists

29.10 References

[1] Penzias, A. A.; Wilson, R. W. (1965). "A Measurement of Excess Antenna Temperature at 4080 Mc/s". *The Astrophysical Journal* **142** (1): 419–421. Bibcode:1965ApJ...142..419P. doi:10.1086/148307.

[2] Smoot Group (28 March 1996). "The Cosmic Microwave Background Radiation". Lawrence Berkeley Lab. Retrieved 2008-12-11.

[3] "Planck reveals an almost perfect Universe". Max Planck Gesellschaft. 21 March 2013. Retrieved 2013-06-03.

[4] Kaku, M. (2014). "First Second of the Big Bang". *How the Universe Works*. Discovery Science.

[5] Fixsen, D. J. (2009). "The Temperature of the Cosmic Microwave Background". *The Astrophysical Journal* **707** (2): 916–920. arXiv:0911.1955. Bibcode:2009ApJ...707..916F. doi:10.1088/0004-637X/707/2/916.

[6] Dodelson, S. (2003). "Coherent Phase Argument for Inflation". *AIP Conference Proceedings* **689**: 184–196. arXiv:hep-ph/0309057. Bibcode:2003AIPC..689..184D. doi:10.1063/1.1627736.

[7] Baumann, D. (2011). "The Physics of Inflation" (PDF). University of Cambridge. Retrieved 2015-05-09.

[8] White, M. (1999). "Anisotropies in the CMB". *Proceedings of the Los Angeles Meeting, DPF 99*. UCLA. arXiv:astro-ph/9903232. Bibcode:1999dpf..conf.....W.

[9] Wright, E.L. (2004). "Theoretical Overview of Cosmic Microwave Background Anisotropy". In W. L. Freedman. *Measuring and Modeling the Universe*. Carnegie Observatories Astrophysics Series. Cambridge University Press. p. 291. arXiv:astro-ph/0305591. ISBN 0-521-75576-X.

[10] The Planck Collaboration, *Planck 2013 results. XXVII. Doppler boosting of the CMB: Eppur si muove*, arXiv:1303.5087, Bibcode:2014A&A...571A..27P, doi:10.1051/0004-6361/201321556

[11] Guth, A. H. (1998). *The Inflationary Universe: The Quest for a New Theory of Cosmic Origins*. Basic Books. p. 186. ISBN 978-0201328400. OCLC 35701222.

[12] Cirigliano, D.; de Vega, H.J.; Sanchez, N. G. (2005). "Clarifying inflation models: The precise inflationary potential from effective field theory and the WMAP data".   *Physical Review D* **71** (10):   77–115. arXiv:astro-ph/0412634. Bibcode:2005PhRvD..71j3518C. doi:10.1103/PhysRevD.71.103518.

[13] Abbott, B. (2007). "Microwave (WMAP) All-Sky Survey". Hayden Planetarium. Retrieved 2008-01-13.

[14] Gawiser, E.; Silk, J. (2000). "The cosmic microwave background radiation". *Physics Reports*. 333–334; 245–267. arXiv:astro-ph/0002044.   Bibcode:2000PhR...333..245G. doi:10.1016/S0370-1573(00)00025-9.

[15] Smoot, G. F. (2006). "Cosmic Microwave Background Radiation Anisotropies: Their Discovery and Utilization". *Nobel Lecture*. Nobel Foundation. Retrieved 2008-12-22.

[16] Hobson, M.P.; Efstathiou, G.; Lasenby, A.N. (2006). *General Relativity: An Introduction for Physicists*. Cambridge University Press. p. 388. ISBN 0-521-82951-8.

[17] Unsöld, A.; Bodo, B. (2002). *The New Cosmos, An Introduction to Astronomy and Astrophysics* (5th ed.). Springer–Verlag. p. 485. ISBN 3-540-67877-8.

[18] Confrontation of Cosmological Theories with Observational Data, M. S. Longair, page 144

[19] Cosmology II: The thermal history of the Universe, Ruth Durrer

[20] Gamow, G. (1948).   "The Origin of Elements and the Separation of Galaxies".   *Physical Review* **74** (4):   505–506.   Bibcode:1948PhRv...74..505G. doi:10.1103/PhysRev.74.505.2.

[21] Gamow, G. (1948). "The evolution of the universe". *Nature* **162** (4122):   680–682.   Bibcode:1948Natur.162..680G. doi:10.1038/162680a0. PMID 18893719.

[22] Alpher, R. A.; Herman, R. C. (1948).   "On the Relative Abundance of the Elements".   *Physical Review* **74** (12):   1737–1742.   Bibcode:1948PhRv...74.1737A. doi:10.1103/PhysRev.74.1737.

[23] Assis, A. K. T.; Neves, M. C. D. (1995). "History of the 2.7 K Temperature Prior to Penzias and Wilson" (PDF) (3). pp. 79–87. but see also Wright, E. L. (2006). "Eddington's Temperature of Space". UCLA. Retrieved 2008-12-11.

[24] Penzias, A. A. (2006). "The origin of elements" (PDF). *Nobel lecture*. Nobel Foundation. Retrieved 2006-10-04.

[25] Dicke, R. H. (1946). "The Measurement of Thermal Radiation at Microwave Frequencies". *Review of Scientific Instruments* **17** (7): 268–275. Bibcode:1946RScI...17..268D. doi:10.1063/1.1770483. PMID 20991753. This basic design for a radiometer has been used in most subsequent cosmic microwave background experiments.

[26] The Cosmic Microwave Background Radiation (Nobel Lecture) by Robert Wilson 8 Dec 1978, p. 474

[27] Dicke, R. H.;   et al.   (1965).   "Cosmic Black-Body Radiation".   *Astrophysical Journal* **142**: 414–419. Bibcode:1965ApJ...142..414D. doi:10.1086/148306.

[28] The history is given in Peebles, P. J. E (1993). *Principles of Physical Cosmology*. Princeton University Press. pp. 139–148. ISBN 0-691-01933-9.

[29] "The Nobel Prize in Physics 1978".   Nobel Foundation. 1978. Retrieved 2009-01-08.

[30] Narlikar, J. V.; Wickramasinghe, N. C. (1967).   "Microwave Background in a Steady State Universe". *Nature* **216** (5110):   43–44.   Bibcode:1967Natur.216...43N. doi:10.1038/216043a0.

[31] McKellar, A.; Kan-Mitchell, June; Conti, Peter S. (1941). "Molecular Lines from the Lowest States of Diatomic Molecules Composed of Atoms Probably Present in Interstellar Space". *Publications of the Dominion Astrophysical Observatory (Victoria, BC)* **7** (6): 251–272.

[32] Peebles, P. J. E.; et al. (1991). "The case for the relativistic hot big bang cosmology". *Nature* **352** (6338): 769–776. Bibcode:1991Natur.352..769P. doi:10.1038/352769a0.

[33] Harrison, E. R. (1970).   "Fluctuations at the threshold of classical cosmology".   *Physical Review D* **1** (10):   2726–2730.   Bibcode:1970PhRvD...1.2726H. doi:10.1103/PhysRevD.1.2726.

[34] Peebles, P. J. E.; Yu, J. T. (1970). "Primeval Adiabatic Perturbation in an Expanding Universe". *Astrophysical Journal* **162**: 815–836. Bibcode:1970ApJ...162..815P. doi:10.1086/150713.

[35] Zeldovich, Y. B. (1972).   "A hypothesis, unifying the structure and the entropy of the Universe". *Monthly Notices of the Royal Astronomical Society* **160** (7–8): 1P–4P. doi:10.1016/S0026-0576(07)80178-4.

[36] Doroshkevich, A. G.; Zel'Dovich, Y. B.; Syunyaev, R. A. (1978) [12–16 September 1977].   "Fluctuations of the microwave background radiation in the adiabatic and entropic theories of galaxy formation".   In Longair, M. S.; Einasto, J. *The large scale structure of the universe; Proceedings of the Symposium*.   Tallinn, Estonian SSR: Dordrecht, D. Reidel Publishing Co.   pp.   393–404. Bibcode:1978IAUS...79..393S. While this is the first paper to discuss the detailed observational imprint of density inhomogeneities as anisotropies in the cosmic microwave background, some of the groundwork was laid in Peebles and Yu, above.

[37] Smooth, G. F.; et al.   (1992).   "Structure in the COBE differential microwave radiometer first-year maps". *Astrophysical Journal Letters* **396** (1): L1–L5. Bibcode:1992ApJ...396L...1S. doi:10.1086/186504.

[38] Bennett, C.L.; et al. (1996). "Four-Year COBE DMR Cosmic Microwave Background Observations: Maps and Basic Results". *Astrophysical Journal Letters* **464**: L1–L4.

arXiv:astro-ph/9601067.   Bibcode:1996ApJ...464L...1B.
doi:10.1086/310075.

[39] Grupen, C.; et al. (2005). *Astroparticle Physics*. Springer.
pp. 240–241. ISBN 3-540-25312-2.

[40] Miller, A. D.; et al. (1999). "A Measurement of
the Angular Power Spectrum of the Microwave Background Made from the High Chilean Andes". *Astrophysical
Journal* **521** (2): L79–L82.   arXiv:astro-ph/9905100.
Bibcode:1999ApJ...521L..79T. doi:10.1086/312197.

[41] Melchiorri, A.; et al. (2000). "A Measurement
of Ω from the North American Test Flight of
Boomerang". *Astrophysical Journal* **536** (2): L63–L66.
arXiv:astro-ph/9911445.   Bibcode:2000ApJ...536L..63M.
doi:10.1086/312744.

[42] Hanany, S.; et al. (2000). "MAXIMA-1: A Measurement of
the Cosmic Microwave Background Anisotropy on Angular
Scales of 10'–5°". *Astrophysical Journal* **545** (1): L5–L9.
arXiv:astro-ph/0005123.   Bibcode:2000ApJ...545L...5H.
doi:10.1086/317322.

[43] de Bernardis, P.; et al. (2000). "A flat Universe
from high-resolution maps of the cosmic microwave
background radiation". *Nature* **404** (6781): 955–959.
arXiv:astro-ph/0004404.   Bibcode:2000Natur.404..955D.
doi:10.1038/35010035. PMID 10801117.

[44] Pogosian, L.; et al. (2003). "Observational constraints on cosmic string production during brane
inflation". *Physical Review D* **68** (2): 023506.
arXiv:hep-th/0304188.   Bibcode:2003PhRvD..68b3506P.
doi:10.1103/PhysRevD.68.023506.

[45] Hinshaw, G.; (WMAP collaboration); Bennett, C. L.; Bean,
R.; Doré, O.; Greason, M. R.; Halpern, M.; Hill, R. S.;
Jarosik, N.; Kogut, A.; Komatsu, E.; Limon, M.; Ode-
gard, N.; Meyer, S. S.; Page, L.; Peiris, H. V.; Spergel,
D. N.; Tucker, G. S.; Verde, L.; Weiland, J. L.; Wol-
lack, E.; Wright, E. L.; et al. (2007). "Three-year
Wilkinson Microwave Anisotropy Probe (WMAP) obser-
vations: temperature analysis". *Astrophysical Journal (Sup-
plement Series)* **170** (2): 288–334. arXiv:astro-ph/0603451.
Bibcode:2007ApJS..170..288H. doi:10.1086/513698.

[46] Guillaume, C.-É., 1896, *La Nature* 24, series 2, p. 234, cited
in "History of the 2.7 K Temperature Prior to Penzias and
Wilson" (PDF)

[47] Eddington, A., The Internal Constitution of the Stars, cited
in "History of the 2.7 K Temperature Prior to Penzias and
Wilson" (PDF)

[48] Weinberg, S. (1972). *Oxford Astronomy Encyclopedia*. John
Wiley & Sons. p. 514. ISBN 0-471-92567-5.

[49] Kragh, H. (1999). *Cosmology and Controversy: The Historical Development of Two Theories of the Universe*. ISBN
0-691-00546-X. "In 1946, Robert Dicke and coworkers at
MIT tested equipment that could test a cosmic microwave

background of intensity corresponding to about 20K in the
microwave region. However, they did not refer to such
a background, but only to 'radiation from cosmic matter'.
Also, this work was unrelated to cosmology and is only mentioned because it suggests that by 1950, detection of the
background radiation might have been technically possible,
and also because of Dicke's later role in the discovery". See
also Dicke, R. H.; et al. (1946). "Atmospheric Absorption
Measurements with a Microwave Radiometer". *Physical Review* **70** (5–6): 340–348. Bibcode:1946PhRv...70..340D.
doi:10.1103/PhysRev.70.340.

[50] George Gamow, *The Creation Of The Universe* p.50 (Dover
reprint of revised 1961 edition) ISBN 0-486-43868-6

[51] Gamow, G. (2004) [1961]. *Cosmology and Controversy:
The Historical Development of Two Theories of the Universe*. Courier Dover Publications. p. 40. ISBN 978-0-
486-43868-9.

[52] Erwin Finlay-Freundlich, "Ueber die Rotverschiebung der
Spektrallinien" (1953) *Contributions from the Observatory,
University of St. Andrews* ; no. 4, p. 96–102. Finlay-
Freundlich also gave two extreme values of 1.9K and 6.0K
in Finlay-Freundlich, E.: 1954, "Red shifts in the spectra of
celestial bodies", Phil. Mag., Vol. 45, pp. 303–319.

[53] Helge Kragh, Cosmology and Controversy: The Historical
Development of Two Theories of the Universe (1999) ISBN
0-691-00546-X. "Alpher and Herman first calculated the
present temperature of the decoupled primordial radiation
in 1948, when they reported a value of 5 K. Although it was
not mentioned either then or in later publications that the radiation is in the microwave region, this follows immediately
from the temperature ... Alpher and Herman made it clear
that what they had called "the temperature in the univerese"
the previous year referred to a blackbody distributed background radiation quite different from sunliight".

[54] Shmaonov, T. A. (1957). "Commentary". *Pribory
i Tekhnika Experimenta* (in Russian) **1**: 83.
doi:10.1016/S0890-5096(06)60772-3.

[55] It is noted that the "measurements showed that radiation intensity was independent of either time or direction of observation ... it is now clear that Shmaonov did observe the
cosmic microwave background at a wavelength of 3.2cm"

[56] Naselsky, P. D.; Novikov, D.I.; Novikov, I. D. (2006). *The
Physics of the Cosmic Microwave Background*. ISBN 0-521-
85550-0.

[57] Helge Kragh, Cosmology and Controversy: The Historical
Development of Two Theories of the Universe

[58] Doroshkevich, A. G.; Novikov, I.D. (1964). "Mean Density of Radiation in the Metagalaxy and Certain Problems in Relativistic Cosmology". *Soviet Physics Dok-
lady* **9** (23): 4292–4298. Bibcode:1999EnST...33.4292W.
doi:10.1021/es990537g.

[59] *Nobel Prize In Physics: Russia's Missed Opportunities*, RIA Novosti, Nov 21, 2006

[60] Sanders, R.; Kahn, J. (13 October 2006). "UC Berkeley, LBNL cosmologist George F. Smoot awarded 2006 Nobel Prize in Physics". UC Berkeley News. Retrieved 2008-12-11.

[61] Kovac, J.M.; et al. (2002). "Detection of polarization in the cosmic microwave background using DASI". *Nature* **420** (6917): 772–787. arXiv:astro-ph/0209478. Bibcode:2002Natur.420..772K. doi:10.1038/nature01269. PMID 12490941.

[62] Readhead, A. C. S.; et al. (2004). "Polarization Observations with the Cosmic Background Imager". *Science* **306** (5697): 836–844. arXiv:astro-ph/0409569. Bibcode:2004Sci...306..836R. doi:10.1126/science.1105598. PMID 15472038.

[63] A. Readhead et al., "Polarization observations with the Cosmic Background Imager", Science 306, 836-844 (2004).

[64] Staff (March 17, 2014). "BICEP2 2014 Results Release". *National Science Foundation*. Retrieved March 18, 2014.

[65] Clavin, Whitney (March 17, 2014). "NASA Technology Views Birth of the Universe". *NASA*. Retrieved March 17, 2014.

[66] Overbye, Dennis (March 17, 2014). "Space Ripples Reveal Big Bang's Smoking Gun". *The New York Times*. Retrieved March 17, 2014.

[67] Overbye, Dennis (March 24, 2014). "Ripples From the Big Bang". *New York Times*. Retrieved March 24, 2014.

[68] Ade, P.A.R. (BICEP2 Collaboration); et al. (June 19, 2014). "Detection of B-Mode Polarization at Degree Angular Scales by BICEP2" (PDF). *Physical Review Letters* **112**: 241101. arXiv:1403.3985. Bibcode:2014PhRvL.112x1101A. doi:10.1103/PhysRevLett.112.241101. PMID 24996078. Retrieved June 20, 2014.

[69] http://www.math.columbia.edu/~||woit/wordpress/?p=6865

[70] Overbye, Dennis (June 19, 2014). "Astronomers Hedge on Big Bang Detection Claim". *New York Times*. Retrieved June 20, 2014.

[71] Amos, Jonathan (June 19, 2014). "Cosmic inflation: Confidence lowered for Big Bang signal". *BBC News*. Retrieved June 20, 2014.

[72] Cowen, Ron (2015-01-30). "Gravitational waves discovery now officially dead". *nature*. doi:10.1038/nature.2015.16830.

[73] Scott, D. (2005). "The Standard Cosmological Model". arXiv:astro-ph/0510731 [astro-ph].

[74] Durham, Frank; Purrington, Robert D. (1983). *Frame of the universe: a history of physical cosmology*. Columbia University Press. pp. 193–209. ISBN 0-231-05393-2.

[75] Assis, A. K. T.; Paulo, São; Neves, M. C. D. (July 1995). "History of the 2.7 K Temperature Prior to Penzias and Wilson" (PDF). *Apeiron* **2** (3): 79–87.

[76] Brandenberger, Robert H. (1995). "Formation of Structure in the Universe". p. 8159. arXiv:astro-ph/9508159. Bibcode:1995astro.ph..8159B.

[77] Bennett, C. L.; (WMAP collaboration); Hinshaw, G.; Jarosik, N.; Kogut, A.; Limon, M.; Meyer, S. S.; Page, L.; Spergel, D. N.; Tucker, G. S.; Wollack, E.; Wright, E. L.; Barnes, C.; Greason, M. R.; Hill, R. S.; Komatsu, E.; Nolta, M. R.; Odegard, N.; Peiris, H. V.; Verde, L.; Weiland, J. L.; et al. (2003). "First-year Wilkinson Microwave Anisotropy Probe (WMAP) observations: preliminary maps and basic results". *Astrophysical Journal (Supplement Series)* **148**: 1–27. arXiv:astro-ph/0302207. Bibcode:2003ApJS..148....1B. doi:10.1086/377253. This paper warns, "the statistics of this internal linear combination map are complex and inappropriate for most CMB analyses."

[78] Noterdaeme, P.; Petitjean, P.; Srianand, R.; Ledoux, C.; López, S. (February 2011). "The evolution of the cosmic microwave background temperature. Measurements of TCMB at high redshift from carbon monoxide excitation". *Astronomy and Astrophysics* **526**: L7. arXiv:1012.3164. Bibcode:2011A&A...526L...7N. doi:10.1051/0004-6361/201016140.

[79] Wayne Hu. "Baryons and Inertia".

[80] Wayne Hu. "Radiation Driving Force".

[81] Hu, W.; White, M. (1996). "Acoustic Signatures in the Cosmic Microwave Background". *Astrophysical Journal* **471**: 30–51. arXiv:astro-ph/9602019. Bibcode:1996ApJ...471...30H. doi:10.1086/177951.

[82] WMAP Collaboration; Verde, L.; Peiris, H. V.; Komatsu, E.; Nolta, M. R.; Bennett, C. L.; Halpern, M.; Hinshaw, G.; et al. (2003). "First-Year Wilkinson Microwave Anisotropy Probe (WMAP) Observations: Determination of Cosmological Parameters". *Astrophysical Journal Supplement Series* **148** (1): 175–194. arXiv:astro-ph/0302209. Bibcode:2003ApJS..148..175S. doi:10.1086/377226.

[83] Hanson, D.; et al. (2013). "Detection of B-mode polarization in the Cosmic Microwave Background with data from the South Pole Telescope". *Physical Review Letters* **111** (14). arXiv:1307.5830. Bibcode:2013PhRvL.111n1301H. doi:10.1103/PhysRevLett.111.141301.

[84] Lewis, A.; Challinor, A. (2006). "Weak gravitational lensing of the CMB". *Physics Reports* **429**: 1–65. arXiv:astro-ph/0601594. Bibcode:2006PhR...429....1L. doi:10.1016/j.physrep.2006.03.002.

[85] Seljak, U. (June 1997). "Measuring Polarization in the Cosmic Microwave Background". *Astrophysical Journal* **482**: 6–16. arXiv:astro-ph/9608131. Bibcode:1997ApJ...482....6S. doi:10.1086/304123.

[86] Seljak, U.; Zaldarriaga M. (March 17, 1997). "Signature of Gravity Waves in the Polarization of the Microwave Background". *Phys. Rev.Lett.* **78** (11): 2054–2057. arXiv:astro-ph/9609169. Bibcode:1997PhRvL..78.2054S. doi:10.1103/PhysRevLett.78.2054.

[87] Kamionkowski, M.; Kosowsky A. & Stebbins A. (March 17, 1997). "A Probe of Primordial Gravity Waves and Vorticity". *Phys. Rev.Lett.* **78** (11): 2058–2061. arXiv:astro-ph/9609132. Bibcode:1997PhRvL..78.2058K. doi:10.1103/PhysRevLett.78.2058.

[88] Zaldarriaga, M.; Seljak U. (July 15, 1998). "Gravitational lensing effect on cosmic microwave background polarization". *Physical Review D.* 2 **58**. arXiv:astro-ph/9803150. Bibcode:1998PhRvD..58b3003Z. doi:10.1103/PhysRevD.58.023003.

[89] "Scientists Report Evidence for Gravitational Waves in Early Universe". Retrieved 2007-06-20.

[90] "Gravitational waves: have US scientists heard echoes of the big bang?". The Guardian. 2014-03-14. Retrieved 2014-03-14.

[91] 'BICEP2 I: Detection Of B-mode Polarization at Degree Angular Scales' on arXiv

[92] "Space Ripples Reveal Big Bang's Smoking Gun". March 17, 2014.

[93] Planck Collaboration Team (19 September 2014). "Planck intermediate results. XXX. The angular power spectrum of polarized dust emission at intermediate and high Galactic latitudes". arXiv:1409.5738.

[94] Overbye, Dennis (22 September 2014). "Study Confirms Criticism of Big Bang Finding". *New York Times*. Retrieved 22 September 2014.

[95] "Big Bang research blunder leaves multiverse theory in ruins, theoretical physicist claims". *The Independent*.

[96] "Polarization detected in Big Bang's echo". *Nature News & Comment*.

[97] ESA Planck (Oct 22, 2013). "Planck Space Mission". Retrieved Oct 23, 2013.

[98] NASA/Jet Propulsion Laboratory (October 22, 2013). "Long-sought pattern of ancient light detected". *ScienceDaily*. Retrieved October 23, 2013.

[99] Hanson, D.; et al. (Sep 30, 2013). "Detection of B-Mode Polarization in the Cosmic Microwave Background with Data from the South Pole Telescope". *Physical Review Letters.* 14 **111**. arXiv:1307.5830. Bibcode:2013PhRvL.111n1301H. doi:10.1103/PhysRevLett.111.141301.

[100] The Polarbear Collaboration (October 2014). "A Measurement of the Cosmic Microwave Background B-Mode Polarization Power Spectrum at Sub-Degree Scales with POLARBEAR" (PDF). *The Astrophysical Journal* **794**: 171. arXiv:1403.2369. Bibcode:2014ApJ...794..171T. doi:10.1088/0004-637X/794/2/171. Retrieved November 16, 2014.

[101] "POLARBEAR project offers clues about origin of universe's cosmic growth spurt". *Christian Science Monitor*. October 21, 2014.

[102] Clavin, Whitney; Harrington, J.D. (21 March 2013). "Planck Mission Brings Universe Into Sharp Focus". *NASA*. Retrieved 21 March 2013.

[103] Staff (21 March 2013). "Mapping the Early Universe". *New York Times*. Retrieved 23 March 2013.

[104] Planck Collaboration (2015). "Planck 2015 results. XIII. Cosmological parameters (See Table 4 on page 31 of pfd).". arXiv:1502.01589. Bibcode:2015arXiv150201589P.

[105] Kogut, A.; Lineweaver, C.; Smoot, G. F.; Bennett, C. L.; Banday, A.; Boggess, N. W.; Cheng, E. S.; De Amici, G.; Fixsen, D. J.; Hinshaw, G.; Jackson, P. D.; Janssen, M.; Keegstra, P.; Loewenstein, K.; Lubin, P.; Mather, J. C.; Tenorio, L.; Weiss, R.; Wilkinson, D. T.; Wright, E. L. (1993). "Dipole Anisotropy in the COBE Differential Microwave Radiometers First-Year Sky Maps". *Astrophysical Journal* **419**: 1–6. arXiv:astro-ph/9312056. Bibcode:1993ApJ...419....1K. doi:10.1086/173453.

[106] Aghanim, N.; Armitage-Caplan, C.; et al. (2013). "Planck 2013 results. XXVII. Doppler boosting of the CMB: Eppur si muove". *Astronomy & Astrophysics* **571** (27): A27. arXiv:1303.5087. Bibcode:2014A&A...571A..27P. doi:10.1051/0004-6361/201321556.

[107] http://antwrp.gsfc.nasa.gov/apod/ap090906.html

[108] http://iopscience.iop.org/1126-6708/2005/07/029/

[109] Inoue, K. T.; Silk, J. (2007). "Local Voids as the Origin of Large-Angle Cosmic Microwave Background Anomalies: The Effect of a Cosmological Constant". *Astrophysical Journal* **664** (2): 650–659. arXiv:astro-ph/0612347. Bibcode:2007ApJ...664..650I. doi:10.1086/517603.

[110] Rossmanith, G.; Räth, C.; Banday, A. J.; Morfill, G. (2009). "Non-Gaussian Signatures in the five-year WMAP data as identified with isotropic scaling indices". *Monthly Notices of the Royal Astronomical Society* **399** (4): 1921–1933. arXiv:0905.2854. Bibcode:2009MNRAS.399.1921R. doi:10.1111/j.1365-2966.2009.15421.x.

[111] Schild, R. E.; Gibson, C. H. (2008). "Goodness in the Axis of Evil". arXiv:0802.3229 [astro-ph].

[112] Bernui, A.; Mota, B.; Rebouças, M. J.; Tavakol, R. (2005). "Mapping the large-scale anisotropy in the WMAP data". *Astronomy and Astrophysics* **464** (2): 479–485. arXiv:astro-ph/0511666. Bibcode:2007A&A...464..479B. doi:10.1051/0004-6361:20065585.

[113] Jaffe, T.R.; Banday, A. J.; Eriksen, H. K.; Górski, K. M.; Hansen, F. K. (2005). "Evidence of vorticity and shear at large angular scales in the WMAP data: a violation of cosmological isotropy?". *The Astrophysical Journal* **629**: L1–L4. arXiv:astro-ph/0503213. Bibcode:2005ApJ...629L...1J. doi:10.1086/444454.

[114] de Oliveira-Costa, A.; Tegmark, Max; Zaldarriaga, Matias; Hamilton, Andrew (2004). "The significance of the largest scale CMB fluctuations in WMAP". *Physical Review D* **69** (6): 063516. arXiv:astro-ph/0307282. Bibcode:2004PhRvD..69f3516D. doi:10.1103/PhysRevD.69.063516.

[115] Schwarz, D. J.; Starkman, Glenn D.; et al. (2004). "Is the low-*l* microwave background cosmic?". *Physical Review Letters* **93** (22): 221301. arXiv:astro-ph/0403353. Bibcode:2004PhRvL..93v1301S. doi:10.1103/PhysRevLett.93.221301.

[116] Bielewicz, P.; Gorski, K. M.; Banday, A. J. (2004). "Low-order multipole maps of CMB anisotropy derived from WMAP". *Monthly Notices of the Royal Astronomical Society* **355** (4): 1283–1302. arXiv:astro-ph/0405007. Bibcode:2004MNRAS.355.1283B. doi:10.1111/j.1365-2966.2004.08405.x.

[117] Liu, Hao; Li, Ti-Pei (2009). "Improved CMB Map from WMAP Data". arXiv:0907.2731v3 [astro-ph].

[118] Sawangwit, Utane; Shanks, Tom (2010). "Lambda-CDM and the WMAP Power Spectrum Beam Profile Sensitivity". arXiv:1006.1270v1 [astro-ph].

[119] Liu, Hao; et al. (2010). "Diagnosing Timing Error in WMAP Data". arXiv:1009.2701v1 [astro-ph].

[120] Tegmark, M.; de Oliveira-Costa, A.; Hamilton, A. (2003). "A high resolution foreground cleaned CMB map from WMAP". *Physical Review D* **68** (12): 123523. arXiv:astro-ph/0302496. Bibcode:2003PhRvD..68l3523T. doi:10.1103/PhysRevD.68.123523. This paper states, "Not surprisingly, the two most contaminated multipoles are [the quadrupole and octupole], which most closely trace the galactic plane morphology."

[121] O'Dwyer, I.; Eriksen, H. K.; Wandelt, B. D.; Jewell, J. B.; Larson, D. L.; Górski, K. M.; Banday, A. J.; Levin, S.; Lilje, P. B. (2004). "Bayesian Power Spectrum Analysis of the First-Year Wilkinson Microwave Anisotropy Probe Data". *Astrophysical Journal Letters* **617** (2): L99–L102. arXiv:astro-ph/0407027. Bibcode:2004ApJ...617L..99O. doi:10.1086/427386.

[122] Slosar, A.; Seljak, U. (2004). "Assessing the effects of foregrounds and sky removal in WMAP". *Physical Review D* **70** (8): 083002. arXiv:astro-ph/0404567. Bibcode:2004PhRvD..70h3002S. doi:10.1103/PhysRevD.70.083002.

[123] Bielewicz, P.; Eriksen, H. K.; Banday, A. J.; Górski, K. M.; Lilje, P. B. (2005). "Multipole vector anomalies in the first-year WMAP data: a cut-sky analysis". *Astrophysical Journal* **635** (2): 750–60. arXiv:astro-ph/0507186. Bibcode:2005ApJ...635..750B. doi:10.1086/497263.

[124] Copi, C.J.; Huterer, Dragan; Schwarz, D. J.; Starkman, G. D. (2006). "On the large-angle anomalies of the microwave sky". *Monthly Notices of the Royal Astronomical Society* **367**: 79–102. arXiv:astro-ph/0508047. Bibcode:2006MNRAS.367...79C. doi:10.1111/j.1365-2966.2005.09980.x.

[125] de Oliveira-Costa, A.; Tegmark, M. (2006). "CMB multipole measurements in the presence of foregrounds". *Physical Review D* **74** (2): 023005. arXiv:astro-ph/0603369. Bibcode:2006PhRvD..74b3005D. doi:10.1103/PhysRevD.74.023005.

[126] Planck shows almost perfect cosmos – plus axis of evil

[127] Found: Hawking's initials written into the universe

[128] Krauss, Lawrence M.; Scherrer, Robert J. (2007). "The return of a static universe and the end of cosmology". *General Relativity and Gravitation* **39** (10): 1545–1550. arXiv:0704.0221. Bibcode:2007GReGr..39.1545K. doi:10.1007/s10714-007-0472-9.

[129] Adams, Fred C.; Laughlin, Gregory (1997). "A dying universe: The long-term fate and evolution of astrophysical objects". *Reviews of Modern Physics* **69** (2): 337–372. arXiv:astro-ph/9701131. Bibcode:1997RvMP...69..337A. doi:10.1103/RevModPhys.69.337.

[130] Cosmic Rebirth Encoded in Background Radiation?

29.11   External links

- Student Friendly Intro to the CMB A pedagogic, step-by-step introduction to the cosmic microwave background power spectrum analysis suitable for those with an undergraduate physics background. More in depth than typical online sites. Less dense than cosmology texts.

- CMBR Theme on arxiv.org

- Audio: Fraser Cain and Dr. Pamela Gay – Astronomy Cast. The Big Bang and Cosmic Microwave Background – October 2006

- Visualization of the CMB data from the Planck mission

- Copeland, Ed. "CMBR: Cosmic Microwave Background Radiation". *Sixty Symbols*. Brady Haran for the University of Nottingham.

Chapter 30

Gravitational wave

Not to be confused with gravity wave.

In physics, **gravitational waves** are ripples in the curvature of spacetime which propagate as waves, travelling outward from the source. Predicted in 1916[1][2] by Albert Einstein to exist on the basis of his theory of general relativity,[3][4] gravitational waves theoretically transport energy as **gravitational radiation**. Sources of detectable gravitational waves could possibly include binary star systems composed of white dwarfs, neutron stars, or black holes. The existence of gravitational waves is a possible consequence of the Lorentz invariance of general relativity since it brings the concept of a limiting speed of propagation of the physical interactions with it. Gravitational waves cannot exist in the Newtonian theory of gravitation, in which physical interactions propagate at infinite speed.

Although gravitational radiation has not been *directly* detected, there is *indirect* evidence for its existence.[5] For example, the 1993 Nobel Prize in Physics was awarded for measurements of the Hulse–Taylor binary system which suggest that gravitational waves are more than mathematical anomalies. Various gravitational wave detectors exist and on 17 March 2014, astronomers at the Harvard–Smithsonian Center for Astrophysics erroneously claimed that they had detected and produced "the first direct image of gravitational waves across the primordial sky" within the cosmic microwave background, providing flawed evidence for inflation and the Big Bang.[5][6][7][8][9][10] On 19 June 2014, lowered confidence in confirming the cosmic inflation findings was reported[11][12][13] and on 19 September 2014, a further reduction in confidence was reported.[14][15] On 30 January 2015, even less confidence yet was reported;[16][17] Nature went as far as publishing a news article entitled "Gravitational waves discovery now officially dead".[18]

30.1 Introduction

In Einstein's theory of general relativity, gravity is treated as a phenomenon resulting from the curvature of spacetime.

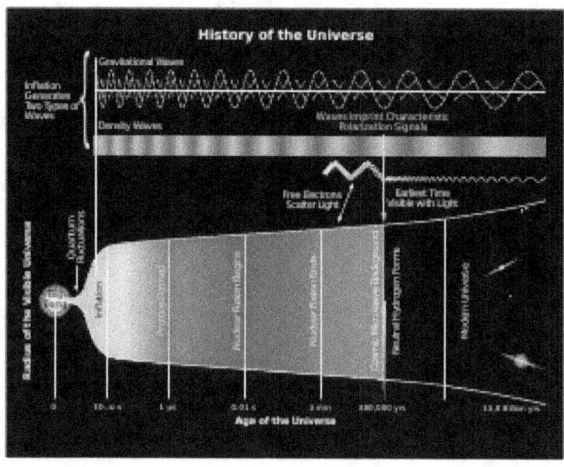

History of the Universe - gravitational waves are hypothesized to arise from cosmic inflation, a faster-than-light expansion just after the Big Bang (17 March 2014).[7][9][10]

This curvature is caused by the presence of mass. Generally, the more mass that is contained within a given volume of space, the greater the curvature of spacetime will be at the boundary of this volume.[5] As objects with mass move around in spacetime, the curvature changes to reflect the changed locations of those objects. In certain circumstances, accelerating objects generate changes in this curvature, which propagate outwards at the speed of light in a wave-like manner. These propagating phenomena are known as gravitational waves.

As a gravitational wave passes a distant observer, that observer will find spacetime distorted by the effects of strain. Distances between free objects increase and decrease rhythmically as the wave passes, at a frequency corresponding to that of the wave. This occurs despite such free objects never being subjected to an unbalanced force. The magnitude of this effect decreases inversely with distance from the source. Inspiralling binary neutron stars are predicted to be a powerful source of gravitational waves as they coalesce, due to the very large acceleration of their masses as they orbit close to one another. However, due to the astronomical distances to these sources the effects when measured on Earth are predicted to be very small, having strains of less than 1 part in 10^{20}. Scientists are attempting to demonstrate the existence of these waves with ever more sensitive detectors. The current most sensitive measurement is about one part in 5×10^{22} (as of 2012) provided by the LIGO and VIRGO observatories.[19] The lack of detection in these observatories provides an upper limit on the frequency of such powerful sources.[20][21] A space based observatory, the Laser Interferometer Space Antenna, is currently under development by ESA.

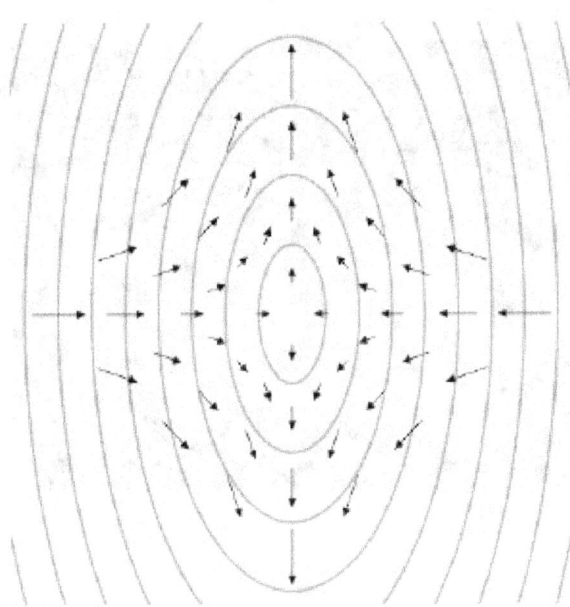

Linearly polarised gravitational wave

Gravitational waves should penetrate regions of space that electromagnetic waves cannot. It is hypothesized that they will be able to provide observers on Earth with information about black holes and other exotic objects in the distant Universe. Such systems cannot be observed with more traditional means such as optical telescopes and radio telescopes. In particular, gravitational waves could be of interest to cosmologists as they offer a possible way of observing the very early universe. This is not possible with conventional astronomy, since before recombination the universe was opaque to electromagnetic radiation.[22] Precise measurements of gravitational waves will also allow scientists to test the general theory of relativity more thoroughly.

In principle, gravitational waves could exist at any frequency. However, very low frequency waves would be impossible to detect and there is no credible source for detectable waves of very high frequency. Stephen W. Hawking and Werner Israel list different frequency bands for gravitational waves that could be plausibly detected, ranging from 10^{-7} Hz up to 10^{11} Hz.[23]

30.2 Effects of a passing gravitational wave

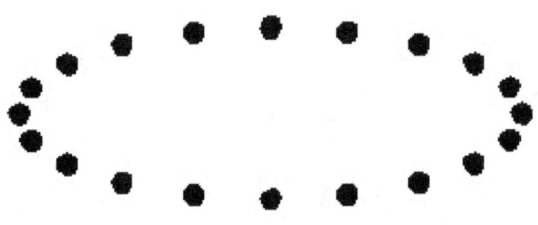

The effect of a plus-polarized gravitational wave on a ring of particles.

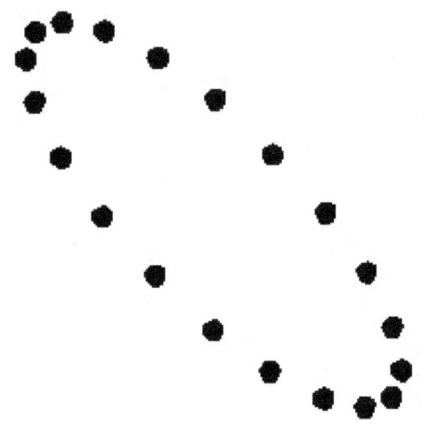

The effect of a cross-polarized gravitational wave on a ring of particles.

The effects of a passing gravitational wave can be visualized by imagining a perfectly flat region of spacetime with a group of motionless test particles lying in a plane (the surface of your screen). As a gravitational wave passes through the particles along a line perpendicular to the plane of the particles (i.e. following your line of vision into the screen), the particles will follow the distortion in spacetime, oscillating in a "cruciform" manner, as shown in the animations. The area enclosed by the test particles does not change and there is no motion along the direction of propagation.

The oscillations depicted here in the animation are exaggerated for the purpose of discussion—in reality a gravitational wave has a very small amplitude (as formulated in linearized gravity). However, they enable us to visualize the kind of oscillations associated with gravitational waves as produced for example by a pair of masses in a circular orbit. In this case the amplitude of the gravitational wave is constant, but its plane of polarization changes or rotates at twice the orbital rate and so the time-varying gravitational wave size (or 'periodic spacetime strain') exhibits a variation as shown in the animation.[24] If the orbit is elliptical then the gravitational wave's amplitude also varies with time according to Einstein's quadrupole formula.[25]

Like other waves, there are a few useful characteristics describing a gravitational wave:

- **Amplitude**: Usually denoted h, this is the size of the wave — the fraction of stretching or squeezing in the animation. The amplitude shown here is roughly $h = 0.5$ (or 50%). Gravitational waves passing through the Earth are many billions times weaker than this — $h \approx 10^{-20}$. Note that this is not the quantity that would be analogous to what is usually called the amplitude of an electromagnetic wave, which would be $\frac{dh}{dt}$.

- **Frequency**: Usually denoted f, this is the frequency with which the wave oscillates (1 divided by the amount of time between two successive maximum stretches or squeezes)

- **Wavelength**: Usually denoted λ, this is the distance along the wave between points of maximum stretch or squeeze.

- **Speed**: This is the speed at which a point on the wave (for example, a point of maximum stretch or squeeze) travels. For gravitational waves with small amplitudes, this is equal to the speed of light, c.

The speed, wavelength, and frequency of a gravitational wave are related by the equation $c = \lambda f$, just like the equation for a light wave. For example, the animations shown here oscillate roughly once every two seconds. This would correspond to a frequency of 0.5 Hz, and a wavelength of about 600,000 km, or 47 times the diameter of the Earth.

In the example just discussed, we actually assume something special about the wave. We have assumed that the wave is linearly polarized, with a "plus" polarization, written h_+. Polarization of a gravitational wave is just like polarization of a light wave except that the polarizations of a

gravitational wave are at 45 degrees, as opposed to 90 degrees. In particular, if we had a "cross"-polarized gravitational wave, $h_\times$, the effect on the test particles would be basically the same, but rotated by 45 degrees, as shown in the second animation. Just as with light polarization, the polarizations of gravitational waves may also be expressed in terms of circularly polarized waves. Gravitational waves are polarized because of the nature of their sources. The polarization of a wave depends on the angle from the source, as we will see in the next section.

30.3 Sources of gravitational waves

In general terms, gravitational waves are radiated by objects whose motion involves acceleration, provided that the motion is not perfectly spherically symmetric (like an expanding or contracting sphere) or cylindrically symmetric (like a spinning disk or sphere). A simple example of this principle is provided by the spinning dumbbell. If the dumbbell spins like wheels on an axle, it will not radiate gravitational waves; if it tumbles end over end like two planets orbiting each other, it will radiate gravitational waves. The heavier the dumbbell, and the faster it tumbles, the greater is the gravitational radiation it will give off. If we imagine an extreme case in which the two weights of the dumbbell are massive stars like neutron stars or black holes, orbiting each other quickly, then significant amounts of gravitational radiation would be given off.

Some more detailed examples:

- Two objects orbiting each other in a quasi-Keplerian planar orbit (basically, as a planet would orbit the Sun) *will* radiate.

- A spinning non-axisymmetric planetoid — say with a large bump or dimple on the equator — *will* radiate.

- A supernova *will* radiate except in the unlikely event that the explosion is perfectly symmetric.

- An isolated non-spinning solid object moving at a constant velocity *will not* radiate. This can be regarded as a consequence of the principle of conservation of linear momentum.

- A spinning disk *will not* radiate. This can be regarded as a consequence of the principle of conservation of angular momentum. However, it *will* show gravitomagnetic effects.

- A spherically pulsating spherical star (non-zero monopole moment or mass, but zero quadrupole moment) *will not* radiate, in agreement with Birkhoff's theorem.

More technically, the third time derivative of the quadrupole moment (or the l-th time derivative of the l-th multipole moment) of an isolated system's stress–energy tensor must be nonzero in order for it to emit gravitational radiation. This is analogous to the changing dipole moment of charge or current necessary for electromagnetic radiation.

30.3.1 Power radiated by orbiting bodies

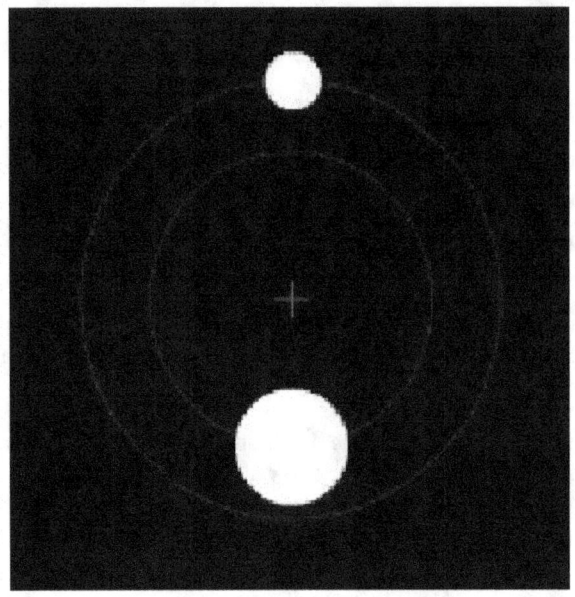

Two stars of dissimilar mass are in circular orbits. Each revolves about their common center of mass (denoted by the small red cross) in a circle with the larger mass having the smaller orbit.

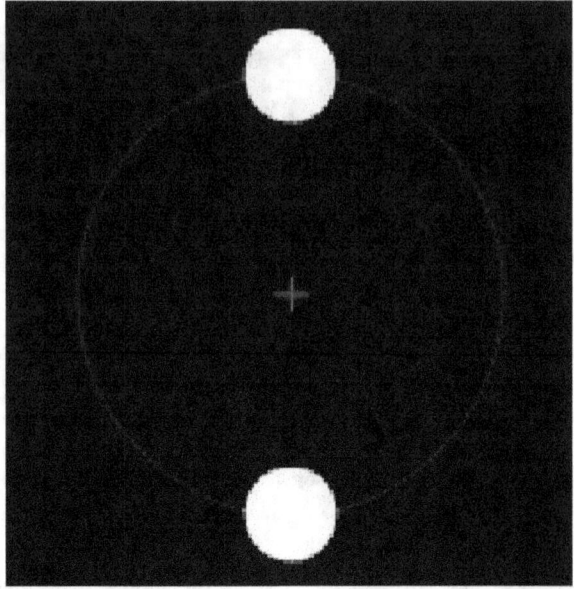

Two stars of similar mass are in circular orbits about their center of mass

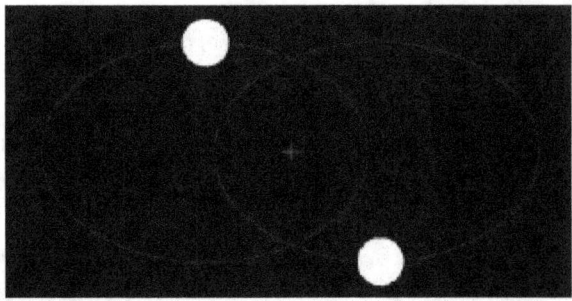

Two stars of similar mass are in highly elliptical orbits about their center of mass

Gravitational waves carry energy away from their sources and, in the case of orbiting bodies, this is associated with an inspiral or decrease in orbit. Imagine for example a simple system of two masses — such as the Earth-Sun system — moving slowly compared to the speed of light in circular orbits. Assume that these two masses orbit each other in a circular orbit in the $x - y$ plane. To a good approximation, the masses follow simple Keplerian orbits. However, such an orbit represents a changing quadrupole moment. That is, the system will give off gravitational waves.

Suppose that the two masses are m_1 and m_2, and they are separated by a distance r. The power given off (radiated) by this system is:

$$P = \frac{dE}{dt} = -\frac{32}{5}\frac{G^4}{c^5}\frac{(m_1 m_2)^2 (m_1 + m_2)}{r^5}, \quad [26]$$

where G is the gravitational constant, c is the speed of light in vacuum and where the negative sign means that power is being given off by the system, rather than received. For a system like the Sun and Earth, r is about 1.5×10^{11} m and m_1 and m_2 are about 2×10^{30} and 6×10^{24} kg respectively. In this case, the power is about 200 watts. This is truly tiny compared to the total electromagnetic radiation given off by the Sun (roughly 3.86×10^{26} watts).

In theory, the loss of energy through gravitational radiation could eventually drop the Earth into the Sun. However, the total energy of the Earth orbiting the Sun (kinetic energy + gravitational potential energy) is about 1.14×10^{36} joules of which only 200 joules per second is lost through gravitational radiation, leading to a decay in the orbit by about 1×10^{-15} meters per day or roughly the diameter of a proton. At this rate, it would take the Earth approximately 1×10^{13} times more than the current age of the Universe to spiral onto the Sun. This estimate overlooks the decrease in r over time, but the majority of the time the bodies are far apart and only radiating slowly, so the difference is unimportant in this example.

A more dramatic example of radiated gravitational energy is represented by two solar mass ($M\odot$) neutron stars orbiting at a distance from each other of 1.89×10^8 m (only 0.63 light-seconds apart). [The Sun is 8 light minutes from the Earth.] Plugging their masses into the above equation shows that the gravitational radiation from them would be 1.38×10^{28} watts, which is about 100 times more than the Sun's electromagnetic radiation.

30.3.2 Orbital decay from gravitational radiation

See also: Two-body problem in general relativity

Gravitational radiation robs the orbiting bodies of energy. It first circularizes their orbits and then gradually shrinks their radius. As the energy of the orbit is reduced, the distance between the bodies decreases, and they rotate more rapidly. The overall angular momentum is reduced however. This reduction corresponds to the angular momentum carried off by gravitational radiation. The rate of decrease of distance between the bodies versus time is given by:[26]

$$\frac{dr}{dt} = -\frac{64}{5}\,\frac{G^3}{c^5}\,\frac{(m_1 m_2)(m_1 + m_2)}{r^3}$$

where the variables are the same as in the previous equation.

The orbit decays at a rate proportional to the inverse third power of the radius. When the radius has shrunk to half its initial value, it is shrinking eight times faster than before. By Kepler's Third Law, the new rotation rate at this point will be faster by $\sqrt{8} = 2.828$, or nearly three times the previous orbital frequency. As the radius decreases, the power lost to gravitational radiation increases even more. As can be seen from the previous equation, power radiated varies as the inverse fifth power of the radius, or 32 times more in this case.

If we use the previous values for the Sun and the Earth, we find that the Earth's orbit shrinks by 1.1×10^{-20} meter per second. This is 3.5×10^{-13} m per year, which is about 1/300 the diameter of a hydrogen atom. The effect of gravitational radiation on the size of the Earth's orbit is unnoticeable over the age of the universe. Actually, this effect is absolutely negligible compared to the increase of Earth's orbit due to losing mass of Sun via radiation (4.7×10^{-9} m/s).[27] This is not true for closer orbits.

A more practical example is the orbit of a Sun-like star around a heavy black hole. Our Milky Way is believed to have a 4 million $M\odot$ black hole at its center in Sagittarius A. Such supermassive black holes are being found in the center of almost all galaxies. For this example take a 2 million

$M\odot$ black hole with a solar-mass star orbiting it at a radius of 1.89×10^{10} m (63 light-seconds). The mass of the black hole will be 4×10^{36} kg and its gravitational radius will be 6×10^9 m. The orbital period will be 1,000 seconds, or a little under 17 minutes. The solar-mass star will draw closer to the black hole by 7.4 meters per second or 7.4 km per orbit. A collision will not be long in coming.

Assume that a pair of 1 $M\odot$ neutron stars are in circular orbits at a distance of 1.89×10^8 m (189,000 km). This is a little less than 1/7 the diameter of the Sun or 0.63 light-seconds. Their orbital period would be 1,000 seconds. Substituting the new mass and radius in the above formula gives a rate of orbit decrease of 3.7×10^{-6} m/s or 3.7 mm per orbit. This is 116 meters per year and is not negligible over cosmic time scales.

Suppose instead that these two neutron stars were orbiting at a distance of 1.89×10^6 m (1890 km). Their period would be 1 second and their orbital velocity would be about 1/50 of the speed of light. Their orbit would now shrink by 3.7 meters per orbit. A collision is imminent. A runaway loss of energy from the orbit results in an ever more rapid decrease in the distance between the stars. They will eventually merge to form a black hole and cease to radiate gravitational waves. This is referred to as the inspiral.

The above equation can not be applied directly for calculating the lifetime of the orbit, because the rate of change in radius depends on the radius itself, and is thus non-constant with time. The lifetime can be computed by integration of this equation (see next section).

30.3.3 Orbital lifetime limits from gravitational radiation

Orbital lifetime is one of the most important properties of gravitational radiation sources. It determines the average number of binary stars in the universe that are close enough to be detected. Short lifetime binaries are strong sources of gravitational radiation but are few in number. Long lifetime binaries are more plentiful but they are weak sources of gravitational waves. LIGO is most sensitive in the frequency band where two neutron stars are about to merge. This time frame is only a few seconds. It takes luck for the detector to see this blink in time out of a million year orbital lifetime. It is predicted that such a merger will only be seen once per decade or so.

The lifetime of an orbit is given by:[26]

$$t = \frac{5}{256}\,\frac{c^5}{G^3}\,\frac{r^4}{(m_1 m_2)(m_1 + m_2)}$$

where r is the initial distance between the orbiting bodies.

This equation can be derived by integrating the previous equation for the rate of radius decrease. It predicts the time for the radius of the orbit to shrink to zero. As the orbital speed becomes a significant fraction of the speed of light, this equation becomes inaccurate. It is useful for inspirals until the last few milliseconds before the merger of the objects.

Substituting the values for the mass of the Sun and Earth as well as the orbital radius gives a very large lifetime of 3.44×10^{30} seconds or 1.09×10^{23} years (that is approximately 10^{13} times larger than the age of the universe). The actual figure would be slightly less than that. The Earth will break apart from tidal forces if it orbits closer than a few radii from the Sun. This would form a ring around the Sun and instantly stop the emission of gravitational waves.

If we use a 2 million $M\odot$ black hole with a solar mass star orbiting it at 1.89×10^{10} meters, we get a lifetime of 6.50×10^{8} seconds or 20.7 years.

Assume that a pair of solar mass neutron stars with a diameter of 10 kilometers are in circular orbits at a distance of 1.89×10^{8} m (189,000 km). Their lifetime is 1.30×10^{13} seconds or about 414,000 years. Their orbital period will be 1,000 seconds and it could be observed by LISA if they were not too far away. A far greater number of white dwarf binaries exist with orbital periods in this range. White dwarf binaries have masses on the order of our Sun and diameters on the order of our Earth. They cannot get much closer together than 10,000 km before they will merge and cease to radiate gravitational waves. This results in the creation of either a neutron star or a black hole. Until then, their gravitational radiation will be comparable to that of a neutron star binary. LISA is the only gravitational wave experiment that is likely to succeed in detecting such types of binaries.

If the orbit of a neutron star binary has decayed to 1.89×10^{6} m (1890 km), its remaining lifetime is 130,000 seconds or about 36 hours. The orbital frequency will vary from 1 revolution per second at the start and 918 revolutions per second when the orbit has shrunk to 20 km at merger. The gravitational radiation emitted will be at twice the orbital frequency. Just before merger, the inspiral can be observed by LIGO if the binary is close enough. LIGO has only a few minutes to observe this merger out of a total orbital lifetime that may be billions of years. The chance of success with LIGO as initially constructed is quite low despite the large number of such mergers occurring in the universe, because the sensitivity of the instrument does not 'reach' out to enough systems to see events frequently. No mergers have been seen in the few years that initial LIGO has been in operation, and it is thought that a merger should be seen about once per several tens of years of observing time with initial LIGO.[28] The upgraded Advanced

LIGO detector, with a ten times greater sensitivity, 'reaches' out 10 times further—encompassing a volume 1000 times greater, and seeing 1000 times as many candidate sources. Thus, the expectation is that detections will be made at the rate of tens per year.

30.3.4   Wave amplitudes from the Earth–Sun system

We can also think in terms of the amplitude of the wave from a system in circular orbits. Let θ be the angle between the perpendicular to the plane of the orbit and the line of sight of the observer. Suppose that an observer is outside the system at a distance R from its center of mass. If R is much greater than a wavelength, the two polarizations of the wave will be

$$h_+ = -\frac{1}{R}\frac{G^2}{c^4}\frac{2m_1 m_2}{r}(1 + \cos^2\theta)\cos\left[2\omega(t-R)\right],$$

$$h_\times = -\frac{1}{R}\frac{G^2}{c^4}\frac{4m_1 m_2}{r}(\cos\theta)\sin\left[2\omega(t-R)\right].$$

Here, we use the constant angular velocity of a circular orbit in Newtonian physics:

$$\omega = \sqrt{G(m_1 + m_2)/r^3}.$$

For example, if the observer is in the x - y plane then $\theta = \pi/2$, and $\cos(\theta) = 0$, so the $h_\times$ polarization is always zero. We also see that the frequency of the wave given off is twice the rotation frequency. If we put in numbers for the Earth-Sun system, we find:

$$h_+ = -\frac{1}{R}\frac{G^2}{c^4}\frac{4m_1 m_2}{r} = -\frac{1}{R}1.7 \times 10^{-10} \text{ m}.$$

In this case, the minimum distance to find waves is $R \approx 1$ light-year, so typical amplitudes will be $h \approx 10^{-26}$. That is, a ring of particles would stretch or squeeze by just one part in 10^{26}. This is well under the detectability limit of all conceivable detectors.

30.3.5   Radiation from other sources

Although the waves from the Earth-Sun system are minuscule, astronomers can point to other sources for which the radiation should be substantial. One important example is the Hulse-Taylor binary — a pair of stars, one of which is a pulsar.[29] The characteristics of their orbit can be deduced from the Doppler shifting of radio signals given off by the pulsar. Each of the stars are about 1.4 $M\odot$ and the size of their orbit is about 1/75 of the Earth-Sun orbit. This means the distance between the two stars is just a few times larger than the diameter of our own Sun. The combination of greater masses and smaller separation means that the energy given off by the Hulse-Taylor binary will be far greater than the energy given off by the Earth-Sun system — roughly 10^{22} times as much.

The information about the orbit can be used to predict just how much energy (and angular momentum) should be given off in the form of gravitational waves. As the energy is carried off, the stars should draw closer to each other. This effect is called an inspiral, and it can be observed in the pulsar's signals. The measurements on the Hulse-Taylor system have been carried out over more than 30 years. It has been shown that the gravitational radiation predicted by general relativity allows these observations to be matched within 0.2 percent. In 1993, Russell Hulse and Joe Taylor were awarded the Nobel Prize in Physics for this work, which was the first indirect evidence for gravitational waves. The orbital lifetime of this binary system before merger is a few hundred million years.[30]

Inspirals are very important sources of gravitational waves. Any time two compact objects (white dwarfs, neutron stars, or black holes) are in close orbits, they send out intense gravitational waves. As they spiral closer to each other, these waves become more intense. At some point they should become so intense that direct detection by their effect on objects on Earth or in space is possible. This direct detection is the goal of several large scale experiments.[31]

The only difficulty is that most systems like the Hulse-Taylor binary are so far away. The amplitude of waves given off by the Hulse-Taylor binary as seen on Earth would be roughly $h \approx 10^{-26}$. There are some sources, however, that astrophysicists expect to find with much larger amplitudes of $h \approx 10^{-20}$. At least eight other binary pulsars have been discovered.[32]

30.4 Astrophysics and gravitational waves

During the past century, astronomy has been revolutionized by the use of new methods for observing the universe. Astronomical observations were originally made using visible light. Galileo Galilei pioneered the use of telescopes to enhance these observations. However, visible light is only a small portion of the electromagnetic spectrum, and not all objects in the distant universe shine strongly in this particular band. More useful information may be found, for example, in radio wavelengths. Using radio telescopes, astronomers have found pulsars, quasars, and other extreme

Two-dimensional representation of gravitational waves generated by two neutron stars orbiting each other.

objects that push the limits of our understanding of physics. Observations in the microwave band have opened our eyes to the faint imprints of the Big Bang, a discovery Stephen Hawking called the "greatest discovery of the century, if not all time". Similar advances in observations using gamma rays, x-rays, ultraviolet light, and infrared light have also brought new insights to astronomy. As each of these regions of the spectrum has opened, new discoveries have been made that could not have been made otherwise. Astronomers hope that the same holds true of gravitational waves.

Gravitational waves have two important and unique properties. First, there is no need for any type of matter to be present nearby in order for the waves to be generated by a binary system of uncharged black holes, which would emit no electromagnetic radiation. Second, gravitational waves can pass through any intervening matter without being scattered significantly. Whereas light from distant stars may be blocked out by interstellar dust, for example, gravitational waves will pass through essentially unimpeded. These two features allow gravitational waves to carry information about astronomical phenomena never before observed by humans.

The sources of gravitational waves described above are in the low-frequency end of the gravitational-wave spectrum (10^{-7} to 10^5 Hz). An astrophysical source at the high-frequency end of the gravitational-wave spectrum (above 10^5 Hz and probably 10^{10} Hz) generates relic gravitational waves that are theorized to be faint imprints of the Big Bang like the cosmic microwave background (see gravitational wave background).[33] At these high frequencies it is potentially possible that the sources may be "man made"[23] that is, gravitational waves generated and detected in the laboratory.[34][35]

30.4.1 Energy, momentum, and angular momentum carried by gravitational waves

Waves familiar from other areas of physics such as water waves, sound waves, and electromagnetic waves are able to carry energy, momentum, and angular momentum. By carrying these away from a source, waves are able to rob that source of its energy as well as its linear and angular momentum. Gravitational waves perform the same function. Thus, for example, a binary system loses angular momentum as the two orbiting objects spiral towards each other—the angular momentum is radiated away by gravitational waves.

The waves can also carry off linear momentum, a possibility that has some interesting implications for astrophysics.[36] After two supermassive black holes coalesce, emission of linear momentum can produce a "kick" with amplitude as large as 4000 km/s. This is fast enough to eject the coalesced black hole completely from its host galaxy. Even if the kick is too small to eject the black hole completely, it can remove it temporarily from the nucleus of the galaxy, after which it will oscillate about the center, eventually coming to rest.[37] A kicked black hole can also carry a star cluster with it, forming a hyper-compact stellar system.[38] Or it may carry gas, allowing the recoiling black hole to appear temporarily as a "naked quasar". The quasar SDSS J092712.65+294344.0 is believed to contain a recoiling supermassive black hole.[39]

30.5 Detecting gravitational waves

30.5.1 Difficulties in detection

Gravitational waves are not easily detectable. This knowledge gap is primarily due to the massive presence of noise in the low frequencies where antennas currently operate. Gravitational waves are expected to have frequencies 10^{-16} Hz $< f < 10^4$ Hz.[41]

30.5.2 Ground-based interferometers

Main article: Gravitational wave detector

Though the Hulse-Taylor observations were very important, they give only *indirect* evidence for gravitational waves. A more conclusive observation would be a *direct* measurement of the effect of a passing gravitational wave, which could also provide more information about the system that generated it. Any such direct detection is complicated by the extraordinarily small effect the waves would produce on a

Evidence of gravitational waves in the infant universe may have been uncovered by the BICEP2 radio telescope. The microscopic examination of the focal plane of the BICEP2 detector is shown here.[7][8][9][10][40] In 2015, however, the BICEP2 findings were confirmed to be the result of dust.[18]

detector. The amplitude of a spherical wave will fall off as the inverse of the distance from the source (the $1/R$ term in the formulas for h above). Thus, even waves from extreme systems like merging binary black holes die out to very small amplitude by the time they reach the Earth. Astrophysicists expect that some gravitational waves passing the Earth may be as large as $h \approx 10^{-20}$, but generally no bigger.[42]

A simple device theorised to detect the expected wave motion is called a Weber bar — a large, solid bar of metal isolated from outside vibrations. This type of instrument was the first type of gravitational wave detector. Strains in space due to an incident gravitational wave excite the bar's resonant frequency and could thus be amplified to detectable levels. Conceivably, a nearby supernova might be strong enough to be seen without resonant amplification. With this instrument, Joseph Weber claimed to have detected daily signals of gravitational waves. His results, however, were contested in 1974 by physicists Richard Garwin and David Douglass. Modern forms of the Weber bar are still operated, cryogenically cooled, with superconducting quantum interference devices to detect vibration. Weber bars are not sensitive enough to detect anything but extremely powerful gravitational waves.[43]

MiniGRAIL is a spherical gravitational wave antenna using this principle. It is based at Leiden University, consisting of an exactingly machined 1150 kg sphere cryogenically cooled to 20 mK.[44] The spherical configuration allows for equal sensitivity in all directions, and is somewhat experimentally simpler than larger linear devices requiring high vacuum. Events are detected by measuring deformation of the detector sphere. MiniGRAIL is highly sensitive in the 2–4 kHz range, suitable for detecting gravitational waves from rotating neutron star instabilities or small black hole mergers.[45]

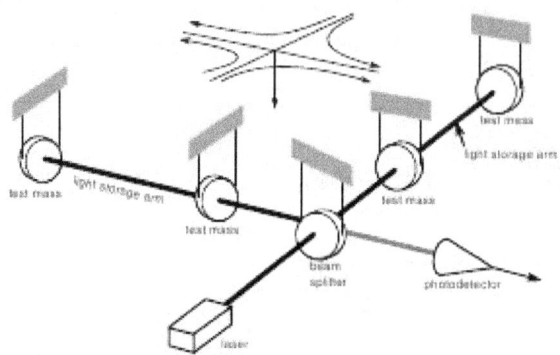

A schematic diagram of a laser interferometer

A more sensitive class of detector uses laser interferometry to measure gravitational-wave induced motion between separated 'free' masses.[46] This allows the masses to be separated by large distances (increasing the signal size); a further advantage is that it is sensitive to a wide range of frequencies (not just those near a resonance as is the case for Weber bars). Ground-based interferometers are now operational. Currently, the most sensitive is LIGO — the Laser Interferometer Gravitational Wave Observatory. LIGO has three detectors: one in Livingston, Louisiana; the other two (in the same vacuum tubes) at the Hanford site in Richland, Washington. Each consists of two light storage arms that are 2 to 4 kilometers in length. These are at 90 degree angles to each other, with the light passing through 1m diameter vacuum tubes running the entire 4 kilometers. A passing gravitational wave will slightly stretch one arm as it shortens the other. This is precisely the motion to which an interferometer is most sensitive.

Even with such long arms, the strongest gravitational waves will only change the distance between the ends of the arms by at most roughly 10^{-18} meters. LIGO should be able to detect gravitational waves as small as $h \sim 5 \times 10^{-20}$. Upgrades to LIGO and other detectors such as Virgo, GEO 600, and TAMA 300 should increase the sensitivity still further; the next generation of instruments (Advanced LIGO and Advanced Virgo) will be more than ten times more sensitive. Another highly sensitive interferometer, KAGRA, is under construction in the Kamiokande mine in Japan. A key point is that a tenfold increase in sensitivity (radius of 'reach') increases the volume of space accessible to the instrument by one thousand times. This increases the rate at which detectable signals should be seen from one per tens of years of observation, to tens per year.[28]

Interferometric detectors are limited at high frequencies by shot noise, which occurs because the lasers produce photons randomly; one analogy is to rainfall—the rate of rainfall, like the laser intensity, is measurable, but the raindrops, like photons, fall at random times, causing fluctuations around the average value. This leads to noise at the output of the detector, much like radio static. In addition, for sufficiently high laser power, the random momentum transferred to the test masses by the laser photons shakes the mirrors, masking signals at low frequencies. Thermal noise (e.g., Brownian motion) is another limit to sensitivity. In addition to these 'stationary' (constant) noise sources, all ground-based detectors are also limited at low frequencies by seismic noise and other forms of environmental vibration, and other 'nonstationary' noise sources; creaks in mechanical structures, lightning or other large electrical disturbances, etc. may also create noise masking an event or may even imitate an event. All these must be taken into account and excluded by analysis before a detection may be considered a true gravitational wave event.

Space-based interferometers, such as LISA and DECIGO, are also being developed. LISA's design calls for three test masses forming an equilateral triangle, with lasers from each spacecraft to each other spacecraft forming two independent interferometers. LISA is planned to occupy a solar orbit trailing the Earth, with each arm of the triangle being five million kilometers. This puts the detector in an excellent vacuum far from Earth-based sources of noise, though it will still be susceptible to shot noise, as well as artifacts caused by cosmic rays and solar wind.

There are currently two detectors focusing on detection at the higher end of the gravitational wave spectrum (10^{-7} to 10^5 Hz): one at University of Birmingham, England, and the other at INFN Genoa, Italy. A third is under development at Chongqing University, China. The Birmingham detector measures changes in the polarization state of a microwave beam circulating in a closed loop about one meter across. Two have been fabricated and they are currently expected to be sensitive to periodic spacetime strains of $h \sim 2 \times 10^{-13}/\sqrt{\text{Hz}}$, given as an amplitude spectral density. The INFN Genoa detector is a resonant antenna consisting of two coupled spherical superconducting harmonic oscillators a few centimeters in diameter. The oscillators are designed to have (when uncoupled) almost equal resonant frequencies. The system is currently expected to have a sensitivity to periodic spacetime strains of $h \sim 2 \times 10^{-17}/\sqrt{\text{Hz}}$, with an expectation to reach a sensitivity of $h \sim 2 \times 10^{-20}/\sqrt{\text{Hz}}$. The Chongqing University detector is planned to detect relic high-frequency gravitational waves with the predicted typical parameters $?_g \sim 10^{10}$ Hz (10 GHz) and $h \sim 10^{-30} - 10^{-31}$.

30.5.3   Using pulsar timing arrays

Pulsars are rapidly rotating stars. A pulsar emits beams of radio waves that, like lighthouse beams, sweep through the sky as the pulsar rotates. The signal from a pulsar can be detected by radio telescopes as a series of regularly spaced pulses, essentially like the ticks of a clock. Gravitational waves affect the time it takes the pulses to travel from the pulsar to a telescope on Earth. A pulsar timing array uses millisecond pulsars to seek out perturbations due to gravitational waves in measurements of pulse arrival times at a telescope, in other words, to look for deviations in the clock ticks. In particular, pulsar timing arrays can search for a distinct pattern of correlation and anti-correlation between the signals over an array of different pulsars (resulting in the name "pulsar timing array"). Although pulsar pulses travel through space for hundreds or thousands of years to reach us, pulsar timing arrays are sensitive to perturbations in their travel time of much less than a millionth of a second.

Globally there are three active pulsar timing array projects. The North American Nanohertz Gravitational Wave Observatory uses data collected by the Arecibo Radio Telescope and Green Bank Telescope. The Parkes Pulsar Timing Array at the Parkes radio-telescope has been collecting data since March 2005. The European Pulsar Timing Array uses data from the four largest telescopes in Europe: the Lovell Telescope, the Westerbork Synthesis Radio Telescope, the Effelsberg Telescope and the Nancay Radio Telescope. (Upon completion the Sardinia Radio Telescope will be added to the EPTA also.) These three projects have begun collaborating under the title of the International Pulsar Timing Array project.

30.5.4   Einstein@Home

Main article: Einstein@Home

In some sense, the easiest signals to detect should be constant sources. Supernovae and neutron star or black hole mergers should have larger amplitudes and be more interesting, but the waves generated will be more complicated. The waves given off by a spinning, aspherical neutron star would be "monochromatic"—like a pure tone in acoustics. It would not change very much in amplitude or frequency.

The Einstein@Home project is a distributed computing project similar to SETI@home intended to detect this type of simple gravitational wave. By taking data from LIGO and GEO, and sending it out in little pieces to thousands of volunteers for parallel analysis on their home computers, Einstein@Home can sift through the data far more quickly than would be possible otherwise.[47]

30.5.5   Primordial gravitational waves

Main article: Primordial gravitational wave

Primordial gravitational waves are gravitational waves observed in the cosmic microwave background. They were allegedly detected by the BICEP2 instrument, an announcement made on 17 March 2014, which was withdrawn on 30 January 2015 ("the signal can be entirely attributed to dust in the Milky Way"[18]).

30.6   Mathematics

Einstein's equations form the fundamental law of general relativity. The curvature of spacetime can be expressed mathematically using the metric tensor — denoted $g_{\mu\nu}$. The metric holds information regarding how distances are measured in the space under consideration. Because the propagation of gravitational waves through space and time change distances, we will need to use this to find the solution to the wave equation.

Spacetime curvature is also expressed with respect to a covariant derivative, ∇ , in the form of the Einstein tensor, $G_{\mu\nu}$. This curvature is related to the stress–energy tensor, $T_{\mu\nu}$, by the key equation

$$G_{\mu\nu} = \frac{8\pi G_N}{c^4} T_{\mu\nu},$$

where G_N is Newton's gravitational constant, and c is the speed of light. We assume geometrized units, so $G_N = 1 = c$.

With some simple assumptions, Einstein's equations can be rewritten to show explicitly that they are wave equations. To begin with, we adopt some coordinate system, like (t, r, θ, ϕ) . We define the "flat-space metric" $\eta_{\mu\nu}$ to be the quantity that — in this coordinate system — has the components we would expect for the flat space metric. For example, in these spherical coordinates, we have

$$\eta_{\mu\nu} = \begin{bmatrix} -1 & 0 & 0 & 0 \\ 0 & 1 & 0 & 0 \\ 0 & 0 & r^2 & 0 \\ 0 & 0 & 0 & r^2 \sin^2\theta \end{bmatrix}.$$

This flat-space metric has no physical significance; it is a purely mathematical device necessary for the analysis. Tensor indices are raised and lowered using this "flat-space metric".

Now, we can also think of the physical metric $g_{\mu\nu}$ as a matrix, and find its determinant, $\det g$. Finally, we define a quantity

$$\bar{h}^{\alpha\beta} \equiv \eta^{\alpha\beta} - \sqrt{|\det g|}\, g^{\alpha\beta}$$

This is the crucial field, which will represent the radiation. It is possible (at least in an asymptotically flat spacetime) to choose the coordinates in such a way that this quantity satisfies the "de Donder" gauge conditions (conditions on the coordinates):

$$\nabla_\beta \bar{h}^{\alpha\beta} = 0,$$

where ∇ represents the flat-space derivative operator. These equations say that the divergence of the field is zero. The linear Einstein equations can now be written[48] as

$$\Box \bar{h}^{\alpha\beta} = -16\pi \tau^{\alpha\beta}$$

where $\Box = -\partial_t^2 + \Delta$ represents the flat-space d'Alembertian operator, and $\tau^{\alpha\beta}$ represents the stress-energy tensor plus quadratic terms involving $\bar{h}^{\alpha\beta}$. This is just a wave equation for the field with a source, despite the fact that the source involves terms quadratic in the field itself. That is, it can be shown that solutions to this equation are waves traveling with velocity 1 in these coordinates.

30.6.1 Linear approximation

The equations above are valid everywhere — near a black hole, for instance. However, because of the complicated source term, the solution is generally too difficult to find analytically. We can often assume that space is nearly flat, so the metric is nearly equal to the $\eta^{\alpha\beta}$ tensor. In this case, we can neglect terms quadratic in $\bar{h}^{\alpha\beta}$, which means that the $\tau^{\alpha\beta}$ field reduces to the usual stress–energy tensor $T^{\alpha\beta}$. That is, Einstein's equations become

$$\Box \bar{h}^{\alpha\beta} = -16\pi T^{\alpha\beta}$$

If we are interested in the field far from a source, however, we can treat the source as a point source; everywhere else, the stress–energy tensor would be zero, so

$$\Box \bar{h}^{\alpha\beta} = 0$$

Now, this is the usual homogeneous wave equation — one for each component of $\bar{h}^{\alpha\beta}$. Solutions to this equation

are well known. For a wave moving away from a point source, the radiated part (meaning the part that dies off as $1/r$ far from the source) can always be written in the form $A(t-r,\theta,\phi)/r$, where A is just some function. It can be shown[49] that — to a linear approximation — it is always possible to make the field traceless. Now, if we further assume that the source is positioned at $r = 0$, the general solution to the wave equation in spherical coordinates is

$$\bar{h}^{\alpha\beta} = \frac{1}{r}\begin{bmatrix} 0 & 0 & 0 & 0 \\ 0 & 0 & 0 & 0 \\ 0 & 0 & A_+(t-r,\theta,\phi) & A_\times(t-r,\theta,\phi) \\ 0 & 0 & A_\times(t-r,\theta,\phi) & -A_+(t-r,\theta,\phi) \end{bmatrix}$$

$$\equiv \begin{bmatrix} 0 & 0 & 0 & 0 \\ 0 & 0 & 0 & 0 \\ 0 & 0 & h_+(t-r,r,\theta,\phi) & h_\times(t-r,r,\theta,\phi) \\ 0 & 0 & h_\times(t-r,r,\theta,\phi) & -h_+(t-r,r,\theta,\phi) \end{bmatrix}$$

where we now see the origin of the two polarizations.

30.6.2 Relation to the source

If we know the details of a source — for instance, the parameters of the orbit of a binary — we can relate the source's motion to the gravitational radiation observed far away. With the relation

$$\Box \bar{h}^{\alpha\beta} = -16\pi \tau^{\alpha\beta}$$

we can write the solution in terms of the tensorial Green's function for the d'Alembertian operator:[48]

$$\bar{h}^{\alpha\beta}(t,\vec{x}) = -16\pi \int G^{\alpha\beta}_{\gamma\delta}(t,\vec{x};t',\vec{x}')\,\tau^{\gamma\delta}(t',\vec{x}')\,dt'\,d^3x'$$

Though it is possible to expand the Green's function in tensor spherical harmonics, it is easier to simply use the form

$$G^{\alpha\beta}_{\gamma\delta}(t,\vec{x};t',\vec{x}') = \frac{1}{4\pi}\delta^\alpha_\gamma\,\delta^\beta_\delta\,\frac{\delta(t \pm |\vec{x}-\vec{x}'| - t')}{|\vec{x}-\vec{x}'|}$$

where the positive and negative signs correspond to ingoing and outgoing solutions, respectively. Generally, we are interested in the outgoing solutions, so

$$\bar{h}^{\alpha\beta}(t,\vec{x}) = -4 \int \frac{\tau^{\alpha\beta}(t-|\vec{x}-\vec{x}'|,\vec{x}')}{|\vec{x}-\vec{x}'|}\,d^3x'$$

If the source is confined to a small region very far away, to an excellent approximation we have:

$$\bar{h}^{\alpha\beta}(t,\vec{x}) \approx -\frac{4}{r} \int \tau^{\alpha\beta}(t-r,\vec{x}')\,\mathrm{d}^3 x'$$

where $r = |\vec{x}|$.

Now, because we will eventually only be interested in the spatial components of this equation (time components can be set to zero with a coordinate transformation), and we are integrating this quantity — presumably over a region of which there is no boundary — we can put this in a different form. Ignoring divergences with the help of Stokes' theorem and an empty boundary, we can see that

$$\int \tau^{ij}(t-r,\vec{x}')\,\mathrm{d}^3 x' = \int x'^i x'^j \nabla_k \nabla_l \tau^{kl}(t-r,\vec{x}')\,\mathrm{d}^3 x'$$

Inserting this into the above equation, we arrive at

$$\bar{h}^{ij}(t,\vec{x}) \approx -\frac{4}{r} \int x'^i x'^j \nabla_k \nabla_l \tau^{kl}(t-r,\vec{x}')\,\mathrm{d}^3 x'$$

Finally, because we have chosen to work in coordinates for which $\nabla_\beta \bar{h}^{\alpha\beta} = 0$, we know that $\nabla_\beta \tau^{\alpha\beta} = 0$. With a few simple manipulations, we can use this to prove that

$$\nabla_0 \nabla_0 \tau^{00} = \nabla_j \nabla_k \tau^{jk}$$

With this relation, the expression for the radiated field is

$$\bar{h}^{ij}(t,\vec{x}) \approx -\frac{4}{r}\frac{\mathrm{d}^2}{\mathrm{d}t^2} \int x'^i x'^j \tau^{00}(t-r,\vec{x}')\,\mathrm{d}^3 x'$$

In the linear case, $\tau^{00} = \rho$, the density of mass-energy.

To a very good approximation, the density of a simple binary can be described by a pair of delta-functions, which eliminates the integral. Explicitly, if the masses of the two objects are M_1 and M_2 , and the positions are $\vec{x}_1$ and $\vec{x}_2$, then

$$\rho(t-r,\vec{x}') = M_1\delta^3(\vec{x}'-\vec{x}_1(t-r)) + M_2\delta^3(\vec{x}'-\vec{x}_2(t-r))$$

We can use this expression to do the integral above:

$$\bar{h}^{ij}(t,\vec{x}) \approx -\frac{4}{r}\frac{\mathrm{d}^2}{\mathrm{d}t^2}\left\{ M_1 x_1^i(t-r)x_1^j(t-r) + M_2 x_2^i(t-r)x_2^j(t-r) \right\}$$

Using mass-centered coordinates, and assuming a circular binary, this is

$$\bar{h}^{ij}(t,\vec{x}) \approx -\frac{4}{r}\frac{M_1 M_2}{R}\, n^i(t-r)n^j(t-r)$$

where $\vec{n} = \vec{x}_1/|\vec{x}_1|$. Plugging in the known values of $\vec{x}_1(t-r)$, we obtain the expressions given above for the radiation from a simple binary.

30.7   See also

- Gravitational wave background

- Cosmic gravitational wave background

- Big Bang Observer (BBO), proposed successor to LISA

- DECIGO "Deci-hertz Interferometer Gravitational wave Observatory", the planned laser interferometric detector in space

- Gravitational field

- Gravitomagnetism

- Graviton

- Gravitational wave astronomy

- Hawking radiation, for gravitationally induced electromagnetic radiation from black holes

- HM Cancri

- LIGO, VIRGO, GEO 600, and TAMA 300 — Gravitational wave detectors

- Linearised Einstein field equations

- LISA the proposed Laser Interferometer Space Antenna

- Peres metric

- pp-wave spacetime, for an important class of exact solutions modelling gravitational radiation

- Spin-flip, a consequence of gravitational wave emission from binary supermassive black holes

- Sticky bead argument, for a physical way to see that gravitational radiation should carry energy

- Tidal force

30.8 References

[1] Einstein, A (June 1916). "Näherungsweise Integration der Feldgleichungen der Gravitation". *Sitzungsberichte der Königlich Preussischen Akademie der Wissenschaften Berlin.* part 1: 688–696.

[2] Einstein, A (1918). "Über Gravitationswellen". *Sitzungsberichte der Königlich Preussischen Akademie der Wissenschaften Berlin.* part 1: 154–167.

[3] Finley, Dave. "Einstein's gravity theory passes toughest test yet: Bizarre binary star system pushes study of relativity to new limits.". Phys.Org.

[4] The Detection of Gravitational Waves using LIGO, B. Barish

[5] "First Second of the Big Bang". *How The Universe Works 3.* 2014. Discovery Science.

[6] http://www.theguardian.com/science/2014/jun/04/gravitational-wave-discovery-dust-big-bang-inflation

[7] Staff (17 March 2014). "BICEP2 2014 Results Release". *National Science Foundation.* Retrieved 18 March 2014.

[8] "First Direct Evidence of Cosmic Inflation". *http://www.cfa.harvard.edu.* Harvard-Smithsonian Center for Astrophysics. 17 March 2014. Retrieved 17 March 2014.

[9] Clavin, Whitney (17 March 2014). "NASA Technology Views Birth of the Universe". *NASA.* Retrieved 17 March 2014.

[10] Overbye, Dennis (17 March 2014). "Detection of Waves in Space Buttresses Landmark Theory of Big Bang". *New York Times.* Retrieved 17 March 2014.

[11] Overbye, Dennis (19 June 2014). "Astronomers Hedge on Big Bang Detection Claim". *New York Times.* Retrieved 20 June 2014.

[12] Amos, Jonathan (19 June 2014). "Cosmic inflation: Confidence lowered for Big Bang signal". *BBC News.* Retrieved 20 June 2014.

[13] Ade, P.A.R. et al. (BICEP2 Collaboration) (19 June 2014). "Detection of B-Mode Polarization at Degree Angular Scales by BICEP2" (PDF). *Physical Review Letters* **112**: 241101. arXiv:1403.3985. Bibcode:2014PhRvL.112x1101A. doi:10.1103/PhysRevLett.112.241101. PMID 24996078. Retrieved 20 June 2014.

[14] Planck Collaboration Team (19 September 2014). "Planck intermediate results. XXX. The angular power spectrum of polarized dust emission at intermediate and high Galactic latitudes". *ArXiv.* arXiv:1409.5738. Bibcode:2014arXiv1409.5738P. Retrieved 22 September 2014.

[15] Overbye, Dennis (22 September 2014). "Study Confirms Criticism of Big Bang Finding". *New York Times.* Retrieved 22 September 2014.

[16] Clavin, Whitney (30 January 2015). "Gravitational Waves from Early Universe Remain Elusive". *NASA.* Retrieved 30 January 2015.

[17] Overbye, Dennis (30 January 2015). "Speck of Interstellar Dust Obscures Glimpse of Big Bang". *New York Times.* Retrieved 31 January 2015.

[18] Cowen, Ron (2015-01-30). "Gravitational waves discovery now officially dead". *nature.* doi:10.1038/nature.2015.16830.

[19] LIGO Scientific Collaboration; Virgo Collaboration (2012). "Search for Gravitational Waves from Low Mass Compact Binary Coalescence in LIGO's Sixth Science Run and Virgo's Science Runs 2 and 3". *Physical Review D* **85**: 082002. arXiv:1111.7314. Bibcode:2012PhRvD..85h2002A. doi:10.1103/PhysRevD.85.082002.

[20] LIGO Scientific Collaboration; Virgo Collaboration (2012). "All-sky search for gravitational-wave bursts in the second joint LIGO-Virgo run". *Physical Review D* **85**: 122007. arXiv:1202.2788. Bibcode:2012PhRvD..85l2007A. doi:10.1103/PhysRevD.85.122007.

[21] LIGO Scientific Collaboration; Virgo Collaboration (2013). "Search for gravitational waves from binary black hole inspiral, merger, and ringdown in LIGO-Virgo data from 2009-2010". *Physical Review D* **87**: 022002. arXiv:1209.6533. Bibcode:2013PhRvD..87b2002A. doi:10.1103/PhysRevD.87.022002.

[22] Krauss, LM; Dodelson, S; Meyer, S (2010). "Primordial Gravitational Waves and Cosmology". *Science* **328** (5981): 989–992. arXiv:1004.2504. Bibcode:2010Sci...328..989K. doi:10.1126/science.1179541. PMID 20489015.

[23] Hawking, S. W. and Israel, W., *General Relativity: An Einstein Centenary Survey,* Cambridge University Press, Cambridge, 1979, 98.

[24] Landau, L. D. and Lifshitz, E. M., *The Classical Theory of Fields.* Fourth Revised English Edition, Pergamon Press., 1975, 356–357.

[25] Einstein, A (1918). "Über Gravitationswellen". *Sitzungsberichte, Preussische Akademie der Wissenschaften* **154**.

[26] Gravitational Radiation

[27] Beatty, Kelly. "Why is the Earth moving away from the sun?". *New Scientist.*

[28] LIGO Scientific Collaboration; Virgo Collaboration (2010). "Predictions for the rates of compact binary coalescences observable by ground-based gravitational-wave detectors". *Classical and Quantum Gravity* **27**: 17300. arXiv:1003.2480. Bibcode:2010CQGra..27q3001A. doi:10.1088/0264-9381/27/17/173001.

[29] Relativistic Binary Pulsar B1913+16: Thirty Years of Observations and Analysis

[30] The discovery of the first binary pulsar

[31] Crashing Black Holes

[32] Binary and Millisecond Pulsars

[33] L. P. Grishchuk (1976), "Primordial Gravitons and the Possibility of Their Observation", Sov. Phys. JETP Lett. 23, p. 293.

[34] Braginsky, V. B., Rudenko and Valentin, N. Section 7: "Generation of gravitational waves in the laboratory", *Physics Report* (Review section of *Physics Letters*), 46, No. 5. 165–200, (1978).

[35] Li, Fangyu, Baker, R. M L, Jr., and Woods, R. C., "Piezoelectric-Crystal-Resonator High-Frequency Gravitational Wave Generation and Synchro-Resonance Detection", in the proceedings of *Space Technology and Applications International Forum (STAIF-2006)*, edited by M.S. El-Genk, American Institute of Physics Conference Proceedings, Melville NY 813: 2006.

[36] Merritt, D.; et al. (May 2004). "Consequences of Gravitational Wave Recoil". *The Astrophysical Journal Letters* **607** (1): L9–L12. arXiv:astro-ph/0402057. Bibcode:2004ApJ...607L...9M. doi:10.1086/421551.

[37] Gualandris, A.; Merritt, D.; et al. (May 2008). "Ejection of Supermassive Black Holes from Galaxy Cores". *The Astrophysical Journal* **678** (2): 780–797. arXiv:0708.0771. Bibcode:2008ApJ...678..780G. doi:10.1086/586877.

[38] Merritt, D.; Schnittman, J. D.; Komossa, S. (2009). "Hypercompact Stellar Systems Around Recoiling Supermassive Black Holes". *The Astrophysical Journal* **699** (2): 1690–1710. arXiv:0809.5046. Bibcode:2009ApJ...699.1690M. doi:10.1088/0004-637X/699/2/1690.

[39] Komossa, S.; Zhou, H.; Lu, H. (May 2008). "A Recoiling Supermassive Black Hole in the Quasar SDSS J092712.65+294344.0?". *The Astrophysical Journal* **678** (2): L81–L84. arXiv:0804.4585. Bibcode:2008ApJ...678L..81K. doi:10.1086/588656

[40] Overbye, Dennis (24 March 2014). "Ripples From the Big Bang". *New York Times*. Retrieved 24 March 2014.

[41] Thorne, Kip S. (1995). "Gravitational Waves". *Cornell University Library*.

[42] David G. Blair (Ed.) (1991). *The detection of gravitational waves*. Cambridge University Press.

[43] For a review of early experiments using Weber bars, see Levine, J. (April 2004). "Early Gravity-Wave Detection Experiments, 1960–1975". *Physics in Perspective (Birkhäuser Basel)* **6** (1): 42–75. Bibcode:2004PhP.....6...42L. doi:10.1007/s00016-003-0179-6.

[44] Gravitational Radiation Antenna In Leiden

[45] de Waard, Arlette; Luciano Gottardi; Giorgio Frossati (July 2000). *Spherical Gravitational Wave Detectors: cooling and quality factor of a small CuAl6% sphere* (PDF). Marcel Grossmann meeting on General Relativity. Rome, Italy: World Scientific Publishing Co. Pte. Ltd. (published December 2002). pp. 1899–1901. Bibcode:2002nmgm.meet.1899D. doi:10.1142/9789812777386_0420. ISBN 9789812777386.

[46] The idea of using laser interferometry for gravitational wave detection was first mentioned by Gerstenstein and Pustovoit 1963 Sov. Phys.–JETP 16 433. Weber mentioned it in an unpublished laboratory notebook. Rainer Weiss first described in detail a practical solution with an analysis of realistic limitations to the technique in R. Weiss (1972). "Electromagetically Coupled Broadband Gravitational Antenna". Quarterly Progress Report, Research Laboratory of Electronics, MIT 105: 54.

[47] Einstein@Home

[48] Thorne, Kip (April 1980). "Multipole expansions of gravitational radiation". *Reviews of Modern Physics* **52** (2): 299–339. Bibcode:1980RvMP...52..299T. doi:10.1103/RevModPhys.52.299.

[49] C. W. Misner, K. S. Thorne, and J. A. Wheeler (1973). *Gravitation*. W. H. Freeman and Co.

30.9   Further reading

- Chakrabarty, Indrajit, "Gravitational Waves: An Introduction". arXiv:physics/9908041 v1, Aug 21, 1999.

- Landau, L. D. and Lifshitz, E. M., The Classical Theory of Fields (Pergamon Press),(1987).

- Will, Clifford M., *The Confrontation between General Relativity and Experiment*. Living Rev. Relativity 9 (2006) 3.

- Peter Saulson, "Fundamentals of Interferometric Gravitational Wave Detectors", World Scientific, 1994.

- J. Bicak, W.N. Rudienko, "Gravitacionnyje wolny w OTO i probliema ich obnarużenija", Izdatielstwo Moskovskovo Universitieta, 1987.

- A. Kułak, "Electromagnetic Detectors of Gravitational Radiation", PhD Thesis, Cracow 1980 (In Polish).

- P. Tatrocki, "On intuitive description of graviton detector", www.philica.com .

- P. Tatrocki, "Can the LIGO, VIRGO, GEO600, AIGO, TAMA, LISA detectors really detect?", www. philica.com .

30.10 Bibliography

- Berry, Michael, *Principles of Cosmology and Gravitation* (Adam Hilger, Philadelphia, 1989). ISBN 0-85274-037-9

- Collins, Harry, *Gravity's Shadow: The Search for Gravitational Waves*, University of Chicago Press, 2004.

- P. J. E. Peebles, *Principles of Physical Cosmology* (Princeton University Press, Princeton, 1993). ISBN 0-691-01933-9.

- Wheeler, John Archibald and Ciufolini, Ignazio, *Gravitation and Inertia* (Princeton University Press, Princeton, 1995). ISBN 0-691-03323-4.

- Woolf, Harry, ed., *Some Strangeness in the Proportion* (Addison–Wesley, Reading, Massachusetts, 1980). ISBN 0-201-09924-1.

30.11 External links

- Gravitational waves at *Encyclopædia Britannica*

-

- Gravitational Waves on *In Our Time* at the BBC. (listen now)

- The LISA Brownbag – Selection of the most significant e-prints related to LISA science

- Astroparticle.org. To know everything about astroparticle physics, including gravitational waves

- Caltech's Physics 237-2002 Gravitational Waves by Kip Thorne **Video plus notes:** Graduate level but does not assume knowledge of General Relativity, Tensor Analysis, or Differential Geometry; Part 1: Theory (10 lectures), Part 2: Detection (9 lectures)

- www.astronomycast.com January 14, 2008 Episode 71: Gravitational Waves

- Laser Interferometer Gravitational Wave Observatory. LIGO Laboratory, operated by the California Institute of Technology and the Massachusetts Institute of Technology

- The LIGO Scientific Collaboration

- Einstein's Messengers – The LIGO Movie by NSF

- Home page for Einstein@Home project, a distributed computing project processing raw data from LIGO Laboratory, searching for gravitational waves

- The National Center for Supercomputing Applications – a numerical relativity group

- Caltech Relativity Tutorial – A basic introduction to gravitational waves, and astrophysical systems giving off gravitational waves

- Resource Letter GrW-1: Gravitational waves – a list of books, journals and web resources compiled by Joan Centrella for research into gravitational waves

- Mathematical and Physical Perspectives on Gravitational Radiation – written by B F Schutz of the Max Planck Institute explaining the significance and background of some key concepts in gravitational radiation

- Binary BH Merger – estimating the radiated power and merger time of a BH binary using dimensional analysis

30.12 Text and image sources, contributors, and licenses

30.12.1 Text

- **Quantum gravity** *Source:* https://en.wikipedia.org/wiki/Quantum_gravity?oldid=690077606 *Contributors:* AstroNomer–enwiki, Matusz, Miguel–enwiki, Roadrunner, Stevertigo, Ubiquity, Bobby D. Bryant, Mcarling, NuclearWinner, Anders Feder, Susurrus, Coren, Charles Matthews, Timwi, Reddi, Tpbradbury, Phys, Bevo, Raul654, BenRG, Frazzydee, Jeffq, Sdedeo, Rholton, Wereon, Ilya (usurped), Seth Ilys, Ancheta Wis, Giftlite, Herbee, Fropuff, Endlessnameless, Malyctenar, Jason Quinn, Finn-Zoltan, YapaTi–enwiki, Lumidek, Marcus2, Joyous!, TJSwoboda, Vitaleyes, Davidclifford, JimJast, Rich Farmbrough, Guanabot, FT2, Masudr, Pjacobi, Pie4all88, David Schaich, Bender235, Clement Cherlin, El C, PhilHibbs, Army1987, Apyule, VBGFscJUn3, PWilkinson, Daniel Arteaga–enwiki, Keenan Pepper, Cjthellama, DonJStevens, Velella, Dabbler, Tycho, Cal 1234, RJFJR, Count Iblis, ThomasWinwood, Anarchimede, Scarykitty, Woohookitty, Igny, ToddFincannon, Mpatel, GregorB, Joke137, Christopher Thomas, Marudubshinki, Graham87, Yurik, Kroggz, Rjwilmsi, Eoghanacht, Jrasowsky, JHMM13, Smithfarm, Ems57fcva, FayssalF, Itinerant1, Lmatt, Chobot, Hmonroe, YurikBot, Hillman, ErkDemon, JocK, SCZenz, Roy Brumback, Bota47, Zunaid, JonathanD, 2over0, Arthur Rubin, Modify, LeonardoRob0t, Caco de vidro, RG2, KasugaHuang, Resolute, SmackBot, Samdutton, Vald, Eskimbot, Hbackman, Onebravemonkey, Gilliam, Chris the speller, Ben.c.roberts, Cthuljew, Silly rabbit, Complexica, Colonies Chris, QFT, Soosed, Theanphibian, Shushruth, Ck lostsword, Yevgeny Kats, DJIndica, Lambiam, Vampus, Vincenzo.romano, Jaganath, JorisvS, Bjankuloski06, RoboDick–enwiki, IronGargoyle, Dicklyon, SirFozzie, Treyp, Twunchy, Piccor, Kurtan–enwiki, Harold f, CalebNoble, Duduong, Paulmlieberman, TVC 15, UncleBubba, TAz69x, Sam Staton, ST47, B, Patrick O'Leary, Epbr123, Koeplinger, Klasovsky, Markus Pössel, Keraunos, Headbomb, Marek69, MichaelMaggs, Tim Shuba, MER-C, ParadiZio, Clementvidal, Perlygatekeeper, VoABot II, Alvatros–enwiki, Bdalevin, SHCarter, Jpod2, DAGwyn, Nucleophilic, LorenzoB, Rickard Vogelberg, DancingPenguin, Rettetast, Victor Blacus, AstroHurricane001, Yonidebot, Acalamari, Mstuomel, Fullmetal2887, NewEnglandYankee, DorganBot, CardinalDan, Idioma-bot, Sheliak, VolkovBot, Pleasantville, Seattle Skier, AlnoktaBOT, TXiKiBoT, Dllahr, Rdekleer, Saibod, Cyberchip, Wikiwikimoore, Carlorovelli, LoreMiles, StevenJohnston, SieBot, LeadSongDog, Bentogoa, Coldcreation, ReluctantPhilosopher, StaticG, GarbagEcol, ClueBot, The Thing That Should Not Be, EoGuy, Polyamorph, Andwor9, Notburnt, Tms9, Alexbot, Resoru, Eeekster, Tamaratrouts, Brews ohare, SchreiberBike, Askahre, BOTarate, Lambtron, DumZiBoT, XLinkBot, Rror, Facts707, SilvonenBot, Theonlydavewilliams, Mhsb, Truthnlove, Ttimespan, Trifonov–enwiki, Addbot, Mortense, Grayfell, Eric Drexler, Gravitophoton, DOI bot, AkhtaBot, CanadianLinuxUser, Frosty726, LaaknorBot, Delaszk, Tassedethe, Tide rolls, Taketa, Titan1129, Krano, Luckas-bot, Yobot, WikiDan61, Pigretational, Wireader, Allowgolf–enwiki, Wiki Roxor, Jim1138, IRP, Sziwbot, Quantity, Materialscientist, Citation bot, ArthurBot, LilHelpa, Amareto2, Ekwos, KrisBogdanov, Rolfguthmann, StealthCopyEditor, 안단테, Dan6hell66, Rabsmith, Hep thinker, Paine Ellsworth, DrArthurRubinPHD, Lagelspeil, Nunc aut numquam, Vacuunaut, Van Speijk, Knowandgive, Craig Pemberton, Udifuchs, Citation bot 2, Citation bot 1, Citation bot 4, Jonesey95, Hirvenkürpa, Tom.Reding, Pmokeefe, Casimir9999, Dac04, Dude1818, Valeriy Pischenko, Follyland, TrueTeargem, N0814444, Earthandmoon, Korepin, DARTH SIDIOUS 2, Musictime4me, RjwilmsiBot, EmausBot, Francophile124, Octaazacubane, Fotoni, Slightsmile, Garfield Salazar, Hhhippo, JSquish, John Cline, Fæ, LostAlone, Brazmyth, Throwmeaway, Arbnos, Ebrambot, Kusername, DanielBurnstein, TonyMath, L Kensington, Maschen, Donner60, Parusaro, Apratim07, Terra Novus, Isocliff, Googledin!, ClueBot NG, SpikeTorontoRCP, Science writer, Preon, Raidr, Jhmmok, 336, Widr, Helpful Pixie Bot, Bibcode Bot, Bardsley Rides a Segway, Apelikedawg, FiveColourMap, Trevayne08, Mr.viktor.stepanov, Brainssturm, BattyBot, Jimw338, Ryanr666, Kryomaxim, Garuda0001, CuriousMind01, Saehry, TwoTwoHello, Sanathdevalapurkar, Andyhowlett, GabeIglesia, Sanathlab, Roiwallace, Spencer.mccormick, Spencerfjase, MrShlongNo1, Marc D. Garrett, D00d00ballz, Gigantmozg, Susan.grayeff, Polytope24, Frinthruit, Anrnusna, Dfyytj, Monkbot, Umut Alihan Dikel, Amortias, Klj1234, Pfpguy, KasparBot, Jespergrimstrup and Anonymous: 297

- **String theory** *Source:* https://en.wikipedia.org/wiki/String_theory?oldid=689563569 *Contributors:* AxelBoldt, Sodium, Mav, Bryan Derksen, Zundark, The Anome, Tarquin, Taw, Eean, Malcolm Farmer, Hephaestos, Olivier, Drseudo, Stevertigo, Spiff–enwiki, Edward, PhilipMW, Michael Hardy, Bewildebeast, Dante Alighieri, Gabbe, Graue, Tgeorgescu, Mcarling, CesarB, Looxix–enwiki, Ahoerstemeier, Theresa knott, Suisui, Angela, Den fjättrade ankan–enwiki, Jdforrester, Julesd, Salsa Shark, Schneelocke, Charles Matthews, Timwi, Bemoeial, Jitse Niesen, 4lex, Greenrd, ErikStewart, Furrykef, Saltine, Phys, Omegatron, Bevo, Topbanana, Trent, Nufy8, Robbot, Craig Stuntz, Fredrik, Chris 73, R3m0t, COGDEN, Mirv, Wjhonson, Sverdrup, Academic Challenger, DHN, Hadal, Khlo, ElBenevolente, HaeB, Tobias Bergemann, Giftlite, DocWatson42, Christopher Parham, Awolf002, Mporter, Amorim Parga, Mikez, Harp, Kim Bruning, Tom harrison, Ferkelparade, Leflyman, Fropuff, No Guru, Anville, Moyogo, Curps, Pashute, Nomad–enwiki, Mboverload, Solipsist, SWAdair, DemonThing, Wmahan, Btphelps, MSTCrow, Decoy, Chowbok, Gadfium, Steuard, Pgan002, Quadell, Carandol–enwiki, Antandrus, Beland, JoJan, Khaosworks, Tothebarricades.tk, Thincat, Tomruen, Shidobu, Icairns, Lumidek, NoPetrol, Avihu, Fanghong–enwiki, Trevor MacInnis, Lacrimosus, Zro, Mike Rosoft, D6, Urvabara, Felix Wan, Jkl, Discospinster, ElTyrant, Rich Farmbrough, Rhobite, Pjacobi, Alien life form, Vapour, Silence, Kzzl, LindsayH, Mani1, Pavel Vozenilek, Paul August, Bender235, Kjoonlee, Mashford, Kelvinc, Perlman10s, Panu–enwiki, Brian0918, Dpotter, Livajo, El C, Laurascudder, Shanes, Zegoma beach, RoyBoy, Causa sui, Bobo192, Directorstratton, Janna Isabot, Smalljim, John Vandenberg, Flxmghvgvk, I9Q79oL78KiL0QTFHgyc, Physicistjedi, Bongoo, 4v4l0n42, Merope, Geschichte, Linuxlad, Phils, Merenta, Alansohn, Gary, JYolkowski, Enirac Sum, Ryanmcdaniel, Arthena, Borisblue, Rd232, Plumbago, Axl, R Calvete, Lightdarkness, Kocio, Bart133, Wtmitchell, Isaac, Tycho, Cal 1234, Fadereu, CloudNine, Sciurinæ, Computerjoe, Kusma, DV8 2XL, Pwqn, Gene Nygaard, Ringbang, Ceyockey, Falcorian, Bobrayner, Joriki, Mel Etitis, Linas, BillC, Jacobolus, HFarmer, Before My Ken, Netdragon, MONGO, GeorgeOrr, Mpatel, Bbatsell, GregorB, 조학현, Joke137, Christopher Thomas, Dysepsion, GSlicer, Jan.bannister, Graham87, Magister Mathematicae, Hillbrand, BD2412, Elvey, Galwhaa, Raymond Hill, JIP, RxS, Athelwulf, Edison, Sjakkalle, Rjwilmsi, Xgamer4, Jake Wartenberg, Arabani, MarSch, TheRingess, Jmcc150, Aero66, Crazynas, Juan Marquez, R.e.b., Bubba73, DoubleBlue, Zelos, AlisonW, Asafavi, Lionelbrits, Conorific, Zunz, Mathbot, Crazycomputers, RexNL, Gurch, Algri, TeaDrinker, Zifnabxar, XAXISx, Erik4, Phoenix2–enwiki, Antimatter15, Ggb667, Chobot, Visor, DVdm, Mhking, VolatileChemical, Bgwhite, Algebraist, Ben Tibbetts, YurikBot, Ugha, Wavelength, Borgx, NuclearFusion–enwiki, Angus Lepper, Hairy Dude, Jimp, Hillman, Cyferx, Wolfmankurd, Pip2andahalf, RussBot, Moronoman, Crazytales, Pippo2001, Bhny, Pigman, SpuriousQ, Branman515, Stephenb, Gaius Cornelius, Eleassar, Bovineone, Cheesus, Shanel, NawlinWiki, Tong–enwiki, Mike18xx, SCZenz, Cleared as filed, Bdiah, Pym98, SColombo, Haemo, FF2010, Closedmouth, Reyk, Brina700, Chris Brennan, Vicarious, Brianlucas, Geoffrey.landis, Hitchhiker89, Spliffy, Pred, ArielGold, Roy Fultun, Ilmari Karonen, Katieh5584, Pentasyllabic, Lunch, DVD R W, WikiFew, That Guy, From That Show!, Street Scholar, AndrewWTaylor, QSquared, Sardanaphalus, Vanka5, MacsBug, Hvitlys, SmackBot, Kurochka, Zazaban, Tom Lougheed, Prodego, KnowledgeOfSelf, Hydrogen Iodide, Melchoir, Vald, Skrewtape, Atomota, Canthusus, GaeusOctavius, Cool3, Andyvn2, Skizzik, RobertM525, Dauto, Bluebot, SSJ 5, Keegan, Aidan Croft, Thumperward, Oli Filth, Silly rabbit, Timneu22, SchfiftyThree, Moshe Constantine Hassan Al-Silverburg, Complexica,

Rediahs, RayAYang, Aero77, Adamstevenson, Ikiroid, Epastore, Baronnet, Ned Scott, Sbharris, Colonies Chris, Konstable, Sct72, Sewlong, Can't sleep, clown will eat me, Timothy Clemans, Onorem, Neilanderson, EvelinaB, TKD, KerathFreeman, Addshore, UU, The tooth, Pepsidrinka, Somebody2292, --=The Doctor=--, Fuhghettaboutit, Cybercobra, Irish Souffle, Nakon, Jdlambert, James McNally, MichaelBillington, Lostart, Insineratehymn, Drphilharmonic, SpiderJon, DMacks, Ihatetoregister, Where, Michael IFA, Yevgeny Kats, Vasiliy Faronov, Byelf2007, Angela26, Visium, Rory096, Zymurgy, Harryboyles, Mdl53711, T-dot, Titus III, Ergative rlt, MagnaMopus, UberCryxic, Vgy7ujm, Linnell, Mgiganteus1, Nonsuch, IronGargoyle, Ckatz, DoItAgain, AstroGod, Kirbytime, Jimbo Mahoney, FredrickS, Invisifan, Ryulong, Ryanjunk, MathStuf, Mike Doughney, Norm mit, Hindol, Dan Gluck, Huntscorpio, Iridescent, K, Sunoco, You? Me? Us?, CzarB, Rabinzkaman, JoeBot, Lottamiata, Tony Fox, Vrkaul, Torrazzo, Gil Gamesh, Areldyb, Courcelles, Tawkerbot2, Gebrah, Shamvil, DKqwerty, Lbr123, Harold f, Heqs, Devourer09, Duduong, Sarvagnya, Dewayne76, JForget, Cg-realms, InvisibleK, CRGreathouse, CmdrObot, Earthlyreason, Van helsing, Olaf Davis, CBM, Rawling, Jibal, Witten Is God, Nunquam Dormio, Giko, KnightLago, Thubsch, Leujohn, SlashDot, TheTito, Karenjc, Myasuda, Emarv, Cydebot, Gmusser, Gogo Dodo, Jkokavec, Kahananite, Quajafrie, Michael C Price, Doug Weller, DumbBOT, Narayanese, AlphaNumeric, SRoughsedge, Vanished User jdksfajlasd, Woland37, Zalgo, Daniel Olsen, UberScienceNerd, Bkazaz, DJBullfish, Thijs!bot, Epbr123, Rwmnau, Babemachine, Pimpin101, Mbell, O. Faigl.ladislav, Ucanlookitup, Andyjsmith, Headbomb, Tcturner2002, Marek69, Brahmajnani, Arthurcprado-enwiki, Y.t., D3gtrd, Babemonkey, Dark dude, Duncan McB, EdJohnston, MichaelMaggs, Ancientanubis, Natalie Erin, Hempfel, Jomoal99, Mmortal03, Mentifisto, Geekdom04, AntiVandalBot, Luna Santin, Seaphoto, Ed270791, Opelio, Doc Tropics, David136a, NithinBekal, Dotdotdotdash, Helicopter, Poshzombie, MontanNito, Dylan Lake, Maximilian77, Shlomi Hillel, Db63376, SamIAmNot, Knotwork, Res2216firestar, Superior IQ Genius, MER-C, Andonic, Sitethief, 100110100, TallulahBelle, Nestamachine, Savant13, Daynightrader, Goldenglove, Charibdis, Acroterion, Ophion, Aigisthos, Editmyhandman, Aruben537, Magioladitis, WolfmanSF, Bongwarrior, VoABot II, Yandman, JamesBWatson, رياض, Qutt, Jespinos, Kevinmon, Aka042, Froid, DAGwyn, Catgut, Panser Born, Ensign beedrill, Perspectival, JJ Harrison, Dirac66, Justanother, Aziz1005, Cpl Syx, ChazBeckett, Teardrop onthefire, WLU, Stephen shenker, Robin S, SkepticVK, Joshua Davis, Mkroh, B9 hummingbird hovering, S3000, Hdt83, MartinBot, FlieGerFaUstMe262, Ytomem, Shimwell, Arjun01, KrishSundaresan, Anaxial, Jay Litman, Alexcalamaro, Andrej.westermann, Smokizzy, LedgendGamer, Cyrus Andiron, Peteryoung144, Tgeairn, Artaxiad, HEL, AlphaEta, J.delanoy, AstroHurricane001, Maurice Carbonaro, Yonidebot, Morris729, M C Y 1008, 69gangsta420, It Is Me Here, Shawn in Montreal, Janus Shadowsong, Bailo26, Fredsie, Madagaskar07, Duchesserin, AntiSpamBot, CHIAGEHYANG, Chiswick Chap, Watsup1313, Belovedfreak, HaloInverse, NewEngland Yankee, Scott1329m, Thesis4Eva, Policron, Jrcla2, WJBscribe, Rnricklefs, Jamesofur, Eyelidlessness, Jonnyk aus, Kvdveer, JavierMC, Izno, Xiahou, CardinalDan, Sheliak, HamatoKameko, Malik Shabazz, Concertmusic, JohnBlackburne, JustinHagstrom, Fences and windows, Wooba doob, Philip Trueman, DoorsAjar, HowardFrampton, TXiKiBoT, Zidonuke, Red Act, Kriak, Calwiki, Technopat, Hqb, Andrius.v, Anonymous Dissident, Crohnie, AlysTarr, Qxz, Vanished user ikijeirw34iuaeolaseriffic, Impunv, Seraphim, Martin451, Don4of4, ABigGreenHippo, Huperphuff, LeaveSleaves, Kaenneth, StringyGuy, Maxim, Erth64net, Meters, Lamro, Rickstauduhar, Enviroboy, Turgan, Anna512, PhysPhD, Northfox, NPguy, Matthew Sanders, Luke Walkerson, Newbyguesses, MissMJ, SieBot, Escher26, J.A.Ireland, BA (IHPST), 4wajzkd02, Robdunst, Dreamafter, Pallab1234, Dbelange, MTHarden, Lemonflash, Kylemew, Yintan, GlassCobra, Wpegden, Likebox, Flyer22 Reborn, Exert, ProGeek314, Arbor to SJ, Babawhitemoose, Caidh, Dhatfield, Audree, Oxymoron83, Pretty Green, Weaselstomp, Manway, Alex.muller, Taco Manipulator, Tschach, Manheat84, Anchor Link Bot, Mikebernstein, ImperialismGo, Nergaal, Ionfield, Ayleuss, Sh4wz0r, Naturespace, Martarius, Phyte, ClueBot, The Thing That Should Not Be, String4d, Illusion96, Polyamorph, Mpd1989, Alexdeburca18, Wiggl3sLimited, Excirial, Kjramesh, Jusdafax, Resoru, WikiZorro, Eeekster, Verum-enwiki, Tamaratrouts, Gtstricky, Humanino, Brews ohare, NuclearWarfare, Cenarium, Arjayay, Razorflame, Scoobey, BOTarate, Sideswiper, Thingg, Capudo, BVBede, Versus22, Introductory adverb clause, MelonBot, SoxBot III, Egmontaz, Notpayingthepsychiatrist, DumZiBoT, BahTab, TimothyRias, Aj00200, Reaperfromhell, Dunkaroo207, XLinkBot, AlexGWU, Impshum, Saeed.Veradi, Little Mountain 5, Guy392, David424, Truthnlove, Qweeveen, Tayste, Addbot, Steven66s, Denali134, Elemented9, Varrey280303, Eric Drexler, Some jerk on the Internet, Fizzycyst, Uruk2008, DOI bot, Jojhutton, AngryBacon, Captain-tucker, Auspex1729, Kongr43gpen, Fgnievinski, Rhetoric Of A Sophist, Ronhjones, CanadianLinuxUser, Cst17, Download, Glane23, Bassbonerocks, Chzz, Favonian, Kronix35, LinkFA-Bot, Udugunit, Aktsu, Tassedethe, Numbo3-bot, Anpecota, Tide rolls, HerpesVirus, SDJ, OlEnglish, Scourge of God, Davidmedlar, Couldbenoway66, Yobot, Maxdamantus, Terrisknickers, Kartano, TaBOT-zerem, Julia W, Unique and proud of it, FireMouseHQ, Terrifictriffid, ArchonMagnus, CinchBug, Synchronism, AnomieBOT, Cleeseheb, 1exec1, Charlesvi, Bigdaddy4x4, Gitman4, Jim1138, IRP, Mintrick, Drweetmola, Ornamentalone, M00npirate, Gautam10, Csigabi, Poli-Psy, Materialscientist, 90 Auto, Citation bot, Teleprinter Sleuth, Vuerqex, Twri, Frankenpuppy, Fuzzy Bob Saget, DirlBot, Georgepowell2008, Heidisql, Cureden, Ekwos, Capricorn42, Gensanders, NFD9001, Anna Frodesiak, Tomwsulcer, A23649, Pra1998, Coretheapple, Ruy Pugliesi, Jagbag2, Vandalism destroyer, Ab1, Omnipaedista, Bandit5005, Shirik, RibotBOT, Waleswatcher, Saalstin, Amaury, Aaron35510, Caz34, Doulos Christos, Sewblon, Born Gay, Capricorn24, SchnitzelMannGreek, A. di M., SpacePyjamas, Kierkkadon, A.amitkumar, Dougofborg, StringLove, Nobelprizewinner, Astiburg, FrescoBot, Fortdj33, Paine Ellsworth, Goodbye Galaxy, HJ Mitchell, Steve Quinn, Vhann, Kwiki, Xhaoz, Citation bot 1, Batong, Gil987, Pinethicket, I dream of horses, Tallboyhoops1991, Three887, Steveo27five, RedBot, Sardinita, Serols, Vhsatheeshkumar, Swisstingle, DeletionUK, Reconsider the static, IVAN3MAN, Remingtonhill1, Orenburg1, Coltonhs, Willy Weazley, Smamaret, Bethovenn, Dinamikbot, Dc987, Oswaldo Zapata, Egemont, Syebo, Alaithiran, Reaper Eternal, Seahorseruler, Ybungalobill, Quaker phil, Specs112, Dr. Aakash Patel, Tbhotch, StormbringerUK, Minimac, Mathgenius3141592, Keegscee, Omgwaffels, Mick le pick, Solancel, Aznhero3793, Dwielark, Afteread, Enauspeaker, EmausBot, MaooaM, Immunize, Az29, Milkocookie, Faolin42, Fotoni, RA0808, RenamedUser01302013, 8digits, Yukiseaside, Slightsmile, Tommy2010, Winner 42, Wikipelli, JonezyKiDx, Joe Gazz84, ZéroBot, Timeitsways, John Cline, Cogiati, Quaqa, Chrispaps2413, Nasulikid, Vollrath2323, Benjamin1414141414141414, Arbnos, Green Lane, A930913, Bamyers99, Azeraphale, H3llBot, Encyclopadia, Danga1988, Ollainen, PoisonGM, Wayne Slam, OnePt618, Knome335, L Kensington, Lulzprotuns, Kranix, Rpcappello, Maschen, Vastly-enwiki, Donner60, CatFiggy, CountMacula, Orange Suede Sofa, Etov, M1k3 101, Bill william compton, Wakabaloola, TERBAFAN, Nickslspride34, NeuralLotus, Isocliff, Brechbill123, Xanchester, ClueBot NG, Martti Muukkonen, KagakuKyouju, Jeff Song, This lousy T-shirt, Satellizer, Name Omitted, Marcdean123, Wiki incorp, Frietjes, O.Koslowski, Alexdamaino9, Dream of Nyx, Blackhall616, Widr, Sashhere, WikiPuppies, Stu181, T00g00d96, Pluma, Storm.sarup, Helpful Pixie Bot, Manzeet, Waffleboy36, HMSSolent, Mikeshelton1, Bibcode Bot, 2001:db8, Phillip.phillipson, Hoaxinator, Lowercase sigmabot, Thor cherubim, Mrshabam, Nishch, Flowerhat15, AvocatoBot, Housegeek224, MahRanch, Benzband, Altaïr, Benhenchdickthomas, Shreyakstring, Sweaty maori sphincter, DaFalk, Dsabo74, Ratanmaitra, MM4EVAH, Steven.w.kowalski, Minsbot, JGallardo2600, Dylanlatham, Myfriendganesha, OCCullens, Likeaboss189, Sean271293, LinusE8, BattyBot, Several Pending, Aldrich2122, CommanderMoka, The Illusive Man, ChrisGualtieri, KoalamaN2, Trevorkid45, Catsloveit07, Alex Modzz, Rustyjamsen, Goh ryangoh, Dexbot, Exolius, Hilander316, Alman1234321, SuperCalzer, LightandDark2000, MeekMelange, BQND, Cdarrai1, Kephir, TheMonkeyboy524, Michael Anon, TwoTwoHello, Mattfat8, Lugia2453, Anruy, Rachel weld, Jamesx12345, AHusain314, BossEditors, Hillbillyholiday, Joeinwiki, Mattninja, Theshadow444, Asaa82, Jakemarz197, Kzhang1025, Epicgenius, Spongbob456789, ⁊, TestMaster,

Ianreisterariola, GrapperJ, Makeitnasty, Moemajdi, I am One of Many, Nualalvy, BAZINGASS, St3fanPC, Eyesnore, Isaac grozd, Jordanis-sexyaf1999, Baruch6525, Mosbruckercj, Ihatedirac2k13, Jonamithy121314, 123physicsquantum, Jt198, RaphaelQS, HeyJude70, AParker628, DimReg, A.k.blaze1, Joshuk, Zenibus, Nianoobasik, Ihelpapplen, Gamo To Apoel, SacredLabyrinth, Ginsuloft, Vampre1122, Dimension10, Howard Wolowitz, AddWittyNameHere, Polytope24, Elysion, Tutun12S, Longerboats5, SimonWombat8, Konveyor Belt, Vtank54, Micheal545, Hck24, Caliae19, Hexafish, Simpick, TheRealTheKoi, Bballbro62, Monkbot, ArmyPath, TheQ Editor, Jtsmith098, Joshmiller1, Hanseer360, XXvPIEvXx, Dbennett 24, Ghikpenos, Nick65633, Saundra03, Thehippothatknows, Sewwgers, Teelaskeletor, Cirksena, Balockaye1234, Plop-pyDoo, Yesufu29, Lumpy2k14, Podayeruma, Abstract92, Sbenfiel, Monkman2k4, Swegwegdgfyetkfoffkkfkfkv, John95541234, Poopman224, ScrapIronIV, Tetra quark, GeneralizationsAreBad, Shivansh2014n, KasparBot, SHUCKYLUCKY, Fabiotheoto, FartGoblin, Joca potato, Josh-cool246, Theoretical Physisist4444, Reg7d88 and Anonymous: 1555

- **Loop quantum gravity** *Source:* https://en.wikipedia.org/wiki/Loop_quantum_gravity?oldid=689285284 *Contributors:* Bryan Derksen, The Anome, AstroNomer–enwiki, RK, Toby Bartels, Miguel–enwiki, Schewek, Ewen, Michael Hardy, TakuyaMurata, Islandboy99, GTBacchus, Mcarling, Looxix–enwiki, Ahoerstemeier, Cyp, Kimiko, Palfrey, Jordi Burguet Castell, Mxn, Charles Matthews, Sanxiyn, Maximus Rex, Phys, Omegatron, Finlay McWalter, Dmytro, Sdedeo, Astronautics–enwiki, Peak, Chris Roy, Mirv, Sverdrup, Kn1kda, Hadal, Jheise, Clementi, Connelly, Giftlite, Sj, Fastfission, Herbee, Anville, Dratman, Curps, JeffBobFrank, Jason Quinn, Gzornenplatz, C17GMaster, DÅugosz, Philo-Vivero, DefLog–enwiki, Gadfium, HorsePunchKid, Sam Hocevar, Lumidek, Tdent, Joyous!, M1ss1ontomars2k4, Eep², Poccil, Rich Farm-brough, Avriette, Pjacobi, Vsmith, MuDavid, Pavel Vozenilek, Bender235, ESkog, Clement Cherlin, Peter M Gerdes, Drhex, John Vandenberg, C S, Cmdrjameson, GTubio, Tweet Tweet, Slicky, Ral315, Lysdexia, Arthena, Xaphan9966, Wtmitchell, Greg Kuperberg, Count Iblis, Egg, Lee-Anne, Kazvorpal, Killing Vector, Linas, Merlinme, HFarmer, Sympleko, Hfarmer, Mpatel, GregorB, J M Rice, Ae7flux, Tjbk tjb, Alienus, Fleisher, Sjö, Rjwilmsi, Nightscream, Zbxgscqf, Bubba73, FlaBot, John Baez, Don Gosiewski, Smithbrenon, Chobot, Spasemunki, Bgwhite, Roboto de Ajvol, YurikBot, Wavelength, RobotE, Rt66lt, Hillman, DanMS, Chaos, Salsb, Welsh, Schmock, Crasshopper, Beanyk, Akashmitra, Bota47, JonathanD, Endomion, Modify, Petri Krohn, Ilmari Karonen, Caco de vidro, Benandorsqueaks, SmackBot, Bayardo, FlashSheridan, Unyoyega, Vald, JMiall, Chris the speller, IvanAndreevich, DHN-bot–enwiki, Colonies Chris, Chlewbot, Pepsidrinka, Chrylis, MegaHasher, TriTertButoxy, Lambiam, Vincenzo.romano, Loadmaster, Konklone, K, G-W, Kurtan–enwiki, Harold f, Will314159, Friendly Neighbour, Vyznev Xnebara, Ian Beynon, Myasuda, Gmusser, Rjm656s, Fournax, Headbomb, Nick Number, MichaelMaggs, Edokter, Byrgenwulf, Knot-work, Arch dude, Igodard, Yill577, WolfmanSF, Tonyfaull, Skylights76, Rickard Vogelberg, Gwern, AltiusBimm, Melamed katz, Vanished user 47736712, WJBscribe, Izno, KittyHawker, Sheliak, Maxximet, AlnoktaBOT, Nxavar, Jackfork, Carlorovelli, Anotherak, SieBot, Keski-val, AS, Robdunst, Hugh16, Senderista–enwiki, Bnsreenath, Caidh, Oxymoron83, Dcattell, Swiebodzice, Sk8hack, Danthewhale, Martarius, Sfan00 IMG, Shaded0, Djr32, CohesionBot, JavierReynaldo, Arjayay, SchreiberBike, Pqnelson, Mjaniec, DumZiBoT, Ianbay, Neuralwarp, XLinkBot, Fastily, Tenner47, Arthur chos, Avoided, Tenderbuttons, Benplusnumber, Balungifrancis, Addbot, DOI bot, 15lsoucy, Tarosic, De-bresser, SamatBot, Yobot, Ibayn, 4th-otaku, AnomieBOT, VanishedUser sdu9aya9fasdsopa, Archon 2488, Francois33, Citation bot, Xqbot, Imushfiq, MIRROR, Pra1998, Dumontierc, Omnipaedista, Franco3450, Rr2000, FrescoBot, Paine Ellsworth, Nunc aut numquam, Martlet1215, Citation bot 1, Jonesey95, Tom.Reding, Schiefesfragezeichen, ROMVLVS, Casimir9999, RobinK, Meier99, Dinamik-bot, Bj norge, ElPeste, Afteread, EmausBot, Detogain, John of Reading, Racerx11, GoingBatty, XinaNicole, Ensabah6, Uploadvirus, ZéroBot, Arbnos, Zueignung, Wa-terCrane, Crown Prince, LaurentRDC, Isocliff, Vodkacannon, Raidr, Helpful Pixie Bot, Titodutta, Bibcode Bot, BG19bot, Spaligo, KateWish-ing, PhnomPencil, Sylvain.maurin, Kecchina, Halfb1t, Brad7777, Fylbecatulous, Jimw338, MyTuppence, Mogism, LTWoods, Andyhowlett, Jawa0, &reasNink, SomeFreakOnTheInternet, Tentinator, EvergreenFir, DimReg, Pedarkwa, Db9199 24, Anrnusna, Notspelly, Ntomlin1996, Monkbot, Isbromberg, Dsprc, YeOldeGentleman, Tetra quark, Wulframm and Anonymous: 334

- **Causal fermion system** *Source:* https://en.wikipedia.org/wiki/Causal_fermion_system?oldid=674718007 *Contributors:* Michael Hardy, Head-bomb, Magioladitis, David Eppstein, Maschen, Wcherowi, OccultZone, Fefinster, Klj1234, Pfpguy and Anonymous: 2

- **Grand Unified Theory** *Source:* https://en.wikipedia.org/wiki/Grand_Unified_Theory?oldid=688904036 *Contributors:* AxelBoldt, Lee Daniel Crocker, Mav, AstroNomer–enwiki, XJaM, Heron, Michael Hardy, Zocky, CesarB, Looxix–enwiki, Emperorbma, Dysprosia, Phys, Omega-tron, Northgrove, Robbot, Securiger, Lowellian, Meelar, Caknuck, Giftlite, Jmnbpt, Herbee, Fropuff, Xerxes314, Golbez, Ary29, Sam Ho-cevar, Lumidek, IcycleMort, M1ss1ontomars2k4, Mike Rosoft, Discospinster, 4pq1injbok, Pjacobi, Silence, JustinWick, Bobo192, Small-jim, John Vandenberg, Apyule, Foobaz, 19Q79oL78KiL0QTFHgyc, Jeodesic, Physicistjedi, Jérôme, Alansohn, Krischik, Sligocki, Mac Davis, GeorgeStepanek, RJFJR, Lee-Anne, DV8 2XL, Falcorian, Simetrical, Linas, Mindmatrix, FeanorStar7, GregorB, Ruziklan, Mekong Bluesman, Ashmoo, Rachel1, Rjwilmsi, Strait, Drrngrvy, FlaBot, Margosbot–enwiki, DannyWilde, Rune.welsh, BradBeattie, Snailwalker, Phoenix2–enwiki, Guanxi, DVdm, YurikBot, Ugha, Hairy Dude, Michael Slone, Gaius Cornelius, CambridgeBayWeather, NawlinWiki, Wiki alf, Joel7687, JocK, Zwobot, IceCreamAntisocial, Ms2ger, Noclip, Moogsi, CWenger, Smurrayinchester, Curpsbot-unicodify, Caco de vidro, Jaysbro, Sbyrnes321, SmackBot, Tom Lougheed, Eskimbot, Dauto, Silly rabbit, DHN-bot–enwiki, Colonies Chris, Scwlong, QFT, Addshore, Wen D House, Dreadstar, Pwjb, Gbinal, Thorsen, Vina-iwbot–enwiki, Arglebargle IV, Rory096, Ben Jos, Mr. Lefty, Ckatz, SirFozzie, Quaeler, Baderyp, Richwhite10, Cydebot, Peripitus, Michael C Price, Tawkerbot4, Thijs!bot, Headbomb, J.christianson, Luna Santin, Alphachimpbot, JAnDbot, Nyq, Satarsa–enwiki, Homy, Mbarbier, Durianking, Danmctaggart, Maliz, JCarlos, AstroHurricane001, Adavidb, Bogey97, Qatter, Jeepday, Econofire, Lseixas, Jaffar33, Eismc2, Alphanon, Praveen pillay, KabbalistPhysicist, PaddyLeahy, Hemadh, Will Scot 55, Datpol, Mof-fitma, ClueBot, DFRussia, James edmiston, Ordinaterr, Djr32, PixelBot, Weysheehai, Sun Creator, Subdolous, Dekisugi, Louis925, AnonySci-entist, TimothyRias, Forbes72, SilvonenBot, Bywater100, Truthnlove, Balungifrancis, Addbot, Micromaster, Favonian, Mohitsridhar, 84user, OlEnglish, WikiDan61, Amirobot, AnomieBOT, Girl Scout cookie, Theunify, Karanmohan, Materialscientist, Citation bot, Obersachsebot, Blennow, Under22Entreprenuer, Dale Ritter, Senouf, Ernsts, FrescoBot, Paine Ellsworth, Steven Avraham Rosten, Ironboy11, Thamntamil, Sła-womir Biały, GreenRoot, Ysyoon, John85, Gil987, Stupidsimple, Tom.Reding, Casimir9999, RobinK, Aknochel, Grandunifier, RjwilmsiBot, Afteread, EmausBot, Slawekb, Arbnos, L Kensington, ClueBot NG, ClaudeDes, Helpful Pixie Bot, Bibcode Bot, Bernard Rementilla, Kkumer, Wer900, Dilaton, Hilander316, Ryanr666, CuriousMind01, Davidyevgeny, Cjean42, Franzl aus tirol, Sagnac, Gabelglesia, Lmboyer04, Ovidiu cupsa, Jwratner1, Gilitejman1, Soumilm, Tetra quark, BuzzBloom, JD Wilcox, KasparBot and Anonymous: 132

- **Quantum field theory** *Source:* https://en.wikipedia.org/wiki/Quantum_field_theory?oldid=686503248 *Contributors:* AxelBoldt, CYD, Mav, The Anome, XJaM, Roadrunner, Stevertigo, Michael Hardy, Tim Starling, IZAK, TakuyaMurata, SebastianHelm, Looxix–enwiki, Ahoerste-meier, Cyp, Glenn, Rotem Dan, Stupidmoron, Charles Matthews, Timwi, Jitse Niesen, Kbk, Rudminjd, Wik, Phys, Bevo, BenRG, North-grove, Robbot, Bkalafut, Gandalf61, Rursus, Fuelbottle, Tobias Bergemann, Ancheta Wis, Giftlite, Lethe, Dratman, Alison, St3vo, Mboverload, DefLog–enwiki, ConradPino, Amarvc, Pcarbonn, Karol Langner, APH, AmarChandra, D6, CALR, Urvabara, Discospinster, Guanabot, Igori-vanov–enwiki, Masudr, Pjacobi, Vsmith, Nvj, MuDavid, Bender235, Pt, El C, Shanes, Sietse Snel, Physicistjedi, KarlHallowell, PWilkinson, He-

Webrider, Woojamon, Gth759k, C0N6R355, SaltyBoatr, Dnarby, SieBot, Timb66, Nubiatech, Hertz1888, Likebox, RadicalOne, Arjen Dijksman, Udirock, Rafimoor, Arthana, Fedosin, Coldcreation, H1nkles, TFCforever, ShajiA, ImageRemovalBot, Martarius, ClueBot, Dypteran, NickCT, IceUnshattered, Plastikspork, Wwheaton, R000t, RYNORT, Flyingcar73242, Glibik, Eric.brasseur, Agge1000, Anaholic, ChandlerMapBot, Oxnard27, RAmesbury, Paulcmnt, Ajoykt, Denveron, PixelBot, Tamaratrouts, Brews ohare, NuclearWarfare, Hans Adler, Hasteur, DS1000, Friedlibend und tapfer, Roger491127, Aitias, John C. Huang, Dkress14, Steve D. Gage, DumZiBoT, Antti29, TimothyRias, RQG, Helixweb, BarretB, XLinkBot, Namreh ekim, Johanley, Rror, WikHead, Cleatus 69, Truthnlove, Mitrg, MTessier, Addbot, Eric Drexler, Some jerk on the Internet, Jojhutton, Smdowney, Haruth, Booger Eating Moron, Fluffernutter, Cst17, EconoPhysicist, Delaszk, Favonian, LinkFA-Bot, Bob K31416, Patton123, Lightbot, ScAvenger, Gameseeker, Blablablob, Luckas-bot, Yobot, Ptbotgourou, LGB, RHB100, KamikazeBot, Szajci, Synchronism, AnomieBOT, Jim1138, 9258fahsflkh917fas, Piano non troppo, Classicalmatter, Materialscientist, The High Fin Sperm Whale, Citation bot, ArthurBot, LilHelpa, Xqbot, Wavgfkl, Tasudrty, Amareto2, Capricorn42, Nasnema, Nanog, Jzhuo, Gap9551, NOrbeck, Almabot, GrouchoBot, Efield, RibotBOT, Waleswatcher, Victamonn, Charvest, Rayclipper, The Wiki ghost, Schekinov Alexey Victorovich, A. di M., Nym6433, Cooguy77, CES1596, FrescoBot, Baz.77.243.99.32, Paine Ellsworth, Ironboy11, Lipsquid, Tzviscarr, Sławomir Biały, Rymmen, Citation bot 1, Jimfarley, Greenrev, Pinethicket, Tsester, Tom.Reding, Zanzapod, Lithium cyanide, Number67c, Tcnuk, Riccardo.fabris, IVAN3MAN, Corinne68, Meier99, Kfchurchill, TobeBot, Callanecc, Cardinality, Iphegenia, Earthandmoon, Sandman888, Andrea105, Mean as custard, RjwilmsiBot, Alph Bot, Ripchip Bot, WildBot, EmausBot, John of Reading, Troubled asset, WikitanvirBot, Docjudith, Jazzalex, Heracles31, Benschopp8, RenamedUser01302013, Slightsmile, Netheril96, K6ka, Coffeefilter, Hhhippo, Empty Buffer, Albertolanzoni71, Kharoutinian, Juan Feo. Araya, Quondum, EWikist, Iiar, Ashell2, Bsmith671, Maschen, Donner60, Bill william compton, Rjowsey, ChuispastonBot, RockMagnetist, NTox, Vadher lalit, Alexandrevennes, Petrb, ClueBot NG, Bloodjam, TehGrauniad, SusikMkr, Cntras, David.bertalan, Braincricket, Richukuttan, Bopomofo, Helpful Pixie Bot, Witherpshins, Anofein, Bibcode Bot, BG19bot, Hewhoamareismyself, Gresavage, Datechanger, Bonginkosi zwane, Afbrat151, F=q(E+v^B), Bobo123456, Harizotoh9, Tremere2, Klilidiplomus, Tabish rafiq, Jffjgfjgfgf, Stigmatella aurantiaca, Khazar2, Ewjohnsonjohnson, JYBot, Dexbot, Twhitguy14, CuriousMind01, TwoTwoHello, 1Todd1, OneToddWon, Shivamshaiv, Mrvd42, Jamesx12345, Shivamshaivpatel, Andyhowlett, Shivam scientist, Gmxian89, NeapleBerlina, Jwratner1, DavRosen, Frinthruit, Anrnusna, Unifiedfields, Mezafo, Macjames4444, Mahusha, Abhishek.m.patel89, Monkbot, JhonASF, TE5ITA, Zachman727, Neeraj Bhakta, Jayache88, ChamithN, The Average Wikipedian, Bondy11u, Corsairio, Tetra quark, Spyridon Vossos, 2TonyTony, KasparBot, Kafishabbir, DamnLetMeLogIn, Trambak J Chall, Harshumarathe and Anonymous: 748

• **General relativity** *Source:* https://en.wikipedia.org/wiki/General_relativity?oldid=688738274 *Contributors:* AxelBoldt, Mav, Bryan Derksen, The Anome, AstroNomer~enwiki, Ap, RK, Andre Engels, XJaM, Chrislintott, JeLuF, Christian List, William Avery, Roadrunner, Kt-square, B4hand, Stevertigo, Frecklefoot, Patrick, Boud, Michael Hardy, Menchi, Ixfd64, Bcrowell, Nimrod~enwiki, TakuyaMurata, Mcarling, Minesweeper, Alfio, Looxix~enwiki, ArnoLagrange, Ellywa, Ahoerstemeier, Stevenj, William M. Connolley, Snoyes, Angela, Mark Foskey, Julesd, Salsa Shark, AugPi, Andres, Evercat, Hectorthebat, Hick ninja, A.Tigges~enwiki, Gingekerr, Jitse Niesen, Gutza, Rednblu, Doradus, Wik, Dragons flight, Tero~enwiki, Phys, Shizhao, Elwoz, BenRG, Banno, Northgrove, Phil Boswell, Robbot, Craig Stuntz, Sdedeo, Bvc2000, Goethean, Altenmann, Romanm, Lowellian, Mayooranathan, Gandalf61, Blainster, Diderot, DHN, Hadal, Alba, Johnstone, Fuelbottle, Isopropyl, Xanzzibar, Carnildo, Tobias Bergemann, Enochlau, Ancheta Wis, Tosha, Giftlite, JamesMLane, Graeme Bartlett, Mikez, BenFrantzDale, Lethe, Tom harrison, Fropuff, Everyking, Physman, Curps, Michael Devore, Jason Quinn, Alvestrand, SWAdair, Glengarry, Bobblewik, Edcolins, DefLog~enwiki, Pgan002, Knutux, GeneralPatton, HorsePunchKid, Robert Brockway, Kaldari, MadIce, Karol Langner, Rjpetti, Rdsmith4, JimWae, Anythingyouwant, Martin Wisse, Thincat, Euphoria, Icairns, Zfr, AmarChandra, Zondor, Econrad, JimJast, Discospinster, Rich Farmbrough, Guanabot, Pak21, ThomasK, Masudr, Pjacobi, Vsmith, Cdyson37, Jowr, Paul August, SpookyMulder, Dmr2, Bender235, Deabrilo, Ground, Ben Standeven, Nabla, Livajo, El C, Worldtraveller, Shanes, Etimbo, Causa sui, Bobo192, Robotje, Smalljim, Rbj, JW1805, ParticleMan, I9Q79oL78KiL0QTFHgyc, Mr2001, Matt McIrvin, PWilkinson, Haham hanuka, Schnolle, Varuna, Jumbuck, Jérôme, Alansohn, Hackwrench, Cctoide, Crebbin, Wikidea, SlimVirgin, Benefros, Alexwg, Wtmitchell, Orionix, CloudNine, Bsadowski1, DV8 2XL, LordLoki, HenryLi, Oleg Alexandrov, Kelly Martin, Linas, FeanorStar7, Sabejias, Moneky, Kzollman, Cleonis, Mpatel, Jok2000, Schzmo, Pdn~enwiki, GregorB, Plrk, Wayward, Joke137, Christopher Thomas, Mandarax, Colodia, Canderson7, Rjwilmsi, WCFrancis, MarSch, Eyu100, JoshuaeUK, JHMM13, Mike Peel, SanitysEdge, R.e.b., Ems57fcva, Bubba73, Gringo300, Ian Pitchford, RobertG, Mishuletz, Arnero, Mathbot, Nihiltres, Vsion, Perfect Tommy~enwiki, Itinerant1, Alfred Centauri, Gparker, Slant, Carrionluggage, Srleffler, Chobot, DVdm, Bgwhite, Dresdnhope, Manscher, Roboto de Ajvol, YurikBot, Wavelength, Bearm1185, Splintercellguy, Hillman, EDG, MattWright, RussBot, Loom91, AVM, KSmrq, DanMS, SpuriousQ, Shawn81, Eleassar, Shanel, Syth, Madcoverboy, Tailpig, Schlafly, Dputig07, Beanyk, Tony1, Dna-webmaster, Enormousdude, 2over0, KGasso, Petri Krohn, GraemeL, Rlove, Sambc, LeonardoRob0t, Geoffrey.landis, HereToHelp, Willtron, Meegs, Bsnd?, Finell, Luk, Sardanaphalus, SmackBot, Kuroçhka, Hydrogen Iodide, Pavlovič, Guangarra, Unyoyega, Nickst, Delldot, Motorneuron, Cessator, Harald88, Edgar181, Shai-kun, Sectryan, Gilliam, Skizzik, Dauto, Saros136, Silly rabbit, Complexica, Colonies Chris, Zven, Abyssal, RProgrammer, Hve, RedHillian, BentSm, Phaedriel, Khoikhoi, Cybercobra, Downwards, Coolbho3000, Nakon, Peterwhy, SkyWriter, DMacks, Nairebis, Henning Makholm, UncleFester, Bidabadi~enwiki, Byelf2007, SashatoBot, Lambiam, Lapaz, Cronholm144, Gizzakk, CPMcE, JorisvS, Goodnightmush, Ckatz, Frokor, Garthbarber, SirFozzie, SandyGeorgia, Midnightblueowl, RichardF, Novangelis, Peter Horn, MTSbot~enwiki, Kvng, JarahE, Licorne, Quaeler, Fan-1967, Editor.singapore, MFago, JoeBot, ShyK, MOBle, RekishiEJ, CapitalR, MD:astronomer, Courcelles, Tawkerbot2, JRSpriggs, Kurtan~enwiki, Harold f, JForget, Sakurambo, Thermochap, Avanu, NickW557, MarsRover, Harrigan, Ian Beynon, Cydebot, Jasperdoomen, WillowW, Fl, MC10, Mato, Pascal.Tesson, Michael C Price, Christian75, DumbBOT, Biblbroks, Omicronpersei8, Crum375, N. Macchiavelli, Epbr123, Fisherjs, Markus Pössel, Martin Hogbin, MrXow, Oliver202, Headbomb, Pjvpjv, Tom Barlow, Davidhorman, D.H, AntiVandalBot, Abu-Fool Danyal ibn Amir al-Makhiri, Tkirkman, Gnixon, VectorPosse, TimVickers, Scepia, Dawz, Billevans~enwiki, Tim Shuba, Rico402, Archmagusrm, Jaredroberts, JAnDbot, Vorpal blade, Hut 8.5, YK Times, Acroterion, Pervect, Magioladitis, Connormah, RogierBrussee, WolfmanSF, JamesBWatson, Swpb, Ling.Nut, Soulbot, Pixel ;-), KConWiki, WhatamIdoing, Eldumpo, Allstarecho, User A1, Mollwollfumble, Chris G, Archen~enwiki, Thompson.matthew, STBot, Mermaid from the Baltic Sea, Shentino, Mschel, CommonsDelinker, Pbroks13, J.delanoy, DrKay, R. Baley, Numbo3, Leafsfan85, Lantonov, M C Y 1008, Mathlabster, Zedmelon, Aboutmovies, C quest000, Tcisco, Marrilpet, Aatomic1, Potatoswatter, Kolja21, Lseixas, Rémih, Caracalocelot, DemonicInfluence, Sheliak, Deor, Part Deux, JohnBlackburne, Philip Trueman, TXiKiBoT, Coder Dan, GimmeBot, Gombo, Hqb, Rei-bot, IPSOS, Qxz, T doffing, Molinogi, Fizzackerly, JhsBot, Leafyplant, Geometry guy, Ilyushka88, Thebigbendizzle, SwordSmurf, Andy Dingley, Gabrielsleitao, Lamro, Antixt, Vector Potential, James-Chin, Arcfrk, Ccheese4, StevenJohnston, Katzmik, YohanN7, Dnarby, SieBot, Tiddly Tom, Work permit, Yintan, RadicalOne, Wizzard2k, SteakNShake, Arbor to SJ, Babareddeer, JSpung, Phil Bridger, Wmpearl, Oxymoron83, Henry Delforn (old), Csloomis, Thehotelambush, Lightmouse, BrightRoundCircle, OpTioNiGhT, The-G-Unit-Boss, Emgg, AWeishaupt, Divinestuff, Coldcreation, Adam Cuerden, Duae Quartunciae, Heptarchy of teh Anglo-Saxons, baby, Randomblue, TFCforever, Danthewhale, Martarius, Sfan00 IMG,

ClueBot, The Thing That Should Not Be, Rjd0060, Metaprimer, Wwheaton, Der Golem, JTBX, TheAmigo42, CounterVandalismBot, Viran, Blanchardb, Rotational, Agge1000, Itzguru, Tanketz, CohesionBot, Eeekster, Stealth500, Brews ohare, NuclearWarfare, PhySusie, SockPuppetForTomruen, SchreiberBike, Another Believer, RubenGarciaHernandez, AC+79 3888, MasterOfHisOwnDomain, He6kd, TimothyRias, Lazyrussian, PseudoOne, Skarebo, NellieBly, JinJian, Truthnlove, Everydayidiot, Tayste, Balungifrancis, Addbot, Mortense, Some jerk on the Internet, Fizzycyst, DOI bot, Mistyocean3, Metagraph, Stariki, Fluffernutter, Schmoolik, MrOllie, Download, EconoPhysicist, Delaszk, Favonian, LinkFA-Bot, Tuition, Tassedethe, Nnedass, Tide rolls, Lightbot, Knutls, Luckas-bot, Ptbotgourou, Legobot II, Julia W, Trickyboarder93, Superamoeba, AnomieBOT, Kristen Eriksen, Giordano.ferdinandi, Jim1138, Jo3sampl, Materialscientist, Wandering Courier, The High Fin Sperm Whale, Citation bot, Xqbot, Stlwebs, Sionus, Amareto2, Unigfjkl, Nickkid5, Stsang, GrouchoBot, Collin21594, RibotBOT, Rucko123, GhalyBot, Acannas, LucienBOT, Paine Ellsworth, Lagelspeil, Steve Quinn, Knowandgive, Pokyrek, Citation bot 1, Citation bot 4, Electrozity8, Pinethicket, LittleWink, Jonesey95, A412, Tom.Reding, Yougeeaw, Barras, Jauhienij, Meier99, Citator, Comet Tuttle, Hughston, Defender of torch, Duoduoduo, Aribashka, Iibbmm, Diannaa, Earthandmoon, Tbhotch, Brambleclawx, Marie Poise, RjwilmsiBot, Aznhero3793, Ripchip Bot, EmausBot, WikitanvirBot, Immunize, Zhaskey, Fly by Night, DuKu, GoingBatty, Jmencisom, Slightsmile, Hhhippo, JSquish, ZéroBot, Cogiati, Stanford96, Empty Buffer, Sanford123456, H3llBot, Quondum, REkaxkjdsc, Monterey Bay, Mr little irish, TonyMath, Brandmeister, Maschen, Puffin, Carmichael, Newstv11, RockMagnetist, Sona11235, WizardofCalculus, Milk Coffee, Whoop whoop pull up, Mjbmrbot, Helpsome, ClueBot NG, Manubot, Hagenfeldt, This lousy T-shirt, SusikMkr, Ggonzalm, Jj1236, Mgvongoeden, Snotbot, Widr, Jamester234, Pluma, Ginger.spice14, Bibcode Bot, Jeraphine Gryphon, Lowercase sigmabot, Quarkgluonsoup, Bolatbek, Marsambe, Amp71, Mark Arsten, Lovepool1220, Marsambe1, Benzband, ENG.F.Younis, 123matt123, DeviantFrog, IrishDevil2, F=q(E+v^B), Egbertus2, Harizotoh9, Doctor Lipschitz, Snow Blizzard, Zoldyick, Roozitaa, BattyBot, Reed07, Vanobamo, JoshuSasori, Stigmatella aurantiaca, Cyberbot II, Abhay ravi, ChrisGualtieri, Maestro814, Deathlasersonline, Plokijnu, Billyshiverstick, Read Blooded, Theeditor6079, Flyer1997, Dexbot, Suffian Akhtar, Kryomaxim, Twhitguy14, CuriousMind01, J0437-4715, Jamesx12345, Among Men, WorldWideJuan, Devinray1991, 1888software, EvergreenFir, Enchantedscience, Mohamed F. El-Hewie, Vai ra'a toa Taina, NeapleBerlina, Jwratner1, Gigantmozg, Ginsuloft, SirKesuma, Anrnusna, JaconaFrere, Osamabin7, Juenni32, Filedelinkerbot, SantiLak, Aryabhatt 21, Willbh15, S11027158, Cjsmith.us, Cris Cyborg, PeterShawhan, Evgeniy E., Sweeeeeeeed, Tetra quark, Praveece, JuanLT2045, Jf2839, GeneralizationsAreBad, KasparBot, Lemonberry622, Pizzaman62, Dgray101, Amrespi2007 and Anonymous: 702

- **Spacetime** *Source:* https://en.wikipedia.org/wiki/Spacetime?oldid=689860450 *Contributors:* Paul Drye, The Cunctator, Dreamyshade, Bryan Derksen, Malcolm Farmer, Josh Grosse, XJaM, Karl Palmen, Stevertigo, Patrick, Infrogmation, Smelialichu, Michael Hardy, Wshun, Pit-enwiki, Deljr, Karada, Mcarling, Looxix-enwiki, Snoyes, Kingturtle, Glenn, Loren Rosen, Hollgor, Adam Bishop, Dcoetzee, Reddi, Jay, E23-enwiki, Omegatron, Fvw, Robbot, Kristof vt, Goethean, Ashley Y, Sverdrup, Blainster, DHN, Papadopc, Tobias Bergemann, Finlander, Matt Gies, Giftlite, ByteCoder, Wolfkeeper, Herbee, Tom Radulovich, Everyking, Snowdog, Michael Devore, Niteowlneils, Yekrats, Eequor, Utcursch, Beland, Karol Langner, Wikimol, JimWae, Karl-Henner, Adashiel, ELApro, Chris Howard, Juan Ponderas, Discospinster, Rich Farmbrough, Cacycle, Ascánder, Dolda2000, Bender235, Ben Standeven, El C, Rgdboer, Lankiveil, Shoujun, Teorth, Che090572, Rbj, Tobacman, I9Q79oL78KiL0QTFHgyc, Como, Obradovic Goran, Free Bear, Keenan Pepper, Sourcer66-enwiki, Riana, Geoff-codes, ReyBrujo, Arag0rn, DonQuixote, Eddie Dealtry, DominicC13, H2g2bob, Loxley-enwiki, Camw, StradivariusTV, TheNightFly, Pkeck, ^demon, Doran, Jeff3000, Mpatel, GregorB, Palica, Graham87, Deltabeignet, Li-sung, Mkn1234, MekaD, Rjwilmsi, KYPark, Kinu, Vary, MarSch, FayssalF, Lebha, Mathbot, Alexjohnc3, Jrtayloriv, Exelban, Pete.Hurd, Tardis, Chobot, Tene, DVdm, VolatileChemical, YurikBot, Wavelength, Splintercellguy, Wolfmankurd, CanadianCaesar, Yamara, NawlinWiki, Mipadi, Trovatore, Schlafly, JocK, Crasshopper, Tony1, T, Zythe, Gadget850, Sahands, Light current, Zzuuzz, StuRat, KGasso, JoanneB, Heathhunnicutt, Anclation-enwiki, RG2, Teply, Mejor Los Indios, Qero, Eigenlambda, Sardanaphalus, SmackBot, RDBury, Formativ, Maksim-e-enwiki, Forteller-enwiki, RaulMiller, Ashill, Kurochka, Lestrade, InverseHypercube, KnowledgeOfSelf, C.Fred, AndreasJS, Jaytan, Alex earlier account, JeffieAlex, Yamaguchi先生, Gilliam, NickGarvey, JMiall, Oli Filth, TheScurvyEye, Silly rabbit, Complexica, Dahigkid, Jerome Charles Potts, Nbarth, Sbharris, Bryan Truitt, Can't sleep, clown will eat me, Tamfang, Chlewbot, Rrburke, Celarnor, Tsop, CanDo, Dylanrush, RaCha'ar, Mtmelendez, Looris, Richard001, Hammer1980, Romanski, Sayden, Kuru, MagnaMopus, Hernoor, Homan2006, 16@r, Loadmaster, Lampman, Hypnosifl, Ace Frahm, Inquisitus, FVP, Shoeofdeath, Newone, Yourstruly, Andrew Hampe, Lxl, Aeons, Xammer, Paolodm, CalebNoble, Robinhw, JForget, Twipie, Blve23, Jsd, Jnoa, WeggeBot, Myasuda, Azakreski, Joshua BishopRoby, Cydebot, AniMate, Kanags, Fl, MC10, Llort, Eu.stefan, Palindromica, Manfroze, DarkLink, Ameliorate!, DBaba, TarquiniusWikipedius, Kylewriter, Raoul NK, Letranova, Thijs!bot, Wikid77, Jedibob5, HappyInGeneral, Gamer007, Headbomb, Vertium, RolanGaros, Pigalle, Washingtonlerias, Ubuthustra, D.H, Nick Number, Klausness, Sam42, DarthNemesis, Northumbrian, Escarbot, WikiSlasher, AntiVandalBot, Seaphoto, Maxibons, Tim Shuba, Braindrain0000, Tempest115, Jrw@pobox.com, Narssarssuaq, Husond, MER-C, Andrewdolby, RogierBrussee, Bongwarrior, VoABot II, Bakken, Appraiser, Faizhaider, Cuardin, Stijn Vermeeren, Trebor1, Catgut, Cardamon, NMarkRoberts, IkonicDeath, MetsBot, Mwasim1, JaGa, GuelphGryphon98, NatureA16, FisherQueen, Flowanda, MartinBot, TechnoFaye, Wikeepeedier, Player 03, Tgeairn, HEL, J.delaney, Bobvinson, Maurice Carbonaro, Foober, 3halfinchfloppy, Lantonov, NewEnglandYankee, LeighvsOptimvsMaximvs, KylieTastic, Ja 62, Vinsfan368, Izno, Idioma-bot, Makewater, 28bytes, VolkovBot, XCelam, JohnBlackburne, AlnoktaBOT, Philip Trueman, Zidonuke, Red Act, Anonymous Dissident, Yilloslime, Fizzackerly, PaulTanenbaum, PhilyG, Wingedsubmariner, Hotmoklet, Eubulides, Zhongsan, SmileToday, Falcon8765, Cubed mass, LachlanSosa, StevenJohnston, Hunter826242, PSSnyder, Hobojaks, YohanN7, SieBot, ShiftFn, Paradoctor, Dawn Bard, Vanished user 82345ijgeke4tg, Flyer22 Reborn, Csblack, Henry Delforn (old), Nuttycoconut, MrWikiMiki, Hjelmerus, Dposte46, Jeanlovecomputers, Mátyás, OKBot, Fedosin, Coldcreation, Fuddle, Mike2vil, Anchor Link Bot, MarkMLl, VanishedUser sdu9aya9fs787sads, ImageRemovalBot, Martarius, De728631, ClueBot, The Thing That Should Not Be, EoGuy, Exploto, Razimantv, IshanAlmazi, Shinpah1, JFlav, Noca2plus, Eeekster, Tam 66 7, DPCU, Cenarium, Mozart21, Mentor364, Themantyke, McXX, Tin Whistle Man, Galor612, MalWilley, NERIC-Security, Rror, Avoided, Whizmd, Addbot, Derivator, Gravitophoton, DOI bot, Gul e, Startstop123, Gustavo José Meano Brito, Vishnava, CanadianLinuxUser, CarsracBot, RTG, Monypool12, DFS454, Glane23, Tod.davidson, Mcsploogerson, AnnaFrance, Favonian, Doniago, West.andrew.g, Numbo3-bot, Lightbot, Zorrobot, Legobot, Yobot, Ht686rg90, THEN WHO WAS PHONE?, Allowgolf-enwiki, Synchronism, AnomieBOT, Jim1138, Materialscientist, Citation bot, Maxis ftw, ChristianH, Expooz, Xqbot, Δζ, Anna Frodesiak, Shindamaru, False vacuum, Frankie0607, RibotBOT, Gsard, MerlLinkBot, Bearnfæder, 7575474087ALBERT, CES1596, FrescoBot, H.W. Clihor, Paine Ellsworth, Dogposter, Tj2691, Majopius, Mouselarry, Haeinous, Vbrcat, Citation bot 1, Deadtotruth, Pinethicket, Hypernovic, Lesath, 10metreh, Tom.Reding, Smuckola, Rushbugled13, A8UDI, RedBot, Sjb13, Elvis633, December21st2012Freak, Fredkinfollower, IVAN3MAN, Double sharp, Euriditi, TobeBot, 0x30114, LogAntiLog, Jonkerz, Dinamik-bot, Capt. James T. Kirk, Aribashka, Brianann MacAmhlaidh, Seahorseruler, Bj norge, The Utahraptor, Scipioafricanus, Reagster, Thslackliner19, Theslackliner19, EmausBot, WikitanvirBot, AlexUT, Pekka.virta, Stewiefool, RA0808, Slightsmile, Wikipelli, Scalable Vector Raccoon, Thecheesykid, Hhhippo, JSquish, Fæ, StringTheory11, Int21hexster, Quondum, Ellie Rickett, Donner60, Nelium12, Maheshbahadur, Ihardlythinkso, AndyTheGrump, RockMagnetist, Winston7, Rmashhadi, ClueBot NG, MelbourneStar, Irisrune, Lanthanum-138, Doorsrocklikerocks, Frietjes, Helpful Pixie Bot, HMSSolent, Gob Lofa, Bibcode Bot, Transscientific, Bobc3, BG19bot, Hz.tiang, Fw0116, Davidiad, Piguy101, Giarcea, Naveedyaykhan, Cadiomals, Kaaalbert, MrBill3, Wiki2103, Penguinstorm300, RiseUpAgain, The1337gamer, Nirmal kumar 9, SteveBM, Stigmatella aurantiaca, Zachhansonhart, U5ard, Nkzf, Khazar2, Ekren, Larskk101, Vanished user 23i4hjwrjfiij4t, Kevinfrank17, Twhitguy14, CuriousMind01, Mike.leivers, Kingcircle, DmVdx, Telfordbuck, Cadillac000, Asad shayan, Epicgenius, Bantennyson, Eachandall, AnthonyJ Lock, Ericgermate, Wedgeline, RogrMexico, Jwratner1, NottNott, Frinthruit, Iliketrains1234567890, Skyshad4w, AnonymousAuthority, TuxLibNit, Da.pro1, Monkbot, BethNaught, Zacwill, 97dc, Neeraj Bhakta, KH-1, Vedic Earthian, Aaronfranke, 39Debangshu, Tetra quark, Isambard Kingdom, Epictacotree, Anand2202, Kbap2002, Djniew, Hriton, KasparBot, Edgar-lausanne, Alout and Anonymous: 553

- **Effective field theory** *Source:* https://en.wikipedia.org/wiki/Effective_field_theory?oldid=629899452 *Contributors:* Maximus Rex, Phys, Giftlite, Xerxes314, Dratman, Lumidek, Frankenschulz, Fwb22, Bluemoose, Chobot, YurikBot, Bambaiah, Salsb, Caco de vidro, Reedy, TimBentley, QFT, Xristy, Mathsci, Xxanthippe, JamesAM, Headbomb, Stephanhartmannde, Gwern, Tarotcards, Fylwind, Irazvi, Snow in quarks-enwiki, Addbot, Mjamja, Perusnarpk, Cydelin, False vacuum, FrescoBot, Ysyoon, Ed Reiner, Puzl bustr, ZéroBot, Asi013, BG19bot, Harimak, Cerabot-enwiki, Mark viking and Anonymous: 15

- **Graviton** *Source:* https://en.wikipedia.org/wiki/Graviton?oldid=677652863 *Contributors:* CYD, Bryan Derksen, Timo Honkasalo, XJaM, Fubar Obfusco, Maury Markowitz, Kaczor-enwiki, Jketola, TakuyaMurata, Eric119, Looxix-enwiki, Glenn, Cyan, Wooster, Charles Matthews, Timwi, Wik, BenRG, Donarreiskoffer, Scott McNay, Stephan Schulz, Arkuat, Chris Roy, Merovingian, Davidl9999, Giftlite, Xerxes314, Jason Quinn, Matt Crypto, CryptoDerk, RetiredUser2, Icairns, Zfr, Lumidek, Ukexpat, Urvabara, Discospinster, Pjacobi, Vapour, Brian0918, El C, Joanjoc-enwiki, Dalf, Army1987, Mpvdm, La goutte de pluie, Physicistjedi, Daniel Arteaga-enwiki, Zenosparadox, Dethtron5000, Keenan Pepper, Viridian, SidP, Falcorian, Skeejay, Simetrical, Dr Archeville, Mpatel, Kyleca, Tmassey, Christopher Thomas, Tevatron-enwiki, Kbdank71, Nightscream, Koavf, Mike Peel, Ems57fcva, FlaBot, RexNL, Chobot, DVdm, Roboto de Ajvol, Spacepotato, Anonymous editor, SnoopY-enwiki, Salsb, Bachrach44, Hyperbrand, NickBush24, Pnrj, RL0919, EEMIV, IslandGyrl, Bota47, C h fleming, Petri Krohn, Mario23, Alias Flood, Tim314, Teply, GrinBot-enwiki, SmackBot, Amcbride, Melchoir, Eskimbot, Gilliam, Skizzik, Timneu22, Complexica, Villarinho, Colonies Chris, V1adis1av, Chlewbot, Xyzzyplugh, Jmnbatista, Fuhghettaboutit, Sadi Carnot, Yevgeny Kats, TenPoundHammer, Lambiam, Zaphraud, JorisvS, Mr Stephen, Ramuman, Quasar Jarosz, Lottamiata, Firewall62, Kurtan-enwiki, CmdrObot, BeenAroundAWhile, WeggeBot, Shultz IV, UncleBubba, Michael C Price, Anthmoo, Thijs!bot, Epbr123, Headbomb, KevinS06, Opelio, Spartaz, JAnDbot, Xoneca, SHCarter, Pikazilla, Robin S, STBot, Kostisl, J.delanoy, Tarotcards, Coppertwig, Wesino, Sava ankit2006, Tygrrr, Idioma-bot, Sheliak, JoAnneThrax, TXiKiBoT, WilliamSommerwerck, Hqb, Anonymous Dissident, Antixt, SieBot, Flyer22 Reborn, Henry Delforn (old), ClueBot, Ergn, Darkicebot, DenverRedhead, Addbot, Eric Drexler, Uruk2008, DOI bot, BrianBop, PJonDevelopment, F Notebook, Legobot, Picturesofnothing, Dov Henis, Alfredschrader, Eric-Wester, AnomieBOT, VanishedUser sdu9aya9fasdsopa, Jim1138, Materialscientist, Citation bot, Tomflaherty, ProtectionTaggingBot, Waleswatcher, FrescoBot, Juto20, LucienBOT, Paine Ellsworth, I dream of horses, Tom.Reding, RedBot, Omar.tigereyes, IVAN3MAN, Ashish.kotwal, Michael9422, D0wnfalle, EmausBot, Octaazacubane, 8digits, Slightsmile, K6ka, Thecheesykid, User10 5, Resprinter123, Orbjeeples, Puffin, Herk1955, ClueBot NG, Raidr, Masssly, Helpful Pixie Bot, Bibcode Bot, BG19bot, Shapoopy178, ServiceAT, PhnomPencil, Trevayne08, Brainssturm, Tjamcclain2, ChrisGualtieri, Ariscod, TheUyulala, LightandDark2000, Jessybun, Makecat-bot, Kryomaxim, JRYon, Andyhowlett, Mark viking, Yorsh07, CensoredScribe, WPratiwi, Monkbot, Bryan Paul Senior, Dr.Begich, Nompynuthead, Jacobflarsen and Anonymous: 196

- **Force carrier** *Source:* https://en.wikipedia.org/wiki/Force_carrier?oldid=651865824 *Contributors:* Camembert, Looxix-enwiki, Bevo, Alison, Urvabara, Mal-enwiki, Euyyn, Matt McIrvin, Keenan Pepper, TenOfAllTrades, Bubba73, Complexica, QFT, Sadi Carnot, Alan.ca, Nethac DIU, Myasuda, Headbomb, Trevyn, Alphachimpbot, Macboots, Maurice Carbonaro, VolkovBot, Antoni Barau, FDominec, SieBot, Jim E. Black, ClueBot, Anon126, WikHead, Truthnlove, Addbot, Mac Dreamstate, Barak Sh, Luckas-bot, Yobot, Flewis, Bci2, ArthurBot, Carlog3, Armando-Martin, JSquish, NatNapoletano, ClueBot NG, Bibcode Bot, The Illusive Man, Monkbot, Tetra quark and Anonymous: 21

- **Dilaton** *Source:* https://en.wikipedia.org/wiki/Dilaton?oldid=647593052 *Contributors:* Michael Hardy, Charles Matthews, Mporter, Lupin, Fropuff, Lumidek, Pjacobi, Bender235, Jérôme, RJFJR, Jeff3000, Mpatel, Kyleca, GregorB, Ems57fcva, FlaBot, Roboto de Ajvol, Wavelength, SmackBot, Pdaoust, Colonies Chris, V1adis1av, QFT, CRGreathouse, Thijs!bot, Keraunos, Headbomb, Jpod2, Anonymous Dissident, Antixt, Paradoctor, Alexbot, Yosefverbin, SkyLined, Addbot, Luckas-bot, AnomieBOT, Hep thinker, MastiBot, Jakeukalane, ZéroBot, Cogiati, Throwmeaway, Ebrambot, TonyMath, Fluctuating metric, Raidr, Helpful Pixie Bot, Bibcode Bot, Jeremy112233 and Anonymous: 12

- **Quantum electrodynamics** *Source:* https://en.wikipedia.org/wiki/Quantum_electrodynamics?oldid=687229117 *Contributors:* Bryan Derksen, DWeir, Stevertigo, Spsprez, Tim Starling, TakuyaMurata, Looxix-enwiki, Ahoerstemeier, Wooster, Joshuabowman, Greenrd, Patrick0Moran, Phys, Ortonme, Robbot, Sanders muc, Diberri, Ancheta Wis, Giftlite, Sj, Wolfkeeper, Lethe, Dratman, JeffBobFrank, Markus Kuhn, Waltpohl, Sukael, Jason Quinn, DefLog-enwiki, LucasVB, Beland, Icairns, AmarChandra, Setfn, Geanyon, Lucidish, Urvabara, Pjacobi, Dbachmann, Mani1, Tooto, Triona, Neilrieck, Harley peters, Viriditas, .:Ajvol:., Matt McIrvin, Photonique, Rje, Helix84, Keenan Pepper, Count Iblis, DV8 2XL, Daubigne, Linas, Mpatel, Joke137, Alan Canon, ObsidianOrder, Magister Mathematicae, Snafflekid, Rjwilmsi, Ian the younger, Strait, Patrick Gill, Salix alba, Rangek, FlaBot, Weihao.chiu-enwiki, Jsldub, Margosbot-enwiki, Srleffler, Chobot, YurikBot, Ugha, RussBot, JabberWok, Salsb, Spike Wilbury, Janke, SCZenz, Larsobrien, Gadget850, Ott2, Light current, Tribaal, Enormousdude, 2over0, Smoggyrob, Aeosynth, Didymos-enwiki, GrinBot-enwiki, Sbyrnes321, Chipef, Artemisfowl3rd, SmackBot, F, Rex the first, Tom Lougheed, Rjtucke, Melchoir, Unyoyega, Hmains, Chris the speller, Complexica, Aolanonawanabe, QFT, Ignirtoq, Voyajer, Aolanaonwaswronglyaccused, Sadi Carnot, DJIndica, Lambiam, MobyDike, EGGS, JorisvS, Zarniwoot, RafaelRGarcia, Phuzion, Eyebum, JMK, Achoo5000, Alberto4d, Retune, CmdrObot, Van helsing, Runningonbrains, Cydebot, Thijs!bot, Headbomb, Speedyboy, Second Quantization, D.H, Nick Number, Dawnseeker2000, Escarbot, Stannered, AntiVandalBot, Stripsi, Krapsin, And4e, Blaine Steinert, Yill577, Hroðulf, Bongwarrior, VoABot II, Bakken, Wormcast, Hekerui, Helianthi, Maliz, Daniel james, Alro, R'n'B, CommonsDelinker, Andrej.westermann, Science5, HEL, Maurice Carbonaro, Cpiral, Speed8ump, Omeganumber, Max woolfson, Squids and Chips, Idioma-bot, Sheliak, VolkovBot, Antoni Barau, Rei-bot, TBond, Richwil, Moose-32, Ivanivanovich, Minestrone Soup, SieBot, Timb66, Graham Beards, Peeter.joot, LeadSongDog, Iameukarya, C0manPayne, Niceguyedc, DnetSvg, Estirabot, Brews ohare, Kmellem, Lazyrussian, Ost316, WikHead, Truthnlove, Jabberwoch, RandomTool2, Addbot, Mathieu Perrin, Gravitophoton, LaaknorBot, CarsracBot, AgadaUrbanit, Tide rolls, Lightbot, Zorrobot, Legobot, Ptbotgourou, TaBOT-zerem, TestEditBot, AnomieBOT, Materialscientist, Citation bot, Northryde, Alexshag, Plumpurple, Br77rino, Pra1998, Salbers, Peter484, A. di M., 路路, FrescoBot, Paine Ellsworth, Kenneth Dawson, UCSD3.14159, Techibun, Jonesey95, Skyerise, LiborX, MondalorBot, RobinK, Merlion444, Jensen.andrew, Dylan1946, EmausBot, Quondum, TonyMath, Brandmeister, Lulzprotuns, Maschen, HCPotter, ClueBot NG, CocuBot, Deep Thought, Helpful

Pixie Bot, Addihockey10 (automated), Bibcode Bot, Mark Arsten, F=q(E+v^B), TrollWizard, Makecat-bot, Paul adrien, Spikitiger, Cltschirhart, Edward.hughes, Mahusha, Dhm4444, Delbert7, Mywalnut, Mario Castelán Castro, Ggf4t, KasparBot, Apricotpie and Anonymous: 156

- **Background independence** *Source:* https://en.wikipedia.org/wiki/Background_independence?oldid=659843309 *Contributors:* The Anome, Time, Phil Boswell, Sdedeo, Merovingian, Jason Quinn, Poceil, Jkl, Pjacobi, Cmdrjameson, Hooperbloob, Lysdexia, Danski14, Axl, Woohookitty, Linas, Opie, Ligulem, Itinerant1, Kjlewis, McGinnis, Jpeob, JonathanD, Caco de vidro, SmackBot, Reedy, AndyZ, Lambiam, JoeBot, CmdrObot, Ian Beynon, Unclenuclear, Hebrides, Thijs!bot, Dogaroon, Headbomb, JustAGal, Lself, Igodard, Autotheist, Markov Chain, Tonyfaull, MujiKha, Katharineamy, Wesino, DorganBot, Ross Fraser, Venny85, MvL1234, Bobathon71, Environnement2100, Addbot, Yobot, Ibayn, AnomieBOT, Citation bot, NOrbeck, Wesino, Wesino, Wesino, Wesino, Walsero, Quondum, Rezabot, Helpful Pixie Bot, ChrisGualtieri, Enyokoyama, DimReg, Monkbot, Loraof and Anonymous: 29

- **Minkowski space** *Source:* https://en.wikipedia.org/wiki/Minkowski_space?oldid=689525478 *Contributors:* XJaM, William Avery, Stevertigo, Patrick, PhilipMW, Michael Hardy, Tim Starling, Karada, Stupidmoron, Hawthorn, Charles Matthews, Kbk, Zoicon5, LMB, Phys, Fvw, Josh Cherry, Jheise, Marc Venot, Decumanus, Giftlite, Gene Ward Smith, BenFrantzDale, Lethe, MathKnight, Fropuff, Dratman, Anythingyouwant, Frau Hitt-enwiki, Hidaspal, Bender235, Syp, Rgdboer, I9Q79oL78KiL0QTFHgyc, Phils, Anthony Appleyard, NukWik, Egg, Ringbang, Japanese Searobin, Stemonitis, Linas, StradivariusTV, Mpatel, Tlroche, Rjwilmsi, Zbxgscqf, R.e.b., Lionelbrits, Mathbot, Chobot, Dylan Thurston, DVdm, Hmonroe, YurikBot, Hairy Dude, Rhythm, Nick, Arthur Rubin, KasugaHuang, That Guy, From That Show!, Sardanaphalus, KnightRider-enwiki, SmackBot, Turbos10-enwiki, Vald, ZerodEgo, Xie Xiaolei, Bluebot, Alexwagner, Complexica, Waprap, Jbergquist, Andrei Stroe, Gregapan, Lambiam, Eliyak, Jim.belk, NongBot-enwiki, Bosons, Xenure, Loadmaster, Dan Gluck, JRSpriggs, CRGreathouse, Cydebot, Michael C Price, Thijs!bot, Martin Hogbin, Headbomb, D.H, Noclevername, Gökhan, JAnDbot, Jyotirmoyb, Sullivan.t.j, First Harmonic, Maurice Carbonaro, TomyDuby, Goutui, WaiteDavid137, Fylwind, Equazcion, XCelam, JohnBlackburne, Red Act, Nxavar, DWP17, PaulTanenbaum, Geometry guy, Zain Ebrahim111, StevenJohnston, YohanN7, Juanmantoya, Paradoctor, Happysailor, DaveBeal, Henry Delforn (old), Udirock, Mr. Stradivarius, Renata500, Martarius, ClueBot, RODERICKMOLASAR, Bastien Sens-Méyé-enwiki, Sun Creator, Carriearchdale, Forbes72, Whizmd, Addbot, Mortense, Kwvan, Gregz08, 84user, Tide rolls, Cesiumfrog, Luckas-bot, Yobot, Amirobot, AnomieBOT, Sfaefaol, Illegal604, Corwin323, ArthurBot, Nanog, NOrbeck, RibotBOT, Ashi009, MeDrewNotYou, Paine Ellsworth, Sławomir Biały, Guitarstud101, Tsester, Serols, Stephen Henry Davies, Tcnuk, Retired user 0001, Dinamik-bot, John of Reading, WikitanvirBot, AlexUT, Netheril96, Hhhippo, Quondum, Thine Antique Pen, JFB80, ClueBot NG, Jack Greenmaven, CaroleHenson, Helpful Pixie Bot, F=q(E+v^B), Aisteco, ChrisGualtieri, Deltahedron, Makecat-bot, Twhitguy14, TwoTwoHello, CsDix, Frinthruit, JaconaFrere, Monkbot, Biblioworm, WillemienH, KasparBot, Arnaud Dorthe and Anonymous: 91

- **Color confinement** *Source:* https://en.wikipedia.org/wiki/Color_confinement?oldid=683221681 *Contributors:* Lorenzarius, Michael Hardy, Schneelocke, Timwi, Doradus, Phys, Secretlondon, Robbot, Ruakh, Isopropyl, Giftlite, Herbee, Xerxes314, Tagishsimon, Lumidek, MuDavid, Jag123, Cortonin, Jon.baldwin, Mpatel, Isnow, Ashmoo, Rjwilmsi, FlaBot, Pfctdayelise, Chobot, Mushin, Bambaiah, Hairy Dude, Ohwilleke, Salsb, E2mb0t-enwiki, Tetracube, WAS 4.250, Banus, SmackBot, Incnis Mrsi, Jjalexand, DHN-bot-enwiki, Sbharris, Colonies Chris, VMS Mosaic, Lambiam, Sinistrum, TaggedJC, Newone, Treue-enwiki, Thijs!bot, Headbomb, Cultural Freedom, Yill577, Natsirtguy, Melamed katz, A Nobody, Spoxjox, Anonymous Dissident, SieBot, FlowerFaerie087, Denisarona, Asher196, Manishearth, Dab240, Mustufailed, Texas Chainstore Manager, Addbot, Dudemanfellabra, Db1101, Eutactic, Luckas-bot, AnomieBOT, Materialscientist, Xqbot, D'ohBot, Jonesey95, RedBot, TobeBot, Trappist the monk, DixonDBot, EmausBot, Peaceray, Maschen, StanS, Helpful Pixie Bot, Bibcode Bot, Orentago, BG19bot, Slinkblot, Jdellamalva, Tikki and Anonymous: 48

- **Perturbation theory** *Source:* https://en.wikipedia.org/wiki/Perturbation_theory?oldid=683335366 *Contributors:* CYD, FlorianMarquardt, Stevertigo, Michael Hardy, Kku, AugPi, Ideyal, Jitse Niesen, Phys, Robbot, Lowellian, Tobias Bergemann, Giftlite, BenFrantzDale, Neilc, Karol Langner, The Land, Igorivanov-enwiki, MuDavid, Bender235, Cmdrjameson, Haham hanuka, Keenan Pepper, RJFJR, Count Iblis, Dirac1933, Mattbrundage, Djsasso, Oleg Alexandrov, Linas, Yansa, SeventyThree, Bubba73, Mathbot, ChrisChiasson, YurikBot, Piet Delport, Tong-enwiki, Joel7687, Dhollm, Tony1, DerHannes, Artemisfowl3rd, SmackBot, Mmernex, Tom Lougheed, Mcld, Chris the speller, Bduke, Complexica, Nbarth, Colonies Chris, MaxSem, Ohconfucius, Nishkid64, Harryboyles, Tomatoman, JorisvS, Frokor, Hiiiiiiiiiiiiiiiiiii, Charles Baynham, Chetvorno, Khromegnome, CBM, Myasuda, Cydebot, Tawkerbot4, Roy W. Wright, Headbomb, Ben pcc, Engelbaet, David Eppstein, Alexei Kopylov, P.wormer, Cuzkatzimhut, Maghnus, Bphillah, Lechatjaune, EverGreg, Vsst, SieBot, JerroldPease-Atlanta, Nancy, Yhkhoo, ClueBot, Warbler271, PtolemyGalen, Mild Bill Hiccup, ZIIvette, CohesionBot, Guiermo, Lacce, Crowsnest, DumZiBoT, Terry0051, Queenmomcat, Download, Yobot, TaBOT-zerem, AnomieBOT, Pownuk, Nfr-Maat, J04n, Resident Mario, Pradameinhoff, FrescoBot, Lotje, Mhilferink, Yger, Qweilun, Mattedia, Zfeinst, ClueBot NG, Wcherowi, LPOG1, Helpful Pixie Bot, Vlos2008, PhnomPencil, Anylai, Dexbot, Andyhowlett, Pdecalculus, Hctrmycss and Anonymous: 77

- **Topological quantum field theory** *Source:* https://en.wikipedia.org/wiki/Topological_quantum_field_theory?oldid=686033948 *Contributors:* TakuyaMurata, Marknen, Charles Matthews, Fuzheado, Phys, Aenar, Hadal, Fropuff, Icaims, RJFJR, BD2412, Ketiltrout, Rjwilmsi, R.e.b., Chris Pressey, John Baez, Wavelength, Conscious, Gspr, David Farris, Ajt, Nbarth, Colonies Chris, Twyder, 345Kai, PKT, Headbomb, Rsocol, YK Times, Sheliak, Camrn86, SieBot, Arjayay, SchreiberBike, Oo09nj76t5, Addbot, Angelobear, Unara, Citation bot, Omnipaedista, Charvest, FrescoBot, Citation bot 1, Orenburg1, EmausBot, John of Reading, Άρτεμις, ZéroBot, Quondum, Maschen, Cntras, Bibcode Bot, Enyokoyama, Kryomaxim, Paritto, AHusain314, Cyrapas, Talithin, K9re11, Broido, Neilwanderson and Anonymous: 34

- **Unruh effect** *Source:* https://en.wikipedia.org/wiki/Unruh_effect?oldid=678224411 *Contributors:* Miguel-enwiki, Maury Markowitz, Paul A., Stevenj, Charles Matthews, Maximus Rex, Phys, Omegatron, Topbanana, Jeffq, Rursus, Wolfkeeper, Icairns, Lumidek, CALR, Pjacobi, Paul August, Bender235, Ben Standeven, Key45, Alamino, Nihil-enwiki, Fwb22, Lysdexia, Count Iblis, Gene Nygaard, StuTheSheep, Isomeme, GregorB, Rjwilmsi, Smithph, Jaraalbe, GangofOne, YurikBot, Dlugosz, A bit iffy, SmackBot, Slashme, IvanAndreevich, Ignirtoq, JRSpriggs, Vaughan Pratt, CmdrObot, Chrumps, Michael C Price, Thijs!bot, Headbomb, Dtgriscom, Escarbot, Luna Santin, David Eppstein, Jaehacker, Leyo, Kenneth M Burke, StevenJohnston, Gerakibot, Schizodelight, Likebox, General Epitaph, Gulmammad, Brews ohare, GHenter, Addbot, Lightbot, Luckas-bot, Yobot, Aldebaran66, Wireader, AnomieBOT, Archon 2488, Citation bot, Gonzalcg, Armbrust, False vacuum, A. di M., Dshrek, Citation bot 1, Redrose64, Jonesey95, Erkluelaneu, Slightsmile, Netheril96, Hhhippo, ZéroBot, Bamyers99, Zueignung, Fluctuating metric, Helpful Pixie Bot, Bibcode Bot, LevCanada, Miljenkomelo, Salamandra85, Enyokoyama, Andyhowlett, Ontics, Jaggee and Anonymous: 50

- **Quantum field theory in curved spacetime** *Source:* https://en.wikipedia.org/wiki/Quantum_field_theory_in_curved_spacetime?oldid=688543325 *Contributors:* TakuyaMurata, Charles Matthews, Phys, Daniel Arteaga-enwiki, BRW, Mpatel, YurikBot, Mets501, David edwards,

Mbell, Fournax, Headbomb, Infophile, Murks, Wesino, Likebox, Martarius, Addbot, Luckas-bot, Freebirth Toad, Kismalac, DrilBot, Duchpawel, Jjspinorfield1, Maschen, ClueBot NG, Raidr, Castncoot, Drewsberry, WikiJuggernaut, Zirus101 and Anonymous: 22

- **Scale relativity** *Source:* https://en.wikipedia.org/wiki/Scale_relativity?oldid=686081154 *Contributors:* The Anome, William Avery, Jordi Burguet Castell, Spoirier-enwiki, MBisanz, I9Q79oL78KiL0QTFHgyc, Woohookitty, BD2412, Rjwilmsi, Vegaswikian, JonathanD, SmackBot, Colonies Chris, Descubes, JForget, CmdrObot, Cydebot, Headbomb, Probios, Clementvidal, Magioladitis, R'n'B, Lantonov, Tmcdanel, Liometopum, Polyamorph, Phi0618, GSMR, Verbal, Yobot, AnomieBOT, FrescoBot, Jonesey95, Heurisko, John of Reading, Klbrain, Hhhippo, Snotbot, Helpful Pixie Bot, KLBot2, BG19bot, Hdembinski, Andyhowlett, Jo-Jo Eumerus, BethNaught, Banana31, Loraof, OliasailO and Anonymous: 18

- **Weinberg–Witten theorem** *Source:* https://en.wikipedia.org/wiki/Weinberg%E2%80%93Witten_theorem?oldid=670075787 *Contributors:* Michael Hardy, Phys, Rholton, Giftlite, Fred Stober, Lumidek, Lubaf, Eb.hoop, StephanKetz, Bender235, Lycurgus, Cmdrjameson, Physicistjedi, Tony Sidaway, GregorB, BD2412, Rjwilmsi, Thekohser, Chobot, Ohwilleke, Conscious, Teply, That Guy, From That Show!, SmackBot, Colonies Chris, Cydebot, Headbomb, Avicennasis, Infovarius, R'n'B, Lantonov, Sinik-enwiki, SchreiberBike, Addbot, Mjamja, Download, OlEnglish, Yobot, AnomieBOT, Bronckobuster, Omnipaedista, Jonesey95, Raidr, Bibcode Bot, Contact '97, Brad7777, Soni, Almaionescu and Anonymous: 11

- **Composite gravity** *Source:* https://en.wikipedia.org/wiki/Composite_gravity?oldid=650403276 *Contributors:* Phys, Lumidek, Mpatel, SmackBot, Colonies Chris, 100110100, PhilKnight, Yobot and Anonymous: 1

- **Induced gravity** *Source:* https://en.wikipedia.org/wiki/Induced_gravity?oldid=684797654 *Contributors:* Kku, Phys, Lumidek, Jag123, HFarmer, Mpatel, GregorB, DVdm, YurikBot, Salsb, Ilmari Karonen, Colonies Chris, QFT, Radagast83, JorisvS, BBuchbinder, Aldis90, Alphachimpbot, SchreiberBike, Addbot, Mortense, Legobot, Charvest, FrescoBot, ZéroBot, Bibcode Bot, CitationCleanerBot, MaiyaH78 and Anonymous: 15

- **Holographic principle** *Source:* https://en.wikipedia.org/wiki/Holographic_principle?oldid=689834128 *Contributors:* Derek Ross, Vicki Rosenzweig, Bryan Derksen, Timo Honkasalo, The Anome, William Avery, Stevertigo, Bcrowell, Mcarling, KAMiKAZOW, Emperor, AugPi, Reddi, Xaven, Phil Boswell, Sverdrup, Tobias Bergemann, Graeme Bartlett, Nat Krause, LeYaYa, Tromer, Jason Quinn, HorsePunchKid, Togo-enwiki, Tdent, Brianhe, Leibniz, Pjacobi, Dbachmann, Bender235, Pink18, Smalljim, I9Q79oL78KiL0QTFHgyc, Scentoni, PWilkinson, Guy Harris, Sligocki, Gpvos, Mindmatrix, StradivariusTV, Mpatel, Joke137, Christopher Thomas, Prothonotar, Rjwilmsi, Koavf, Jfraatz, Utopos, Chobot, Roboto de Ajvol, Wavelength, Bhny, Chris Capoccia, JocK, Bmdavll, Addps4cat, Gzabers, Kermit2, 2over0, Closedmouth, Sardanaphalus, SmackBot, ZerodEgo, Armeria, Mirokado, Bluebot, TimBentley, Nick Levine, Wen D House, Pwjb, LoveEncounterFlow, Byelf2007, Lambiam, JorisvS, Ckatz, PEiP, Hypnosifl, Stephen B Streater, Jynus, UncleDouggie, Lahiru k, Vyznev Xnebara, Foresee, Mhs5392, Peterdjones, Michael C Price, Alexnye, Doug Weller, Johnfn, Qwyrxian, Al Lemos, Headbomb, Davidhorman, Mmortal03, Escarbot, Knotwork, Len Raymond, JAnDbot, Kungfoofairy, Mikhailfranco, Kentucho, Mange01, Maurice Carbonaro, NerdyNSK, Cpiral, McSly, Potatoswatter, Borat fan, Sheliak, Satani, Philip Trueman, TXiKiBoT, PhysPhD, AlleborgoBot, Cowlinator, Richard1968, Fcady2007, Mark Germine, M.D., Emfetz, Likebox, Robertcurrey, MrWikiMiki, Cmcelwain, Ln2069, Hamiltondaniel, Kallog, Quinling, ClueBot, Drmies, F-j123, HenrikErlandsson, M4gnum0n, Wndl42, SchreiberBike, DumZiBoT, TimothyRias, Andrw, DOI bot, Simonm223, Crus4d3, LaaknorBot, Tassedethe, Verbal, Legobot, Flash.starwalker, Luckas-bot, Yobot, Systemizer, Pcap, AnomieBOT, Materialscientist, Citation bot, ArthurBot, Coreetheapple, RibotBOT, Charvest, Anthropodeus, Giddeon Fox, Prari, FrescoBot, Citation bot 2, Citation bot 1, DrilBot, Tom.Reding, MastiBot, Cjrcl, Lightlowemon, Dr. Salvia, WikitanvirBot, Eekerz, GoingBatty, Netheril96, Wikipelli, Italia2006, ZéroBot, Cogiati, KzLzawlin, Isocliff, ClueBot NG, Raidr, Helpful Pixie Bot, Bibcode Bot, Gordonben, Bmusician, CitationCleanerBot, Caypartisbot, BattyBot, Jimw338, Epicgenius, Christophe1946, Polytope24, JaconaFrere, Almaionescu, Monkbot, Stringer63, Hardkhora, Atreus57, AnimikhRoy967, Toes1111111231111123111123, Nøkkenbuer, Christos Theopoulos, Npmats, The 1editr and Anonymous: 124

- **Phenomenological quantum gravity** *Source:* https://en.wikipedia.org/wiki/Phenomenological_quantum_gravity?oldid=635793052 *Contributors:* DragonflySixtyseven, Gregbard, Omnipaedista, FreeKnowledgeCreator, Nunc aut numquam and Anonymous: 1

- **Quantum foam** *Source:* https://en.wikipedia.org/wiki/Quantum_foam?oldid=681335431 *Contributors:* Stevertigo, Angela, MD87, BenRG, Peak, Wile E. Heresiarch, Ancheta Wis, Curps, Sword-enwiki, Rich Farmbrough, The bellman, Robotje, Smalljim, Cherlin, 99of9, Pearle, BRW, RJFJR, Redvers, Kazvorpal, Linas, Slocombe, GregorB, Dysepsion, BD2412, Platypus222, Spacepotato, Conscious, Shaddack, ErkDemon, Asarelah, 2over0, Groyolo, SmackBot, Bbq332, Scwlong, Dreadstar, Btg2290, Hypnosifl, CapitalR, George100, Legis, Mbell, VolkovBot, Walor, Anonymous Dissident, M-streeter97, OlavN, Pierre-Alain Gouanvic, Biglebouski, MikeGogulski, SEppley, Martarius, ClueBot, DragonBot, Alexbot, CaptainVideo890, Addbot, Physicman123, Lightbot, JackieBot, Xqbot, GrouchoBot, Amaury, Flaviusvulso, FrescoBot, Paine Ellsworth, Mhusainluvsbooks, Pinethicket, Three887, RedBot, Serols, Spinachwrangler, Deanmullen09, EmausBot, Physics16, Uploadvirus, Spoonman61494, AsmundErvik, Terraflorin, ClueBot NG, Joefromrandb, Raidr, Widr, Theopolisme, Davut360, Osiris, ChrisGualtieri, YFdyhbot, EuroCarGT, CuriousMind01, Paulstelian97, Mirvios, Fixature, Becauseican8, Melcous and Anonymous: 78

- **Cosmic microwave background** *Source:* https://en.wikipedia.org/wiki/Cosmic_microwave_background?oldid=689986403 *Contributors:* AxelBoldt, Bryan Derksen, The Anome, Tarquin, AstroNomer-enwiki, AdamW, XJaM, Roadrunner, SimonP, Jaknouse, Youandme, Tedernst, Boud, Michael Hardy, EddEdmondson, Modster, Dominus, Loisel, Alfio, Egil, Stevenj, Ec5618, Timwi, Reddi, Cmbant, Chuunen Baka, Donarreiskoffer, Robbot, Jredmond, Peak, Mirv, Rursus, Mark Krueger, Carlj7, JerryFriedman, Kevin Saff, Graeme Bartlett, Harp, Art Carlson, Guanaco, Gzornenplatz, SWAdair, Bobblewik, TerokNor, Quadell, Beland, Karol Langner, Oneiros, MFNickster, Infradig, Sam Hocevar, Lumidek, Iantresman, Tsemii, Burschik, Mschlindwein, Deglr6328, Flyhighplato, Safety Cap, Moverton, Rich Farmbrough, Guanabot, Pjacobi, Wrp103, Vsmith, Jonathanischoice, RJHall, JustinWick, Livajo, Pt, Edward Z. Yang, Art LaPella, Army1987, Mtruch, MITalum, Svenax, I9Q79oL78KiL0QTFHgyc, Physicistjedi, (aeropagitica), Haham hanuka, Alansohn, Anthony Appleyard, 119, Free Bear, Eric Kvaalen, Andrew Gray, Proxide, JHG, GeorgeStepanek, Kdau, EAi, Count Iblis, Cmapm, DV8 2XL, Gene Nygaard, Ceyockey, Falcorian, Oleg Alexandrov, Ian Moody, Kelly Martin, TheNightFly, Pol098, Yukawa-enwiki, Jok2000, Uris, Wdanwatts, Joke137, Tevatron-enwiki, Rnt20, Graham87, Grammarbot, Edison, Josh Parris, Sjö, Drbogdan, Rjwilmsi, Zbxgscqf, Strangethingintheland, Marasama, Hjb26, Mike s, Mike Peel, Ttwaring, Fragglet, Phoenix2-enwiki, Bgwhite, YurikBot, Wavelength, Jimp, RussBot, Gaius Cornelius, Eleassar, David R. Ingham, NawlinWiki, DragonHawk, Grafen, Chrisbrl88, Deckiller, FF2010, Smkolins, Orioane, Jules.LT, CWenger, QmunkE, ArielGold, Kungfuadam, Profero, Teply, GrinBot-enwiki, BenWandelt, SmackBot, Esradekan, Ashill, Saravask, Onsly, Benjaminevans82, Hmains, RobertKennedy, Lindosland, Andyzweb, Bluebot, Kurykh, H2ppyme, Myxsoma, Silly rabbit, Sangrolu, DHN-bot-enwiki, Scwlong, Modest Genius, Wikipedia

brown, Rrburke, LouScheffer, Aldaron, Wen D House, Cybercobra, Bowlhover, Hgilbert, DenisRS, Zadignose, Ligulembot, Rossp, Sashato-
Bot, Robomaeyhem, JzG, UberCryxic, Hypnosifl, Wwagner, Johnny 0, KJS77, Iridescent, Pathosbot, Mssgill, Chetvorno, Friendly Neighbour,
CWY2190, Makuabob, Cydebot, Peripitus, Hsxavier-enwiki, Tbird1965, Alaibot, ZombieLoffe, הסרפת, Astrophysics Kid, Headbomb, Peter
Gulutzan, Davidhorman, Raphiq, Gioto, Widefox, Orionus, TexMurphy, Rico402, Arch dude, Hut 8.5, .anacondabot, Antelan, VoABot II,
Catslash, Pcp071098, Bubba hotep, First Harmonic, Allstarecho, LorenzoB, Kornfan71, Davidjcmorris, Keith D, R'n'B, Rrostrom, Yonide-
bot, Tgebbie, Jpwest, Migran, Александр Сигачёв, Austin512, Novis-M, Tarotcards, Rominandreu, Wesino, DorganBot, Epistemenical, Sheliak, VolkovBot, Svmich, Sporti, Craigheinke, TXiKiBoT, MusicScience, Anonymous Dissident, Michael H 34, Broadbot, SwordSmurf, James
McBride, Kbrose, Biscuittin, SieBot, Hertz1888, Csmart287, Wing gundam, Zbvhs, Csblack, Mimihitam, Jdaloner, RMB1987, Duae Quartunciae, Anchor Link Bot, Wyattmj, Martarius, GarbagEcol, ClueBot, The Thing That Should Not Be, Niceguyedc, Agge1000, ChandlerMapBot, I
am a violinist, Excirial, Homonihilis, Nymf, Alexbot, Jefflayman, SolomonFreer, PixelBot, Bob108, Telekenesis, Tnxman307, Mastertek, Natty
sci-enwiki, BOTarate, Panos84, Aitias, Nakomaru, Jonverve, DumZiBoT, BarretB, XLinkBot, DCCougar, BodhisattvaBot, Gwark, Dthom-
sen8, ErgoSum88, Ich42, Addbot, Dryphi, DOI bot, Ronhjones, Chotabe, Ka Faraq Gatri, Proxima Centauri, Ehrenkater, Astro-norte, Lightbot, OlEnglish, Zorrobot, Ben Ben, Luckas-bot, Yobot, Ptbotgourou, Legobot II, Aldebaran66, Amble, Wireader, Azcolvin429, AnomieBOT,
Stuffed cat, Captain Quirk, Hunnjazal, Citation bot, Xqbot, Plastadity, Seb.mag, Nnivi, Cydelin, Srich32977, Lithopsian, J04n, GrouchoBot,
EqualMusic, Frankie0607, Kyng, Amaury, Mnmngb, Bigger digger, Fotaun, CES1596, GliderMaven, Nagualdesign, FrescoBot, LucienBOT,
Paine Ellsworth, Binrdow, Citation bot 2, HamburgerRadio, Citation bot 1, HRoestBot, MoonGirl78, Jonesey95, Tom.Reding, Lithium cyanide,
Pmokeefe, RedBot, IVAN3MAN, RockSolidCosmo, TobeBot, Trappist the monk, Comet Tuttle, Michael9422, LI995, Earthandmoon, Tbhotch,
Marie Poise, Wikiborg4711, Siranmichel, DexDor, Cwsavage78, Mathewsyriac, EmausBot, WikitanvirBot, Immunize, Quantanew, GoingBatty,
Snorgway, Italia2006, Grondilu, ZéroBot, Medeis, Quondum, AManWithNoPlan, Miguelzuma, Iiar, Pumpkinking0192, Tbgriswold, Hang Li
Po, ChiZeroOne, DASHBotAV, ClueBot NG, Ulflund, Factorial8, Helpful Pixie Bot, Bibcode Bot, BG19bot, Omegafold, AvocatoBot, Socal212,
Ninney, Altaïr, Natalia.missana, Sparkie82, Fivemusketeers, U-95, ChrisGualtieri, JYBot, Dexbot, Neicdk, Manjolis, LightandDark2000, An-
tunesi, Reatlas, Rfassbind, User74-enwiki, Qmgsobserver, Praemonitus, Zlelik2000, OxygenBlueDansk, AbiLtoCen, Johndric Valdez, Exoplan-
etaryscience, Jlarks73, Monkbot, Filedelinkerbot, Falcon9v1.1, Unatnas1986, Trackteur, Werzaz, Anthul, SkyFlubbler, Samoniel1, Tullyojr,
SwagYolo420ilovethis, Tetra quark, Anand2202, GeneralizationsAreBad, Freakcrane870, Feelthhis, Outedexits and Anonymous: 292

- **Gravitational wave** *Source:* https://en.wikipedia.org/wiki/Gravitational_wave?oldid=689815079 *Contributors:* Bryan Derksen, The Anome,
Toby Bartels, PierreAbbat, Roadrunner, Panairjdde-enwiki, Olivier, Stevertigo, Tim Starling, Stormwriter, Voidvector, GTBacchus, Gbleem,
Looxix-enwiki, William M. Connolley, Bogdangiusca, Cherkash, Timwi, Wikiborg, Reddi, Dfeuer, DonPaolo, Doradus, Patrick0Moran, Mi-
terdale, Raul654, BenRG, Sanders muc, Henrygb, Unyounyo, Giftlite, Graeme Bartlett, Harp, Herbee, Curps, Eequor, Zeimusu, Beland,
Zantolak, Rdsmith4, Mikko Paananen, Tomruen, Sam Hocevar, Urhixidur, Deglr6328, Freakofnurture, Rich Farmbrough, Mgtoohey, Emtilt,
DanP, RJHall, Pt, El C, Alereon, Iridia, Harley peters, Cmdrjameson, Rbj, Mpvdm, I9Q79oL78KiL0QTFHgyc, Nk, Alazoral, Pearle, Fatphil,
Mpeisenbr, Daniel Arteaga-enwiki, Mac Davis, Swift, BRW, Peter McGinley, DV8 2XL, Gene Nygaard, Falcorian, Oleg Alexandrov, Postrach,
Gmaxwell, Mindmatrix, Duncan.france, Trapolator, Mpatel, Tabletop, Dhs, Pdn-enwiki, Trevor Andersen, TotoBaggins, GregorB, BlaiseFE-
gan, Waldir, Joke137, Halcatalyst, Qwertyus, Drbogdan, Rjwilmsi, Strait, TheRingess, Mike Peel, Miserlou, Mikedelsol, Erns57fcva, Bubba73,
TiagoTiago, RobyWayne, Goudzovski, DVdm, Hmonroe, Wavelength, Hairy Dude, Hillman, Hellbus, Eleassar, Długosz, Ccook, Beanyk,
2over0, C h fleming, Anclation-enwiki, Wbrameld, Kungfuadam, Benandorsqueaks, Finell, KasugaHuang, Eigenlambda, Sardanaphalus, Smack-
Bot, Lestrade, Mgreenbe, Brossow, Herbm, Markov, Izehar, Movementarian, Gnuthomson, Kostmo, DHN-bot-enwiki, Timmothias, Colonies
Chris, Sct72, MaxSem, Nick Levine, WinstonSmith, Coolv, Ephogy, Wen D House, Lucretius-enwiki, Lambiam, JorisvS, Profjim, Coke-
laer-enwiki, Lottamiata, Newone, MOBle, Achoo5000, Mssgill, Kurtan-enwiki, CRGreathouse, Floridi-enwiki, Vyznev Xnebara, Syphondu,
Bmk, Myasuda, Cydebot, HermanFinster, Raoul NK, Headbomb, Rriegs, Timalfred, Cerrigno, CTZMSC3, Seaphoto, Zachwoo, Spartaz,
Pkk1253, DagosNavy, JAnDbot, CosineKitty, Hut 8.5, Chagai, WolfmanSF, Paul Niquette, BatteryIncluded, LorenzoB, Laur2ro, Vssun, R'n'B,
Sicanjal, Lantonov, Austin512, Ranemanoj, Wesino, Potatoswatter, Joshua Issac, Treisijs, Yecril, Idioma-bot, Sheliak, Michael H 34, Ng.j,
BotKung, SwordSmurf, Lamro, Klappspatier, SieBot, Aeronaut6, Rsmoling, Corrado7mari, Csblack, Momo san, Aaarnooo, Duae Quartun-
ciae, Chrisjthornhill, Hamiltondaniel, Levine2, Martarius, Trojancowboy, The Thing That Should Not Be, Balashpersia-enwiki, Djr32, Alexbot,
JerryBomb, Michaellreilly, PhySusie, Cavaglia-enwiki, Aitias, DumZiBoT, TimothyRias, InternetMeme, Northwesterner1, Truthnlove, Kb-
dankbot, Addbot, DOI bot, Proxima Centauri, ChenzwBot, Tassedethe, 84user, Lightbot, Cesiumfrog, SPat, Quantumobserver, Micke, Yin-
weichen, Yobot, Quasar1826, AnomieBOT, VanishedUser sdu9aya9fasdsopa, Oracelau, Piekhorn, Materialscientist, Citation bot, Persistent76,
LilHelpa, Obersachsebot, Xqbot, Gap9551, RibotBOT, Curtmed, Privalov, CES1596, FrescoBot, Indomei, Sae1962, Citation bot 1, PigFlu
Oink, Tom.Reding, Toncho11-enwiki, RedBot, Jhbuk, SkyMachine, IVAN3MAN, Meier99, Trappist the monk, Salvatorac, Dinamik-bot,
Earthandmoon, Cayuela, EmausBot, 7daysahead, Elementaro, Slightsmile, K6ka, Hhhippo, A2soup, Stanford96, JosJuice, Colin.campbell.27,
Maschen, Zueignung, Jodoka, Sethking28, Herk1955, Spicemix, Whoop whoop pull up, Henry maxfield, Xonqnopp, ClueBot NG, Hagen-
feldt, Kyu54636, Antichristos, Rezabot, Iggamuffin2445654, Megalobingosaurus, Helpful Pixie Bot, IgglesPickles, LeidenLorentz, Bibcode Bot,
BG19bot, Curlingbug, Malyszkz, OdessaCamp, Jordatech, Brainssturm, Pawprintoz, TheInfernoX, Ronin712, Rhlozier, Yaaxbalam, Mygskr,
Caroline1981, Beatgiant, LightandDark2000, Zodiark111, BooFred, IWPCHI, Neither Pan nor Pettigrew, Comp.arch, Rocket Eddy, Frinthruit,
Man of Steel 85, Seabuckthorn, Monkbot, Neeraj Bhakta, Sometree, Anand2202, Emilyinorbit, KasparBot, Feelthhis and Anonymous: 257

30.12.2 Images

- **File:AdS3.svg** *Source:* https://upload.wikimedia.org/wikipedia/commons/4/47/AdS3.svg *License:* CC BY-SA 3.0 *Contributors:* This file was
derived from: AdS3 (new).png
Original artist:

- derivative work: Alex Dunkel (Maky)

- **File:AdditionComplexes.svg** *Source:* https://upload.wikimedia.org/wikipedia/commons/6/67/AdditionComplexes.svg *License:* CC BY-
SA 3.0 *Contributors:* Own work *Original artist:* <a href='//commons.wikimedia.org/wiki/User:-Strogoff-' title='User:-Strogoff-'>*Frédéric
MICHEL*</a>

- **File:Albert_Einstein_portrait.jpg** *Source:* https://upload.wikimedia.org/wikipedia/en/f/f7/Albert_Einstein_portrait.jpg *License:* PD-US
Contributors:

http://images.google.com/hosted/life/628e99ef2e26233d.html *Original artist:*
E. O. Hoppe. (1878-1972) Published on LIFE

- **File:Ambox_important.svg** *Source:* https://upload.wikimedia.org/wikipedia/commons/b/b4/Ambox_important.svg *License:* Public domain *Contributors:* Own work, based off of Image:Ambox scales.svg *Original artist:* Dsmurat (talk · contribs)

- **File:BH_LMC.png** *Source:* https://upload.wikimedia.org/wikipedia/commons/5/5e/BH_LMC.png *License:* CC BY-SA 2.5 *Contributors:* Own work *Original artist:* User:Alain r

- **File:Black_Hole_Merger.jpg** *Source:* https://upload.wikimedia.org/wikipedia/commons/d/d1/Black_Hole_Merger.jpg *License:* Public domain *Contributors:* Taken from http://www.space.com/imageoftheday/image_of_day_060203.html credit is listed to NASA. *Original artist:* NASA

- **File:Black_Hole_Milkyway.jpg** *Source:* https://upload.wikimedia.org/wikipedia/commons/c/cd/Black_Hole_Milkyway.jpg *License:* CC BY-SA 2.5 *Contributors:* Gallery of Space Time Travel *Original artist:* Ute Kraus, Physics education group Kraus, Universität Hildesheim, Space Time Travel, (background image of the milky way: Axel Mellinger)

- **File:CERN_LHC_Tunnel1.jpg** *Source:* https://upload.wikimedia.org/wikipedia/commons/f/fc/CERN_LHC_Tunnel1.jpg *License:* CC BY-SA 3.0 *Contributors:* Own work *Original artist:* Julian Herzog (Website)

- **File:Calabi-Yau-alternate.png** *Source:* https://upload.wikimedia.org/wikipedia/commons/5/55/Calabi-Yau-alternate.png *License:* CC BY-SA 2.5 *Contributors:* Transferred from en.wikipedia to Commons by Lunch. *Original artist:* The original uploader was Lunch at English Wikipedia

- **File:Calabi-Yau.png** *Source:* https://upload.wikimedia.org/wikipedia/commons/d/d4/Calabi-Yau.png *License:* CC BY-SA 2.5 *Contributors:* own work by Lunch
  http://en.wikipedia.org/wiki/Image:Calabi-Yau.png (english Wikipedia) *Original artist:* Lunch

- **File:Calabi_yau.jpg** *Source:* https://upload.wikimedia.org/wikipedia/commons/f/f3/Calabi_yau.jpg *License:* Public domain *Contributors:* Mathematica output, created by author *Original artist:* Jbourjai

- **File:Clebsch_Cublic.png** *Source:* https://upload.wikimedia.org/wikipedia/commons/7/7c/Clebsch_Cublic.png *License:* CC BY-SA 3.0 *Contributors:* I created this on my own computer using the free software Surfer *Original artist:* Fly by Night

- **File:Cmbr.svg** *Source:* https://upload.wikimedia.org/wikipedia/commons/c/cd/Cmbr.svg *License:* Public domain *Contributors:* Own work *Original artist:* Quantum Doughnut

- **File:Commons-logo.svg** *Source:* https://upload.wikimedia.org/wikipedia/en/4/4a/Commons-logo.svg *License:* ? *Contributors:* ? *Original artist:* ?

- **File:Compactification_example.svg** *Source:* https://upload.wikimedia.org/wikipedia/commons/f/f5/Compactification_example.svg *License:* CC BY-SA 4.0 *Contributors:* Brian Greene (2004). The Elegant Universe (DVD). Part II (String's the thing): WGBH Boston Video. Event occurs at 43:55. OCLC 54019786 *Original artist:* Alex Dunkel (Maky)

- **File:Compton_Scattering.svg** *Source:* https://upload.wikimedia.org/wikipedia/commons/8/8a/Compton_Scattering.svg *License:* CC BY-SA 3.0 *Contributors:* Own work *Original artist:* Dylan1946

- **File:Compton_qed.jpg** *Source:* https://upload.wikimedia.org/wikipedia/commons/5/51/Compton_qed.jpg *License:* CC BY-SA 3.0 *Contributors:* Own work (Original text: *I (Pra1998 (talk)) created this work entirely by myself.*) *Original artist:* Pra1998 (talk)

- **File:Crab_Nebula.jpg** *Source:* https://upload.wikimedia.org/wikipedia/commons/0/00/Crab_Nebula.jpg *License:* Public domain *Contributors:* HubbleSite: gallery, release. *Original artist:* NASA, ESA, J. Hester and A. Loll (Arizona State University)

- **File:D3-brane_et_D2-brane.PNG** *Source:* https://upload.wikimedia.org/wikipedia/commons/8/88/D3-brane_et_D2-brane.PNG *License:* Public domain *Contributors:* Image:D-brane.PNG, oeuvre personnelle. *Original artist:* <a href='//commons.wikimedia.org/w/index.php?title=User:Rogilbert,,&,,action=edit,,&,,redlink=1' class='new' title='User:Rogilbert (page does not exist)'>Rogilbert</a>

- **File:De_Raum_zeit_Minkowski_Bild.jpg** *Source:* https://upload.wikimedia.org/wikipedia/commons/c/c5/De_Raum_zeit_Minkowski_Bild.jpg *License:* Public domain *Contributors:* scan from original book *Original artist:* **Hermann Minkowski**

- **File:Dirac_3.jpg** *Source:* https://upload.wikimedia.org/wikipedia/commons/7/7d/Dirac_3.jpg *License:* Public domain *Contributors:* http://www-history.mcs.st-andrews.ac.uk/PictDisplay/Dirac.html *Original artist:* Cambridge University, Cavendish Laboratory [1]

- **File:E6GUT.svg** *Source:* https://upload.wikimedia.org/wikipedia/commons/9/9c/E6GUT.svg *License:* CC BY-SA 3.0 *Contributors:* Own work, Created from Garret Lisi's Elementary Particle Explorer *Original artist:* Cjean42

- **File:Earth-moon.jpg** *Source:* https://upload.wikimedia.org/wikipedia/commons/5/5c/Earth-moon.jpg *License:* Public domain *Contributors:* NASA [1] *Original artist:* Apollo 8 crewmember Bill Anders

- **File:Edit-clear.svg** *Source:* https://upload.wikimedia.org/wikipedia/en/f/f2/Edit-clear.svg *License:* Public domain *Contributors:* The *Tango!* Desktop Project. *Original artist:*
  The people from the Tango! project. And according to the meta-data in the file, specifically: "Andreas Nilsson, and Jakub Steiner (although minimally)."

- **File:Edward_Witten.jpg** *Source:* https://upload.wikimedia.org/wikipedia/commons/9/97/Edward_Witten.jpg *License:* Public domain *Contributors:* Own work *Original artist:* Ojan

- **File:Einstein_cross.jpg** *Source:* https://upload.wikimedia.org/wikipedia/commons/c/c8/Einstein_cross.jpg *License:* Public domain *Contributors:* http://hubblesite.org/newscenter/archive/releases/1990/20/image/a/ *Original artist:* NASA, ESA, and STScI

30.12.3 Content license